SCANNING PROBE MICROSCOPY AND SPECTROSCOPY

Second Edition

SCANNING PROBE MICROSCOPY AND SPECTROSCOPY

Theory, Techniques, and Applications

Second Edition

Edited by

Dawn A. Bonnell

A JOHN WILEY & SONS, INC., PUBLICATION

New York · Chichester · Weinheim · Brisbane · Singapore · Toronto

For ordering and customer service, call 1-800-CALL-WILEY.

Library of Congress Cataloging-in-Publication Data:

Scanning probe microscopy and spectroscopy: theory, techniques, and applications / edited by Dawn Bonnell.—2nd ed.
    p. cm.
   Rev. ed. of: Scanning tunneling microscopy and spectroscopy. c1993.
   ISBN 0-471-24824-X (cloth : alk. paper)
   1. Scanning probe microscopy. 2. Tunneling spectroscopy. I. Bonnell, Dawn A. II. Scanning tunneling microscopy and spectroscopy.

QH212.S33 S38 2000
502′.8′25—dc21   00-036821

Printed in the United States of America.

10 9 8 7 6 5 4 3 2 1

CONTENTS

PREFACE

Microscopies based on the scanning tunneling and atomic force microscopes have been continuously evolving over the last decade. These developments are hardly unexpected. In fact, the purpose of the first edition of this book, *Scanning Tunneling Microscopy and Spectroscopy: Theory, Technique, and Applications* was to provide introductory concepts, tutorial applications, a background in fundamentals, and examples in the context of recent research. The premise was that this would allow users to follow subsequent developments in the field. The sales history of this book attests to the fact that the approach achieved the original goal. With the commercialization of many new imaging modes and the application to new fields, the needs of the average user of scanning probe microscopy (SPM) have changed. For example imaging modes based on magnetic forces and on surface potential are now routine, rather than specialty tools. The extended use of intermittent contact modes in and out of liquids has greatly expanded applications in life sciences. The purpose of this second edition is to address this broader field with the expectation that readers will again be positioned to use current tools and to exploit the advances of the next decade.

The average user of scanning probe microscopy (SPM) no longer builds a microscope. We have, therefore, eliminated content related to details of microscope design. The concepts necessary to microscope operation and installation remain and are augmented by illustrations of basic data analyses. Introductory descriptions of the physics of tunneling and interatomic forces are added to Chapter 2. The chapters based on fundamental concepts remain essentially unchanged but include updated examples. This applies to Chapters 3–6. New chapters have been dedicated to techniques that have become routine since the mid-1990s: electrostatic and magnetic force microscopies (Chapter 7) and near-field optical microscopies (Chapter 11). Applications to buried interfaces (Chapter 8), biology (Chapter 9), nanomechanics (Chapter 10), and electrochemistry (Chapter 12) have been substantially updated. The predominance of new probe-based imaging tools has motivated a change in the title of the book. As with the previous edition, the book is appropriate for advanced undergraduate- or early graduate-level courses, technical short courses, or as a tutorial or general reference. Readers are also encouraged to refer to periodical research monographs and review papers for the most current research results.

The preparation of this edition met with many practical obstacles. I am extremely grateful for the patience of the contributors, J. Tersoff, R. Hamers, D. Padowitz, W. Unertle, G. Rohrer, R. Smith, R. Colton, N. Burnham, W. Kaiser, L. Bell, M. Hecht, A. Bard, F. Fan, S. Kalinin, B. Huey, D. Higgins, E. Mei, L. Davis, and S. Lindsay, during this process. In addition, I appreciate the considerable practical assistance of S. Kalinin and J. Smith.

In 1993 I acknowledged inspiration provided by my early mentors, for which I am still grateful. With this edition I am acutely aware of the contributions of my students and postdocs over the last decade. Although not all are represented explicitly in the book, all have contributed to my understanding of the field. For that I continue to be grateful.

DAWN A. BONNELL

PREFACE
TO THE FIRST EDITION

Scanning tunneling microscopy (STM), tunneling spectroscopy (STS), and atomic force microscopy (AFM) have evolved into routine surface characterization tools. The question arises: from where can one learn the basic operating principles and techniques of scanned probe microscopies? Hundreds of research papers are published each year which either focus on a particular aspect of the measurement or use STM techniques in passing. These papers are usually directed towards the expert and, consequently, are not very helpful to the novice. Even if this were not the case, the prospective STM user would face the daunting task of sifting through nearly 1000 papers to glean the introductory concepts. Assistance might be found in several of the excellent review articles published in the last eight years, such as those by P. Hansma and J. Tersoff, Y. Kuk and J. Silverman, and H. Rohrer, but even these papers cannot address all of the relevant background required for a beginner.

As the use of STM became more common through access to commercial microscopes, several approaches to teaching STM and AFM have been taken. Scientific societies (the Materials Research Society, the American Vacuum Society, etc.) offer one- or two-day courses on the applications of STM/AFM and, of course, microscope companies supply some instruction with their instruments. Aspects of both STM and AFM are currently introduced in upper level courses offered at universities in physics, materials science, and chemical engineering departments. At present, no text exists upon which to base such courses or to develop alternative instructional formats.

Monographs do exist which serve as overviews of the field at the time of their publication. Notable examples include books edited by R. J. Bohm, N. Garcia, and H. Rohrer (1989) and R. Wiessendanger and H. J. Güntherodt (1992). In addition, proceedings from International Scanning Tunneling Microscopy Conferences held annually between 1984 and 1991 also provide overviews of current knowledge and, the early ones in particular, include much detailed information about instrumentation design. (*See Surf. Sci.* **181** (1987) 1–412. *J. Vac. Sci. and Technol.* **A 6** (1988) 259–556. *J. Vac. Sci. and Technol.* **B 9** (1991) 403–411.) However, like the review articles, these books lack the introductory materials necessary for the novice. A source that contains both basic principles and advanced applications is clearly needed.

In this book we attempt to combine a presentation of basic concepts and operating principles with the more advanced theoretical descriptions, illustrated with examples from recent research. Thus, the book is appropriate for students, scientists and technicians with some background in elementary science, as well as those more fluent in physics and chemistry. The book is written at a level that can be used as an upper level undergraduate or first year graduate course, or as additional materials for a surface science course. This book is not a textbook in the traditional sense; the dynamics of the field preclude a final summary of many of the topics. It is intended to be a bridge between monographs or review type coverages and the potential STM/AFM experimentalist or practicioner. While the fundamental aspects of STM are treated in depth, many general surface science concepts could not be included. Readers are referred to one of the comprehensive texts on surface science for a more general treatment of surface physics; Zangwill, *Physics at Surfaces*, (1988); Adamson, *Physical Chemistry of Surfaces*, (1982), and Prutton, *Electronic Properties of Surfaces*, (1984).

After a review of historical developments leading to the invention of STM (Chapter 1), the basic operating principles of the tunneling microscope are outlined in Chapter 2. Chapter 3 considers the theoretical basis of STM and STS and explicitly addresses the issue of image interpretation. Both the principles and the procedures of tunneling spectroscopy are covered in Chapter 4. Practical issues dealt with in Part II include the expected structures of solid surfaces (Chapter 5) and tip and sample preparation (Chapter 6). Special applications and techniques that have evolved from STM are treated in Part III. Chapter 7 provides a description of atomic force microscopy, including fundamental forces, instrumentation design and image interpretation. Ballistic electron emission microscopy is described in Chapter 8. Applications to electrochemistry (Chapter 9) and biology (Chapter 10) follow.

The significance of the contributions of the coauthors will be self-evident, however, in addition I would like to thank the contributors, J. Tersoff, R. Hamers, W. Inertl, G. Rohrer, R. Colton, N. Burnham, W. Kaiser, L. Bell, M. Hecht, A. Bard, F. Fan, and S. Lindsay, for their cooperation and patience during the editing of this book. I have received helpful comments from J. C. H. Spence, Y. Liang, and Q. Zhong who have read one or more chapters and offered suggestions which resulted in clearer presentation. I am also grateful to G. Hsieh for extensive assistance in editing and for preparing figures, and to D. Gilbert for assistance with typing.

Finally, I would like to acknowledge the good fortune of having worked with T. Y. Tien who directed me towards science, with M. Ruhle who introduced me to the microscopic world, and with D. R. Clarke who first provided me the opportunity to become involved with STM at the emergence of this exciting new field.

DAWN A. BONNELL
January 1993

CONTRIBUTORS

ALLAN J. BARD, Department of Chemistry and Biochemistry, The University of Texas, Austin, TX 78712

L. D. BELL, Center for Space Microelectronics Technology, Jet Propulsion Laboratory, California Institute of Technology, Pasadena, CA 91109

DAWN A. BONNELL, Department of Materials Science, The University of Pennsylvania, Philadelphia, PA 19104

NANCY A. BURNHAM, Department of Physics, Worcester Polytechnic Institute, 100 Institute Road, Worcester, MA 01609-2280

RICHARD J. COLTON, Naval Research Laboratory, Washington, DC 20375-5000

L. C. DAVIS, Scientific Research Laboratory, Ford Motor Company, Dearborn, MI 48121

FU-REN F. FAN, Department of Chemistry and Biochemistry, The University of Texas; Austin, TX 78712

ROBERT J. HAMERS, Department of Chemistry, The University of Wisconsin, Madison, WI 53706

M. H. HECHT, Center for Space Microelectronics Technology, Jet Propulsion Laboratory, California Institute of Technology, Pasadena, CA 91109

DANIEL A. HIGGINS, Department of Chemistry, Kansas State University, Manhattan, KS 66506

BRYAN D. HUEY, Department of Materials, Oxford University, Parks Road, OX1 3PH, UK

W. J. KAISER, Center for Space Microelectronics Technology, Jet Propulsion Laboratory, California Institute of Technology, Pasadena, CA 91109

SERGEI V. KALININ, Department of Materials Science, The University of Pennsylvania, Philadelphia, PA 19104

STUART LINDSAY, Molecular and Cellular Biology Program, Arizona State University Tempe, AZ 85287-1504

ERWEN MEI, Department of Chemistry, Kansas State University, Manhattan, KS 66506

DAVID F. PADOWITZ, Department of Chemistry, Amherst College, Amherst, MA 01002

GREGORY ROHRER, Department of Materials Science and Engineering, Carnegie Mellon University, Pittsburgh, PA 15213

RICHARD L. SMITH, Department of Materials Science and Engineering, Massachusetts Institute of Technology, Cambridge, MA 02139

JERRY TERSOFF, IBM Research, P.O. Box 218, Yorktown Heights, NY 10598

WILLIAM N. UNERTL, Sawyer Research Center, The University of Maine, Orono, ME 04401

PART I

FUNDAMENTALS OF OPERATION

1

INTRODUCTION

Dawn A. Bonnell

In the two decades since the invention of the scanning tunneling microscope (STM), probe microscopies in general have made a dramatic impact in fields as diverse as materials science, semiconductor physics, biology, electrochemistry, tribology, biochemistry, surface thermodynamics, organic chemistry, catalysis, micromechanics, and medical implant technology. The reason for the nearly instantaneous acceptance of scanning probe microscopy (SPM) is that it provides three-dimensional, real-space images of surfaces at high spatial resolution. Images are based on detecting the local interaction between a small probe tip and a surface. Depending on the particular SPM, the images can represent physical surface topography, electronic structure, electric or magnetic fields, or a number of other local properties. When the sample is clean and flat, even atoms can be imaged. These capabilities will be demonstrated throughout the book. In the case of some materials, such as biological tissue and large organic molecules, SPM allows imaging at unprecedented levels of resolution without destroying samples. This is in contrast to other characterization techniques, including scanning electron microscopy and transmission electron microscopy, that require relatively high vacuum conditions. In the case of crystalline solids high-resolution transmission electron microscopy can be used to examine bulk atomic structure; however, this technology does not provide the local information about bonding that is obtained from tunneling spectroscopy. Thus the SPM brings a unique and complimentary probe of atomic structure and local properties to the arsenal of characterization tools.

Although its effect has been sudden and dramatic, SPM did not emerge from an intellectual vacuum. Consider as an illustration the STM, which is one application of the concept of electron tunneling.[1-10] Our understanding of electron tunneling begins in the early 1920s with the advent of quantum mechanics. Shortly thereafter, in 1928, Fowler and Nordheim described field emission from a free metal, in which electrons overcome the potential barrier and leave the surface. Electron tunneling was the focus of occasional attention throughout the 1930s and 1940s; however, as cryogenic and vacuum technologies advanced, new extensions of theory became accessible to experiment. Of particular import was the invention of the transistor and subsequent development of electronic devices, which motivated significant experimental and theoretical efforts in the early 1960s. Giaver used tunneling to measure the superconductor energy gap in metals. Bardeen used a transfer Hamiltonian to incorporate many body effects in the theoretical formulation for tunneling between two electrodes, and Simmons included the effects of image forces on the tunneling barrier. These advances provided a framework within which to understand the operation of Zener diodes, Esaki diodes, and the field ion microscope. One of the interesting applications of the tunneling phenomenon was introduced by Jaklevic and Lambe, who observed inelastic tunneling losses owing to excitation of molecular vibrations in bulk tunnel junctions. This observation led to energy loss spectroscopy across planar junctions, referred to as inelastic tunneling spectroscopy (IETS), which developed into a field by itself during the early 1970s. These developments are summarized in Table 1.1.

In 1971, Young, Ward, and Scire observed vacuum tunneling and transition to field emission in the point-plane geometry that presaged the STM. Binnig, Rohrer, and Gerber, inventors of the STM, observed tunneling through a controlled vacuum gap in 1982 and imaged single atom steps. In 1983, the silicon $(111)-(7 \times 7)$ reconstruction was imaged in real space for the first time and a new field was born. Although this short historical review is by no means comprehensive, it does emphasize that it was the massive amount of both experimental and theoretical work stimulated by the electronics industry in the 1960s that allowed almost

Table 1.1 Summary of progress in electron tunneling

Year	Investigators	Advancement
1928	Fowler, Nordheim	Explanation of field emission
1934	Zener	Theory of interband tunneling
1937	Muller	Field emission microscope
1958	Esaki	Tunneling in p–n junctions
1960	Gaiever	Measurement of superconducting energy gap
1961	Bardeen	Many body effects in tunneling theory
1963	Simmons	Image forces in tunneling theory
1966	Jaklevic, Lambe	Inelastic tunneling spectroscopy
1971	Young, Ward, Scire	Vacuum tunneling in point-plane geometry
1982	Binning, Rohrer, Gerber	Atomic resolution STM

immediate understanding and application of the STM. From a patent in 1981 to a few papers in 1982, the field has experienced a geometric expansion with the consequence that STM use spread worldwide within one decade.

A similar summary could be made regarding atomic forces. The modern history begins with Derjagin et al.,[10] who described interparticle forces and colloidal interactions in the 1960s, and includes the invention of the surface forces apparatus by Israelavicchi[11] in the 1970s and the invention of the atomic force microscope (AFM) by Bennig et al.[12] in the 1980s.

The impact of SPM now extends far beyond a few probe techniques. The development of the scanning probe technology is essentially a method for making localized measurements of any property for which a detection scheme can be devised. Detection strategies are now available for probing first-order, second-order, and even third-order contributions to interactions. Thus it is now possible to detect, for example, thermal gradients, magnetic forces, photon emission, photon absorption, piezo electric strain, electrostriction, magnetostriction, and optical reflection with localized probes. Subsequent chapters describe how these measurements can be exploited in various fields of endeavor.

REFERENCES

1. J. Bardeen, *Phys. Rev. Lett.* **6**, 57 (1961).
2. G. Binnig and H. Rohrer, *Helv. Phys. Acta* **55**, 726 (1982).
3. L. Esaki, *Phys. Rev.* **109**, 603 (1958).
4. R. H. Fowler and L. Nordheim, *Proc. R. Soc.* **A119**, 173 (1928).
5. I. Gaiver, *Phys. Rev. Lett.* **5**, 147 (1960).
6. R. C. Jaklevid and J. Lambe, *Phys. Rev. Lett.* **17**, 1139 (1966).
7. J. G. Simmons, *J. Appl. Phys.* **34**, 1793 (1963).
8. R. Young, J. Ward, and R. Scire, *Rev. Sci. Instrum.* **43**, 999 (1972).
9. C. Zener and Z. *Phys.* **106**, 541 (1937).
10. B. V. Derjaguin, N. V. Churaev, and V. M. Muller in J. A. Kitchener, ed., *Surface Forces* (trans. V. I. Kisin), Consultants Bureau, New York, 1987.
11. J.N. Israelachvili, *Intermolecular and Surface Forces*, Academic Press, New York, 1992.
12. G. Binnig, C. F. Quate, and C. Gerber, *Phys. Rev. Lett.* **56**, 930 (1986).

2

BASIC PRINCIPLES OF SCANNING PROBE MICROSCOPY

Dawn A. Bonnell and Bryan D. Huey

2.1 THE LOCAL PROBE APPROACH

Scanning probe microscopy (SPM) is unique among imaging techniques in that it provides three-dimensional (3-D) real-space images and among surface analysis techniques in that it allows spatially localized measurements of structure and properties. Under optimum conditions subatomic spatial resolution is achieved. These capabilities are accomplished with an instrument that is simple in concept and design. A scanning tunneling microscope (STM) is similar to a profilometer in that the image is obtained by scanning a tip over the surface of a sample. The primary dierences between the two is that the motion control of the scanner is at the angstrom level and the tip does not make contact with the sample in a STM, rather it is maintained at a height of 5 to 50 Å above the sample. Other SPM techniques are based on the same scanner principles but use various detection schemes to access a wide range of properties, including atomic forces, electrostatic and magnetic forces, thermal gradients, and optical intensity. The basic principles associated with these techniques are outlined in this chapter; the specific details and applications are described in subsequent chapters. Starting with STM, electron tunneling is described, then interatomic interactions of tips and solids are discussed. Typical instrumentation is illustrated and issues important to good performance are summarized. Finally, several common data analysis procedures are demonstrated. Subsequent chapters will address fundamental issues in more detail and illustrate a wide variety of applications of all variants of scanning probe microscopy. For additional insight into the fundamentals on which the discussions in this chapter are based, the reader is referred to several classic textbooks.[1-3] Several research monographs are also available that review applications of SPM.[4-7]

2.2 PRINCIPLES OF ELECTRON TUNNELING

Scanning tunneling microscopy and spectroscopy rely on electron tunneling, a phenomenon that is based on quantum mechanics in that it is not allowed in classical mechanics. In a metal or semiconductor, electrons exist within an energy range, designated by the shaded areas in the 1-D diagram in Figure 2.1. At the interface between that material and an insulator, or indeed a vacuum, there is an energy barrier (Fig. 2.1a). If a second metal or semiconductor is placed near the first and a voltage imposed between the two, then the shape of the energy barrier is changed and there is a driving force for electrons to move across the barrier (Fig. 2.1b). However, in classical mechanics, the electron cannot travel across the barrier unless its energy is raised to a value larger than E_{b1}, i.e., it must go over the barrier not through it. Quantum mechanics allows a finite, albeit small, number of electrons to traverse the barrier if the thickness z is small. The solutions to Schroedinger's equation inside the barrier have the form:

$$\psi(z) = \psi(0)e^{-\kappa z} \tag{2.1}$$

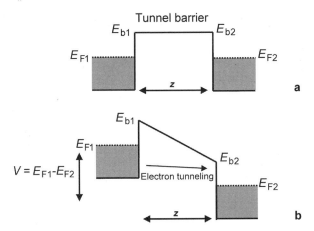

Figure 2.1. The energy levels in two solids separated by an insulating or vacuum barrier (**a**) with no bias applied between the solids and (**b**) with an applied bias. Energies of the electrons in the solids are indicated by the shaded areas up to E_{F1} and E_{F2}, which are the Fermi levels of the respective materials. The applied bias V is $E_{F1}-E_{F2}$, and z is the distance between the two solids.

where

$$\kappa = \frac{\sqrt{2m(V-E)}}{\hbar}$$

where m is the mass of an electron, $\hbar$ is Planck's constant, E is the energy of electron, and V is the potential in the barrier. The probability that an electron will cross the barrier is the tunneling current (I), and it decays exponentially with the barrier width z as

$$I \propto e^{-2\kappa z} \qquad (2.2)$$

Only electrons with energies between the Fermi levels of the two materials can tunnel, due to the constraint that electrons must exist in a filled state at the energy in the negative material and an unfilled state must exist at the energy in the positive material.

A more realistic depiction of the voltage dependence of barrier function includes the rounding near the surfaces owing to the presence of image forces developed in the material. The voltage dependence of a tunneling barrier is illustrated in Figure 2.2. The shape of the barrier changes with sample-tip separation and with applied bias. Note the asymmetry in the barrier.

The energy diagram in Figure 2.1 is general and is usually first thought of in terms of two macroscopic flat metal electrodes (parallel plates). These energy relations also apply to the configuration of a STM, i.e., a tip-plane geometry. The STM imaging process is shown in Figure 2.3. A sharp metal tip, often made of Pt or W, is brought within proximity of a planar sample surface. When the two

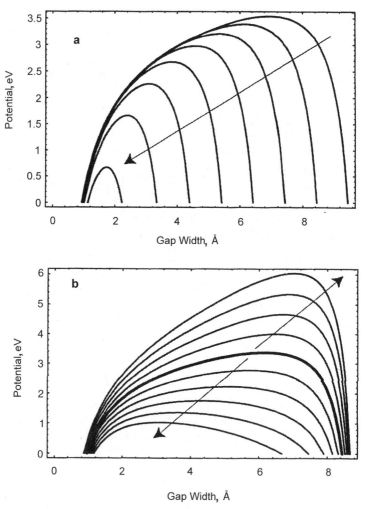

Figure 2.2. A more realistic depiction of the tunneling barrier, including the effect of image charge on the shape of the barrier. **a**, As the distance between the solids is decreased (*arrow*), the size and shape of the barrier change. **b**, The barrier also changes when a voltage is applied, the arrows show the response to opposite signs of bias.[25]

surfaces are sufficiently close that the wave functions overlap the resulting current, I—given in Eq. (2.3) and developed in more detail in Chapter 3—is

$$I = C\rho_t\rho_s e^{z*k^{1/2}} \tag{2.3}$$

where z is the sample-tip separation, ρ_t is the tip electronic structure, ρ_s is the sample electronic structure, and C is a constant. The current is exponentially dependent on the sample–tip separation, i.e., a 1 Å decrease in the sample–tip separation will increase the current by one order of magnitude and a decrease

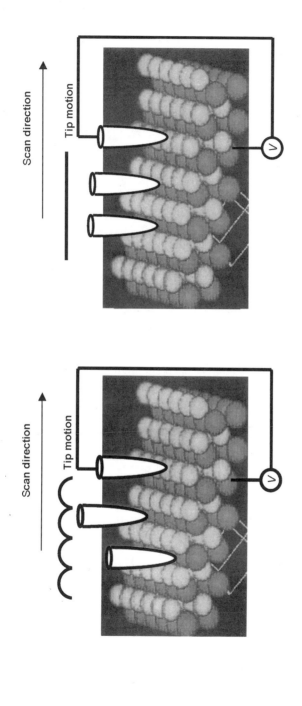

(a) Constant current imaging

(b) Constant height imaging

Figure 2.3. STM. A bias is applied between the sample and tip. As the tip is scanned from left to r᾿ᾰt, (**a**) either the tip is moved vertically to keep current constant or (**b**), the vertical position is held constant and the current varies.

of 2 Å will increase the current by two orders of magnitude. With this level of sensitivity, the tunneling current can be used to control the sample–tip separation with high vertical resolution.

To produce images, the STM can be operated in two modes. In constant current imaging, a feedback mechanism is enabled that maintains a constant current while a constant bias is applied between the sample and tip (Fig. 2.3**a**). As seen in Eq. (2.3), these conditions require a constant sample–tip separation. As the tip is scanned over the sample, the vertical position of the tip is altered to maintain that constant separation. The motion in all three directions (x, y, and z) is controlled by piezoelectric elements. Simple voltage ramps applied to a piezoelectric tube cause the tip to scan the surface, and the voltage signal from a comparison circuit is directed to the z piezoelectric element. The signal required to alter the vertical tip position is the image, which represents a constant charge density contour of the surface. The image is recorded as a 2-D array of integers representing heights at a specific x, y positions. Altering the level of the current set point or the applied bias produces contours of different charge densities. An alternative imaging mode is constant height operation in which constant height and constant applied bias are simultaneously maintained (Fig. 2.3**b**). A variation in current results as the tip scans the sample surface because topographic structure varies the sample–tip separation. In this case, the current is the image and can be related to charge density. There are advantages inherent in both modes of operation; the former produces contrast directly related to electron charge density profiles, whereas the latter provides for faster scan rates not being limited by the response time of the vertical driver. Atomic resolution images are possible only under optimized sample and tip conditions. Larger sample–tip separations and blunt tips have the effect of smearing the localized structure and produce topographic images with somewhat lower resolution. One illustration of the spatial resolution is shown in the constant current image of Figure 2.4; many more will be presented in subsequent chapters.

The contrast in any STM image is a convolution of the electronic structure of the sample, and the electronic structure of the tip, described in detail in Chapter 3. It is only when the latter contribution is minimal that the variation of contrast in the image can be attributed to electronic properties of the sample. Fortunately, this condition can be easily met on many materials; therefore, Eq. (2.3), in which this assumption is implicit, is often a reasonable approximation.

Tunneling spectroscopy (TS) performed in the STM provides information about the electronic structure of the sample by probing the sample density of states as a function of energy. The concept is similar to traditional sandwich junction spectroscopy (inelastic tunneling spectroscopy, IETS), the primary dierence being that a vacuum gap exists between the two electrodes in scanning tunneling spectroscopy (STS), whereas an insulating thin film exists between planar electrodes in IETS. There are also two methods of performing STS. The first, referred to as point spectroscopy, involves moving the tip to a feature of interest, disengaging the feedback mechanism, modulating the tip bias, and recording the resulting variation in current. The idealized band structures in Figure 2.5 illustrate current flow in this

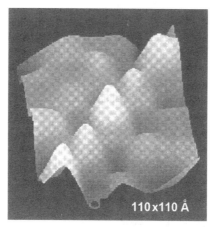

Figure 2.4. A constant current image of SrTiO$_3$ (100) in a region that exhibits 1 × 2 symmetry. Images such as this are acquired in UHV with biases in the range of 0.5 to 2.0 V and currents of about 0.5 nA. (Courtesy of X. Lin and D. Bonnell.)

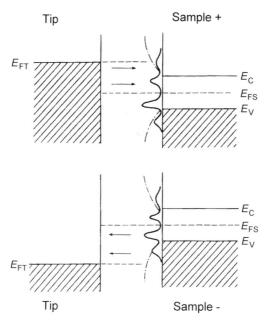

Figure 2.5. Tunneling spectroscopy using idealized band structures in which the tip is modeled as a metal with continuous occupation to the Fermi level, and the sample is imaged as a semiconductor with surface states. Shaded areas refer to energy regions in which states are occupied.[26]

situation. As the bias is ramped, either positively or negatively, the current varies in response to the changing electron density in the energy window. Ramping in both directions probes both the occupied and unoccupied energy states, with the magnitude of the current at a specific voltage directly related to the density of states of the sample at that energy. Because the constant, C in Eq. (2.4) is a linear function of voltage, spectroscopy data can be related to the sample electronic structure D_s in the form of conductance:

$$I/V = C * D_s e^{-1.025*Z*\phi^{1/2}} \tag{2.4}$$

and voltage derivatives thereof. The second method of obtaining spectroscopic information involves simultaneously collecting images at various biases. This can be accomplished by modulating the bias at a high frequency (with respect to the time constant of the feedback controller) and recording the current at several discreet values of applied bias or using the same sample and hold method of point spectroscopy at every image point. The simultaneous collection of multiple current images, or collection of spectra at each image pixel, is sometimes referred to as current imaging tunneling spectroscopy (CITS). A comparison of the periodic contrast in such images with atomic charge superposition and total energy calculations indicates the spatial position of energy states with respect to the lattice.

Note from Figures 2.1 and 2.2 that the tunneling barrier is asymmetric. The implication is that at the same magnitude of bias, a forward tunneling electron experiences a different barrier than does a reverse tunneling electron. The consequence to tunneling spectra is that the negative half of a spectrum is sometimes higher in intensity than the positive half. In other words, the tip contribution to the spectra is smaller in one direction than in the other. The basis for analyzing spectra and applications to surface defects and adsorption are presented in Chapter 4. A program for calculating tunneling spectra is given in Appendix III.

2.3 PRINCIPLES OF ATOMIC FORCES

A large number of SPM variants have developed based on the detection of forces between the tip and sample. A variety of tip–surface interactions may be measured by an atomic force microscope (AFM), depending on the separation between tip and sample. Obviously, during contact with the sample the tip predominantly experiences repulsive Van der Waals forces. When lifted above the surface, however, long-range interactions, notably electric or magnetic forces, may dominate the interaction with the tip.

2.3.1 Short-Range Interactions

A comprehensive treatment of surface and intermolecular forces can be found in an excellent book by Israelachevili.[8] The fundamental interactions at short distances are the van der Waals interactions, which are responsible for the formation of solids, wetting, etc. At distances of a few nanometers, van der Waals forces are sufficiently strong to move macroscopic objects such as AFM cantilevers. Van der Waals interactions consist of three components: polarization, induction, and dispersion. Polarization refers to permanent dipole moments such as exist in water molecules or in $BaTiO_3$. Induction refers to the contribution of induced dipoles. Dispersion is due to instantaneous fluctuations of electrons, which occur at the frequency of light causing optical dispersion, thus the name.

To see how the potential, and therefore the force, is related to materials properties, a simple case is considered for illustration. In a simple system, an analytical expression for the van der Waals potential, for example for two identical gas molecules, is

$$U_{vdw} = -\left(\frac{1}{4\pi\varepsilon_o}\right)^2 \left(\frac{\mu^4}{3kT} + 2\mu^2\alpha + \frac{3}{4}\alpha^2\hbar\omega\right)\frac{1}{z^6} \tag{2.5}$$

where μ is the dipole moment, T is temperature, α is the polarizability, ε_o is the permittivity of free space and $\hbar\omega$ is the ground state energy of the electrons. For more complex systems, a reasonable approximation of the interaction is obtained assuming that dispersion dominates and is isotropic, additive, and nonretarded. Under these assumptions, the van der Waals potential between two atoms is $-C/z^6$, between an atom and a flat surface is $-C/z^3$ between two planes is $-A/12\rho z^2$, and between a sphere and a plane is $-AR/6z$; where z is the distance between objects, R is the radius of the sphere, A is the Hamaker constant, and C is a constant known as the London coefficient. The complete interaction must include the repulsive as well as attractive terms (Fig. 2.6). The Hamaker constant is the term that characterizes the properties of the materials, including collective interactions and polarization if calculated from dielectric and optical properties. It is

$$A \approx \pi^2 C\rho_1\rho_2 = \pi^2\rho_1\rho_2\left(\frac{3\alpha^2\hbar\omega}{4(4\pi\varepsilon_o)^2}\right)$$

or, more precisely,

$$A = \frac{3kT}{4}\left(\frac{\varepsilon_1 - \varepsilon_3}{\varepsilon_1 + \varepsilon_3}\right)\left(\frac{\varepsilon_2 - \varepsilon_3}{\varepsilon_2 + \varepsilon_3}\right)$$
$$+ \frac{3\hbar\omega}{8\sqrt{2}}\left[\frac{(n_1^2 - n_3^2)(n_2^2 - n_3^2)}{(n_1^2 + n_3^2)^{1/2}(n_2^2 + n_3^2)^{1/2}\left[(n_1^2 + n_3^2)^{1/2} + (n_2^2 + n_3^2)^{1/2}\right]}\right] \tag{2.6}$$

where k is the Boltzmann constant, T is temperature, ρ is the density, ε_i, is the dielectric constant in medium i (i.e., sample (1), tip (2), or intervening material (3)), and n_i is refractive index. The Hamaker constant is on the order of 10^{-19} J for most solids.

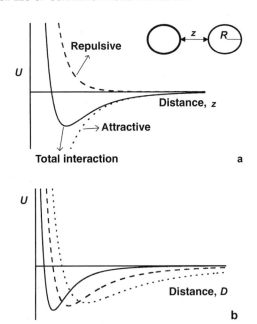

Figure 2.6. (**a**)The potential from an attractive van der Waals interaction (*dotted line*) and a hard wall repulsive interaction (*dashed line*) for two points showing the total interaction (*solid line*) as a sum of the two (**b**). The effect of geometry on the total potential is illustrated by comparing a point-point interactions (two molecules; *solid line*), a point-plane interaction (a molecule and a surface; *dashed line*) and a sphere-plane interaction (an AFM tip and a surface; *dotted line*). The power law decreases with larger components.

Remembering the following useful relationships makes a link between potential, force (measurable in the AFM), and force gradient (also measurable in the AFM).

		For a Spring	For a Sphere-plane, $z \ll R$
Potential	$= U$	$= 1/2\,kz^2$	$\propto 1/z$
Force	$= -dU/dz$	$= -kz$	$\propto -1/z^2$
Force gradient	$= -d^{-2}U/dz^2$	$= -k$	$\propto -1/z^3$

These relationships illustrate what happens in an AFM. The force between any two materials in a vacuum is attractive, the force between identical materials is attractive, the force between different materials can be repulsive, and exchanging the sample and tip materials does not change A. Also, the sensitivity or resolution of the images is related to slope of the potential curve. Atomic resolution is achieved only at small sample–tip separations. Examples of the magnitude of forces due to van der Waals interactions alone are compared in Table 2.1 for Cu spheres in vacuum. Nanometer scale force detection is easily accessed in AFM. In fact, estimates of the minimum detectable force are in the range of 10^{-18} N, and forces as small as 10^{-15} N have been measured.

Table 2.1   Forces between copper spheres in vacuum

Sphere Diameter, nm	1-nm Separation	10-nm Separation	100-nm Separation
1	10^{-9} N	10^{-11} N	10^{-13} N
10	10^{-8} N	10^{-10} N	10^{-12} N
100	10^{-7} N	10^{-9} N	10^{-11} N

2.3.2   Long-Range Interactions

In many situations, long-range forces act in addition to short-range forces between two surfaces. Examples of long-range interactions include electrostatic attraction or repulsion, current-induced or static-magnetic interactions, and capillary forces due to the surface energy of water condensed between the sample and tip. Close to the surface, these forces are much smaller than those due to van der Waals interactions and usually contribute little to the signal. Farther from the surface, the van der Waals interactions decay rapidly to the point of being negligible. In this regime, long-range forces are still significant. This difference in decay length provides a means to distinguish the two types of interactions.

The general relations describing the force experienced by a tip above a *homogeneous* surface for electrostatic and magnetic interactions are described in Eqs. (2.7) and (2.8), where ΔV is the difference in potential between sample and tip, C is the tip–sample capacitance as a function of separation z, B_{sample} is the magnetic field emanating from the sample surface, and m is the magnetic dipole of the tip. Conducting and/or magnetic tips are obviously necessary to access electric or magnetic fields.

$$F_{electrostatic} = -\frac{1}{2}(\Delta V)^2 \frac{\partial C}{\partial z} \tag{2.7}$$

$$F_{magnetostatic} = \nabla(m \cdot B_{sample}) \tag{2.8}$$

These relations are oversimplifications but suffice to describe the operating principles of electrostatic force imaging (EFM) and magnetic force imaging (MFM). A more rigorous treatment is given in Chapter 7 along with several examples of electric and magnetic field studies, instrumentation approaches to optimizing imaging, and challenges in image interpretation.

2.3.3   Force Detection and AFM Imaging

The underlying principle of AFM is that the interactions between the end of a probe tip that is mounted on a cantilever result in a response in the cantilever, notably a deflection. There are several modes of imaging based on atomic forces. In principle, one can measure the deflection of the cantilever associated with the force. Keeping the deflection constant by varying the vertical position of the tip produces a constant force image analogous to constant current STM. When done sufficiently close to the surface that van der Waals interactions dominate, the image represents the topographic structure of the surface. Deflection is not the most sensitive measurement, having a relatively small signal-to-noise ratio. An alternative more sensitive measurement uses the vibrational characteristics of the cantilever. The mechanical resonant frequency of the cantilever is determined by the dimensions of the structure and the properties of materials from which it is made. This resonant frequency (distinct from the scanner resonance discussed later in this chapter) is related to the cantilever spring constant according to Eq. (2.9), where the cantilever is conceptually treated as a classical, 1-D, lightly loaded, "fixed-free" beam:

$$\omega_o = \frac{t}{\ell^2}\sqrt{\frac{E}{\rho}} = \sqrt{\frac{k}{m_{\text{eff}}}} \tag{2.9}$$

The properties of some typical cantilevers are listed below:

Vendor	Cantilever	Resonant Frequency	Spring Constant
Olympus	Etched Si	200–400 kHz	42 N/m
Digital Instruments	Tapping mode Si	200–400 kHz	20–100 N/m
IBM	Super cone	275–450 kHz	50–200 N/m
Thermomicroscopes	Ultralever	30–300 kHz	0.2–50 N/m

The resonance is located by oscillating the cantilever with various frequencies and measuring the rms response (Fig. 2.7). At the resonant frequency of the cantilever the oscillation is amplified. When the cantilever is placed in a force gradient, the response is damped, shifting the resonant frequency. The vibration amplitude A detected at a given frequency ω changes as a function of the force gradient as shown in Eq. (2.10).

$$A(\omega = \omega'_o) = a_o Q\left(\frac{\omega_o}{\omega'_o}\right) \approx a_o Q\left(1 - \frac{\partial F/\partial z}{2k}\right)^{-1} \tag{2.10}$$

where Q is the quality factor and k is the spring constant of the cantilever. Note that by measuring the change in amplitude both the magnitude and sign of the force gradient can be determined for the case where the cantilever is initially oscillated away from resonance. Varying the vertical position of the tip such that the amplitude of oscillation at a particular frequency is constant produces a constant force gradient image. This ac mode has a larger signal-to-noise ratio than does the dc (deflection) mode. In addition, ac AFM is generally considered to be less damaging to soft surfaces (polymers) and to samples with poor substrate adhesion (nanotubes or DNA on silicon).

It is also possible to image based on the frequency or the phase of the cantilever oscillation. Extending the ac method to improve signal-to-noise further, an additional lock-in amplifier allows the phase or resonant frequency of the oscillat-

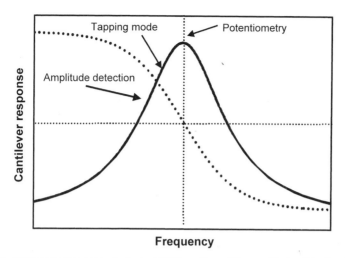

Figure 2.7. Frequency response of a typical AFM cantilever. The amplitude of the tip vibration as a function of driving frequency exhibits a peak that indicates the tip resonant frequency. The second line shows the phase of the cantilever oscillation.

ing cantilever to be the feedback signal. This is useful because any force gradient acting on the tip shifts the cantilever resonant frequency by decreasing the effective cantilever spring constant according to Eq. (2.11) (where c is a constant, k is the spring constant, and F' is a force gradient normal to the surface):

$$\Delta\omega_o = c\sqrt{k} - c\sqrt{k - F'} \qquad (2.11)$$

This frequency shift is actually responsible for the change in amplitude described by Eq. (2.10). By either maintaining a constant phase shift through adjustments of the cantilever driving frequency ("frequency modulation"), or by measuring the change in phase as a function of sample position, high-resolution maps of a variety of forces are obtained. Appropriate frequency's set points for various imaging modes are shown in Figure 2.7.

As discussed earlier, the strength of various force interactions depend on the tip–sample separation and are related to surface topography. To distinguish the topography from electrostatic or magnetic forces, a combination contact and noncontact measurement is made. Figure 2.8 illustrates the difference in topography based on short-range interactions and a force image based on long-range interactions. The example describes a magnetically or electrostatically inhomogeneous material with a rough surface. Typically, the topography is first obtained using dc or ac AFM, after which the tip is separated a preset distance from the surface (chosen so that short-range forces contribute negligibly compared to the electrostatic or magnetic forces). At this preset sample–tip separation, long-range interactions are measured, usually with an ac technique. This procedure results in the tip retracing the topography, but with a large sample–tip separation, effectively eliminating the

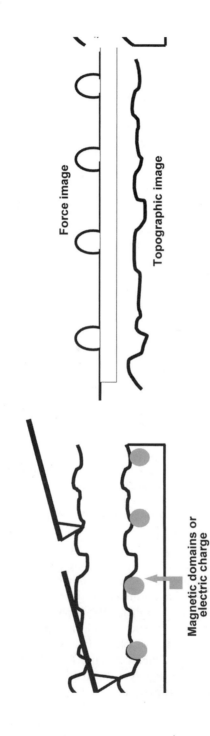

Force image

Topographic image

Magnetic domains or electric charge

Figure 2.8. Long-range force imaging. The tip acquires topographic structure at short range and retraces the topography above the surface. The process subtracts topographic variations to first order. Above the surface only long-range forces are detected. A material with inhomogeneous magnetic or electrostatic features would produce the topographic and force images at the left.

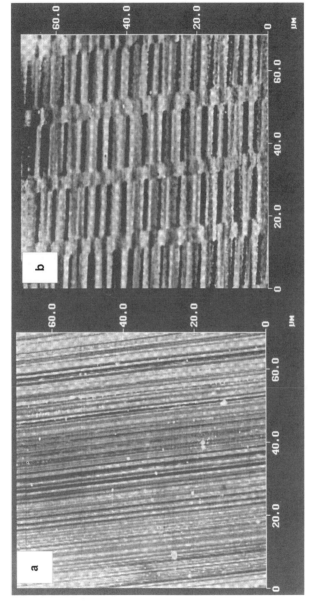

Figure 2.9. (a), Topographic structure and (b) magnetic force image of a compact disk. (Courtesy of R. Alvarez)

topographic contribution from the force image. The process can be done pixel by pixel, line by line, or scan by scan; an example of the second is shown in Figure 2.9. The topographic structure of the CD disk results from surface preparation and exhibits striations from a polishing process. Digital information is stored on the disk as small magnetic domains that are unrelated to surface topography. These are easily imaged in MFM. Various possible complications and artifacts, as well as several other examples of local property measurements, are discussed in detail in Chapter 7.

2.4 SYSTEM DESIGN

All SPMs consist of five parts: the tip, the scanner, the detector, the electronic control system, and a vibration isolation system. For a STM, the tip is effectively a metal wire, whereas for an AFM it is generally a physical tip attached to (or etched from) a cantilever. Details of tip construction and properties are given in Chapter 6. The detector for a STM is simply a voltmeter; for an AFM it, is commonly a laser and photodiode. Control systems and vibration isolation are similar for all SPMs.

2.4.1 STM Instrumentation

Although originally most STMs were made with tripod scanners, modern micro-scopes are almost invariably based on a tube configuration. The principles of piezoelectric behavior are the basis of operation of the tube scanners. In this geometry, the inner and outer surfaces of a tube of piezoelectric material are coated with a thin metal electrode. The outside is separated into four sections, which are electrically isolated from each other. For the lateral motion of scanning the voltage is applied across the tube; for vertical motion, the voltage is applied between the inside and outside of the tube. The segmented tube scanners are compact and have higher resonant frequencies, usually 12 to 20 kHz; however there is the potential for more cross talk and this configuration is inherently nonlinear.

Scanners are characterized by their scan range and resonant frequency. The scan range is a function of the piezoelectric material, the scanner dimensions, and the applied voltage range. To increase the scan range it is better to increase the piezo-electric constant rather than the size of the scanner. The resonant frequency is the most important parameter in the design of a STM, because it limits the scan rate, determines the stability with respect to mechanical vibrations, and affects electronic noise reduction strategy. The higher resonant frequency is advantageous for all considerations.

The tip or the sample can be mounted on the scanner. The sensitivity of the sample–tip approach mechanism is critical to successful operation of the STM. The criterion is that it must be possible to bring the sample and tip to within 50 Å of each other without making contact. This requires that the sample–tip approach mechanism have step resolution of 50 Å while exhibiting a dynamic range on the order of centimeters. Several types of mechanisms have been used, including electrostrictive devices, electromagnetic walkers, and enhanced sensitivity owing to a large mechanical disadvantage based on levers or differential springs.

A flow chart for the electronic control system is shown in Figure 2.10.[9] High voltage power supplies connected to the scanner usually have variable gains and are programmed with coordinated ramp functions to cause scanning. The low voltage power supply provides the sample–tip bias and is programmable for spectroscopy measurements. As the magnitude of tunneling currents is in the range of nanoamps it is necessary to amplify this signal by 7 to 10 orders of magnitude. This is done as close as possible to the tunnel junction, because any noise incorporated into the signal before this amplifier will be amplified with the signal. The output of the current amplifier enters the feedback controller, which compares the value to a reference and outputs a signal to the z voltage supply that will alter the tip position appropriately. The current preamplifier should have a gain variable over three orders of magnitude, the feedback circuit should have variable filter time constant and variable comparison gain. This is necessary because the sensitivity and response time of the feedback loop and the tendency of the circuit to become unstable and oscillate depend on the filter time constant and the total gain of the circuit. Because the gain at the sample–tip junction is not known and usually varies, it is an immense practical advantage to have control over these parameters.

2.4.2 AFM Instrumentation

Scanners in AFM are of the same design as those for STM, though they are available in a broader range of sizes to accommodate the larger scans that are of practical interest with AFM. A well-equipped SPM laboratory generally has three such scanners of approximately 1, 10, and 100 μm maximum lateral scan dimensions to accommodate the wide variety of samples. Both piezoelectric constants and dimensions are adjusted to achieve these larger scan ranges. Depending on the design, cantilevers or samples can be mounted on the scanner.

The quality of a ac-AFM cantilever, known as the Q-factor, is defined as the ratio of peak height at resonance to peak width at the full width at half maximum point.[10] This Q-factor is clearly tip dependent and can be modified by the surrounding medium (oscillating the tip in particular fluids or in vacuum can alter Q by several orders of magnitude). As Q improves, so does the signal-to-noise ratio, but a high Q cantilever has a longer response time. The driving frequency dependence of both amplitude and phase for a typical "diving-board" cantilever is shown in Figure 2.7.

The historical development cantilever motion detection involved using tunneling, capacitance, and eventually optical reflection as the signal.[11,12] Most AFMs are now based on the latter, in which a laser reflects off of the back of the cantilever and impinges on a position-sensitive photo diode. Any cantilever motion causes a detectable change in the laser's path (Fig. 2.11). A four-segment diode can quantify vertical and/or lateral motion by summing the signals. For example, vertical motion is determined as follows: $Z = (A + C) - (B + D)$, whereas lateral motion is $X = (A + B) - (C + D)$. The system can feedback on Z to produce AEM images. X is usually considered to be a measure of local friction, as discussed in Chapter 10.

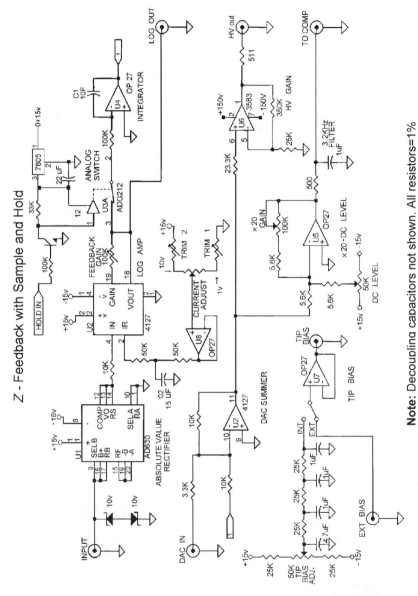

Note: Decoupling capacitors not shown. All resistors=1%

Figure 2.10. STM electronic systems. (Courtesy of R. Colton)

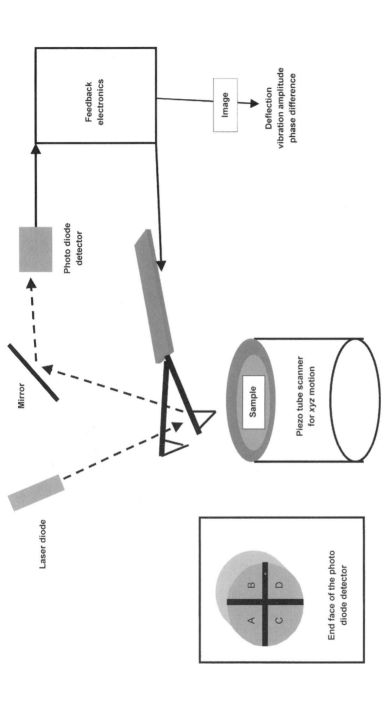

Figure 2.11. Operating principle of an AFM. The sample is mounted on a scanner. The cantilever and tip are positioned near the surface with a macroscopic positioning device, e.g., small motors or inchworm electronic devices. Cantilever deflection is detected with a photo diode that records the position of a laser beam that has been reflected off the top of the cantilever. System electronics operate a feedback circuit that access different signals to control tip oscillation and position, depending of the imaging mode. If the photo diode is segmented (*insert*) then variations in tip oscillation and position are quantified by the ratios of signals in each quadrant.

25

Specular reflection of the laser from the surface can sometimes interfere with the signal reflected from the cantilever, especially when imaging in fluids, although this difficulty is usually overcome by tilting the cantilever roughly 10° with respect to the surface.

Cantilever deflection has been measured using an integrated piezoelectric sensor as well.[13] This method has the obvious advantage of minimizing the need for external optical components, which is especially attractive for the miniaturization of force probes and the development of low-cost cantilever-based sensors. In fact, piezo-sensors, piezo-actuators, and feedback electronics have all been fully integrated onto a single cantilever.[14]

2.4.3 Vibration Isolation

The desired resolution in SPM imaging of 0.1 Å vertically and 1.0 Å laterally, imposes the requirement that noise from any source be < 0.01 Å in z and 0.1 Å in x and y. Mechanical vibrations are a large component of the noise, and it is particularly challenging to reduce these to less than a hundredth of an angstrom in the vertical direction. Vibrations that reach the sample–tip junction originate from the building, the table/chamber system on which the microscope sits, and possibly from components of the microscope itself.

Typical building vibrations are in the range of 5 to 100 Hz. Frames, walls, and floors of buildings undergo shear and bending vibrations, which are continuous excitations driven by machines running at or below line frequency (50 to 60 Hz). The maximum resonances occur at frequencies between 10 and 25 Hz, which are subharmonics of the driving frequencies. The exact frequency of the floor response is directly related to the maximum floor load. The amplitudes of the vibrations depend on the distance from major floor supports and distance from the ground floor, i.e., the farther away from a node that occurs at the floor support, the larger the vibration amplitude. Higher floors usually exhibit larger amplitudes than the ground floor. In addition to the continuously driven excitations, attention must also be paid to the floor response of irregular motion such as that caused by people walking, door closings, and elevator operation. Even some human voices and radios are in the right frequency range to drive microscope components. The typical floor response is shown in Figure 2.12 along with some variations possible within a building. Amplitudes can be as large as 100s of μms.

The table/chamber can have additional resonances in the range of 30 to 100 Hz, which can be driven by acoustic waves as well as the sources discussed above. Acoustic noise arising from machines operating at 50 to 60 Hz can couple to a typical vacuum chamber to produce a broad peak near 35 to 55 Hz, depending on the shape, mass, and center of gravity of the system. Resonance of simple system components can be estimated and can be designed to avoid those that are likely to couple acoustically.

Because so much of the design of the SPM, and ultimately its success, depends on damping vibration, it is useful to consider a few basic principles of vibration isolation. The response of the STM to vibrations can be understood by separating the tunneling assembly and the isolation assembly and modeling them as coupled oscillators having masses m_t, m_i ($m_i \gg m_t$) and spring constants k_i, k_t.[15] The discussion is restricted to vibrations of floors in a purely vertical mode. The equations of motion for the system with no damping are given by:

$$m_i x_i + k_i x_i + k_i(x_i - x_t) = k_i x_b \sin(\omega t) \tag{2.12}$$

$$m_t x_t + k_t(x_t - x_i) = 0 \tag{2.13}$$

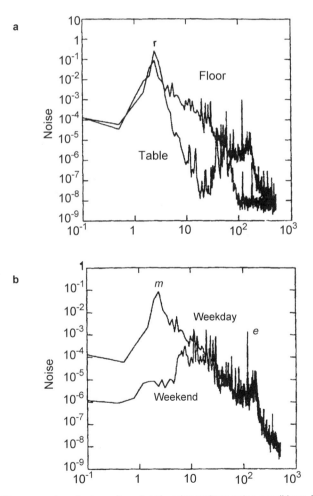

Figure 2.12. Power spectra of a tunneling signal under various noise conditions. The functions are the sum of Fourier coefficients squared. Sharp peaks labeled e are electrical noise, broader peaks labeled m are mechanical noise. The resonance of the air legs is labeled r. Reduced building activity affects the low-frequency noise significantly.

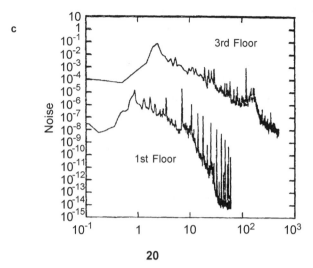

Figure 2.12. *(Continued)*

where x_i and x_t are the displacements of the isolation and tunneling assemblies, respectively, and $x_b \sin(\omega t)$ is the displacement of the floor, i.e., the forcing function. The transfer function defined for the vibration isolation assembly versus the floor is

$$Z_i = 20 \log \left[\frac{x_i}{x_b} \right] \qquad (2.14)$$

and that for the tunneling assembly versus the vibration isolation assembly is

$$Z_t = 20 \log \left[x_t - \frac{x_i}{x_i} \right] \qquad (2.15)$$

The overall transfer function, then is

$$Z = 20 \log \left[x_t - \frac{x_i}{x_b} \right] \qquad (2.16)$$

Figure 2.13 illustrates the system response by comparing the transfer between the floor and the vibration isolation assemble (*I*); the tunneling assembly and the vibration isolation assembly (*II*); and the sum, which is the transfer between the floor and the tunneling assembly (*III*). The response in curves *I* and *II* have three regions, one frequency range where the system has no effect on vibrations, an eigen frequency at which vibrations are actually amplified, and a large frequency regime where vibrations are attenuated. The eigen frequencies are associated with the

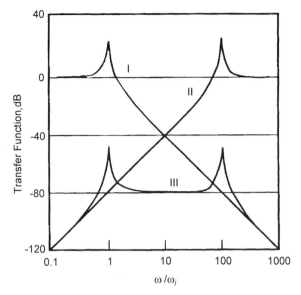

Figure 2.13. Calculated transfer function for the vibrations isolation assembly vs. the floor (*I*), for the tunneling assembly vs. the vibrations isolation assembly (*II*), and for the tunneling assembly vs. the floor (*III*). The resonance of the vibration isolation assembly (*R*) and that of the tunneling assembly (*R'*) are marked. This calculation is for the case in which the resonance of the tunneling assembly is 100 times larger than the isolation assembly. (Redrawn from Ref. 17.)

resonance of the isolation assembly (*I*) and the tunneling assembly itself (*II*). The total response of the system contains two resonances and a center plateau region. The design strategy is to reduce the first resonance as low as possible, increase the frequency of the second resonance, and reduce the plateau region. The overall isolation performance is improved by making the difference in the two eigen frequencies larger and the slope of curve *I* above the eigen frequency steeper. This corresponds to the highest resonance possible for the scanner and the lowest possible for the isolation assembly.

The optimum design of the SPM system depends on the particular vibration characteristics of the environment and those of the scanner. The vibration isolation system that is optimized for the seventh floor of a building on the beach in Santa Barbara, California, may not be the same as one optimized for the first floor of a building on a hill in upstate New York or in the basement of a building in downtown Philadelphia. In addition, an ultrahigh vacuum chamber with Auger, low-energy electron diffraction (LEED), ultraviolet photoemission spectroscopy (UPS), and in situ deposition capabilities will couple differently to acoustic vibrations than will the small canister often used for operation in air. Although the final design may be different, the strategy for vibration isolation is the same for all conditions and can be divided into three frequency regimes. The low-frequency regime is <20 Hz, the medium-frequency range is 20 to 200 Hz, and the high-frequency range is >200 Hz.

The low-frequency regime contains the vibrations from the building and the resonances of certain system components. An effective way to damp these vibrations is to suspend the microscope on long tension wires. This can be accomplished by suspending a massive table from the ceiling by bungy cords, where the length of the cord, the mass of the table, and the spring constant of the cords can be varied to achieve maximum attenuation at the frequencies of concern. The microscope itself could be suspended with tension spring, where again the length and spring constant could be chosen in the most effective range. A second approach to damping low-frequency vibrations is to levitate the system on an air table. The damping characteristics of a typical air table are given in Figure 2.14. In this case, the resonant frequency of the legs is usually set, but the performance, in terms of the magnitude of attenuation, can be improved by increasing the mass of the system. These two approaches affect the amplitude of the vibration that reaches the tunneling/AFM signal. It is also possible to affect the contribution of this noise to the image by increasing the scan rate. It is often possible to scan fast enough that the low-frequency noise that does get into the image is in the form of a curved background, which can be easily eliminated in image analysis. The scan rate is limited by the resonant frequency of the scanner, which, as discussed earlier, should be as high as possible.

Vibrations in the medium-frequency range result from mechanical vibrations from motors, resonances of the table and chamber that are acoustically excited, and unattenuated acoustic noise. The strategies that have been almost universally employed to damp vibrations in this frequency range are to mount the scanner assembly on a stack of materials of different elastic moduli or to suspend the scanner on tension springs with eddy current damping. High-frequency vibrations originating as resonances of the piezo elements can be eliminated from the signal with electrical filters without altering the part of the signal that represents the

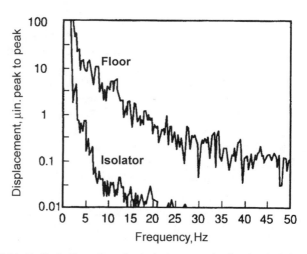

Figure 2.14. Vertical attenuation of a typical pneumatic vibration isolation system.

structural information; however, the design strategy is to have the scanner resonance so high that it is not excited during operation.

Some examples of system designs that yield good performance are shown in Figure 2.15. The UHV chamber of the STM system in Figure 2.15**a** sits on a table that is mounted on air legs (under the cabinet). The pumps are below table level, helping to lower the center of mass of the system. The STM itself is suspended on springs and damped by an eddy current device (Fig. 2.15**b**). This combination results in routine atomic resolution in most locations. In many studies of the atomic structures of surfaces, information in addition to STM analysis is beneficial. Microscope systems based on UHV chambers can be expanded to include any number of complementary analytical tools such as Auger electron spectroscopy (AES), LEED, energy loss spectroscopy (EELS), and mass spectrometry. UHV STMs have even been installed on molecular beam epitaxy film growth chambers. Atomic force microscopes are shown in Figure 2.15, along with control systems. These are ambient microscopes that usually rely on optical benches for vibration isolation. Ambient systems can be more flexible in terms of accepting a variety of sample configurations. Variable-temperature STM and AFM are now commercially available, allowing analyses from liquid He temperatures to 1000°C. Temperature variation introduces an additional instrument challenge in the form of thermal drift. It

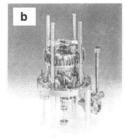

Figure 2.15. SPM systems. **a**, A UHV variable-temperature STM system, and **b**, the microscope assembly. (Courtesy of Omicron Vakuumphysik GmbH.) **c**, An ambient AFM system with control electronics, and **d**, a small stand along with the table top AFM/STM. (Courtesy of Digital Electronics, Inc.)

is relatively straightforward to cool or heat samples in vacuum chambers, even doing so without introducing mechanical vibrations. But to maintain atomic resolution, thermal drift must be less than 1 Å/min.

2.5 DATA ANALYSIS

To extract information from image or spectroscopy data, acquisition artifacts must be removed. The strategies for data analysis have been developed in many fields in which data are in the form of images, including astronomy and medicine, as well as conventional and electron beam–based microscopies. Advances in computer capability allow sophisticated image processing on regular computers. Standard routines for image processing and spectral analysis are regularly included by SPM manufacturers for commercial microscopes. It is also usually possible to save images and spectra in standard formats and transport them to external analysis programs. The strategies for routine SPM data analysis are outlined in this section.

2.5.1 Image Processing

The format of image data is usually a square array of numbers in which the array position corresponds to x, y coordinates and the number represents the signal intensity at that position. There is often a header and footer to the file that contain instrument specific information in a format specific to the manufacturer. When conducting image processing within a commercial SPM package, the formatting details of the header and footer can be ignored, because they are automatically accounted for in the analysis routines. (If you export this data array to an independent image analysis program, the information in the header file will likely be necessary for calibration.) Typical image arrays are 128×128, 256×256, 512×512, and sometimes 1024×1024. The lateral information density is then the scan size divided by the number of pixels. A 124×124 array of a 10-μm× 10-μm area yields a step size on the order of 100 nm, whereas the same array size of a 10 nm × 10 nm area has subangstrom resolution. The information density in the vertical dimension, i.e., the number in the array, depends on the feature height (or force gradient strength). For example, using a 16-bit processor there are 2^{15} possible values for height, with an additional bit indicating the sign. For a scanner with a vertical range of 5 μm, this corresponds to a resolution of 0.75 Å, whereas for a 1-μm vertical range the resolution is 0.15 Å. The first step in data analysis is to confirm that the data have sufficient information density to resolve features of interest.

Image-processing routines can be separated into two categories: Those that compensate for instrument defects and those that quantify surface information. It is virtually impossible to bring a tip into contact with a sample plane so that they are exactly perpendicular; therefore, all images include a plane containing this deviation. This background plane is calculated from a least squares fit and then simply subtracted from the data, leaving $z' = A*(x) + B*(y)$, where A and B are constants. Sometimes nonlinearities in the scanner or very low frequency vibrations give rise to a curvature in the background. This is compensated by using a polynomial of order n for the subtraction rather than a plane, so that $z' = A*(x^n) + B*(y^n)$, where n of 4 or less is usually sufficient. It should be noted that subtracting a plane is not the same mathematical process as rotating the plane. The difference is noted in that a stepped structure becomes a sawtooth on plane subtraction. Consequently, if the directions between surface and edge planes are important, these subtleties need to be considered. Commercial SPMs often offer several mechanisms for subtracting the plane, including whole plane, line by line, or partial plane subtractions. A plane in the x direction can also be subtracted independently of a plane in the y direction. The character of the data dictates which routine is appropriate.

Other instrumental artifacts include tip jumps and incompletely suppressed high-frequency noise. These are often eliminated with simple filters. High pass and low pass filters multiply each pixel by a matrix that weights nearby pixels and averages the values. Relative weighting factors for adjacent pixels with high pass and low pass filters might be

$$
\begin{array}{ccc}
1 & 1 & 1 \\
1 & 9 & 1 \\
1 & 1 & 1
\end{array}
\qquad
\begin{array}{ccc}
5 & 5 & 5 \\
5 & 1 & 5 \\
5 & 5 & 5
\end{array}
$$

These filters have the effect of reducing the measured heights of features that occur over a single pixel and are nonphysical. Figure 2.16 illustrates the effects of processing routines on an AFM image of graphite. The actual image data (Fig. 2.16a) have dark contrast at the top and bottom with a bright region in the center, typical of nonlinearity in the scanner when used near the maximum scan range. This is removed with a third-order background subtraction (**b**) or a flattening (**c**). The resulting image has relatively low contrast, i.e., all of the intensity is in a small range of gray. The data are stretched in the z direction by a contrast enhancement (**d**), which reveals that background curvature in the x direction is present. This is removed with an additional third-order background subtraction in the x direction (**e**), and finally the contrast is enhanced again so that the data fill the vertical scale. The final image displays the features clearly, and these operations have removed only artifacts, not image content.

It is also possible to choose a frequency to eliminate from the image, which can be useful when intermediate frequency electrical noise penetrates the data. This is done by calculating a 2-D Fourier transform of the data and selecting the data to be retained or the noise to be removed (depending on the program), then inverse transforming the data. The approach is extremely effective for periodic data and hence atomic resolution images. The obvious caution is that image content can be simultaneously removed, knowingly or unknowingly, introducing artifacts into the data. In fact, it is possible to filter white noise to the point of producing periodic image features. Great care must be taken in applying these filters, and the details should always be reported in presentations and publications.

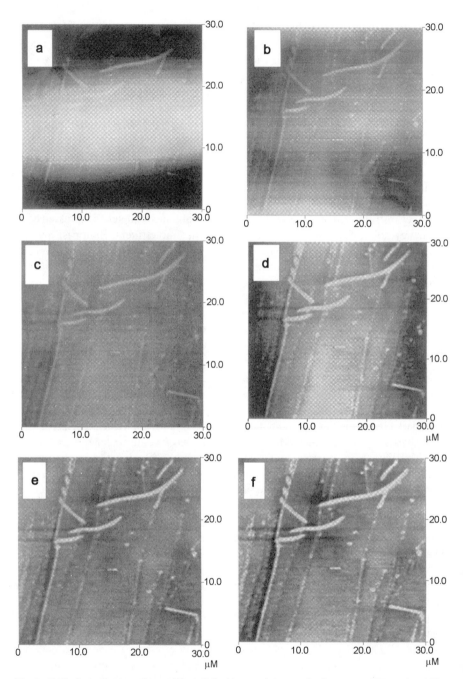

Figure 2.16. Image processing. **a**, The original image data acquired on a graphite surface. The image after a third-order background subtraction (**b**), after a line-by-line plane subtraction or flattening procedure (**c**), and after flattening and a low-pass filter (**d**), **e**, The same image after an additional final third-order background subtraction in the *x* direction. **f**, The image with a final contrast expansion. (Courtesy of R. Alvarez.)

The second category of data processing involves procedures aimed at quantifying aspects of surface structure. An obvious measurable is surface roughness. Quantification of macroscopic surface roughness is a well-established field, and the reader is referred to any of several texts for introductory concepts and advanced discussion.[16,17] Analyses of scanning probe images rely on the same mathematics for roughness quantification. Roughness can be reported as the average peak-to-peak height variation, the root mean square of vertical variation, or in more sophisticated analyses as the autocovariance or power spectral density. Keep in mind that roughness is a scale-dependent parameter and cannot be rigorously represented as a single number. A power spectrum, which is related to a Fourier transform, contains roughness information at all length scales in reciprocal space. A conceptionally straightforward approach to describe roughness at all length scales is to use variational RMS measurements.[18] The root mean square (RMS) roughness of a small subset of the image is calculated. The process is repeated with increasingly larger subsets of the image. The data are then plotted as a function of the size of the image subset. An example is shown in Figure 2.17 for a silicon fracture surface.

In principle, the slope in Figure 2.17**b** is related to the surface character, and any changes in slope are related to characteristic length scales on the surface. In this case, the slope change between 500 nm and 2 μm is related to the striations observed in the image. At some limit, the roughness will become independent of length scale. This value, referred to as maximum roughness, should be that reported if only a single number will be compared. Figure 2.17 compares the variational roughness with the power spectrum (calculated from a Fourier transform) and autocovariance of the same surface. The slopes of both the variational roughness and the power spectrum can be related to the surface fractal dimension. Variational roughness has the advantage that it displays the scale dependence of roughness in real space; however, although RMS is usually included in commercial programs, the variation usually must be performed manually.

When surface features are textured, having some lateral directionality, then a 2-D Fourier transform or autocorrelation function can be used to quantify this aspect of the structure. The width of the feature in Figure 2.17**d** ranges from 700 nm to 2 mm, indicating that the characteristic length of the topographic structure varies with direction. Figure 2.18 compares the image and autocovariance of the surface of a ceramic fiber used in structural composites. The asymmetry in the autocorrelation function quantifies the directionality of the features.

It is important to note that the image processing procedures illustrated in Figure 2.19 affect the roughness calculated from the image. If the image contains a large background plane, there will be a component to the roughness that has a maximum at the image size. The effects of post-acquisition image processing on roughness are illustrated on data from a calibration grid with 10 μm spacing. The grid spacing is identified as a characteristic length in all cases, but the roughness at all scales above the grid dimension depends on image processing. In this case, curve D represents most accurately the real surface. If, on the other hand, the background slope were part of the structure then curve A would better represent the real surface. This example further illustrates the subjective nature of image processing and the care that must be taken to avoid misinterpretation.

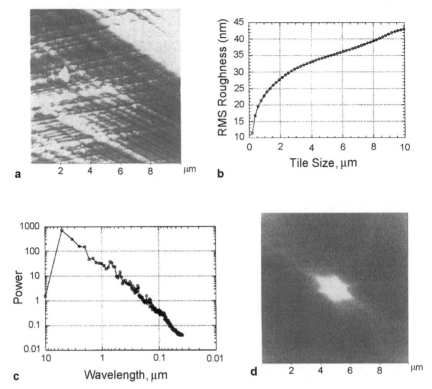

Figure 2.17. AFM image of (**a**) silicon fracture surface, (**b**) the variations in RMS roughness, (**c**) the power spectrum, and (**d**) the 2-D auto correlation function. (Ref. 18)

2.5.2   STM Image Simulations

To make a correlation of atomic positions and features in a SPM image, it is necessary to compare the image with a model. For example, the contrast in a STM image is actually related to charge density (see Eq. (2.3) and Chapters 4 and 5), which is a complicated function of atomic core position and local bonding interactions. In the cases of semiconductors and metals, the experimental images can sometimes be compared to first principles calculations. (Examples can be found in Figures 3.3, 3.4, 3.6 and throughout many of the subsequent chapters.) Numerous examples of studies that combine experimental STM images and calculations of surface charge density for metals, graphite, semiconductors, and even oxides are found in the literature. These approaches to STM image calculation are the most rigorous, and therefore most correct, but can require extensive computational resources and certainly require the services of a knowledgeable and willing theorist. A comparison to first principles calculations becomes problematic when surfaces consist of reconstructions with large unit cells or are made of materials with complex bonding, such as oxides. In these cases, a more approximate approach has been

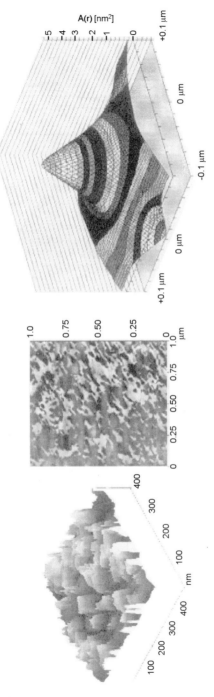

Figure 2.18. AFM image and 2-D autocorrelation function of a ceramic fiber surface.

37

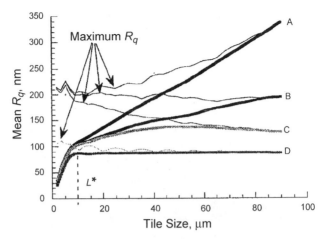

Figure 2.19. The effect of post-acquisition image processing on the calculated RMS roughness of a calibration grid with 10 μm spacing. Curve *A* is the raw data, curve *B* is plane subtracted in the fast scan direction, curve *C* is plane subtracted in both directions, and curve *D* is plane subtracted and flattened. L^* indicates the length determined from the slope change at about 10 μm. (Ref. 18)

found useful in interpreting contrast in STM images to first order. Several authors have suggested procedures, including tight binding approaches and valence sum calculations.[19–21] A simple and general routine is illustrated here for the case of a binary compound with mixed ionic/covalent bonding.[22]

The treatment essentially follows that of Chapter 3 and contains all the assumptions implied by Eq. (3.7). Instead of determining the wave functions in this expression directly through energy minimization, however, Slater-type[23] orbitals are fit to first principle solutions of empty state densities (ESD) for surfaces with high symmetry. This strategy is applied to determine the atomic structure of a reduced $TiO_2(110)$ surface. STM images of TiO_2 have been observed only while accessing empty states in the conduction band that have an energy between E_F and $E_F + eV$. The results of first principle studies show that the density of states (DOS) in this energy range is dominated by Ti 3d orbitals, but there is a finite contribution from O 2p as well. Therefore, the model ESD of the squares of parametric Slater-type Ti 3d and O 2p orbitals is built according to the following expansion:

$$\mathrm{ESD}(\vec{r}) = \sum_i A_i \phi_{Ti_i}^2(\vec{r} - \vec{r}_i) + \sum_j A_j \phi_{O_j}^2(\vec{r} - \vec{r}_j) \qquad (2.17)$$

where the sums are over all Ti sites i and O sites j, the $A_{i,j}$ are the coefficients in the expansion, and the parametric Slate-type orbitals are denoted by ϕ. Each orbital is composed of the product of a radial term and an angular term that are solutions to the Schrodinger equation for a generic spherically symmetric potential in a one electron atom. Only the radial portion of these orbitals is adjusted to fit first principles calculations of reference surfaces. The Ti 3d and O 2p the radial functions are

$$R_{\mathrm{Ti}}(\vec{r}) = \frac{4}{81\sqrt{30}} \left(\frac{3}{2}\alpha_{\mathrm{Ti}} \right)^{7/2} r^2 e^{-\alpha_{\mathrm{Ti}} r/2}$$

and

$$R_{\mathrm{O}}(\vec{r}) = \frac{1}{2\sqrt{6}} (\alpha_{\mathrm{O}})^{5/2} r e^{-\alpha_{\mathrm{O}} r/2} \tag{2.18}$$

The orbitals are fully normalized. The fitting parameters are the A_i, A_j, and α_{Ti}, α_{O}. The coefficients in the expansion represent the average fraction of the total density of states that each type of orbital retains over the energy range relevent to tunneling. The parameters α_{Ti} and α_{O} are inversely related to the nuclear potential for each ion. Altering these parameters changes the effective radii of the Slater orbitals. Note that there are a relatively small number of parameters in the fit. For the example case of highly symmetric $TiO_2(110)$ surfaces only four parameters are adjusted to match reference structures.

The experimental STM image of the surface is shown in Figure 2.20 with inserts of the two possible surface atomic structures. Subtraction of the experimental and calculated images allows quantification of the similarity. In this case, it is clear that the upper model represents the structure of the surface. It is also clear that further refinement is necessary to determine the details of the structure. More complex calculations are required to refine the structure; however, the relatively simple and computationly quick simulations were sufficient to distinguish between two atomic models of the surface. A computer program to implement these calculations is provided in Appendix III.

2.5.3   STS Calculations

One also obtains quantitative information from a comparison of experimental tunneling spectra to models. In the case where a surface has a high density of surface states and the Fermi level is pinned, the derivative or logarithmic derivative of the tunneling spectrum is plotted. This presentation highlights the peaks associated with localized states and determination of the energies is straightforward. This is discussed in detail in Chapter 4. For materials with low surface state density and low electrical conductivity, considerable band bending occurs during the acquisition of the spectra.[24] This situation applies to wide band gap semiconductors. In this case, tunneling spectra manifest properties of the sample such as dielectric constant, carrier concentration, effective mass of the electrons, band gap, and surface charge. If some of these parameters are constant during a measurement, then variations in others can be determined. Figure 2.21 shows the effect of some of these parameters on the tunneling spectrum of a semiconductor with a 3 V band gap (which applies, e.g., to ZnO, TiO_2, and $SrTiO_3$).[25] Similar calculations have been

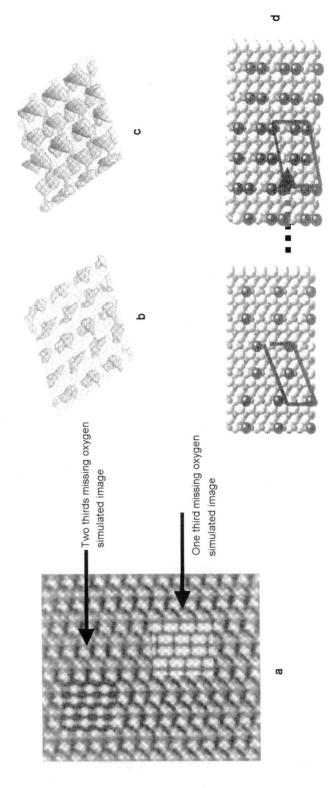

Figure 2.20. STM image simulation. Simulated images of two possible atomic structures are superimposed on the experimental STM image of a TiO₂ surface. The difference images (i.e., experimental image—the simulated image) for the potential atomic structures are compared. The reconstruction is determined to result from the loss of two thirds of the oxygen in the surface bridging rows.

One third missing oxygen simulated image

Two thirds missing oxygen simulated image

a

b

c

d

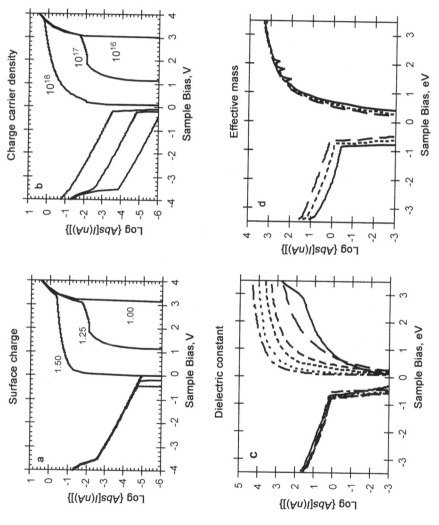

Figure 2.21. The effect of materials properties on tunneling spectra for a material with a 3.2 eV band gap.

41

useful in determining the surface charge associated with adsorption or segregation to surfaces.[26] A computer program to implement these calculations is provided in Appendix IV.

REFERENCES

1. J. D. Jackson, *Classical Electrodynamics*, John Wiley & Sons, New York, 1998.
2. S. M. Sze, *Physics of Semiconductor Devices*, John Wiley & Sons, New York, 1981.
3. A. Zangwill, *Physics at Surfaces*, Cambridge University Press, Cambridge, UK, 1988.
4. Ed. H. J. Guentherodt and R. Wiesendanger, *Scanning Tunneling Microscopy I*, Springer-Verlag, Berlin, 1992.
5. J. Chen, *Scanning Tunneling Microscopy*, 1994.
6. C. Bai, *Scanning Tunneling Microscopy and Its Application*, Springer-Verlag, Berlin, 1992.
7. D. Sarid, *Scanning Force Microscopy*, Oxford University Press, New York, 1991.
8. J. N. Israelachvili, *Intermolecular and Surface Forces*, Academic Press, New York, 1992.
9. D. Di Lellam J. Wandass, R. Colton, and C. Marrian, *Rev. Sci. Instrum.* **60**, 997 (1989).
10. A. Nowick, B. Berry, in *Anelastic Relaxation in Crystalline Solids*, Academic Press, New York, 1972, pp. 621–631.
11. R. Erlandsson, et al., *J. Vac. Sci. Technol A* **6**, 226 (1988).
12. C. Schonenberger and S. F. Alvarado, *Rev. Sci. Instrum.* **60**, 3131 (1989).
13. R. Linnemann, et al., *J. Vac. Sci. Technol. B* **14**, 856 (1996).
14. N. Blanc, et al., *J. Vac. Sci. Tech. B* **14**, 901 (1996).
15. M. Okano, K. Kajimura, S. Wakiyama, et al., *J. Vac. Sci. Technol.* **A5**, 3313 (1987).
16. John. C. Russ, *Surface Roughness* (1988).
17. T. R. Thomas, *Rough Surfaces*, Longmans, London, 1982.
18. J. D. Kiely and S. A. Bonnell, *J. Vac. Sci. Technol. B* **15**, 1483 (1997).
19. R. L. Smith, G. S. Rohrer, K. S. Lee, et al., *Surf. Sci.* **367**, 87 (1996).
20. R. Smith, W. Lu, and G. S. Rohrer, *Surf. Sci.* **322**, 293 (1995).
21. O. Gülseren, R. James, and D. W. Bullett, *Surf. Sci.* **377–379**, 150 (1997).
22. J. R Smith and D. A. Bonnell, *Physical Review B*, **62**, 2000.
23. J. C. Slater, *Quantum Theory of Molecules and Solids*. Vol. 2: *Symmetry and Energy Bands in Crystals*, McGraw-Hill Book Co., New York, 1963.
24. R. Feenstra and J. Stroscio, *J. Vac. Sci. Technol. B* **5**, 923 (1987); R. Feenstra, *Phys. Rev. B*. **50**, 4561 (1994).
25. A. Frye, Ph.D. thesis, University of Pennsylvania, 1999.
26. D. A. Bonnell, *J. Am. Cer. Soc.* **81**, 3049 (1998)
27. P. Thibado, Ph.D. thesis, University of Pennsylvania, 1994.

3

THEORY OF SCANNING TUNNELING MICROSCOPY

Jerry Tersoff

3.1   INTRODUCTION

This chapter reviews the current understanding of scanning tunneling microscopy (STM).[1-4] Chapter 2 introduced the basic principles of STM, which are sufficient for understanding nanometer-scale topographic imaging. In the case of atomic resolution, a microscopic quantum-mechanical theory is needed, and this is presented in Section 3.2. The theory of Section 3.2 relates the STM image to the wavefunctions of the surface. It turns out that, depending on the surface electronic structure, two qualitatively different types of behavior can occur. For some surfaces (in particular metals), the STM image may correspond quite closely to a topograph of the surface, even on an atomic scale. The interpretation of STM images then becomes rather simple. The analysis of images in this case is treated in Section 3.3.

For other surfaces, including many semiconducting surfaces, the STM image may differ drastically from a simple topograph on the atomic scale. In these cases, the dependence of the image on tunneling voltage provides a crucial clue in the interpretation. The analysis of STM images for such surfaces is discussed in Section 3.4. Finally, Section 3.5 touches on the possible distortion of STM images by mechanical interactions between the surface and the tip.

3.2   THEORY OF STM

3.2.1   Beyond Topography

As long as the features resolved are on the nanometer scale or larger, interpretation of the STM image as a surface topograph (complicated by local variations in work function) is generally adequate. But soon after the invention of STM,[1] Binnig et al.[5] reported the first atomic resolution images. On the atomic scale, it is not even clear what one would mean by a topograph. The most reasonable definition would be that a topograph is a contour of constant surface charge density. However, there is no reason why STM should yield precisely a contour of constant charge density, because only the electrons near the Fermi level contribute to the tunneling, whereas all electrons below the Fermi level contribute to charge density. Thus, on some level, the interpretation of STM images as surface topographs must fail. The following sections describe a more precise interpretation of STM images, applicable even in the case of atomic resolution.

The tunneling process is illustrated in Figure 2.1. In principle, one could directly calculate the transmission coefficient for an electron incident on the vacuum barrier between a surface and tip, and some early theoretical studies took this approach.[6] However, such a calculation is not feasible for a realistic model of the tip and surface. Fortunately, for typical tip-sample separations (of order $10\,\text{Å}$ nucleus-to-nucleus) the coupling between tip and sample is weak, and the tunneling can be treated with first-order perturbation theory. Because the problem is otherwise rather intractable except in simplified models of the surface,[6–8] I restrict the remaining discussion to this weak-coupling limit.

3.2.2   Tunneling Hamiltonian Approach

In first-order perturbation theory, the current is

$$I = \frac{2\pi e}{\hbar} \sum_{\mu,\nu} [f(E_\mu) - f(E_\nu)] |M_{\mu\nu}|^2 \delta(E_\nu + V - E_\mu) \tag{3.1}$$

where $f(E)$ is the Fermi function, V is the applied voltage, $M_{\mu\nu}$ is the tunneling matrix element between states ψ_μ and ψ_μ of the respective electrodes, and E_μ is the energy of ψ_μ. For most purposes, the Fermi functions can be replaced by their zero-temperature values, i.e., unit step functions, in which case the term in brackets has magnitude zero or unity. In the limit of small voltage, this expression further simplifies to

$$I = \frac{2\pi}{\hbar} e^2 V \sum_{\mu,\nu} |M_{\mu\nu}|^2 \delta(E_\nu - E_F)\delta(E_\nu - E_F) \tag{3.2}$$

These equations are quite simple. The only real difficulty is in evaluating the tunneling matrix elements $M_{\mu\nu}$. Bardeen[9] showed that, under certain assumptions, these can be expressed as

$$M_{\mu\nu} = \frac{\hbar^2}{2m} \int dS \cdot (\psi_\mu^* \nabla \psi_\nu - \psi_\nu \nabla \psi_\mu^*) \tag{3.3}$$

where the integral is over any surface lying entirely within the barrier region. If we choose a plane for the surface of integration and neglect the variation of the potential in the region of integration, then the surface wavefunction at this plane can be conveniently expanded in the generalized plane wave form

$$\psi = \int d\mathbf{q}\, a_q e^{-\kappa_q z} e^{i\mathbf{q} \cdot \mathbf{x}} \tag{3.4}$$

Here z is height measured from a convenient origin at the surface, $\mathbf{q}$ is the Fourier (spatial frequency) wavevector, $a_{\mathbf{q}}$ is the corresponding expansion coefficient, and the decay constant from Eq (2.3) is

$$\kappa_q^2 = \kappa^2 + |\mathbf{q}|^2 \tag{3.5}$$

A similar expansion applies for the other electrode, replacing a_q with b_q, $\mathbf{z}$ with $z_t - z$, and x with $x - x_t$. Here x_t and z_t are the lateral and vertical components of the position of the tip.

Then, substituting these wavefunctions into Eq. (3.3), one obtains

$$M_{\mu\nu} = -\frac{4\pi^2 \hbar^2}{m} \int d\mathbf{q}\, a_{\mathbf{q}} b_{\mathbf{q}}^* \kappa_q e^{-\kappa_q z_t} e^{i\mathbf{q} \cdot \mathbf{x}_t} \tag{3.6}$$

Thus, given the wavefunctions of the surface and tip separately, i.e., a_q and b_q, one has a reasonably simple expression for the matrix element and tunneling current.

3.2.3 Modeling the Tip

To calculate the tunneling current, and hence the STM image or spectrum, it is first necessary to have explicitly the wavefunctions of the surface and tip, for example, in the form of Eq. (3.4) for use in Eq. (3.6). Unfortunately, the actual atomic structure of the tip is generally not known.[10] Even if it were known, the low symmetry would probably make accurate calculation of the tip wavefunctions infeasible.

One must, therefore, adopt a reasonable but somewhat arbitrary model for the tip. To motivate the simplest possible model for the tip, consider what would be the ideal STM.[11] First, one wants the maximum possible resolution, and therefore the smallest possible tip. Second, one wants to measure the properties of the bare surface, not of the more complex interacting system of surface and tip. Therefore, the ideal STM tip would consist of a mathematical point source of current. In that case, Eq. (3.2) for the current at small voltage reduces to

$$
\begin{aligned}
I &\propto \sum_{\nu} |\psi_\nu(r_t)|^2 \delta(E_\nu - E_F) \\
&\equiv \rho(\mathbf{r}_t, E_F)
\end{aligned}
\tag{3.7}
$$

because any contributions to the matrix element are proportional to the amplitude of ψ_ν at the tip.

Thus the ideal STM would simply measure $\rho(\mathbf{r}_t, E_F)$. This is a familiar quantity, the local density of states at E_F, i.e., the charge density from states at the Fermi level, for the bare surface at the position of the tip. Thus within this model, STM has quite a simple interpretation as measuring a property of the bare surface, without reference to the complex tip–sample system.

It is important to see how far this interpretation can be applied for more realistic models of the tip. Tersoff and Hamann[11] showed that Eq. (3.7) remains valid, regardless of tip size, as long as the tunneling matrix elements can be adequately approximated by those for an s-wave tip wavefunction. The tip position $\mathbf{r}_t$ must then be interpreted as the effective center of curvature of the tip, i.e., the origin of the s-wave that best approximates the tip wavefunctions.

One can also to some extent go beyond the s-wave tip approximation, while still getting useful analytical results. A discussion of the contribution of wave function components of higher angular momentum was given by Tersoff and Hamann,[11] who showed that these made little difference for the observable Fourier components of typical STM images. This issue was raised again by Chen,[12] who extended Tersoff and Hamann's analysis. Chen noted that in more recent images of close-packed metals the relevant Fourier components are high enough that higher angular momentum components of the tip wavefunction could indeed affect the image substantially. However, it is important to recognize that such deviations from the behavior expected for an s-wave tip would be large only on this special class of surfaces. If a tip were to have a purely d_z wavefunction, then the corrugation for, say, Al(111) would be drastically increased, but the effect on surfaces observable with typical STM resolution would be relatively modest. Whether such tips exist is in any case an open question.

Sacks et al.[13] also treated a simple but informative model for a free-electron tip. More recently,[14] it was shown that the s-wave tip model should accurately describe STM images for a free-electron tip (aside from effects of gross tip shape), except in the case of tunneling to band edge states, e.g., to semiconducting surfaces at low voltage; and that even then, none but the lowest Fourier component of the image should differ much from the s-wave result. This, of course, neglects the obvious effects of tip geometry, e.g., double tips.[15]

To model the tip more realistically, one must turn to numerical calculations of wavefunctions for a specific tip. Lang[16–19] performed several illuminating studies in this vein, modeling the tip (and the sample) as an atom adsorbed on a model free-electron metal substrate (i.e., the "jellium" model). His results[16] for the STM images of a sodium and a sulfur atom on jellium (with a sodium atom tip) are shown in Figures 3.1 and 3.2.

In the case where both tip and surface are Na on jellium, and everything is highly metallic, the image in Figure 3.1 is almost indistinguishable from the local density of states (LDOS), in striking confirmation of the s-wave tip model. In addition, both of these look like the total charge density, i.e., the image corresponds closely to a surface topograph. For a Na-on-jellium tip and a sulfur-on-jellium surface (Fig. 3.2), the image is again well reproduced by the LDOS, but these are distinctly different from the total charge density. So although the simple LDOS interpretation Eq. (3.7) remains valid, the image does not correspond to a topograph in this case.

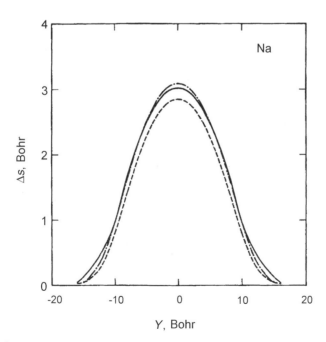

Figure 3.1. Comparison of theoretical STM image for Na adatom sample and Na adatom tip, with contours of constant Fermi-level local density of states (LDOS) and constant total charge density. [—, tip displacement; ·—·—·, Fermi-level LDOS; - - - - -, total density.] (Reprinted with permission from Ref. 16.)

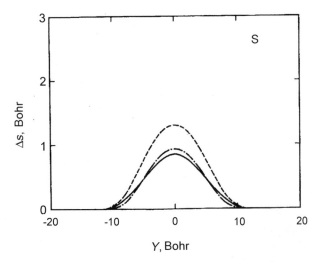

Figure 3.2. Comparison of theoretical STM image for S adatom sample and Na adatom tip, with contours of constant Fermi-level local density of states (LDOS) and constant total charge density. [—, tip displacement; ·····, Fermi-level LDOS; - - - -, total density.] (Reprinted with permission from Ref. 16.)

3.3   METAL SURFACES: STM AS SURFACE TOPOGRAPHY

3.3.1   Calculation of the LDOS

To interpret STM images quantitatively, just as with other experimental techniques, it is often necessary to calculate the image for a proposed structure, or set of structures, and compare with the actual image. The next two chapters discuss some examples and general features of such calculations, as well as the qualitative interpretation of STM images without detailed calculations.

For simple metals, there is typically no strong variation of the local density of states or wavefunctions with energy near the Fermi level. For purposes of STM, the same is presumably true for noble and even transition metals, because the d shell apparently does not contribute significantly to the tunneling current.[18] It is, therefore, convenient in the case of metals to ignore the voltage dependence and consider the limit of small voltage (Eq. 3.7). (Effects of finite voltage are discussed in Section 3.4.)

This is particularly convenient, because we then require only the calculation of the LDOS $\rho(r, E_f)$, a property of the bare surface. In principle, any standard method for calculating electronic structure can be used to obtain $\rho(r, E_f)$. However, this calculation is demanding numerically, and there are technical problems that make many common methods poorly suited. One early example of a calculation for a real metal surface and comparison with experiment is the case of Au(110) 2×1 and 3×1. The LDOS calculated for these Au surfaces is shown in Figure 3.3.

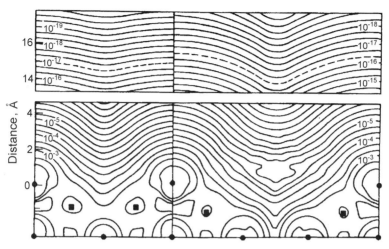

Figure 3.3. Calculated $\rho(r, E_F)$ for Au (110)-(2 × 1) (*left*) and (3 × 1) (*right*) surfaces. The figure shows the (1$\bar{1}$0) plane through outermost atoms. Positions of nuclei are indicated by *solid circles* (in plane) and *solid squares* (out of plane). Contours of constant ρ are labeled in units of a.u.$^{-3}$ eV^{-1}. Note the break in the vertical distance scale. Assuming a 0.9-nm tip radius in the *s*-wave tip model, the center of curvature of the tip is calculated to follow the dashed line. (Reprinted with permission from Ref. 11.)

Even for these relatively simple surfaces, the calculation was at the limits of feasibility at that time. Note in particular the unphysical detail in the deep trough for the 3 × 1 case, which was attributed to incomplete convergence of the basis set (i.e., to Gibbs oscillations), owing to computer memory limits. However, with more powerful computers and more computationally economical scheme,[20] it has become more common to calculate STM images for use in the analysis of experimental data.

3.3.2 Atom-Superposition Modeling

The real strength of STM is that, unlike diffraction, it is a local probe, and so can be readily applied to large complex unit cells, or even to disordered surfaces or to isolated features such as defects. However, while the accurate calculation of $\rho(r, E_f)$ is difficult even for Au(110) 3 × 1, it is out of the question for surfaces with large unit cells and *a fortiori* for disordered surfaces or defects. It is, therefore, highly desirable to have a method, however approximate, for calculating STM images in these important but intractable cases. Such a method was suggested and tested by Tersoff and Hamann.[11] It consists of approximating the LDOS (Eq. (3.7)) by a superposition of spherical atomic-like densities

$$\rho(r, E_F) \propto \sum_R \frac{e^{-2\kappa|r-R|}}{E_0\kappa|r-R|}. \tag{3.8}$$

Here each term is an atomic-like density centered on the atom site. The choice of an s-wave Hankel function allows convenient analytical manipulations and provides an accurate description even at large distances. E_0 is an energy that makes the formula dimensionally correct and is typically of order $0.5-1.0\,\text{eV}$.

This approach is expected to work well for simple and noble metals and was tested in detail for Au(110)[11]. The success of the method relies on the fact that the

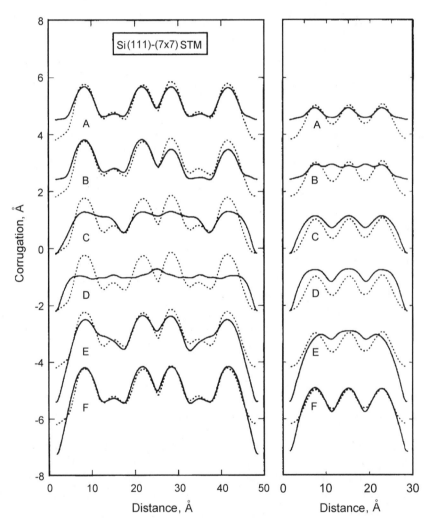

Figure 3.4. Comparison between experimental STM image of Si(111)-(7 × 7) (*dotted line*) and the images calculated with the atom-superposition model (Eq. (3.8)), for six different proposed structures of this surface. *Left panel* the line scan along the long diagonal of the (7 × 7) cell; *right panel* the line scan along the short diagonal. Curve *A* is for the adatom model proposed by Binnig et al,[5] Curve *F* is for the model of Takayanagi et al,[22] other curves are discussed by Tromp et al.[21] (Adapted with permission from Ref. 21.)

model density, by construction, has the same analytical properties as the true density, so that if the model is accurate near the surface, it will automatically describe accurately the decay with distance.

As one example of where this approach can be useful, the image expected for Au(110)-(3 × 1) was calculated for two plausible models of the structure, differing only in the present or absence of a missing row in the second layer.[11] The similarity of the model images at distances of interest suggested that the structure in the second layer could not be reliably inferred from experimental images. Quantifying the limits of valid interpretation in this way is an essential part of the analysis of STM data.

Although this method is intended primarily for metals, Tromp et al.[21] applied it to Si(111)-(7 × 7) with remarkable success. They simulated the images expected for a number of different proposed models of this surface and compared them with an experimental image (Fig. 3.4). The so-called dimer-adatom-stacking fault model of Takayanagi et al.[22] gives an image that agrees almost perfectly with experiment, and a simple adatom model is also rather close. Other models, although intended to be consistent with the STM measurements, lead to images with little similarity to experiment. Thus the usefulness of such image simulations must not be under-estimated.

3.4 SEMICONDUCTING SURFACES: ROLE OF SURFACE ELECTRONIC STRUCTURE

3.4.1 Voltage-Dependent Images

At small voltages, the s-wave approximation for the tip led to the simple result Eq. (3.7). At larger voltages, one might hope that this could be easily generalized to give a simple expression such as

$$ I \sim \int_{E_F}^{E_F+V} \rho(r_t, E)\, dE \qquad (3.9) $$

This is not strictly correct for two reasons. First, the matrix elements and the tip density of states are somewhat energy dependent, and any such dependence is neglected in Eq. (3.9). Second, the finite voltage changes the potential, and hence the wavefunctions, outside the surface. Nevertheless, Eq. (3.9) is a reasonable approximation for many purposes,[19] as long as the voltage is much smaller than the work function. I shall, therefore, use this approximation in discussing STM images of semiconductors at modest voltages.

Unlike metals, semiconductors show a strong variation of LDOS with energy. In particular, this quantity changes discontinuously at the band edges. With negative sample voltage, current tunnels out of the valence band, whereas for positive voltage, current tunnels into the conduction band. The corresponding images, reflecting the spatial distribution of valence and conduction-band wavefunctions respectively, may be qualitatively different.

A particularly simple and illustrative example, which has been studied in great detail, is GaAs(110). It was proposed[11] that, because the valence states are preferentially localized on the As atoms and the conduction states on the Ga atoms, STM images of GaAs(110) at negative and positive sample bias should reveal the As and Ga atoms respectively. Such atom-selective imaging was confirmed by direct calculation of LDOS, and was subsequently observed experimentally by Feenstra et al.[24]

In a single image of GaAs(110), whether at positive or negative voltage, Feenstra et al.[24] found that one simply sees a single "bump" per unit cell (Fig. 3.5). In fact, the images at opposite voltage look quite similar. It is, therefore, crucial to obtain both images simultaneously, so that the dependence of the absolute position of the "bump" on voltage can be determined. Although neither image alone is very informative, by overlaying the two images the zigzag rows of the (110) surface can be clearly seen. Thus in this case, as with many nonmetallic surfaces, voltage-dependent imaging is essential for the meaningful interpretation of STM images on an atomic scale.

Even in this simple case, however, the interpretation of the voltage-dependent images as revealing As or Ga atoms directly is a bit simplistic. Figure 3.6 shows a line scan from a measured GaAs image, as well as theoretical images for two cases: the ideal surface formed by rigid truncation of the bulk; and the real surface, where the As atom buckles upward, and the Ga downward. In each case, two images are

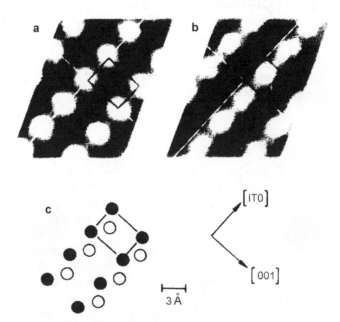

Figure 3.5. Gray-scale STM images of GaAs(110) acquired at sample voltages of (**a**) +1.9 and (**b**) −1.9 V. (**c**), Top view of the GaAs surface structure. The As and Ga atoms are shown as *open* and *solid circles*, respectively. The *rectangle* indicates a unit cell, whose position is the same in all three figures. (Reprinted with permission from Ref. 24.)

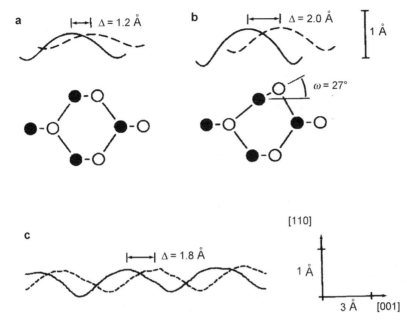

Figure 3.6. Contour of constant LDOS in ($1\bar{1}0$) cross section for occupied (*dashed line*) and unoccupied (*solid line*) states. Absolute height of each curve is arbitrary. Δ is the lateral distance between peak positions for occupied and unoccupied states. **a**, Theoretical results for ideal (unbuckled) surface. **b**, Theoretical results for surface with 27° buckling. The side view of the atomic structure is also shown, with *open* and *solid circles* indicating As and Ga, respectively. **c**, Experimental results. (Reprinted with permission from Ref. 24.)

shown, corresponding to positive and negative bias. For the ideal surface, at both biases the bumps in Figure 3.6 are almost directly over the respective atoms, supporting the simple interpretation of the image. For the buckled surface, though, the apparent positions of the atoms in the images deviate significantly from the actual positions. The separation between the Ga and As atoms in the (001) direction after buckling is <1.3 Å, whereas the separation of the bumps in the image in this direction is 2.0 Å; so the distance in the image differs from that of the atoms by >50%. Qualitatively, one might say that the bumps correspond to the positions of the respective dangling bonds; but such an interpretation is difficult to quantify.

This deviation of the bumps from the atom positions could be viewed as an undesirable complication, because it makes the image even less like a topograph. Alternatively, it is possible to take advantage of this deviation. The apparent positions of the atoms turn out to be rather sensitive to the degree of buckling associated with the (110) surface reconstruction. As a result, it is possible to infer the surface buckling quantitatively from the apparent atom positions.[24] Thus the images are actually quite rich in information, but the quantitative interpretation requires a more detailed analysis than is often feasible. Even for a qualitative analysis, we cannot sufficiently emphasize the importance of voltage-dependent imaging, to help separate electronic from topographic features.

In tunneling to semiconductors, there are additional effects not present in metals. There may be a large voltage drop associated with band bending in the semiconductor, in addition to the voltage drop across the vacuum gap.[26] This means that the tunneling voltage may be substantially less than the applied voltage, complicating the interpretation. More interestingly, local band bending associated with defects or adsorbates on the surface can lead to striking nontopographic effects in the image,[26] and localized states in the band gap lead to fascinating voltage-dependent images.[23] Although beyond the scope of this chapter, these effects are of great interest for anyone involved in STM of semiconductor surfaces.

3.4.2   Imaging Band-Edge States

A particularly interesting situation can arise in tunneling to semiconducting surfaces at low voltages, and also to some semimetals such as graphite.[25,27] At the lowest possible voltages, only states at the band edge participate in tunneling. These band-edge states typically (though not necessarily) fall at a symmetry point at the edge of the surface Brillouin zone. In this case, the states that are imaged have the character of a standing wave on the surface.

This standing-wave character leads to an image with striking and peculiar properties.[25,28] The corrugation is anomalously large; and unlike the normal case, it does not decrease rapidly with distance from the surface. This gives the effect of unusually sharp resolution. For example, in the case of graphite, the unit cell is easily resolved despite the fact that it is only 0.2 nm across. This effect was also seen on Si(111)-(2 × 1).[27]

In most such cases, the image can be adequately described by a universal form,[25] consisting of an array of dips with the periodicity of the lattice. (The dips, however, are weakened by a variety of effects, and in any case may not be well resolved because of the limits of instrumental response.)

Such an image can be extremely misleading. When there is one bump per unit cell, it is tempting to infer that there is one topographic feature per unit cell. However, in the case of Si(111)-(2 × 1), like GaAs(110), it is necessary to combine images from both positive and negative voltage to get a reasonable picture of the surface. In extreme cases, such as certain images of charge-density-wave materials,[29] the image may in fact carry no information whatever regarding the distribution of atoms within the unit cell.

3.5   MECHANICAL TIP–SAMPLE INTERACTIONS

Ideally, in STM the tip and surface are separated by a vacuum gap and are mechanically noninteracting. However, sometimes anomalies are observed, which are most easily explained by assuming a mechanical interaction between the tip and surface. In particular, since the earliest vacuum tunneling experiments of Binnig et al., it has been observed that for dirty surfaces, the current varies less rapidly than expected with vertical displacement of the tip. Coombs and Pethica[30] pointed out that this behavior can be explained by assuming that the dirt mediates a mechanical interaction between surface and tip.

As discussed above, the current is expected to vary with tip height z as $I \propto e^{-2\kappa z}$, where $(h^2 \kappa^2 / 2m) = \phi$, ϕ being the work function. Thus, in principle, ϕ can be determined from $d \ln I / dz$. For dirty surfaces, $d \ln I / dz$ is smaller than expected, leading to an inferred work function that is unphysically small. For graphite in air, this "effective work function" is often <0.1 eV.

Coombs and Pethica[30] suggested that some insulating dirt (e.g., oxide) is squeezed between the surface and tip, acting in effect as a spring. The actual point of tunneling could be a nearby part of the tip that is free of oxide, or an asperity poking through the dirt. As the tip is lowered, the dirt becomes compressed, pushing down the surface or compressing the tip if either is sufficiently soft. As a result, the surface-tip separation does not really decrease as much as expected from the nominal lowering of the tip, and so the current variation is correspondingly less.

This issue gained renewed importance with the observation of giant corrugations in STM of graphite.[31,32] Ridiculously large corrugations were sometimes observed, up to 10 Å or more vertically within the 2 Å wide unit cell of graphite. Soler et al.[31] at first attributed these corrugations to direct interaction of the tip and surface, but a detailed study by Mamin et al.[32] suggests that in fact the interaction is mediated by dirt, consistent with the earlier proposal. Clean UHV measurements apparently give corrugations of 0.3 Å or less.[33]

For the model of direct interaction,[31] there is a complex nonlinear behavior. But for the dirt-mediated interaction model, a linear treatment is appropriate.[30] In this case, assuming the mechanical interaction has negligible corrugation (e.g., because of the large interaction area), the image seen is simply the ideal image, with the vertical axis distorted by a constant scale factor.

This lack of a complete understanding is pointed out by the possible role of forces even in certain clean situations. In observations of close-packed metal surfaces, in which individual atoms were resolved,[34,35] it was suggested that these cases, like graphite, may represent an enhancement of the corrugation by mechanical interactions between tip and surface. Apparent work functions as small as 1 eV are not unusual even in ultraclean UHV experiments, in which atomic-resolution images would seem to be incompatible with direct mechanical contact. Such effects may be associated with a deformable tip (e.g., a "whisker" or a weakly adsorbed atom). Alternatively, such low apparent work functions on clean surfaces might be unrelated to mechanical interactions, reflecting, e.g., the reduction of the real barrier at small tip–surface separation.[36] In any case, the question of what forces exist in clean vacuum experiments, and their effect on STM images, remains relatively unexplored. It therefore seems prudent to constantly monitor the quantity $d \ln I / dz$, e.g., by modulating the tip height slightly at high frequency, as one simple check on whether unexpected interactions are occurring.

ACKNOWLEDGMENTS

It is a pleasure to acknowledge the many people who have contributed to my understanding of scanning tunneling microscopy and to my enjoyment of working in this exciting field. I am particularly grateful to R. M. Feenstra, D. R. Hamann, R. J. Hamers, P. K. Hansma, N. D. Lang, H. Rohrer, and R. M. Tromp.

REFERENCES

1. G. Binnig, H. Rohrer, Ch. Gerber, and E. Weibel, *Phys. Rev. Lett.* **49**, 57 (1982); G. Binnig and H. Rohrer, *Rev. Mod. Phys.* **59**, 615 (1987); P. K. Hansma and J. Tersoff, *J. Appl. Phys.* **61**, R1 (1987); J. Tersoff, in R. J. Behm, N. Garcia, and H. Rohrer, eds., *Scanning Tunneling Microscopy and Related Methods*, NATO ASI Series E vol. **184**, Kluwer Academic Publishers, Dordrecht, 1990.

2. R. M. Feenstra, W. A. Thompson, and A. P. Fein, *Phys. Rev. Lett.* **56**, 608 (1986); J. A. Stroscio, R. M. Feenstra, D. M. Newns, and A. P. Fein, *J. Vac. Sci. Technol. A* **6**, 499 (1988).

3. R. S. Becker, J. A. Golovchenko, D. R. Hamann, and B. S. Schwartzentruber, *Phys. Rev. Lett.* **55**, 2032 (1985).

4. R. J. Hamers, R. M. Tromp, and J. E. Demuth, *Phys. Rev. Lett.* **56**, 1972 (1986).

5. G. Binnig, H. Rohrer, C. Gerber, and E. Weibel, *Phys. Rev. Lett.* **50**, 120 (1983).

6. N. Garcia, C. Ocal, and F. Flores, *Phys. Rev. Lett.* **50**, 2002 (1983); E. Stoll, A Baratoff, A. Selloni, and P. Carnevali, *J. Phys. C* **17**, 3073 (1984).

7. N. D. Lang, *Phys. Rev. B* **36**, 8173 (1987).

8. N. D. Lang, A. Yacoby, and Y. Imry, *Phys. Rev. Lett.* **63**, 1499 (1989).

9. J. Bardeen, *Phys. Rev. Lett.* **6**, 57 (1961).

10. For a unique exception, see Y. Kuk, P. J. Silverman, and H. Q. Nguyen, *J. Vac. Sci. Technol. A* **6**, 524 (1988).

11. J. Tersoff and D. R. Hamann, *Phys. Rev. B* **31**, 805 (1985) and *Phys. Rev. Lett.* **50**, 1998 (1983).

12. C. J. Chen, *J. Vac. Sci. Technol. A* **6**, 319 (1988) and *Phys. Rev. Lett.* **65**, 448 (1990).

13. W. Sacks, S. Gauthier, S. Rousset, and J. Klein, *Phys. Rev. B* **37**, 4489 (1988).

14. J. Tersoff, *Phys. Rev. B* **41**, 1235 (1990).

15. S. Park, J. Nogami, and C. F. Quate, *Phys. Rev. B* **36**, 2863 (1987); H. A. Mizes, S. Park, and W. A. Harrison, *Phys. Rev. B* **36**, 4491 (1987).

16. N. D. Lang, *Phys. Rev. Lett.* **56**, 1164 (1986).

17. N. D. Lang, *Phys. Rev. Lett.* **55**, 230 (1985).

18. N. D. Lang, *Phys. Rev. Lett.* **58**, 45 (1987).

19. N. D. Lang, *Phys. Rev. B* **34**, 5947 (1986).

20. J. Tersoff, *Phys. Rev. B* **40**, 11990 (1989).

21. R. M. Tromp, R. J. Hamers, and J. E. Demuth, *Phys. Rev. B* **34**, 1388 (1986).

22. K. Takayanagi, Y. Tanishiro, M. Takahashi, and S. Takahashi, *J. Vac. Sci. Technol. A* **3**, 1502 (1985), and references therein.

23. R. M. Feenstra, *Phys. Rev. Lett.* **63**, 1412 (1989); P. N. First, J. A. Stroscio, R. A. Dragoset, D. T. Pierce, and R. J. Celotta, *Phys. Rev. Lett.* **63**, 1416 (1989).

24. R. M. Feenstra, J. A. Stroscio, J. Tersoff, and A. P. Fein, *Phys. Rev. Lett.* **58**, 1192 (1987).

25. J. Tersoff, *Phys. Rev. Lett.* **57**, 440 (1986).

26. R. M. Feenstra and J. A. Stroscio, *J. Vac. Sci. Technol. B* **5**, 923 (1987); J. A. Stroscio and R. M. Feenstra, *J. Vac. Sci. Technol. B* **6**, 1472 (1988); R. M. Feenstra and P. Martensson, *Phys. Rev. B* **39**, 7744 (1989).

27. J. A. Stroscio, R. M. Feenstra, and A. P. Fein, *Phys. Rev. Lett.* **57**, 2579 (1986).

28. J. Tersoff, *Phys. Rev. B* **39**, 1052 (1989).

29. R. V. Coleman, B. Giambattista, P. K. Hansma, et al., *Adv. Phys.* **37**, 559 (1988).

30. J. H. Coombs and J. B. Pethica, *IBM J. Res. Develop.* **30**, 455 (1986).

31. J. M. Soler, A. M. Baro, N. Garcia, and H. Rohrer, *Phys. Rev. Lett.* **57**, 444 (1986).

32. H. J. Mamin, E. Ganz, D. W. Abraham, et al., *Phys. Rev. B* **34**, 9015 (1986).

33. R. J. Hamers, in G. Chiarotti, ed., *Physics of Solid Surfaces*, Landolt-Bornstein, p. 363 (1992); S. Gwo and C. K. Shih, *Phys. Rev. B* **47**, 13059 (1993).

34. V. M. Hallmark, S. Chiang, J. F. Rabolt, et al., *Phys. Rev. Lett.* **59**, 2879 (1988).

35. J. Wintterlin, J. Wiechers, H. Brune, et al., *Phys. Rev. Lett.* **62**, 59 (1989).

36. N. D. Lang, *Phys. Rev. B* **37**, 10395 (1988).

METHODS OF TUNNELING
SPECTROSCOPY
WITH THE STM

Robert J. Hamers and David F. Padowitz

4.1 INTRODUCTION

One of the most exciting aspects of scanning tunneling microscopy (STM) is that in addition to providing information on the local topographic structure of the surface, the images inherently contain a large amount of data relating to the local electronic structure of the surface. Tunneling spectroscopy has its origins in some of the earliest STM work of Binnig et al.,[1] who found evidence for voltage-dependent changes in the appearance of some of the atoms within the (7×7) unit cell and correctly noted that such effects might arise from tunneling through discrete electronic states of the sample. These voltage-dependent changes in appearance are clearly seen in Figure 4.1, which shows STM images of $Si(111)-(7 \times 7)$ acquired with a negative bias applied to the sample and with a positive bias on the sample.

Clear differences in the apparent height of many of the atoms are observed, which, as we shall see, arise from the energy-dependent changes in the density of states of the sample. Perhaps the most immediate conclusion that should be drawn from these observations is that it is common to interpret the features observed in scanning tunneling microscope images as "atoms." This superficial interpretation can often be misleading. A proper interpretation of scanning tunneling microscopy images requires an understanding of the tunneling mechanism together with some knowledge of the relevant variables (such as the density of states of sample and tip and the voltage-dependent tunneling probability) that determine the experimental result.

The images obtained in STM often strongly depend on the sample-tip bias voltage in a nontrivial manner.[3-5] Although early STM studies focused on the application of STM as a microscope for observing topographic structure, the sensitivity to bias voltage also means that the STM is sensitive to the energy states of the sample and tip. This has two main implications: tunneling microscopy does not reveal the positions of atoms themselves; and using the STM, it is possible to obtain spectroscopic information with atomic spatial resolution.

Compared to other surface spectroscopy techniques such as ultraviolet photoemission spectroscopy (UPS), inverse photoemission spectroscopy (IPS), electron energy loss spectroscopy (EELS), and infrared reflection-absorption spectroscopy (IRRAS), the STM has a unique advantage. Whereas the other techniques average over a large region of the surface, the tunneling current in STM flows through a region only ~5 Å in diameter,[6] so that spectroscopic information can be obtained on an *atom-by-atom* basis. This is an important advance, because many chemical and physical phenomena at surfaces are associated with active sites such as dopants, impurities, steps, or defects that occupy only a small fraction of the total

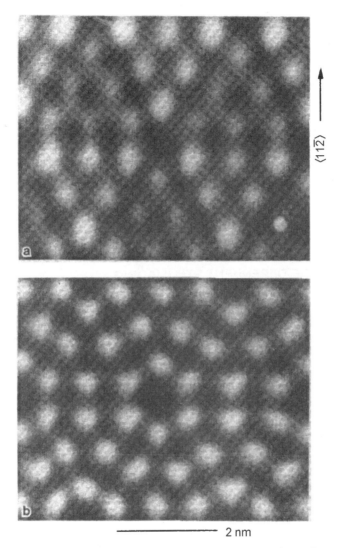

$\langle 11\bar{2}\rangle$

2 nm

Figure 4.1. STM images of Si(111)−(7 × 7) acquired at positive and negative sample bias. **a**, Sample biased at −2 V, showing stacking fault-related asymmetries. **b**, Sample biased at +2 V, revealing symmetric-appearing unit cell. (Reprinted with permission from Ref. 2.)

surface area. With the STM, it is possible to directly measure variations in local electronic structure associated with specific microscopic features and to examine the electronic structure of chemically inequivalent atoms in complex ordered structures such as the (7 × 7) reconstruction of Si(111).

The theory of tunneling in one dimension is well developed as a result of extensive studies performed using planar bulk tunnel junctions. Several excellent books on both theoretical and experimental aspects of bulk tunneling phenomena are available.[7–9] These theories of tunneling in bulk solids, together with more recent work (discussed in Chapter 3) specifically applied to the STM,[10–17] provide a basis for understanding tunneling spectroscopy from a theoretical standpoint. From a practical standpoint, there are many limitations on our ability to measure and/or control the variables affecting the tunneling process. Nevertheless, through careful experimental technique and an understanding of the factors that influence the tunneling, it is possible to glean much new information about the electronic structure of surfaces on a local scale. Here we describe the experimental techniques commonly used for tunneling spectroscopy measurements, discuss the interpretation of the data, and point out some of the advantages and disadvantages of the experimental techniques. We also address the measurement of the tunneling barrier height, owing to the importance of the tunneling transmission probability to the interpretation of tunneling spectroscopy results.[6]

4.2 VOLTAGE-DEPENDENT STM IMAGING

Voltage-dependent STM imaging is the simplest way of obtaining spectroscopic information, by acquiring conventional STM topographic information at different applied voltages and comparing the results. When a voltage V is applied to the sample (with the tip at ground), only states lying between E_F and $E_F + eV$ participate in tunneling. The sign and magnitude of the applied voltage, then, determine which states contribute to the resulting topographic images. This is shown in Fig. 4.2. In many cases, conventional constant-current topographs (CCTs) show bias-dependent changes in appearance, which can be related to the spatial symmetry and energy of the surface electronic states. Such changes are most apparent on semiconductors[3,2,5,18,19] and semimetals such as graphite,[20] for which the different electronic states are spatially separated and strongly localized.

4.2.1 Experimental Technique

One systematic method of identifying these bias-dependent changes is to acquire complete images at different bias voltages over the same nominal region of the surface. In this method, surface inhomogeneities, such as point defects, can be used for alignment. For example, Figure 4.3 shows two sequential images of the Si(001) surface, obtained at negative and positive sample bias.[22] Here, naturally occurring point defects were used as markers to ensure proper registration of the two complementary images. A close examination of Figure 4.3 shows that on the Si(001) surface there are different kinds of characteristic defects, each with unique electronic properties. Even on the nearly perfect regions, however, Figure 4.3a and 4.3b show differences arising from the different local symmetry of the occupied and unoccupied states of the dimers making up the ideal Si(001) surface.

A second method that is sometimes useful is to change the applied bias on a line-by-line basis during the raster-scanning of the tip, providing two (or more) inter-

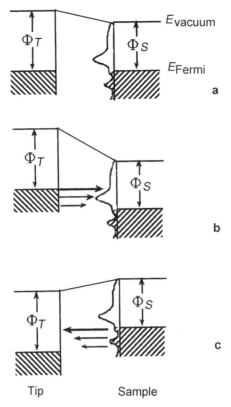

Figure 4.2. Energy levels involved in tunneling. **a**, Zero bias, no current. **b**, Positive sample bias, current from occupied tip states into unoccupied sample states. **c**, Negative sample bias, current from occupied sample states into unoccupied tip states.

leaved images at different bias voltages. Although it seems that this method should provide better (and easier) registration of the images than the first method, in practice one must be careful to realize that as the bias voltage is changed, it is possible for the spatial symmetry of the tip electronic states to change as well. This may lead to shifts and distortions in the images that are different at different bias voltages, as demonstrated by Tromp et al.[23] Such spatially separated, localized electronic states on the tip can result when the tip picks up material from a semiconducting sample by accidental contact or through field-induced migration.

4.2.2 Interpretation of Voltage-Dependent STM Images

Under the assumptions that (1) the tip has uniform density of states, (2) the voltage is low (less than $\simeq 10\,\text{mV}$), (3) the temperature is low, and (4) only s-wave tip wave functions are important, the Tersoff-Hamann theory predicts that the tunneling

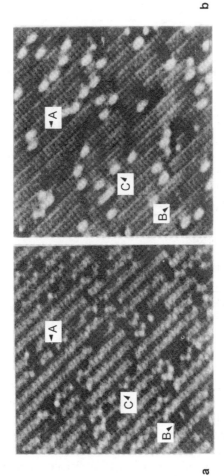

Figure 4.3. STM images of Si(001) surface with characteristic defects. **a**, Image acquired at negative sample bias, tunneling from occupied surface states. **b**, Image acquired at positive sample bias, tunneling into unoccupied surface states. (Reprinted with permission from Ref. 21.)

current can be expressed as

$$I = 32\pi^3\hbar^{-1}e^2V\phi^2R^2\kappa^{-4}e^{2\kappa R}D_t(E_F)\rho(r_0, E_F) \tag{4.1}$$

where D_t is the density of states per unit volume of the tip, R is the tip radius, and $\rho(r_0, E_F)$ is the Fermi-level density of states of the sample, measured at the position corresponding to the center of curvature of the tip.

This important result shows that, under idealized conditions, the STM image reflects the sample electronic structure. For constant tunneling current, it can be seen from this formula that the contour followed by the tip is essentially a contour of constant Fermi-level density of states of the sample, measured at the center of curvature of the tip.

The low-voltage approximation is often violated in practice, because many STM experiments are conducted at bias voltages of between 1 and 3 V. Under these conditions, the tunneling current arises from a range of states lying within eV of the Fermi level. For interpretation of such high-bias results, it is useful to consider the predictions of simple planar tunneling models using the Wentzel-Kramers Brillouin (WKB) approximation. The WKB theory predicts that the tunneling current is given by

$$I = \int_0^{eV} \rho_s(r, E)\rho_t(r, -eV + E)T(E, eV, r)\,dE \tag{4.2}$$

where $\rho_s(r, E)$ and $\rho_t(r, E)$ are the density of states of the sample and tip at location r and the energy E, measured with respect to their individual Fermi levels. For negative sample bias, $eV < 0$ and for positive sample bias, $eV > 0$. The tunneling transmission probability $T(E, eV, r)$ for electrons with energy E and applied bias voltage V is given by

$$T(E, eV) = \exp\left(-\frac{2Z\sqrt{2m}}{\hbar}\sqrt{\frac{\phi_s + \phi_t}{2} + \frac{eV}{2} - E}\right) \tag{4.3}$$

where Z is the distance from sample to tip, and ϕ are work functions. At constant tunneling current I, the contour followed by the tip is a function of the density of states of both sample and tip, together with the tunneling transmission probability. One important point regarding the transmission probability should be noted. Examination of $T(E, eV)$ shows that if $eV < 0$ (i.e., negative sample bias), the transmission probability is largest for $E = 0$ (corresponding to electrons at the Fermi level of the sample). Similarly, if $eV > 0$ (positive sample bias), the probability is largest for $E = eV$ (corresponding to electrons at the Fermi level of the tip). Thus we see that the tunneling probability is always largest for electrons at the Fermi level of whichever electrode is negatively biased.

4.2.3 Applications of Voltage-Dependent STM Imaging

Elemental Semiconductors   In the case of ordered surfaces, voltage-dependent imaging can provide information on the relative spatial locations of the various electronic states at the surface. Some of the surfaces that have been extensively studied by this method include Si(001),[24-26] Ge(001),[27] Si(111)– (7 × 7),[3, 2, 28] the metastable cleaved Si(111)–(2 × 1) surface,[19, 29, 30] GaAs(110),[3, 31-32] GaAs(l00),[33, 34] and graphite.[20, 35]

Figure 4.4 shows the atomic rearrangements involved in the reconstruction of the Si(001) surface. Figure 4.5 shows CCTs of the Si(001) surface obtained at negative and positive bias, reflecting the spatial distribution of filled and empty surface states, respectively. The geometric structure of this surface is well understood[26, 22, 36, 37] and consists of pairs of silicon atoms, each of which is bonded to two Si atoms in the next lower atomic layer and its dimer partner; the resulting (2 × 1) unit cell is outlined in Figure 4.5. At negative sample bias (Fig 4.5a) the STM-CCT's show bean-shaped structures, whereas at positive sample bias (Fig. 4.5b) the images show a weak minimum between the dimer rows and a deep trough along the center of the dimer row. These differences can be seen more quantitatively in the corrugation profiles shown in Figure 4.6. These differences demonstrate a spatial separation of the filled and electronic states on the Si(001) surface.

These differences can be understood based on electronic structure calculations.[36, 37] These show that the electronic structure of the dimers can be described in terms of a π-bonding state slightly below E_F, and a π-antibonding state slightly above E_F, with occupied σ and unoccupied σ^* states lying far from E_F, outside the energy range accessible by tunneling. The occupied π state is predicted to be symmetric with respect to reflection through a mirror plane bisecting the dimer bond (A_1 symmetry), whereas the π^*-antibonding state is antisymmetric with respect to this reflection (B_1 symmetry) and therefore must have a node in the wavefunction at the center of the dimer bond.

The STM contours at negative bias reflect the contours of the occupied π state, whereas at positive bias, the STM tip follows the contour of the empty π^* state. The deep trough observed in the positive-bias STM images corresponds to the location of the node in the wavefunction of the π^* antibonding state. Thus the STM images directly reflect the spatial symmetry of these surface-state wavefunctions.

STM images of Si(111) surfaces also strongly depend on the bias voltage. The (2 × 1) reconstruction of the cleaved Si(111) surface[19, 30] shows a shift between negative- and positive-bias STM images owing to the different spatial location of

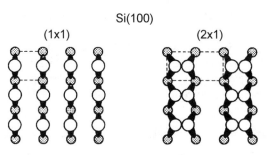

Figure 4.4. Representation of the (2 × 1) reconstruction of the Si(001) surface, with the formation of two-atom dimers in the outermost layer of surface atoms.

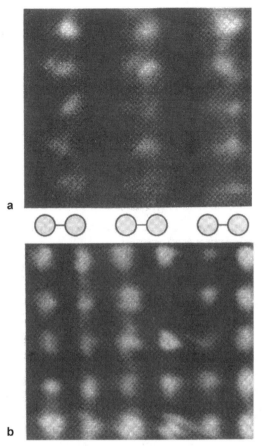

Figure 4.5. High-resolution image of Si(001) acquired at (**a**) negative sample bias and (**b**) positive sample bias. At negative bias, electrons tunnel from an occupied π-state, which has a maximum at the center of the Si-Si dimer bond. At positive bias, electrons tunnel into an empty π*-antibonding state, which has a node in the wavefunction at the center of the Si-Si dimer bond. (Reprinted with permission from Ref. 34.)

occupied and unoccupied electronic states. This shift provides strong confirmation for the π-bonded chain model proposed by Pandey.[38]

More complicated voltage dependence is observed for the Si(111)−(7 × 7) surface. This surface undergoes an extensive reconstruction extending several layers into the bulk. The atomic arrangements were determined[39] based in part on early STM results.[1] At most positive-bias voltages, STM images reveal 12 adatoms of equal height in each unit cell, as shown in Figure 4.1b. In a narrow voltage range around + 1.4 eV sample bias, the two triangular subunits in each unit cell appear to have slightly different heights. Differences between the two triangular subunits can be attributed to a difference in the stacking sequence of the underlying atomic planes, leading to a stacking "fault" in one half, but not the other. The difference in stacking sequence leads to different interactions between the Si atoms in the first

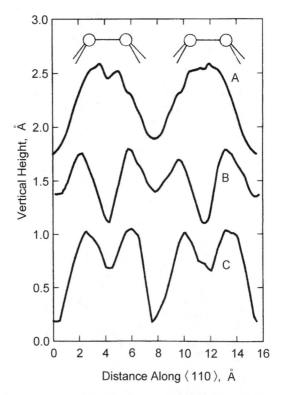

Figure 4.6. Differing appearance of the (2 × 1) reconstruction of Si(001) Corrugation profiles on Si(001) at negative (*A*) and positive bias (*B*), together with the corrugation profile (*C*) after dosing with NH_3; the symmetry in all three cases remains (2 × 1). (Reprinted with permission from Ref. 21.)

full atomic layer and those lying 4.6 Å below. Around + 1.4 eV sample bias, the "unfaulted" half of the unit cell appears higher (indicating a higher density of states) than the faulted half. This situation reverses at negative sample bias voltages, as in Figure 4.1**a**, where the adatoms in the faulted half of the unit cell appear higher than those in the unfaulted half. In addition, the adatoms nearest the deep corner holes appear higher than the central adatoms. These latest differences are attributed to the different environment of the corner adatoms (which are next to one dimer and one rest atom) and the central adatoms (which are next to two dimers and two rest atoms). Such voltage-dependent STM images,[2] together with local tunneling spectroscopy measurements to be discussed later[3, 24] have provided a great deal of insight into the electronic structure of this complicated reconstruction.

Compound Semiconductors In compound semiconductors, voltage-dependent contrast changes can be directly related to charge transfer between surface atoms. The (110) surfaces of III-V semiconductors, such as GaAs(110), contain equal numbers of cations (Ga) and anions (As) in the surface layer; in GaAs, they are arranged in a chainlike structure. Charge transfer from Ga to As results in an occupied state centered at the As atoms and an empty state centered at the Ga atoms, and also results in a slight vertical displacement of the Ga atoms with respect to the As atoms. STM images[5] reveal a shift between the peaks observed at positive and negative bias. The peaks at negative bias correspond to the positions of the As atoms, and those at positive bias reveal the Ga atoms. A comparison of this bias-dependent shift with theoretical calculations also allowed the tilt angle between Ga and As atoms at the surface to be quantitatively determined.[5, 40]

Adsorbates and Defects   Voltage-dependent STM imaging is not restricted to ordered surfaces as above, but is also useful for identifying and studying the properties of impurities and defects at surfaces. On semiconductor surfaces, local charging owing to adsorbates[32, 41] and defects[42] can shift the surface-state bands with respect to the Fermi level, resulting in apparent height changes that correspond to the electrostatic screening length. For strongly charged adsorbates and/or lightly doped semiconductors, these effects can dominate the appearance of defects, as in the case of oxygen on n-type GaAs(110).[41] Adsorbates on metals and covalently bonded adsorbates on semiconductors give rise to local changes in the surface electronic structure that allow them to be imaged.

Changes in the spatial distribution of the surface-state wavefunctions can often be induced by chemisorption. Dosing the Si(001) surface with NH_3 significantly changes the negative-bias STM images, producing a slight depression in the center of the dimer bond, where images on the clean surface showed a maximum.[25, 22] These changes were attributed to hydrogen atoms produced by NH_3 dissociation interacting with the dimers. On the clean surface each surface Si atom is double-bonded via both a σ- and π-bond to its dimer partner, whereas on the hydrogen-covered surface each surface Si atom is bonded via a single σ-bond to its dimer partner and via another σ-bond to a hydrogen atom. The changes in the negative-bias images induced by the reaction with NH_3 result from the different spatial distribution of the Si-Si π-state and the Si-H σ-state. Thus the STM images alone can be used to distinguish dimers that have reacted with NH_3 from those that have not.

Metal overlayers on semiconductors provide rich voltage-dependent structure; those studied include Ag/Si(111),[43] Al/Si(111),[44] and Cu/Si(111).[45] Arsenic-terminated Si(111) and Si(100) surfaces have proven to be interesting systems as well and have been the subject of several studies by Becker and co-workers.[46, 47] Most of these metal-on-silicon systems show strongly voltage-dependent changes in constant-current STM topography resulting from complex electronic structure. For such systems, an understanding of the dependence of the STM images on applied bias is crucial for understanding the atomic features that give rise to the STM topographic structure.

The ability to differentiate between chemically inequivalent atoms is demonstrated in Figures 4.7 and 4.8. Figure 4.7 shows the atomic configuration; the surface consists of a bulk-like (111) surface of Si, with a random distribution of aluminum and silicon adatoms. Figure 4.8 shows the STM images at positive (part **a**, + 1.2 V) and negative (part **b**, −1.2 V) bias.[42] At positive bias, each bright

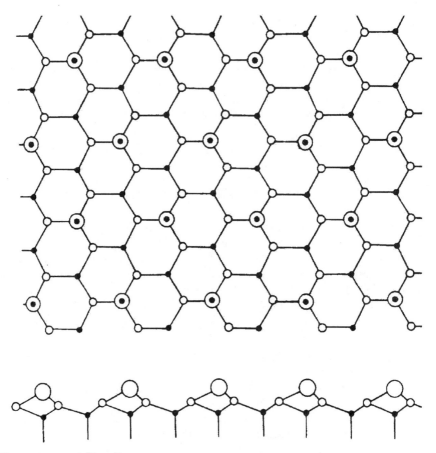

Figure 4.7. The ($\sqrt{3} \times \sqrt{3}$) reconstruction of Al/Si(111). Al adatoms sit in T_4 sites atop a bulk-like Si(111) lattice.

protrusion in the ordered structure is an Al adatom. In many of the locations where an Al adatoms would be expected, the unit cell appears darker but still has a protrusion in the center. These darker unit cells consist of Si adatoms, rather than the expected Al adatoms, situated atop a bulk-like Si(111) surface. Switching the bias to -2 eV causes the contrast to reverse so that the Si adatoms appear $\simeq 1$ Å higher than the Al adatoms. By choosing the bias polarity, then, either the Al adatoms or the Si adatoms can be selectively imaged. The origin of these contrast changes is revealed from theoretical calculations,[48] which show that the Al adatoms have an unoccupied p orbital, which in Si is shifted below E_F. These changes in the state density directly give rise to contrast in the STM constant-current topographs.

Images of molecular adsorbates show pronounced voltage dependences that are not yet fully understood. Adsorbed molecules may be imaged directly by resonant tunneling into or out of molecular orbitals, but typically the unperturbed states of the molecule would be outside the range of energies accessible to tunneling.

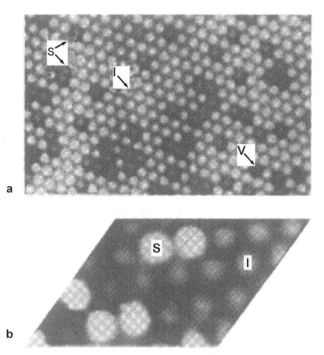

Figure 4.8. STM images of partial ($\sqrt{3} \times \sqrt{3}$) reconstruction of Al/Si(111) (**a**) at positive and (**b**) at negative bias, demonstrating atom-selective imaging. The ($\sqrt{3} \times \sqrt{3}$) overlayer is incomplete, so that half the surface atoms are aluminum adatoms (as in Fig. 4.7) and half are silicon adatoms. **a**, Bright atoms are Al adatoms and dark atoms are Si. **b**, Si atoms appear bright, and Al atoms appear dark. (Reprinted with permission from Ref. 44.)

Adsorption may shift the molecular levels, or the molecule may be imaged indirectly by alteration of the substrate electronic structure or by influencing the tunneling transmission probability. Images of molecules on metals such as benzene on Rh(111) are likely owing to new states around E_F induced by adsorption.[49, 50]

Long-chain alkanes physisorbed on graphite have been extensively studied. Remarkably, the STM may penetrate a thick liquid-crystalline film to image only the layer of molecules directly in contact with the substrate. At biases below 100 meV, the underlying graphite substrate is frequently imaged through the adsorbate. At higher biases the adsorbate emerges.[51] Various functional groups may appear higher or lower than purely topographic relief owing to electronic effects.[52, 53]

Covalently bound organic molecules on Si(100) also show marked voltage dependence. Images at low bias reveal the underlying reacted Si surface. Sample biases below −1 V are within the bulk bandgap, so that only the surface states are accessible. Chemisorption removes the π state of the Si-Si dimer, as in the previous example of NH_3. The adsorbate then images as a depression. At sample biases of −2 to −3 V, the molecules show enhanced tunneling relative to the bare silicon and appear higher than the unreacted silicon.[54] This may reflect a shift

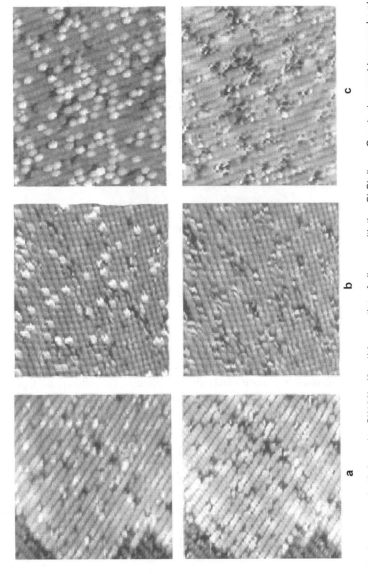

a   b   c

Figure 4.9. Cycloalkenes covalently bound to Si(100)–(2 × 1) by reaction of alkenes with the Si-Si dimer. Constant-current topographs at two biases were taken simultaneously by changing bias on alternate scan lines. At sample bias of −1 V, the adsorbed molecules image as depressions. At −2.5 V, the molecules appear as protrusions. (Reprinted with permission from Ref. 54.)

in the valance-band edge owing to bonding. In addition, there may be increased transmission probability for tunneling through the molecule.[55, 56] In the case of ethylene adsorption on Si(100), the interaction of the tip-field with the adsorbate-silicon system may have to be explicitly considered to account for the observed bias dependence.[57–59]

4.3 MODULATION TECHNIQUES: STS

Constant-current topographs often reveal electronic structure information, but separating the contributions of electronic and geometric structure is not straight-forward. One way of obtaining quantitative spectroscopic information in the STM is the use of modulation techniques.

4.3.1 Experimental Techniques

The most common modulation technique is scanning tunneling spectroscopy, or STS, first used by Binnig and Rohrer.[18, 60] The STS technique involves superimposing a small high-frequency sinusoidal modulation voltage V_{mod} ω_{mod} on top of the constant dc bias V_{dc} between sample and tip. The ac component of the tunneling current is measured with a lock-in amplifier, with the in-phase component directly giving $dI/dV|_{V=V_{DC}}$ simultaneously with the sample topography. To probe the density of states as a function of energy, the lock-in output is measured as a function of V_{DC}. The modulation frequency ω_{mod} must be faster than the closed-loop bandwidth of the STM feedback system (typically 1 to 2 kHz). If the modulation frequency is too low, the feedback control will attempt to compensate for the applied modulation by changing the gap spacing. As the frequency is increased the sample–tip capacitance contributes an additional displacement current 90° out-of-phase with the applied modulation, which becomes larger as the modulation frequency is increased. In practice, the optimal modulation frequency is slightly above the cuto frequency of the feedback loop and can be found by identifying the frequency range where the induced modulation dI/dV is independent of frequency.

4.3.2 Interpretation

High Voltage: Barrier State Spectroscopy   When the applied dc bias V_{dc} is larger than the work functions of sample and tip, structure in dI/dV arises primarily from image potential states or barrier resonances.[18, 28, 61] These states arise from the interaction between the surface electrons and the polarization they induce in the bulk. The interaction traps the electron in the near-surface region while leaving it free to translate parallel to the surface.[62] These states are essentially standing-wave states formed between the sample and the potential barrier between sample and tip. Their energy depends on the shape of the tunneling barrier, which reflects the work functions of sample and tip. The oscillatory electron wavefunctions lead to constructive or destructive interference, depending on the phase-matching conditions at the boundary. Because the shape of the boundary is determined in part by the applied voltage, changing the voltage causes the wavefunctions to alternate between constructive and destructive interference, leading to oscillations in the tunneling probability. Binnig et al.[18] first observed these barrier resonances, using a gold tip on a Ni(100) surface. Becker et al.[28] performed similar studies on tungsten surfaces and showed that through a numerical integration of Schrödinger's equation, the oscillations in dI/dV could be used to determine an absolute sample-tip separation, which is difficult to determine otherwise.

Low Voltages: Surface State Spectroscopy   At lower bias voltages (when V_{dc} is lower than the sample and tip work functions), structure in dI/dV as a function of V_{dc} is associated with the surface density of states. Structure in the surface density of states can arise from critical points in the surface-projected bulk band structure or it may arise from true surface states, which are generally associated with surface reconstructions. The interpretation of these low-bias dI/dV measurements is generally based on the WKB approximation for the tunneling current, given as Eq. (4.2). Differentiating that equation gives

$$\frac{dI}{dV} = \rho_s(r, eV)\rho_t(r, 0)T(eV, eV, r)$$
$$+ \int_0^{eV} \rho_s(r, E)\rho_t(r, E - eV) \frac{dT(E, eV, r)}{dV} dE \qquad (4.4)$$

The first term in Eq. (4.4) is the product of the density of states of the sample, the density of states of the tip, and the tunneling transmission probability T. The second term contains the voltage dependence of the tunneling transmission factor.

Although the tunneling transmission probability T is usually unknown, under the WKB approximation Eq. (4.3) gives a smooth, monotonically increasing function of the applied voltage V. Thus at any fixed location the voltage dependence of the tunneling transmission probability contributes a smooth background on which the spectroscopic information is superimposed. Structure in dI/dV as a function of V can usually be assigned to changes in the state density via the first term of Eq. (4.4). As a result, these measurements can provide a measure of the density of states as a function of energy at any particular location on the surface.

Often, one measures dI/dV at some fixed energy simultaneously with the sample topography, resulting in an image of dI/dV. Changes in dI/dV can be produced by localized electronic states or by changes in the transmission probability. In addition, there is yet another complicating contribution of topographic origin. Consider the state density contours depicted in Figure 4.10. According to the Tersoff-Hamann theory,[11, 12] the tip is expected to follow a contour of constant state density. At vertical separations that are large compared to the distance between the individual atoms on the surface, the atomic corrugations die away and the sample becomes laterally isotropic. For this to occur, the effective decay length of the wavefunctions above a local peak in the topography $(\kappa_p - 1)$ must be shorter than those above a local valley $(\kappa_v - 1)$. This effect can be seen more quantitatively using an

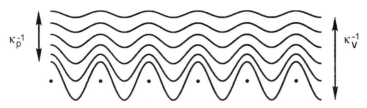

κ_p^{-1}

κ_v^{-1}

Figure 4.10. Exponential decay of equipotential contours for a corrugated surface. The contours decay more quickly above a protrusion than above a depression.

approximation given by Tersoff[11] for the Z-dependent corrugation $\Delta(Z)$,

$$\Delta(Z) \simeq \frac{2}{\kappa} \exp\left(-\frac{\pi^2 Z}{a^2 \kappa} \right) \qquad (4.5)$$

where a is the lattice constant and κ is the average inverse decay length. A simple analysis based on this formula leads to the approximate expression;

$$\kappa_v \simeq \kappa_p - \frac{2\pi^2}{a^2 \kappa} \exp\left(\frac{-\pi^2 \overline{Z}}{\kappa a^2} \right) \qquad (4.6)$$

All other factors being equal, the effective decay length (and the tunneling transmission factor T) will be greater when measured at a location corresponding to a valley in the topography than when measured over a peak. This spatial variation in the transmission probability shows up in measurements of dI/dV as a background that is essentially an inverted topography. Thus images showing the spatial variation of dI/dV obtained under conditions of constant average tunneling current always contain some topographic information convoluted in with the electronic information.

STS measurements of dI/dV at constant average tunneling current also contain a strongly voltage-dependent background whose origin can be seen by first writing the tunneling current as

$$I = \int_0^{eV} \rho_s(r, E) \rho_t(r, E - eV) \exp\left(-\frac{2Z\sqrt{2m}}{\hbar} \sqrt{\phi + \frac{eV}{2} - E} \right) dE \qquad (4.7)$$

For simplicity, we can assume that the density of states ρ is constant for both sample and tip. Then, neglecting the influence of the applied voltage on the tunneling barrier (a valid assumption for $eV \ll \phi$) the derivative dI/dV is given by

$$\frac{dI}{dV} = e\rho_s \rho_t \exp\left(-A\sqrt{\phi - \frac{eV}{2}} \right) Z \qquad (4.8)$$

where $A = 2\hbar^{-1}\sqrt{2m}$. In modulation experiments conducted under conditions of average constant average tunneling current the sample–tip separation Z increases as the dc voltage V is increased; this variation in Z must be explicitly included. In the low-bias limit, Eq. (4.7) reduces to

$$\bar{I} = \rho_s \rho_t \bar{V} \exp\left(-A\sqrt{\phi}Z\right) \tag{4.9}$$

Solving this for Z and substituting the result into Eq. (4.8) gives $dI/dV = \bar{I}/\bar{V}$. Under conditions of constant average tunneling current $\bar{I}$, the quantity dI/dV diverges like $1/V$ as V approaches zero and presents a background term on which the desired spectroscopic information is superimposed.

The $1/V$ divergence problem can be significantly improved by instead operating under conditions of constant resistance. This type of operation was first used by Kaiser and Jaklevic[63] to observe surface states on clean Au(111) and Pd(111) surfaces. On these most of the variation in the surface density of states arises from critical points in the bulk density of states. This can be seen in Figure 4.11; which shows Kaiser and Jaklevic's constant resistance dI/dV spectra for Pd(111) (part **a**), together with the projected bulk band structure (part **b**) and the results of ultraviolet photoemission spectroscopy measurements (part **c**). The peaks near $+1.0$ and -1.3 V in dI/dV closely correspond with critical points in the two dimensional

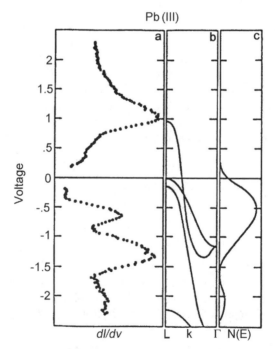

Figure 4.11. Tunneling spectroscopy data on (**a**) the Pd(111) surface obtained under conditions of constant resistance compared to (**b**) the bulk band structure projected onto the (111) surface. Most of the features observed in the tunneling spectroscopy correspond with critical points in the bulk-projected band structure, (**c**) where the density of states is expected to be high. (Reprinted with permission from Ref. 63.)

(2-D) projected band structure, and the dI/dV peak near $-0.6\,\text{eV}$ is in reasonable agreement with a surface state observed in photoemission measurements. At constant average tunneling current, this information would likely have been obscured by the $1/V$ background.

4.3.3 Applications of STS

The modulation-based STS technique has been widely applied. Garcia and co-workers[61, 64] first applied it to clean and oxidized Ni(100) surfaces, where they observed barrier resonances at high voltages but also found evidence for true surface states at lower voltages. Salvan et al.[65] applied this technique to the $(\sqrt{3} \times \sqrt{3})$ R30° overlayer of Au on Si(111) and observed a strong peak near $1\,\text{eV}$ above E_F associated with a surface state, as well as field emission resonances at higher voltages. Becker et al.[66] applied this technique to the famous (7×7) reconstruction of Si(111) and found two unoccupied surface states, at $+1.5$ and $+2.8\,\text{eV}$ above E_F, and also observed differences in the energy of the $1.5\,\text{eV}$ surface state between the faulted and unfaulted halves of the unit cell. In later work,[67] they applied STS to the study of surface states on various alloys of Si, Ge, and Sn and observed structure in dI/dV associated with the surface states of these alloys. Other STS studies investigated the electronic structure of GaSe and stepped Ni(111)/H surfaces.[68, 69]

Both surface state and image potential resonances have been used to distinguish atomic-scale domains of different metals on the basis of work function.[70] Resonances above the work function have lower spatial resolution than determination of surface-state energies, but the intensity and energy separation of the resonance peaks is greater.

Although the standard STS technique is relatively simple to apply, many of the published studies demonstrated two important disadvantages to the standard technique as conventionally applied at constant average tunneling current. The first is the $1/V$ dependence of dI/dV, which makes it difficult to observe structure at low voltages. The second (and more severe) limitation is that at lower voltages, the tip moves toward the surface to maintain constant tunneling current. This effect is particularly dramatic on semiconductor surfaces, because then the density of states near the Fermi level is typically low and the tip must push in even closer to the surface to draw the demanded tunneling current at low voltages. If the density of states is too low, the tip will crash into the surface. Becker and co-workers showed[66, 67] the tip displacement vs. voltage at constant tunneling current for Si(111)−(7 × 7).

The modulation STS technique works well at comparatively large bias voltages, but less well near the Fermi level. Because the electronic structure near the Fermi level is often of particular interest on semiconductors, this may be a significant limitation. In the case of conductors and superconductors, however, it may not be a problem. As in the case of voltage-dependent STM imaging, a complete mapping of the surface electronic structure as a function of energy and position requires many repeated measurements over the same area. This procedure is tedious at best, and is often unsuccessful owing to instability in the tip as well as thermal drifts in the microscope.

4.4 LOCAL *I-V* MEASUREMENTS

Tunneling spectroscopy data can be simultaneously obtained over a wider range of applied voltages by acquiring complete curves of I vs. V, and later analyzing the results. This is the approach taken in early work by Feenstra et al.[30] on the Si(111)−(2 × 1) surface. In such early studies, blunt tips were used and it was not possible to accurately establish the tip position. To take full advantage of the spatial resolution of the STM, the I-V curves need to be measured with "atomic" resolution and at well-defined locations on the surface to be able to correlate the surface "topography" with the local electronic structure.

4.4.1 Experimental Methods

Atomically resolved I-V measurements were first performed by Hamers et al.[3] by rapidly acquiring I-V curves (each at a fixed sample–tip separation) while simultaneously slowly scanning the tip position. The method is denoted current imaging tunneling spectroscopy (CITS) because in addition to the conventional image of the surface height as a function of position, spectroscopic information is obtained at each location, thereby permitting the local electronic characteristics to be directly imaged from the current measurements. This technique was used to map out the complete electronic structure of the Si(111)−(7 × 7) unit cell.

Several methods of acquiring such local I-V information are available. They differ in the details of the data acquisition hardware, but all provide essentially identical information. Typically a sample-and-hold circuit interrupts the feedback loop for a few hundred microseconds to maintain the tip position while the voltage is ramped for the I-V measurement. After a brief delay to damp capacitive transients, feedback resumes at the stabilization voltage V_{stab}. By acquiring the I-V curves rapidly compared to the scan speed of the tip, both the sample topography and spatially resolved tunneling I-V characteristics are measured at each location in a 2-D raster scan over the surface. The high speed eliminates both vertical and lateral drift of the tip position, providing a one-to-one correspondence between each point in the topography and an I-V curve.

One major experimental difficulty with obtaining constant-separation I-V curves is that a wide dynamic range of current measurement is required, because the current varies almost exponentially with the applied voltage. In practice, this creates limitations in the ability to obtain electronic information over a wide voltage range at constant separation. One way around this is to allow the gap spacing to change in a measured, controlled manner, and then correcting the tunneling current for this change in sample–tip separation using the known tunneling barrier height (which is directly related to the inverse decay length). Using this idea, Feenstra and Stroscio[71] measured I-V spectra on GaAs(100), which, when corrected for changes in sample–tip separation, eectively corresponded to performing measurements over eight orders of magnitude in tunneling current. Such measurements at variable sample–tip separation require measuring both the tunneling current and the sample–tip separation as a function of *V*, and also require measuring the tunneling barrier height to determine the decay rate of the wavefunctions in the vacuum.

The stabilization voltage V_{stab}, the bias voltage during normal feedback operation between *I-V* scans, plays a special role by determining the contour that the tip follows as it scans across the surface, and usually must be chosen arbitrarily. The choice of V_{stab} affects the spatially dependent *I-V* curves, because the sample–tip separation may be dierent at each location. Berghaus et al.[72] and Feenstra[30] explicitly demonstrated the changes in local *I-V* spectra on Si(111)−(7 × 7) and Si(111)−(2 × 1) surfaces at different stabilization voltages. Fortunately, the influence of the choice of stabilization voltage can be almost completely eliminated by proper normalization of the spectroscopy data to eliminate the *z*-dependence, using a normalization procedure suggested by Feenstra et al.,[30] which will be discussed in more detail below.

4.4.2 Analysis and Interpretation of *I-V* Data

Analysis of the *I-V* information can be performed in a number of ways. For example, the tunneling current *I* resulting from a particular voltage *V* can be directly imaged—a current image. These images depend on the contour that the tip follows, and so cannot be directly interpreted in a quantitative fashion. However, the symmetry of the images provides information about the different spatial symmetry of the electronic states of the surface. For example Figure 4.12 shows current images on Si(111)−(7 × 7) at various bias voltages between +2 and −2 V. Distinct changes in the symmetry of the images occur at −0.8 and +1.4 V. Analysis of the tunneling spectroscopy data shows that these voltages coincide with steep increases in the local conductance and also match the energies of the surface states known from photoemission studies.[3, 73, 74] In some other cases, however, changes in the tunneling probability can dominate the electronic structure information.[19, 72] The degree to which the choice of stabilization voltage influences the CITS images can be easily assessed by comparing data acquired at different stabilization voltages. On the Si(100) surface, Hamers et al.[24] showed the equivalent information was obtained either at negative or positive sample bias. In any case, interpretation of the CITS images must be done carefully and must be corroborated by studying the *I-V* curves at selected locations or by normalizing the data to remove the *z*-dependence, as will be discussed below.

Identifying the surface-state energies is usually done by analyzing the *I-V* curves at selected locations or averaged over a large region. Figure 4.13**a** shows plots of the conductance (I/V) vs. *V* measured at different specific locations within the Si(111)−(7 × 7) unit cell, together with the surface states observed in ultraviolet photoemission[75] and inverse photoemission[74] spectroscopies. The conductance curves in Figure 4.13**a** show steep "onsets" at particular voltages corresponding to the energies of the surface states. The energies of these onsets correspond exactly

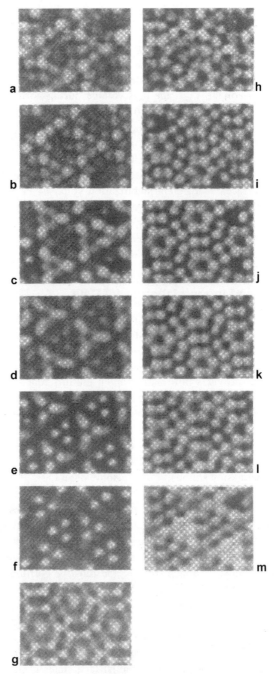

Figure 4.12. Images showing the spatial distribution of tunneling current in the Si(111)−(7 × 7) unit cell as a function of applied bias. Although difficult to interpret directly, such current images show pronounced changes in symmetry at particular voltages, which can be related to the symmetry of the surface states contributing to tunneling. (Reprinted with permission from Ref. 24.)

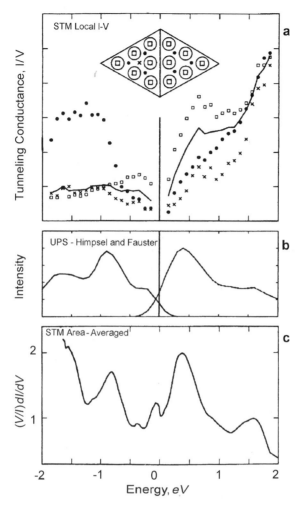

Figure 4.13. **a**, Spatially localized tunneling spectra on the Si(111)–(7 × 7) surface. **b**, Compared with the photoemission/inverse photoemission spectra of Himpsel and Fauster (Reprinted with permission from Refs. 73 and 74. **c**, Normalized tunneling spectra area-averaged over entire (7 × 7) unit cell. (Reprinted with permission from Ref. 42.)

with the energies of the surface states determined by photoemission shown in Figure 4.13**b** and correspond to the energies at which symmetry changes are observed in current images. The atomically resolved tunneling spectroscopy measurements directly reveal the atomic origins of the various electronic states. The states near −0.35 and +0.5 eV arise from the 12 adatoms within each unit cell, while the state near −0.8 eV arises from the 6 "rest" atoms. The states near −0.35 and +1.4 eV also appear to have some contribution from underlying layers, since they show an asymmetry between the faulted and unfaulted halves of the (7 × 7) unit cell.

Because dI/dV is small except at the "onset" energies, spatial maps of dI/dV at these energies can be loosely interpreted as "images" of the surface states themselves. However, just as in modulation experiments, the spatial maps of dI/dV also contain a background contribution in the form of an "inverted topography." Only when dI/dV is large and corresponds to the energy of a surface state can surface states be imaged in this way. This is the case, for example, in the surface-state images of Si(111)$-$(7 $\times$ 7),[3] where images of dI/dV were presented only at energies where the rapid increase in dI/dV overwhelmed this background effect.

Extracting quantitative information about the sample density of states is difficult because the density of the states of the tip ρ, and voltage-dependent tunneling transmission probability $T(E, eV)$ are almost always unknown. This can create confusion when comparing spectroscopy results obtained at different lateral positions, because the sample–tip separation (and consequently, the transmission probability) depends on the contour that the tip follows, which is determined by the feedback stabilization voltage V_{stab}. The voltage dependence of the transmission probability may also vary as a function of position owing to variations in the local work function, band-bending effects, and other phenomena. In the case of superconductors, the energy range of interest is small enough that the voltage dependence of the tunneling barrier is only a minor effect. In the study of surface states of metals and semiconductors, however, the energy range of interest usually extends several electron volts on either side of E_F. Nevertheless, since $T(E, eV)$ is generally a slowly varying function, valuable semiquantitative electronic structure information can still be obtained.

Feenstra et al.[30] argued that normalization of dI/dV by I/V reduces the data to a form like

$$\frac{dI/dV}{I/V} = \frac{d(\log I)}{d(\log V)} = \frac{\rho_s(eV)\rho_t(0) + A(V)}{B(V)} \tag{4.10}$$

This quantity is equal to unity at $V = 0$. The background term $A(V)$ contains the influence of the electric field in the gap on the decay length κ, whereas $B(V)$ normalizes the tunneling transmission probability over the density of states (DOS). Assuming that $A(V)$ and $B(V)$ vary slowly with voltage, structure in $(dI/dV)/(I/V)$ reflects $\rho_s(eV)$.

Figure 4.13c shows the results of this analysis for area-averaged tunneling I-V data for Si(111)$-$(7 $\times$ 7).[76] A comparison of this curve with the atomic-resolution conductivity measurements in Figure 4.13a and the photoemission results in Figure 4.13b shows a close correspondence, with surface-state peaks occurring at -1.5, -0.8, -0.2, $+0.45$, and $+1.55$ eV. Surprisingly, even the relative intensities of the tunneling data and the photoemission data appear in be similar, except for the state near -1.5, which appears quite small in the tunneling data.

One major advantage of this normalization procedure is that it tends to eliminate the distance dependence of the tunneling probability. Feenstra et al.[30] acquired I-V curves at different sample–tip separations on Si(111)$-$(2 $\times$ 11) and verified that this normalization minimizes the influence of the sample–tip separation z, as shown in Figure 4.14. Figure 4.14**b** shows normalized Si(111)−(2 × 1) tunneling spectra measured at five different sample–tip separations (indicated by open triangles, solid triangles, open squares, solid squares, and solid circles). The normalized data overlay one another, demonstrating that the normalization procedure minimizes the eect of changing sample–tip separation. A comparison of these data with predictions of the electronic structure based on Pandey's π-bonded chain model is shown in the Figure 4.14**c**, demonstrating good agreement between the experiment and theory for this particular model. Similar analysis has been done by Kuk and Silverman[77] on Au(100) surfaces.

While calculating $d(\log I)/d(\log V)$, or equivalently, $(dI/dV)/(I/V)$ provides a *convenient* normalization, a number of researchers have incorrectly stated that this "tunneling density of states" is proportional to the true density of states. This is certainly not true. First, simple calculus shows that, because the tunneling current is a continuous function of the applied voltage, $(dI/dV)/(I/V)$ must equal unity at $V = 0$. Thus the above procedure tends to normalize the spectra by the density of states at the Fermi level, which will be different at different locations on the surface. The relative intensities of the peaks in such normalized tunneling spectra also show significant variations from those in the actual state density. The origin of these differences will be discussed below.

The main goal of the (I/V) normalization is to minimize the effect of the voltage dependence of the tunneling barrier. For gapless surfaces where the entire *I-V* curve can be approximated by a simple exponential function of *V*, the normalization works well. If there is a surface-state band gap, however, artificial discontinuities arise in this normalization owing to the strong voltage dependence of the $B(V)$ term in the denominator of Eq. (4.10), particularly near the band edges. An alternative procedure is to extrapolate (I/V) from measurements at voltages well outside the surface-state band gap, or to fit the entire *I-V* curve to an exponential function. In the presence of a surface-state gap, this fitted *I-V* curve more accurately represents the smoothly increasing tunneling transmission probability than the experimental one does. Then the experimental (dI/dV) can be divided by the smoothed *I-V* curve. This provides compensation for voltage-dependent changes in the tunneling barrier but prevents the denominator in Eq. (4.8) from vanishing.[31] If the structure at the edges of the surface state bands are of primary importance, direct plots of *I* vs. *V* are satisfactory.[22]

Thus, although $(dI/dV)/(I/V)$ provides a convenient normalization of the tunneling spectroscopy data, it does not directly yield the sample DOS. Care must be taken in quantitative comparisons, particularly for $I = V$ curves that show significant deviations from a simple exponential form.

4.4.3   Comparison of Normalized *I-V* Data with True DOS

Numerical Integration of Tunneling Equations   To show how the tunneling spectra resulting from this normalization compare with the local state density, an artificial density of states function has been created and then, using Simmons's[78]

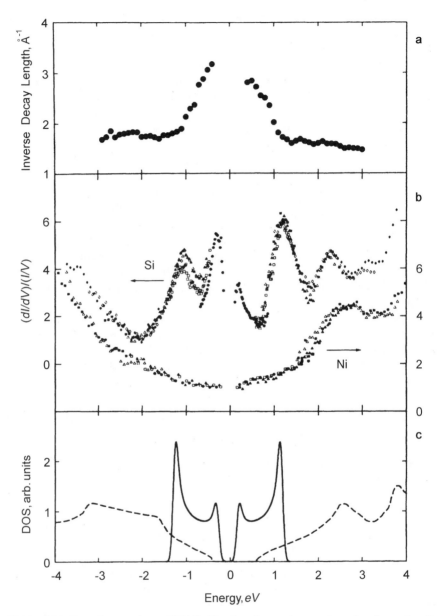

Figure 4.14. Tunneling spectra on Si(111)–(2 × 1) surface. **a,** Inverse decay length as a function of applied bias. **b,** Normalized tunneling spectra on Si(111)–(2 × 1) acquired at a number of different sample–tip separations, along with spectra measured on a Ni film. **c,** Calculated density of surface (*solid line*) and bulk (*dotted line*) states for the π-bonded chain model of Pandey. (Reprinted with permission from Ref. 30.)

formulas (which also include effects of the image potential), numerically integrated to obtain the *I-V* curve predicted from tunneling theory. This *I-V* curve was numerically differentiated and normalized in the same way as experimental data. The shapes and locations of the peaks in the initial input density of states function were chosen so that the calculated $d(\log I)/d(\log V)$ was in reasonable agreement with the experimental curve for $Si(111)-(7 \times 7)$, shown in Figure 4.13**c**. The other parameters used in the calculation were $\phi_{sample} = 2.5\,eV$, $\phi_{tip} = 2.5\,eV$, and $z = 8.0\,Å$.

Figure 4.15 shows the results of this simulation. The DOS function shown in Figure 4.15**a** was used with Simmons's formulas (assuming a constant DOS for the tip) to generate the *I-V* curve expected for this distribution, shown in Figure 4.15**b**. This was then numerically differentiated to give the plot of dI/dV vs. *V* shown in Figure 4.15**c**, and normalized by the static conductivity to produce the normalized spectrum shown in Figure 4.15**d**. Comparing this normalized spectrum in the

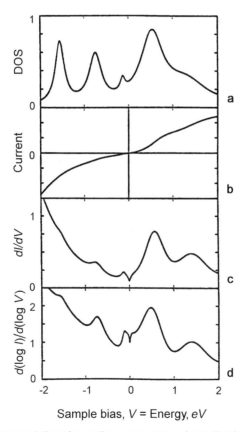

Sample bias, *V* = Energy, *eV*

Figure 4.15. Computer modeling of tunneling *I-V* spectra. **a**, Input density of states function. **b**, Tunneling current calculated by numerical integration of Simmons's equation. **c**, First derivative of tunneling current. **d**, Normalized tunneling spectrum using logarithmic-derivative normalization.

bottom panel with the starting DOS function shown in the top panel shows that they have peaks in nearly the same position. However, the intensities of the peaks are quite different. Unoccupied states of the sample are observed much more clearly and with higher intensity in the normalized tunneling spectra than are occupied states, which is a result of the fact that most of the tunneling electrons arise from states near the Fermi energy of the negatively biased electrode. This numerical simulation demonstrates that the intensities of the peaks observed in $(dI/dV)/(I/V)$ are not proportional to the surface density of states, particularly at negative sample bias.

Relative Intensities of s, p, and d-Band Contributions A second significant difference between the density of states detected by STM and the "true" density of states occurs for transition metals. Lang[79] predicted that for transition metals with partially filled d-bands near the Fermi energy, the contribution of these bands to the tunneling current is usually greatly exceeded by the s- and p-band contributions. This has two sources. First, for equivalent principal quantum numbers, d-states tend to be more tightly bound to the nucleus than s- and p-states. Second, for transition metals the occupied d-bands generally have a smaller principal quantum number than the nearby s- and p-bands (4d vs. 5s, 5p for Mo, for example). The smaller principal quantum number states decay faster in the vacuum, and so their overlap with the wavefunctions of the tip is often negligible. Lang's calculations indicate that occupied d-states will be nearly invisible to the STM, except for metals near the left edge of the period table (such as Ca), for which the d-states lie at or above the Fermi energy.

The predicted inability of tunneling spectroscopy to observe occupied d-states has been confirmed in several studies. Brodde et al.[80] conducted studies on Au(111) but were not able to observed any structure in the tunneling I-V curves associated with the onset of the d-bands at $\sim 2\,\text{eV}$ below E_F. Kuk and Silverman[77] likewise conducted studies of Fe clusters on Au(100) and did not find any observable structure associated with either occupied or unoccupied d-bands previously observed in photoemission and inverse photoemission spectroscopies at 0.6 below and 1.6 eV above E_F.[81] This insensitivity to d-band contributions is unfortunate, as it significantly limits the ability of the STM to distinguish between various transition metals.

Band-Structure Effects When comparing tunneling spectroscopy results with theoretical calculations, it is also important to recognize the importance of the band structure of the solid. The band structure has an effect on tunneling spectroscopy measurements because the rate of decay of the wavefunctions of the sample and tip in the vacuum region is determined by the momentum perpendicular to the surface. In Eq. (4.3), the energy E of the state enters the tunneling probability because the effective height of the tunneling barrier depends on the electron energy. More correctly, it depends on the momentum perpendicular to the surface. Using

$$E = \frac{k^2}{2m} = \left(\frac{k_x^2 + k_y^2 + k_z^2}{2m} \right) \tag{4.11}$$

we can include the effects of nonzero parallel momentum $k_\parallel$ and rewrite the inverse decay length of the wavefunctions as

$$\kappa = \sqrt{2m\bar{\phi}/\hbar^2 + k_\parallel^2} \tag{4.12}$$

where the average barrier height is given by

$$\bar{\phi} = \sqrt{\left(\frac{\phi_s + \phi_t}{2}\right) + \frac{eV}{2} - E} \tag{4.13}$$

and $k_\parallel = \sqrt{k_x^2 + k_y^2}$ is the parallel momentum.

Comparing this expression with the expression given in Eq. (4.3) shows that this essentially amounts to replacing the energy E by

$$E - \hbar^2\left(\frac{k_\parallel^2}{2m}\right)$$

Also, it is apparent from this equation that the wavefunctions have the longest decay length when the parallel momentum is zero, corresponding to the Γ point in the surface Brillouin zone. Thus tunneling spectroscopy measurements have a built-in preference to probe states at Γ. Detecting tunneling from other locations in the Brillouin zone is most readily done for semiconductors where either the high-lying occupied state disperses upward in energy from Γ, or where the lowest energy unoccupied state disperses downward in energy from Γ. Under these circumstances, energy conservation restricts tunneling at lower voltages to states with nonzero parallel momentum; at higher voltages, tunneling from all states will be possible but will occur preferentially from Γ. For very sharp tips, it is predicted that this preferential tunneling from Γ will be counteracted by momentum broadening through the uncertainty principle owing to the strong lateral confinement of the tunneling electrons. This topic will be discussed further below.

4.4.4   Applications of Local *I-V* Spectroscopy

In addition to Si(111)−(7 × 7) work discussed above, the electronic properties of many other clean surfaces and ordered overlayers have been studied using local *I-V* spectroscopy. Not all of these are reviewed here; rather, we simply highlight some of the more typical measurements to illustrate the capabilities of tunneling spectroscopy.

Semiconductors: Band Structure and Defects   Tunneling measurements on Si(111)−(2 × 1)[19, 29, 30, 82] show a close correspondence with the theoretical band structure for the π-bonded chain model proposed by Pandey.[83] Experimental observations of band structure effects have also been made. Feenstra et al.[30] found that the effective decay length of the wavefunctions varied strongly as a function of voltage (Fig. 4.14a). Interesting behavior results from dispersion of the surface-state bands toward E_F as the parallel momentum $k_\parallel$ increases. At low voltages, tunneling can only occur from states with large $k_\parallel$, which have a short decay length. At higher voltages, tunneling occurs from states with $k_\parallel = 0$, which have the slowest decay. From measurements of the inverse decay length as a function of voltage, the dispersion of the surface-state bands could be inferred from the tunneling measurements.

A second example comes from ordered overlayers of aluminum on silicon, which also have a downward-dispersion unoccupied state. Figure 4.16 shows both tunneling spectroscopy data as well as theoretical and experimental determinations of the surface band structure. The experimental band structure determination shows an unoccupied state that disperses downward in energy from $+1.4\,\text{eV}$ at Γ to $0.8\,\text{eV}$ at K' (Fig. 4.16**b**, solid circles). The tunneling spectroscopy measurements show two rather well-defined empty states at $+1.0$ and $+1.4\,\text{eV}$ (Fig. 14.16**a**). The tunneling spectroscopy peak near $+1.0\,\text{eV}$ corresponds closely to the bottom of the empty band observed near K' in the surface Brillouin zone, whereas the second

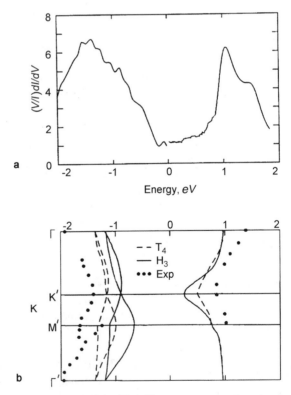

Figure 4.16. a, Tunneling spectra on $(\sqrt{3} \times \sqrt{3})$–Al/Si(111). **b,** Calculated and experimental surface-state band-structures for $(\sqrt{3} \times \sqrt{3})$–Al. (Reprinted with permission from Ref. 84.) The double peak for the unoccupied state observed in STM likely arises from a band-structure effect.

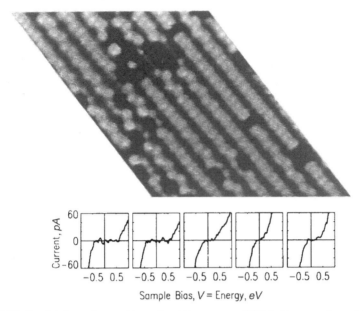

Figure 4.17. Spatial dependence of tunneling *I-V* curves on Si(001). Far away from defect, *I-V* curve shows a clear band gap. Above the defect, the *I-V* curve is metallic in nature. The spatial extent of this metallic character is ∼7 Å. (Reprinted with permission from Ref. 14.)

peak near $+1.4\,\mathrm{eV}$ corresponds to the band-edge at Γ ($k_\parallel = 0$). Here also, the effect is visible mainly because the unoccupied band disperses downward from Γ.

Figure 4.17 shows the capability of atomic-resolution *I-V* measurements to directly probe the electronic properties of defects. The figure shows *I-V* curves acquired at different distances from a small defect, measured simultaneously with the constant-current topograph (at $-1.5\,\mathrm{eV}$).[85] Far away from the defect (10e,10f), the tunneling *I-V* curve shows a clear gap as revealed from the sharp turn-on of the tunneling current near -0.45 and $+0.25\,\mathrm{eV}$ (in agreement wit the gap edges observed in area-averaged, logarithmically differentiated spectra[21]). As the defect is approached, the sharp gap edges disappear and directly over the defect (10c), the tunneling curve exhibits a strong exponential increase both above and below E_F, demonstrating that there is a high density of states at E_F at the defect site.[85] Such spatially dependent measurements also provide a direct measure of the spatial extent of the wavefunctions associated with defects and impurities. Single-atom defects occurring in Al overlayers on Si(111) have also been studied.[42, 76, 84] Figure 4.16a shows tunneling *I-V* measurements made on A1 adatoms and Figure 4.16b shows the *I-V* measurements on the Si adatoms, compared with a theoretical calculation[48] of the band structure for Al adatoms on Si(111). Substituting Si for Al shifts the energy of a p_z state from $+1.1$ to $-0.35\,\mathrm{eV}$, which is readily detectable in the *I-V* spectra aid also leads to large contrast changes in constant-current topographs that were noted earlier and shown in Figure 4.8. Unique localization effects are also observed in tunneling spectra at these defects, which are effectively iso-lated from one another by the large surface atom spacing in the $\sqrt{3}$ structure.[42, 84]

Local Electronic Structure and Chemical Reactivity   The ability to probe the local electronic structure of defects and chemically in-equivalent atoms at surfaces has great potential for the study of surface chemical reactivity on an atom-by-atom basis. An example is the decomposition of NH_3 on Si(111)–(7 × 7) studied by Avouris and Wolkow.[86, 87] The Si(111)–(7 × 7) structure has several types of chemically inequivalent surface atoms exposed. Figure 4.18 combines voltage-dependent STM topographs with local *I-V* measurements before and after exposing the surface to NH_3. Variations in chemical reactivity between the various types of chemically inequivalent atoms are seen. Thus local *I-V* spectroscopy provides a way of directly correlating chemical reactivity with local electronic structure.

Identification of specific surface electronic states involved in a reaction is ex-emplified in the reaction of NH_3 with Si(001)–(2 × 1).[25, 22] Normalized spectra shown in Figure 4.19 reveal a surface-state band gap of $\simeq 0.7$ eV, with peaks at -0.85 and $+0.35$ eV corresponding to the dimer π-bonding and π^*-antibonding states discussed earlier, in agreement with photoemission results.[73, 74] Tunneling

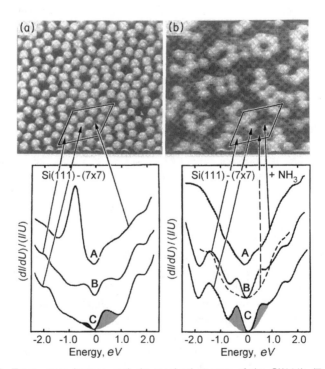

Figure 4.18. Empty state images and site-resolved spectra of the Si(111)–(7 × 7) surface before and after reaction with NH_3. Negative energies are occupied states. (Reprinted with permission from Ref. 86.)

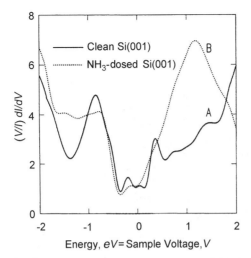

Figure 4.19. Normalized tunneling spectra on clean Si(001) and Si(001) after dosing with NH_3. The clean surface has an occupied state near -0.35 eV and an empty state near $+0.5$ eV, after dosing, these states are eliminated and replaced with a single, Si-H state at $+1.5$ eV. (Reprinted with permission from Ref. 21.)

measurements show that exposure of this surface to NH_3 eliminates these states and replaces them with an intense state lying 1.1 eV above E_F associated with Si-H antibonding states[22] owing to dissociation of the NH_3. Voltage-dependent imaging also provides a contrast between reacted and unreacted dimers, owing to changes in the spatial distribution of occupied states on H adsorption.

Nanostructures   The combination of imaging and local *I-V* measurement has become uniquely valuable in probing the electronic properties of nanostructured materials. Techniques that average over bulk samples cannot pinpoint the relationships between nanometer-scale structure and properties that are fundamental to these materials. Direct measurements of the conductivity of proposed "molecular wires" have been carried out.[88, 89] Coulomb blockade, only apparent at low temperatures in macroscopic materials, has frequently been reported in STM *I-V* measurements on metal nanoparticles. STM images combined with local *I-V* measurements on carbon nanotubes have shown direct correlation between the local density of states (LDOS) and the diameter and helicity of single-walled carbon nanotubes (Fig. 4.20).[90–92] The nanotubes may be semiconducting or metallic depending on their stucture. Carbon nanotubes have also shown evidence of local Schottky diode behavior, postulated to be the result of defects.[93]

Superconductors   Tunneling spectroscopy has been an important probe of superconductivity since Giaever's experiments on tunneling in a planar metal-oxide-superconductor junction confirmed the BCS theory.[94] That work led to Bardeen's transfer hamiltonian theory for tunneling, which in turn provided the

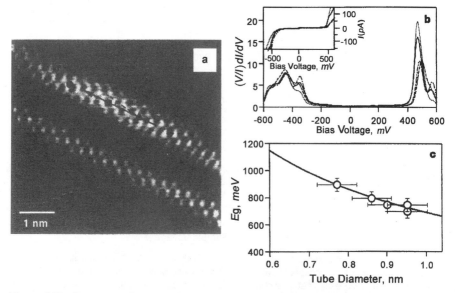

Figure 4.20. Structure and spectra of single-walled carbon nanotubes. **a**, Constant current image at 150 pA and 300 mV. **b**, Spectra taken at the marked points on the image. **c**, Correlation between diameter and band gap for semiconducting nanotubes. (Reprinted with permission from Ref. 91.)

first theoretical paradigm for STM.[95] Atomic spatial resolution and meV energy resolution has made scanning tunneling spectroscopy a favored tool in studies of new superconductors.[96]

Perhaps one of the most dramatic applications of STM has been the work of Hess et al.,[97–100] who used local tunneling spectroscopy to image the electronic state density around a superconducting flux core in $NbSe_2$. Figure 4.21 shows voltage-dependent variations in the vortex lattice. Figure 4.22 shows spectra taken at several points along a line through a flux core.

Several studies have reported spatial variations in superconducting energy gaps. Fein et al.[101] and Kirtley et al.[75] measured $I\text{-}V$ curves at each location while scanning over granular superconductors and later analyzed the data to provide an image of the spatial dependence of the energy gap. Many other recent studies of superconductors have concentrated on the various new high-temperature supercon-ductors,[102–105] as well as organic superconductors.[106] Surface quality is a critical issue. The tunneling spectra of high-Tc materials has been seen to depend on the topmost atomic layer of the surface. Reproducible spectra can be achieved by in situ surface preparation under inert atmospheres.[107, 108]

4.5 GENERAL FEATURES OF TUNNELING SPECTRA

4.5.1 Dynamic Range

To a first approximation, tunneling $I\text{-}V$ curves are exponential in voltage for $V \geq 100$ mV. As a result of this exponential dependence, one difficulty commonly

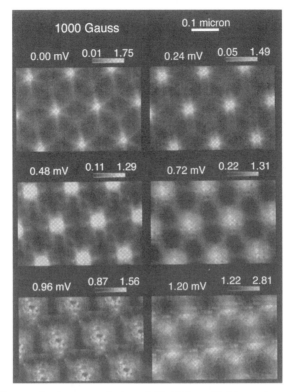

Figure 4.21. Changes in the vortex lattice of 2H-NbSe$_2$ with bias voltage. (Reprinted with permission from Ref. 99.)

encountered in tunneling spectroscopy measurements is that of dynamic range. Assuming a constant density of states for sample and tip, the tunneling current is roughly given by

$$I \propto \int_0^V \exp\left(-AZ\sqrt{\bar{\phi} - V}\right) \tag{4.14}$$

where $A = 1.025 \, \mathrm{eV}^{1/2} \, \mathrm{\mathring{A}}^{-1}$. Figure 4.23 shows the result of numerically integrating this equation for several different values of average work function ϕ. In each case, the sample-tip distance has been adjusted such that the tunneling current is 0.85 nÅ at $V = 2.0$ V. Here, we see that the sample with the highest work function has an I-V curve that is somewhat more linear than the sample with the lowest work function. Qualitatively, this is easy to understand. For a material with a high work function, increasing the applied voltage leads to only a relatively small change in the argument of the exponent, and the integral tends to be comparatively linear. For a material with a low work function, however, increasing the applied voltage makes the exponential factor significantly greater, and the integral increases as a rapid exponential function. In each case, however, even in the limit of constant density of states one must have at least three orders of magnitude dynamic range in the

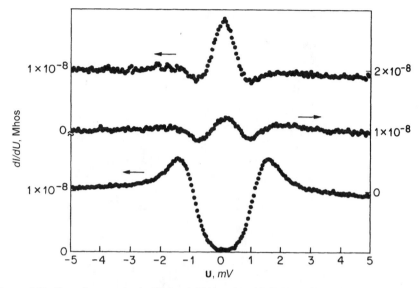

Figure 4.22. Tunneling spectra for $NbSe_2$ 1.85 K and 0.02 T field. *Top curve*, On a vortex; *middle curve*, 75 Å away; *bottom curve*, 2000 Å away. (Reprinted with permission from Ref. 97.)

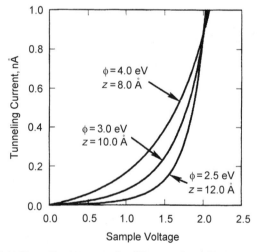

Figure 4.23. Calculated tunneling *I-V* curves for different values of barrier height and sample–tip separation. The spectra were all calculated such that $I_{tunnel} = 0.85$ nÅ at a sample voltage of 2 V.

tunneling current measurement to record the electronic characteristics between about 0.1 and 2 eV.

Within a surface-state band gap the density of states is low, and the problem may be even more acute. One way around this difficulty has been described previously: allowing the tip–sample distance to vary while measuring both the current and the sample–tip separation as a function of voltage. By measuring the tunneling barrier height, it is then possible to effectively normalize the data to a constant sample–tip separation.

4.5.2 Resolution

The resolution of tunneling spectroscopy is determined primarily by the range of electron energies that can contribute to the tunneling current. For inelastic tunneling experiments at low voltages, Hansma[7] determined an effective resolution of $5.4\,kT$, or approximately 140 mV at room temperature. For electronic spectroscopy at higher voltages, the effective resolution is primarily determined by the fact that the electrons involved in tunneling arise from a broad range of energies, determined by the voltage difference between sample and tip. The width of the electron energy distribution can be seen from Eq. (4.3) for the transmission probability $T(E, eV)$. Because the work functions of sample and tip are large, the argument of the exponent is relatively insensitive to E, resulting in a wide energy distribution for the tunneling electrons. Thermal effects can be included by multiplying the tunneling transmission factor by the Fermi function $f(E)$, which describes the probability that a state with energy E is occupied. The definition of $f(E)$ is

$$f(E) = \left[1 + \exp\left(\frac{E - E_F}{kT} \right) \right]^{-1} \tag{4.15}$$

Becuase electrons must tunnel from an occupied state to an unoccupied state, the overall tunneling transmission probability must be multiplied by two Fermi functions

$$T(E, eV, kT) = T(E, eV)f(E)[1 - f(eV - E)] \tag{4.16}$$

The first factor is the temperature-independent transmission probability $T(E, eV)$ defined in Eq. (4.3), the second factor is the probability that the state with energy E on the first electrode (either sample or tip) is occupied, and the third factor is the probability that the state with energy $eV-E$ on the second electrode (either tip or sample) is unoccupied; energy conservation is implied, because the Fermi level of sample and tip are different by an amount eV.

Figure 4.24 shows the relative transmission probability as a function of energy assuming that $\phi_s = \phi_t = 4.0\,eV$, $= 8.0\,\text{Å}$, and the applied bias $V = 2.0\,V$. The *dotted line* is the calculation neglecting thermal broadening, and the *solid line* is calculated including thermal broadening for $T = 300\,K$. It can be seen that the main effects of including finite temperature are to broaden out the Fermi edge and to shift the electron energy distribution to slightly lower energies. In either case, it can be seen that most of the tunneling electrons come from within 300 mV of the Fermi level of the negatively biased electrode. The asymmetric form of $T(E, eV)$, with the sharp increase at E_F, helps make the effective resolution of the STM somewhat higher when probing empty states of the sample than when probing filled states.

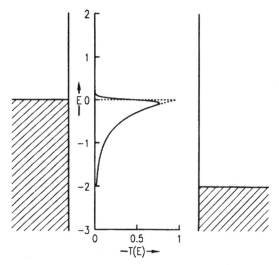

Figure 4.24. Energy distribution of tunneling electrons calculated using Eq. (4.3) and assuming $\phi_s = \phi_t = 4.0\,\text{eV}$, $z = 8.0\,\text{Å}$, and $V = 2.0\,\text{V}$. *Dotted line*, neglects thermal broadening; *solid line*, includes thermal broadening for $T = 300\,\text{K}$.

An additional broadening mechanism may arise from the Heisenberg uncertainty principle, for extremely sharp tips. The localization of the tunneling current to region $\Delta\omega$ of approximately $5\,\text{Å}$ diameter implies an uncertainty of $\sim 1/\Delta\omega = 0.2\,\text{Å}^{-1}$ in the parallel momentum $k_{\parallel}$ of the tunneling electrons. Since the two-dimensional Brillouin zone for the Si(111)–(7 × 7) surface, for example, has a maximum length of $0.155\,\text{Å}^{-1}$ (along the $\Gamma K'$ direction), this can represent a significant averaging over various portions of the surface Brillouin zone. For highly dispersive states, this will tend to contribute an additional energy broadening mechanism since tunneling will no longer be strongly preferred from the Γ point.

4.5.3 Tip Electronic Structure

Many of the present limitations of tunneling spectroscopy arise from the inability to easily prepare well-defined tips of known electronic structure. Tunneling spectra are always a convolution of the electronic structures of the sample and tip. At a minimum, it is necessary to compare spectra acquired at different locations on the surface to distinguish the spatially invariant contribution of the tip and the spatially varying contribution from the sample. Ideally, the geometric and electronic structure of the tip, which will differ from the bulk material, should be known. A number of groups[109–112] have calculated the electronic structure of the tip and its effects on the tunneling current and tunneling spectra. DeParga et al.[113] found that two distinct forms occured for experimental tunneling spectra of a Cu surface obtained with W tips. They then showed that calculated spectra using two different stable clusters to model the W tip matched the experimental results.

Although uncharacterized tips remain the rule, progress continues to be made in the area by combining field ion microscopy and field evaporation techniques with STM.[114,115] Ultimately, it may be possible to perform tunneling spectroscopy experiments with tips of known geometric and electronic structure. Carbon nanotubes, dendrimers, and organic conductors have all been proposed as molecularly well-defined tips.

4.5.4 Anomalies

Just as in conventional scanning tunneling microscopy, there are a number of anomalies and artifacts that can arise in tunneling spectroscopy experiments. Many of these anomalies arise because of the general lack of control over the geometry and electronic structure of the tip. One of the most interesting analysis of tip effects in tunneling spectroscopy was performed by Klitsner et al.[116] On a tip that (on the basis of topographic images) was composed of at least two "microtips," they acquired tunneling spectroscopy data as a function of sample tip separation and at different locations on the sample. At particular locations, tunneling from one microtip was strongly favored over the other. Through a detailed analysis of these tunneling spectra acquired at different locations on a nominally homogeneous sample, it was possible to distinguish the contributions from each microtip.

In general, tip-related artifacts cannot be easily ruled out in any tunneling spectroscopy measurement. When making comparisons between spectra at two different locations on the surface (with the same tip), the electronic structure of the tip is known to be at least constant. In making absolute measurements, however, it is important to perform measurements using a variety of different tips and samples to validate the results.

Negative Differential Resistance   Simple one-dimensional (1-D) tunneling theory between electronically equivalent electrodes generally predicts that tunneling transmission probability increases monotonically with the applied voltage. Because in most cases structure in the DOS is relatively weak, the net result is that the tunneling current generally increases with the applied voltage. However, this is not mandated by the tunneling equations. It is quite possible (over certain limited ranges of applied voltage) for the tunneling current to decrease as the applied voltage is increased. This phenomenon is referred to as negative differential resistance (NDR) because the differential resistance dI/dV is negative in these regions.

The observation of negative differential resistance over localized areas with the scanning tunneling microscope was first reported by Hamers and Koch,[117] who reported NDR in localized regions of $\sim$5 to 10 Å diameter on partially-oxidized silicon surfaces (Figure 4.25). Here the tunneling curve increases normally at low voltage; but at approximately 1 V sample bias, the tunneling current exhibits a clear decrease. The corresponding first derivative becomes negative (Fig. 4.25**b**).

Negative differential resistance can have several origins. First, it can occur if the tunneling transmission function is not a monotonically increasing function of $|V|$.

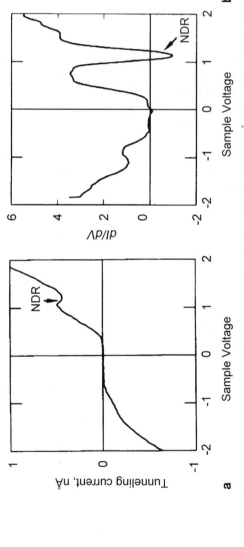

Figure 4.25. a, Tunneling *I–V* curve demonstrating NDR acquired at an oxide trap of approximately 5 Å, diameter on oxidized Si(001). **b,** First derivative of the tunneling *I–V* curve. (Reprinted with permission from Ref. 117.)

This is the case for partially oxidized silicon surfaces which form localized trap states throughout the band gap region. These trap states give rise to the anomalous *I-V* curves and negative differential resistance observed in Figure 4.25. In this case, NDR is observed at positive sample bias, when electrons tunnel from the tip to the sample. The origin of this behavior is illustrated in Figure 4.26. Here the trap energy lies well above the Fermi energy of the sample. At low bias voltages, the trap is unoccupied and the tunneling *I-V* curve is comparatively normal. At a sufficiently positive sample bias, however, electrons from the tip can tunnel through the gap and occupy this trap state. One unique characteristic of these trap states is that they can capture an electron for a long time (easily in the millisecond regime).[118] The negative charge associated with the electron captured in this localized trap state acts as a "Coulomb blockade," preventing further electrons from tunneling until the electron relaxes from the trap state to the Fermi energy. As long as the lifetime of the electron in the trap state is longer than 10^{-10} sec (the average time between tunneling events for a current of 1 nA), then it can easily give rise to strong negative differential resistance. As a result, the tunneling transmission function increases at low voltages (Fig. 4.26a), decreases sharply when the trap is near the Fermi energy, becomes occupied by tunneling electrons (Fig. 4.26b), and then increases again at higher bias voltages.

A second cause of NDR can be the existence of energetically narrow states on either the sample or tip. If the density of states of sample or tip has sufficiently sharp structure, then over a particular range of bias voltages the decrease in the DOS near the Fermi energy can overwhelm the increase in the voltage-dependent tunneling transmission probability, resulting in a net decrease in tunneling current. Bedrossian et al.[119] and Lyo et al.[120] reported NDR on highly B-doped Si samples and attributed it to the presence of energetically localized states. This NDR can also occur from localized states on the tip, because sample and tip play nearly equivalent roles in STM.

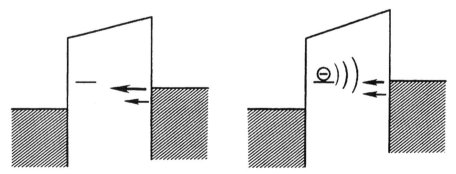

Figure 4.26. Origin of negative differential resistance owing to trap states on partially oxidized silicon surfaces. At low voltage, tunneling electrons do not have sufficient energy to occupy the trap state, and tunneling is normal. At slightly higher voltage, the trap becomes occupied and exerts a Coulomb repulsion, which prevents further tunneling and gives rise to NDR.

Tip-Induced BandBending In typical STM measurements, there is a very high electric field between sample and tip. On semiconductor surfaces, Weimer et al.[121] suggested that the electric field between tip and sample can change the bandbending, as illustrated in Figure 4.27. The significance of this effect depends to a great extent on the density of states within $\sim kT$ of the Fermi energy, because any tendency to change the bandbending will change the occupation of these states. On Si(111)−(7 × 7), the presence of an adatom-related surface state at the Fermi energy keeps the Fermi level well pinned at a constant position. Thus tunneling spectra[3] show surface state energies in good agreement with the values obtained using other surface science techniques. On most other semiconductors, however, the surface the reconstruction results in a surface-state bandgap. As a result, the Fermi-level position at most semiconductor surfaces is determined by defects or the tails of the bulk bands. On Si(001), for example, one particular kind of common defect has a high density of states and pins the Fermi level. With comparatively low numbers of midgap states available to pin the Fermi level, it is quite possible for the tip-induced electric field to affect the band energies. If this occurs, then the full sample–tip potential drop does not occur across the vacuum, but occurs partially in the bulk. As a result, the tunneling spectra will spread out, with all peaks appearing farther away from the Fermi energy than anticipated. On some surfaces the tunneling spectra have been shown to be in good agreement with the results of other surface science techniques,[3,30] indicating that tip-induced bandbending is negligible. On wider bandgap material and unpinned surfaces, it can be significant.

Kaiser et al.[122] showed that on hydrogen-terminated silicon surfaces, the very low density of gap states leaves the Fermi level unpinned. Under those conditions, the electric field between tip and sample penetrates a significant distance into the bulk of the sample, and thereby modifies the energies of the valence and conduction bands. As illustrated in Figure 4.27, on an *n*-type sample a midgap Fermi level implies upward bending of the bands at the surface; as a result, at positive sample bias the bandbending will be increased, and at negative bias the bandbending will

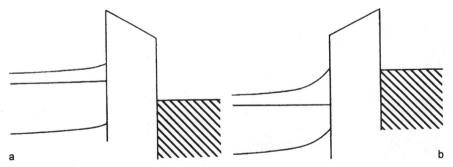

Figure 4.27. Tip-induced bandbending at unpinned, *n*-type semiconductor surfaces. **a,** At negative sample bias, the Schottky barrier is reduced (forward-biased), and the tunneling current is high. **b,** At positive sample bias, the Schottky barrier is increased, and the tunneling current is reduced. On *p*-type samples, the situation is reversed.

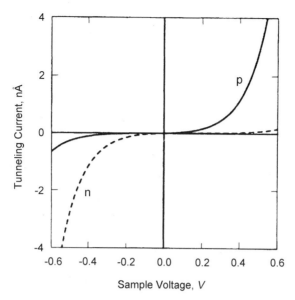

Figure 4.28. Effect of tip-induced bandbending on tunneling current. On n-type samples, tunneling current is high at negative sample bias and low at positive bias. On p-type samples, the situation is reversed. (Modified from Ref. 21.)

be decreased. On a p-type sample the situation is reversed. As shown in Figure 4.28, Kaiser et al. found that the bandbending produced pronounced differences between the general shape of the I-V curves on n-type and p-type material. One result of tip-induced bandbending is that even in the absence of discrete surface-state related electronic structure, the tunneling I-V curves will be asymmetric at positive and negative bias and will be different on n-type and p-type material. As demonstrated by Kaiser et al., this bandbending can be modeled as a Schottky barrier problem.

4.6 OTHER SPECTROSCOPIES

4.6.1 Inelastic Tunneling Spectroscopy

The electronic spectroscopy discussed thus far provides some chemical contrast but is not capable in most cases of actually identifying chemical species adsorbed on a surface. For such purposes, it would be more useful to measure the vibrational spectra of adsorbed species, because vibrational spectra are usually sharp and show features that are characteristic of particular molecular functional groups. Theoretical calculations[14, 123–125] have predicted that under certain circumstances, changes in dI/dV as large as 10% might be observed in tunneling measurements. Observing such structure is not an easy task because the vibrational features are so sharp that a low-temperature STM is required to avoid thermal broadening of the Fermi levels.

Hansma[7] estimated an effective resolution of $5.4\,kT$ for inelastic tunneling, although vibrational features are typically only a few meV wide; as result, measurement of vibrational losses must be performed at liquid helium temperature or lower.

Early experimental studies reported inelastic losses associated with phonons on graphite[126] and sorbic acid vibrations adsorbed on graphite,[127] both obtained on a surface that was completely immersed in liquid helium. In the case of graphite phonons, a good correspondence was observed between peaks in d^2I/d^2V vs. V and the known phonon energies determined from various other methods. For sorbic acid adsorbed on graphite, peaks were observed in the first derivative spectrum instead of the expected second derivative spectrum. In addition, the peaks were intense and the energies of the peaks were different from those measured in bulk tunnel junctions. The origins of these discrepancies are not yet resolved and may arise from a strong coupling with graphite bulk states.

Gregory[128] clearly observed vibrational-inelastic features in a "self-assembled tunnel junction." Although not strictly a "scanning" tunneling microscope, this junction is formed by two fine crossed wires with a layer of adsorbed argon between, forming a tunnel junction of atomic dimensions. Small amounts of hydrocarbon contamination are present in the argon, which can be detected through inelastic losses characteristic of carbon-hydrogen molecular vibrations. Figure 4.29 shows the second derivative of the tunneling current, d^2I/dV^2 obtained by

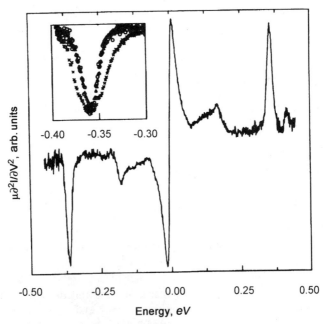

Figure 4.29. Inelastic tunneling spectra obtained on self-assembled tunnel junction. Characteristic vibrational losses are observed near 359 and 173 mV associated with hydrocarbon vibrations. (Reprinted with permission from Ref. 128.)

Gregory. Clear inelastic features are observed at $359\,mV$ ($2900\,cm^{-1}$) and in a range near $173\,mV$ ($1400\,cm^{-1}$). The former, higher energy peak corresponds closely with the energy expected for C-H stretching modes, whereas the second corresponds to a broad range of energies typically associated with C-H bending modes. The spectral widths of the observed features were typically on the order of $20\,mV$ ($\sim160\,cm^{-1}$), owing to a combination of both homogeneous and inhomogeneous line broadening mechanisms. The failure to observe similar vibrational spectra in conventional low-temperature scanning tunneling microscopes remains a question of great interest and importance.

4.6.2 Barrier Height Spectroscopy

In conventional tunneling spectroscopy the tunneling current is measured as a function of the applied voltage, providing a measure of the sample density of states. One of the major uncertainties in such experiments is that the effective height and width of the tunneling barrier are generally unknown. However, it is possible to directly probe the effective tunneling barrier height. Although this technique has received, comparatively little attention, it provides information complementary to tunneling I-V spectroscopy. Barrier height spectroscopy involves measuring the dependence of the tunneling current on the sample–tip separation, at constant applied voltage. In its usual implementation, a constant bias is applied between the sample and tip. A small modulation (typically a few hundredths of an angstrom) is applied to the z-piezo, and the resulting modulation in the tunneling current is measured using a lock-in amplifier. The frequency of the modulation must be faster than the closed-loop bandwidth of the STM feedback electronics, but must be small compared to the mechanical resonance frequency of the piezoelectric scanners. Typically the modulation frequency is chosen to be just slightly faster than the feedback electronics.

In the WKB approximation, the tunneling transmission probability can be directly related to the local work function ϕ_s of the sample. At low bias voltages, the tunneling transmission probability Eq. (4.3) simplifies as

$$T = \exp\left(-\frac{2Z\sqrt{2m}}{\hbar} \sqrt{\frac{\phi_s + \phi_t}{2}} \right) \tag{4.17}$$

The DOS contributions to the current in Eq. (4.2) do not depend on z so $I(z) \propto T(z)$ and

$$\frac{d\ln I}{dz} = \frac{1}{I}\frac{dI}{dz} = \frac{2\sqrt{2m}}{\hbar} \sqrt{\frac{\phi_s + \phi_t}{2}} \tag{4.18}$$

Thus by modulating the z-piezo and measuring the induced modulation (dI/dz) of the tunneling current I, the average work function of sample and tip can be determined. An apparent barrier height can be defined

$$\Phi_a[eV] \equiv 0.952 \left(\frac{\partial \ln I}{\partial z[\text{Å}]} \right)^2 \tag{4.19}$$

Because the work function of the tip is constant, lateral variations in the measured barrier height can be attributed to changes in the local sample work function. In practice, many experimental determinations of barrier heights using this tech-nique have obtained unreasonably small values. At the time of this writing, it appears that several effects may be entering, depending on the experimental conditions.

The first reason for anomalously low barrier heights is the detailed form of the potential function between the tip and the sample. As a result of the close proximity of tip and sample, there is an image potential that has been predicted to modify the potential. However, as Binnig and Rohrer,[129] Coombs et al.,[130] and Lang[131] have pointed out, even though the form of the potential function includes image potential terms that vary as $1/z^4$, in determining the tunneling barrier height these terms cancel, leaving only terms that vary like $1/z^2$. At the distances encountered in STM, the image potential is expected to have negligible effect on the measured barrier heights; this has been further confirmed by Coombs et al.,[130] who numerically integrated Simmons's tunneling equations[78] and concluded that the effect of the image potential was negligible under normal STM conditions.

Lang[131] pointed out that the potential function may be regarded as the sum of an electrostatic potential $v_{es}(z)$ and an exchange-correlation potential $v_{xc}(z)$. In the local density approximation, the exchange-correlation potential varies as the cube root of the electron density. Thus, although the electron density in the vacuum region varies like $n \sim \exp(-\alpha z)$, the exchange-correlation potential contribution to the total potential varies like $v_{xc} \sim \exp(-(1/3)\alpha z)$. Lang performed numerical calculations of the tunneling current as a function of distance using this local density approximation, and used the results to generate an effective barrier height. As shown in Figure 4.30, the results indicate that for sample–tip separations of <12 Bohr (about 6.5 Å), the measured barrier height will be substantially below the true work function.

Another reason for anomalous barrier heights is the mechanical interaction between sample and tip. This is illustrated with the help of Figure 4.31, which shows Chen and Hamers's[132] data on the measured barrier height as a function of sample–tip separation using a tungsten tip and a clean Si(111)–(7 × 7) surface under ultrahigh vacuum conditions. At large distances, the measured barrier height of 3.6 eV is in good agreement with the value anticipated from the work functions of tungsten and silicon. As the sample–tip separation is decreased, the barrier height is constant at first; it then shows a small increase, and shortly thereafter plunges toward zero. At extremely small distances, the sample and tip are in obvious physical contact. Clearly, when the sample and tip are in physical contact, then as the z-piezo scanner tries to push the tip toward the sample the only effect is to slightly compress the sample–tip physical junction, with little eect on the resistance of the junction; thus the barrier height must tend toward zero when the sample and tip are in physical contact.

As the tip is moved toward the sample from larger distances, Figure 4.31 shows that the measured barrier height actually increases slightly. One explanation for this effect was proposed by Chen[16] as indicative of an attractive force between the sample and tip leading to mechanical deformation. The initial assumption is that

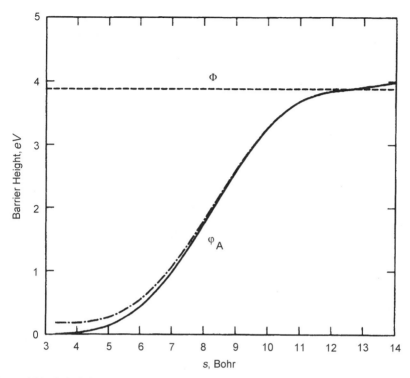

Figure 4.30. Calculation by Lang of the effective barrier height as a function of sample–tip separation. The comparatively slow approach to the equilibrium work function value is primarily owing to the exchange-correlation part of the atomic potential. (Reprinted with permission from Ref. 131.)

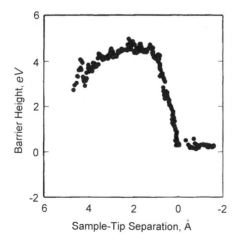

Figure 4.31. Experimental measurement of the barrier height as a function of sample–tip separation on a clean Si(111)–(7 × 7) surface with a tungsten tip. At close distances (corresponding to strong mechanical contact) the barrier drops to nearly zero. (Reprinted with permission from Ref. 132.)

the sample–tip interaction can be modeled as a Morse-type potential. When the sample–tip separation corresponds to the steep, attractive part of the potential, then as the z-piezo pushes the tip toward the surface (during each period of the applied z-piezo modulation), the atom at the end of the tip experiences a greater attraction toward then surface than it does when it is 180° out of phase, in effect "stretching" the tip slightly. The net result of this is that the separation between the sample and the atom at the end of the tip actually changes by slightly more than the z-piezo modulation, leading to an barrier height that is slightly larger than the true work function when the sample–tip interaction is attractive. The prediction of this analysis is that the measured barrier height will be too large when the sample–tip separation corresponds to being on the attractive part of the Morse potential, it will be exactly equal to the true value when the sample–tip separation corresponds to the minimum of the Morse potential, and it will be smaller than the true work function at smaller distances when the potential becomes repulsive (in eect, compressing the tip and/or sample). Chen successfully fit the detailed shape of this function, assuming only that the mechanical interaction can be described with a Morse potential with two free parameters.

Atomic scale geometric contributions to the measured barrier height arise from a "smoothing" of the equipotential contours in the vacuum. As discussed earlier (Fig. 4.10), the effective decay length of the wavefunctions above a surface protrusion must be shorter than that above a depression, because the potential corrugations always decrease at large distances from the sample. Because a shorter decay length is essentially equivalent to a high effective barrier, this means that (all other factors being equal) the barrier height measured above a protrusion will be larger than that measured above a depression.

Finally, it must be remembered that the equation used for the barrier height assumes that the separation between sample and tip is modulated by δz. On tilted samples or samples in which the surface normal is tilted by θ with respect to the z-direction (as defined by the direction of expansion of the z-piezo), the modulation of the gap spacing is $\delta z_{\mathrm{gap}} = \cos(\theta)\,\delta z_{\mathrm{piezo}}$. For surfaces that are not atomically smooth, this convolutes a strong geometric factor into the barrier height measurement proportional to the cosine of the local sample tilt. In particular, this can lead to artifacts in apparent barrier heights at steps or grain boundaries.[133]

4.7 SUMMARY

In this chapter, several of the methods for acquiring and interpreting tunneling spectroscopy data were presented, together with examples of how these techniques have been applied to study the electronic structure of surfaces. The unique ability of the STM to probe the electronic structure of surfaces directly has opened yet another dimension in our understanding of surfaces by allowing us to study the energetics of the electronic states at the surface on an atom-by-atom basis and to directly correlate the geometric positions of the atoms with the resulting electronic structure. Advances in the acquisition and interpretation of tunneling spectroscopy data continue to improve these capabilities. Crucial to the effective use of this powerful technique is proper analysis of tunneling spectroscopy data.

ACKNOWLEDGMENTS

R.J.H. expresses his appreciation to Norton Lang, Jerry Tersoff, Randy Feenstra, Julian Chen, Joe Demuth, and David Cahill, all of whom have had a continuing interest in tunneling spectroscopy. D.F.P. thanks the Trustees of Amherst College. In addition, we would like to acknowledge the U.S. Office of Naval Research and the Petroleum Research Fund for partial support of this work.

REFERENCES

1. G. Binnig, H. Rohrer, C. Gerber, and E. Weibel, *Phys. Rev. Lett.* **50**, 120 (1983).
2. R. Tromp, R. Hamers, and J. Demuth, *Phys. Rev. B* **34**, 1388 (1986).
3. R. Hamers, R. Tromp, and J. Demuth, *Phys. Rev. Lett.* **56**, 1972 (1986).
4. R. M. Feenstra, W. A. Thompson, and A. P. Fein, *Phys. Rev. Lett.* **56**, 608 (1986).
5. R. Feenstra, J. Stroscio, J. Tersoff, and A. Fein, *Phys. Rev. Lett.* **58**, 1192 (1987).
6. N. Lang, *IBM J. Res. Dev.* **30**, 374 (1986).
7. P. Hansma, *Tunneling Spectroscopy: Capabilities, Applications and New Techniques*, Plenum Press, New York, 1982.
8. E. Wolf, *Electron Tunneling Spectroscopy*, Oxford University Press, Oxford, UK, 1986.
9. C. Duke and W. Ford, *Surf. Sci.* **111**, L685 (1981).
10. N. Garcia, F. Flores, and F. Guinea, *J. Vac. Sci. Technol.* **6**, 323 (1988).
11. J. Tersoff and D. Hamann, *Phys. Rev. B* **31**, 805 (1985).
12. J. Tersoff and D. Hamann, *Phys. Rev. Lett.* **50**, 1998 (1983).
13. A. Baratoff, *Physica* **127B**, 143 (1984).
14. B. Persson and A. Baratoff, *Phys. Rev. Lett.* **59**, 339 (1987).
15. N. Lang, *Phys. Rev. B* **34**, 5947 (1986).
16. C. J. Chen, *J. Vac. Sci. Technol.* **6**, 319 (1988).
17. R. Wiesendanger and H.-J. Güntherodt, eds., *Scanning Tunneling Microscopy III*, Springer-Verlag, Berlin, 1996.
18. G. Binnig, et al., *Phys. Rev. Lett.* **55**, 991 (1985).
19. J. Stroscio, R. Feenstra, and A. Fein, *J. Vac. Sci. Technol. A* **5**, 838 (1987).
20. A. Selloni, P. Carnevali, P. Tosatti, and C. Chen, *Phys. Rev. B.* **31**, 2602 (1985).
21. R. Hamers, P. Avouris, and F. Bozso, *Phys. Rev. Lett.* **59**, 2071 (1987).
22. R. Hamers and U. Koehler, *J. Vac. Sci. Technol. A* **7**, 2854 (1989).
23. R. Tromp, E. V. Loenen, J. Demuth, and N. Lang, *Phys. Rev. B* **37**, 9042 (1988).
24. R. Hamers, R. Tromp, and J. Demuth, *Surf. Sci.* **181**, 346 (1987).
25. R. Hamers, P. Avouris, and F. Bozso, *J. Vac. Sci. Technol.* **6**, 508 (1988).

26. R. Hamers, R. Tromp, and J. Demuth, *Phys. Rev. B* **34**, 5343 (1987).

27. J. Kubby, J. Griffith, R. Becker, and J. Vickers, *Phys. Rev. B* **36**, 6079 (1987).

28. R. Becker, J. A. Golovchenko, and B. Swartzentruber, *Phys. Rev. Lett.* **55**, 987 (1985).

29. J. Stroscio, R. Feenstra, and A. Fein, *Phys. Rev. Lett.* **57**, 2579 (1986).

30. R. Feenstra, J. Stroscio, and A. Fein, *Surf. Sci.* **181**, 295 (1987).

31. R. Feenstra and P. Martensson, *Phys. Rev. Lett.* **61**, 447 (1988).

32. J. Stroscio, R. Feenstra, and A. Fein, *Phys. Rev. Lett.* **58**, 1668 (1987).

33. D. Biegelsen, R. Bringans, J. Northrup, and L. Swartz, *Phys. Rev. B* **41**, 5701 (1990).

34. D. K. Biegelsen, L. J. Swartz, and R. D. Bringans, *J. Vac. Sci. Technol. A* **8**, 280 (1990).

35. B. Reihl, J. Gimzewski, J. Nicholls, and E. Tosatti, *Phys. Rev. B* **33**, 5770 (1986).

36. J. Appelbaum, G. Baraff, and D. Hamann, *Phys. Rev. Lett.* **35**, 11 (1975).

37. J. Chadi, *Phys. Rev. Lett.* **43**, 43 (1979).

38. K. Pandey, *Phys. Rev. Lett.* **47**, 1913 (1981).

39. K. Takayanagi, Y. Tanishiro, M. Takahashi, and S. Takahashi, *J. Vac. Sci. Technol. A* **3**, 1502 (1985).

40. J. Tersoff, R. Feenstra, J. Stroscio, and A. Fein, *J. Vac. Sci. Technol. A* **6**, 497 (1988).

41. J. Stroscio, R. Feenstra, and A. Fein, *Phys. Rev. B* **36**, 7718 (1987).

42. R. Hamers, *J. Vac. Sci. Technol. B* **6**,1462 (1988).

43. E. van Loenen, J. Demuth, R. Tromp, and R. Hamers, *Phys. Rev. Lett.* **58**, 373 (1987).

44. R. Hamers, *Phys. Rev. B* **40**, 1657 (1989).

45. J. Demuth, U. Koehler, R. Hamers, and P. P. Kaplan, *Phys. Rev. Lett.* **62**, 641 (1989).

46. R. Becker, T. Klitsner, and J. Vickers, *J. Microsc.* **152**, 157 (1988).

47. R. Becker and J. Vickers, *J. Vac. Sci. Technol. A* **8**, 226 (1990).

48. J. Northrup, *Phys. Rev. Lett.* **53**, 683 (1984).

49. H. Ohtani, R. J. Wilson, S. Chiang, and C. M. Mate, *Phys. Rev. Lett.* **60**, 2398 (1988).

50. F. P. Netzer and K.-H. Frank, *Phys. Rev. B* **40**, 5223 (1989).

51. W. Mizutani, M. Shigeno, M. Ono, and K. Kajimura, *Appl. Phys. Lett.* **56**, 1974 (1990).

52. C. L. Claypool, et al., *J. Phys. Chem. B* **101**, 5978 (1997).

53. F. Fagolioni, C. L. Claypool, N. S. Lewis, and W. A. Goddard III, *J. Phys. Chem. B* **101**, 5996 (1997).

54. D. F. Padowitz and R. J. Hamers, *J. Phys. Chem. B* **102**, 8541 (1998).

55. K. V. Mikkelsen and M. A. Ratner, *Chem. Rev.* **87**, 113 (1987).

56. P. Sautet and C. Joachim, *Surf. Sci.* **271**, 387 (1992).

57. W. Widdra, C. Huang, S. I. Yi, and W. H. Weinberg, *J. Chem. Phys.* **105**, 5605 (1996).

58. H. Ness and A. J. Fisher, *Phys. Rev. B* **55**, 10081 (1997).

59. H. Ness and A. J. Fisher, *Phys. Rev. B* **56**, 12469 (1997).

60. G. Binnig and H. Rohrer, *Surf. Sci.* **157**, L373 (1985).

61. N. Garcia, *IBM J. Res. Dev.* **30**, 533 (1986).

62. E. Louis, F. Flores, and P. Echenique, *Phys. Scripta* **37**, 359 (1988).

63. W. Kaiser and R. Jaklevic, *IBM J. Res. Dev.* **30**, 411 (1985).

64. R. Garcia, J. Saenz, and N. Garcia, *Phys. Rev. B* **33**, 4439 (1986).

65. E. Salvan, H. Fuchs, A. Baratoff, and G. Binnig, *Surf. Sci.* **162**, 634 (1985).

66. R. Becker, B. Swartzentruber, and J. Vickers, *J. Vac. Sci. Technol.* **6**, 472 (1988).

67. R. Becker, T. Klitsner, and J. Vickers, *Phys. Rev. B* **38**, 3537 (1988).

68. A. Humbert, F. Salvan, and C. Mouttet, *Surf. Sci.* **181**, 307 (1987).

69. G. V. de Walle, H. V. Kempen, P. Wyder, and C. Flipse, *Surf. Sci.* **181**, 27 (1987).

70. F. Himpsel, T. Jung, R. Schlittler, and J. K. Gimzewski, *Jpn. J. Appl. Phys. Part 1* **35(6B)**, 3695 (1996).

71. R. Feenstra and J. Stroscio, *J. Vac. Sci. Technol. B* **5**, 923 (1987).

72. T. Berghaus, A. Brodde, H. Neddermeyer, and S. Tosch, *Surf. Sci.* **193**, 235 (1988).

73. F. Himpsel and T. Fauster, *J. Vac. Sci. Technol. A* **2**, 815 (1984).

74. T. Fauster and F. Himpsel, *J. Vac. Sci. Technol. A* **1**, 1111 (1983).

75. J. Kirtley, S. Raider, R. Feenstra, and A. Fein, *Appl. Phys. Lett.* **50**, 1607 (1987).

76. R. Hamers and J. Demuth, *J. Vac. Sci. Technol. A* **6**, 512 (1988).

77. Y. Kuk and P. Silverman, *J. Vac. Sci. Technol. A* **8**, 289 (1990).

78. J. Simmons, *J. Appl. Phys.* **34**, 1793 (1963).

79. N. Lang, *Phys. Rev. Lett.* **58**, 45 (1987).

80. A. Brodde, S. Tosch, and H. Neddermeyer, *J. Microsc.* **152**, 441 (1988).

81. J. Callaway and C. Wang, *Phys. Rev. B* **16**, 2095 (1977).

82. R. Feenstra, W. Thompson, and A. Fein, *Phys. Rev. Lett.* **56**, 608 (1986).

83. K. Pandey, in *Proceedings of the 17th International Conference on the Physics of Semiconductor*, Springer-Verlag, New York, 1985.

84. R. Hamers and J. Demuth, *Phys. Rev. Lett.* **60**, 2527 (1988).

85. J. Demuth, E. van Loenen, R. Tromp, and R. Hamers, *J. Vac. Sci. Technol. B* **5**, 1528 (1987).

86. P. Avouris and R. Wolkow, *Phys. Rev. B* **39**, 5091 (1989).

87. R. Wolkow and P. Avouris, *Phys. Rev. Lett.* **60**, 1049 (1988).

88. A. Dhirani, et al., *J. Chem. Phys.* **106**, 5249 (1997).

89. L. Bumm, et al., *Science* **271**, 1705 (1997).

90. C. H. Olk and J. P. Heremans, *J. Mater. Res.* **9**, 259 (1994).

91. T. W. Odum, J.-L. Huang, P. Kim, and C. M. Leiber, *Nature* **391**, 62 (1998).

92. J. W. G. Wildöer, et al., *Nature* **391**, 59 (1998).

93. P. G. Collins, et al., *Science* **278**, 100 (1997).

94. I. Giaever, *Phys. Rev. Lett.* **5**, 147, 464 (1960).

95. J. Bardeen, *Phys. Rev. Lett.* **6**, 57 (1961).

96. U. Hartmann, *Appl. Phys. A Solids Surf.* **A59**, 41 (1994).

97. H. Hess, et al., *Phys. Rev. Lett.* **62**, 214 (1989).

98. H. Hess, et al., *J. Vac. Sci. Technol. A* **8**, 450 (1990).

99. H. Hess, *Physica* **185–189**, 259 (1991).

100. H. Hess, R. B. Robinson, and J. V. Waszczak, *Physica* **169**, 422 (1991).

101. A. Fein, J. Kirtley, and R. Feenstra, *Rev. Sci. Inst.* **58**, 10 (1987).

102. M. Tanaka, et al., *J. Vac. Sci. Technol. A* **8**, 475 (1990).

103. M. Gallagher and J. Adler, *J. Vac. Sci. Technol. A* **8**, 464 (1990).

104. T. Oshio, et al., *J. Vac. Sci. Technol. A* **8**, 468 (1990).

105. T. Endo, et al., *J. Vac. Sci. Technol. A* **8**, 468 (1990).

106. H. Bando, et al., *J. Vac. Sci. Technol. A* **8**, 479 (1990).

107. P. Mallet, et al., *Phys. Rev. B* **54**, 13324 (1996).

108. K. Kitazawa, *Science* **271**, 313 (1996).

109. K. Kobayashi and M. Tsukada, *J. Vac. Sci. Technol. A* **8**, 170 (1990).

110. S. Ciraci, A. Baratoff, and I. P. Batra, *Phys. Rev. B* **41**, 2763 (1990).

111. D. Lawunmi and M. C. Payne, *J. Phys.* **2**, 3811 (1990).

112. J. Tersoff, *Phys. Rev. B* **41**, 1235 (1990).

113. A. L. V. de Parga, et al., *Phys. Rev. Lett.* **80**, 357 (1997).

114. Y. Kuk and P. Silverman, *Appl. Phys. Lett.* **48**, 1597 (1986).

115. T. Sakurai, et al., *J. Vac. Sci. Technol. A* **8**, 324 (1990).

116. T. Klitsner, R. Becker, and J. Vickers, *Phys. Rev. B* **41**, 3837 (1990).

117. R. Hamers and R. Koch, *Physics and Chemistry of SiO and the Si-SiO Interface*, Plenum Press, New York, 1989.

118. R. Koch and R. Hamers, *Surf. Sci.* **181**, 333 (1987).

119. P. Bedrossian, D. Chen, K. Mortensen, and J. Golovchenko, *Nature* (London) **342**, 258 (1989).

120. I. W. Lyo and P. Avouris, *Science* **245**, 1369 (1989).

121. M. Weimer, J. Kramar, and J. Baldeschwieler, *Phys. Rev. B* **39**, 5572 (1989).

122. W. Kaiser, L. Bell, M. Hecht, and F. Grunthaner, *J. Vac. Sci. Technol. A* **6**, 519 (1988).

123. G. Binnig, N. Garcia, and H. Rohrer, *Phys. Rev. B* **32**, 1336 (1985).

124. B. Persson and J. Demuth, *Solid State Commun.* **57**, 769 (1986).

125. A. Baratoff and B. Persson, *J. Vac. Sci. Technol.* **6**, 331 (1988).

126. D. Smith, G. Binnig, and C. Quate, *Appl. Phys. Lett.* **49**, 1641 (1986).

127. D. Smith, M. Kirk, and C. Quate, *J. Chem. Phys.* **86**, 6034 (1987).

128. S. Gregory, *Phys. Rev. Lett.* **64**, 689 (1990).

129. G. Binnig and H. Rohrer, *Rev. Mod. Phys.* **59**, 615 (1987).

130. J. Coombs, M. Welland, and J. Pethica, *Surf. Sci.* **198**, L353 (1988).

131. N. Lang, *Phys. Rev. B* **37**, 10395 (1988).

132. C. Chen and R. J. Hamers, *J. Vac. Sci. Technol. B* **9**, 230 (1991).

133. R. Wiesendanger, *Scanning Probe Microscopy and Spectroscopy*, Cambridge University Press, Cambridge, UK, 1994.

PART II

TIPS AND SURFACES

5

THE SURFACE
STRUCTURE
OF CRYSTALLINE SOLIDS

William N. Unertl

The purpose of this chapter is to provide an overview of the topic of surface structure and of the role of scanned probe microscopies in the study of surface structure. Emphasis is on the atomic scale structures that form on crystal surfaces, because the vast majority of research has focused on these systems. One goal of the chapter is to give a systematic presentation of the language and notation used to describe the atomic scale structure. Another is to survey the wide variety of structure types, including defects that are found on real surfaces. From an experimental viewpoint, unambiguous determination of the structure of any surface usually requires information from several complementary techniques. Thus to balance the book as a whole, information obtained by techniques other than scanned probe microscopies is also emphasized. In particular, the chapter concludes with a discussion of some of the pros and cons of some of the most useful experimental techniques for surface structure studies.

The spatial arrangements of atoms near surfaces and interfaces determine the chemical, electronic, and mechanical properties of materials in ways that cannot be simply predicted from knowledge of bulk properties. For example, processes such as epitaxial growth of thin films and catalytic reactions are sensitive functions of surface structure and composition.

Most of our atomic scale understanding surfaces has been acquired from studies of the surfaces of highly crystalline metals and semiconductors. Consequently, these materials are fairly well understood. However, significant gaps remain in understanding other materials and, in particular, real surfaces, such as those of noncrystalline materials like polymers. The surface properties of oxides and carbides are dominated by the presence of defects and are only beginning to be understood. Furthermore, real surfaces are often formed in metastable structural states rather than in thermodynamic equilibrium. With the recent introduction of direct imaging techniques like scanning tunneling microscopy (STM) and low-energy electron microscopy (LEEM), the last decade has seen a rapid increase in research directed toward understanding the noncrystalline aspects of surface structure. Surface defects range from atomic scale defects, such as impurities and vacancies, to macroscopic imperfections, such as grain boundaries. Diffusion, nucleation, growth, and chemical reactivity are sensitive to the presence of surface defects of both mesoscopic and microscopic extent. This chapter places more emphasis on surface defects than can be found in many other introductory discussions of surface structure. The reader is assumed to be familiar with the basic concepts of bulk thermodynamics, bulk crystallography, and bulk condensed matter physics.[1] More in-depth reviews of surface structure can be found elsewhere.[2]

5.1 IDEAL SURFACES OF CRYSTALLINE MATERIALS

We begin with a useful conceptual model: the *ideal surface*. Ideal surfaces have translational symmetry in directions parallel to the surface and contain no imperfections.

The simplest ideal surfaces are called *singular surfaces*. They have the structures that would form if a perfect bulk crystal could be cut to expose the atoms of an {*hkl*} plane at the surface without allowing them to adjust to their new environment by relaxing from their bulk positions. Because their atoms have the same arrangement as bulk planes, singular surfaces are also called (1×1) surfaces. Singular surfaces do not exist in nature but the structures of the close-packed surfaces of many face-centered-cubic (FCC) and body-centered-cubic (BCC) metals have lateral structures that are nearly singular. A singular surface is identified by its Miller indices (*hkl*). For example, Figure 5.1 shows geometric models of the (111), (110), and (100) surfaces of FCC and BCC crystals. In the case of FCC surfaces, the density of surface atoms is highest for (111), lower for (100), and lowest for (110). For BCC surfaces, the (110) surface is most densely packed, (100) is more open, and (111) is very open—atoms in the third layer are exposed. Determination of the singular surface structures of other crystal types, such as the diamond and hexagonal-close-packed (HCP) lattices, is left as an exercise.

The structure of an (*hkl*) singular surface is classified according to its *two-dimensional (2-D) Bravais lattice* type. A 2-D Bravais lattice consists of an infinite planar array of points, each of which can be transformed into any other by the translation operator

$$\mathbf{T} = h\mathbf{a}_1 + k\mathbf{a}_2 \qquad (5.1)$$

where h and k are the 2-D Miller indices. The surface analogue of the bulk unit cell is the unit mesh; it is defined by its unit mesh vectors $\mathbf{a}_1$ and $\mathbf{a}_2$. Figure 5.2 shows the unit meshes of the five possible Bravais lattices which, when combined with $\mathbf{T}$, have the property that they can completely cover a plane. The set of individual atoms located inside each unit mesh is called the basis of the surface structure. Although it is not necessary to choose the coordinate system so that one of the atoms of the basis coincides with the origin, it is often convenient to do so. In the case of the simple surface structures shown in Figure 5.1, each basis consists of a single atom. More complex structures are also common. For example, the (100) surface of the NaCl structure has a two-atom basis (Fig. 5.3)—in this case chosen with one atom at the origin of the unit mesh and the other at its center. The distribution of atoms in the unit mesh may also have additional symmetry properties under 1-, 2-, 3-, 4-, and 6-fold rotations, reflections, and glide lines. However, it can be proven that there are only 17 unique combinations of these symmetry operations with the five Bravais lattices; these are called the 2-D space groups.[3] In other words, all possible surface crystal structures must belong to one of these 17 space groups.

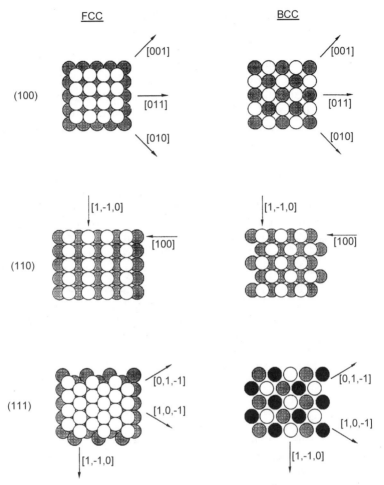

Figure 5.1. Structural models of low-index singular surfaces of FCC and BCC crystals. The *open circles* represent atoms in the surface layer, the *shaded circles* are atoms in the second layer, and the *solid circles* are atoms in the third layer. The *arrows* indicate directions in each surface.

5.1.1 Relaxation

Atoms near a surface are located in an asymmetric environment. Thus the bulk positions no longer correspond to the lowest energy state. This drives the surface atoms into new locations that correspond to the state of minimum free energy. In the simplest cases, only small displacements from singular positions are needed to achieve the equilibrium structure. The most common displacement is *relaxation*. In this case, the structure of each plane of atoms parallel to the surface remains singular, but the spacings between the outermost few planes change from the bulk value.[4-7] Relaxations are not usually detectable with STM or AFM but must be studied with another technique.

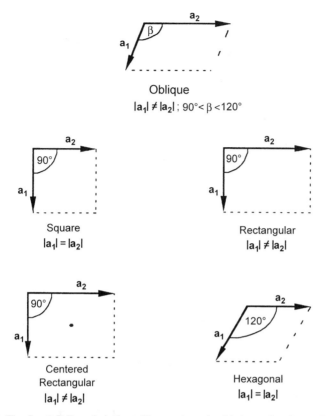

Figure 5.2. The five 2-D Bravais lattices. The angle γ should always be chosen to be $>90°$.

5.1.2   Structures Owing to Adsorption and Segregation

When atoms or molecules are placed on a surface either by adsorption from a gas phase or by segregation from the underlying bulk phase, it is common for highly ordered 2-D crystalline structures to be formed. This was noted as early as 1927 in the original surface structure studies of Davisson and Germer.[8] Often the adsorbed species form a separate layer on top of substrate atoms; Figure 5.4 shows a ball model example for H adsorbed on the Ni(111) surface.[9] In other cases, such as O adsorbed on Ta(100) the adsorbed atoms can mix into the top layer of the substrate (Fig. 5.4b)[10] or, as is the case for N on Ti(001) form a new layer below the surface (Fig. 5.4c).[11] In other systems, adsorption causes major structural rearrangements in the substrate atoms. Hydrogen adsorption on Ni(110) is one example.[12]

The amount of a foreign material adsorbed onto a surface is expressed as a *coverage* Θ where Θ is usually defined as the number of adsorbed atoms or molecules per unit area divided by the number of substrate surface atoms per unit area of the clean surface. Using this convention, the H coverage in Figure 5.4a is $\Theta = 0.25$, and the O coverage in Figure 5.4b is $\Theta = 1/3$.

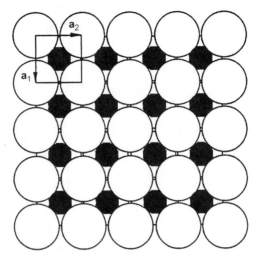

Figure 5.3. The (100) surface of a crystal with the NaCl structure. The primitive unit mesh is outlined in the upper left corner.

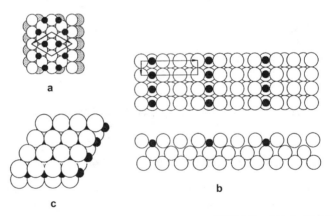

Figure 5.4. Examples of ordered overlayer structures. **a,** Ni(111)-2 × 2-H. **b,** Ti(100)-(1 × 3)-O. **c,** Ti(001)-(1 × 1)-N.

As for the clean surface, the structure of any ordered overlayer can be characterized by its 2-D Bravais lattice (Fig. 5.2), its unit mesh vectors b_1 and b_2, and by the positions of the atoms within the unit mesh. An overlayer unit mesh is normally defined with respect to the unit mesh (a_1, a_2) of the substrate surface. Overlayer unit mesh vectors are illustrated in Figure 5.4a for the H on Ni structure. A notation developed by Wood is used in such simple cases, in which the angle between b_1 and b_2 is the same as the angle between a_1 and a_2.[13] The general form for this notation for an adsorbate Q on the (*hkl*) surface of substrate A with overlayer mesh vectors

$|\mathbf{b}_1| = n|\mathbf{a}_1|$ and $|\mathbf{b}_2| = m|\mathbf{a}_2|$, is

$$A(hkl)\text{-}(n \times m)R(\theta)\text{-}Q$$

where $R(\theta)$ indicates the angle θ (measured in degrees) between $\mathbf{b}_1$ and $\mathbf{a}_1$. If θ is zero, the $R(\theta)$ term is dropped. The structure in Figure 5.4a is thus denoted as a Ni(111)-(2 ×)2-H structure with a two-atom basis. A few other examples of the Wood notation are shown in Figure 5.5. Figure 5.5a is an overlayer on a square substrate mesh such as the (100) surface of an FCC or BCC material. It has $|\mathbf{b}_1| = \sqrt{5}|\mathbf{a}_1|, |\mathbf{b}_2| = \sqrt{5}|\mathbf{a}_2|$, and $\theta = 26.57°$ so that its Wood notation is $(\sqrt{5} \times \sqrt{5})R(26.57)$. Some structures have several forms in the Wood notation. One often encountered in practice is shown in Figure 5.5b. The mesh on the left

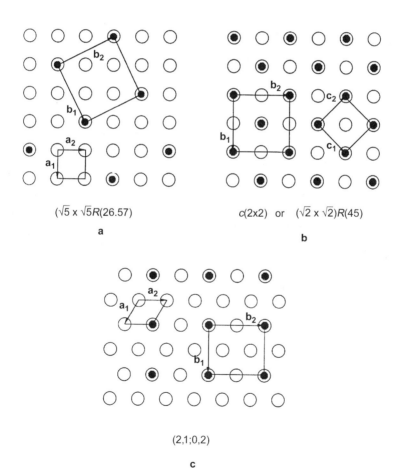

$(\sqrt{5} \times \sqrt{5}R(26.57)$

a

$c(2\times2)$ or $(\sqrt{2} \times \sqrt{2})R(45)$

b

$(2,1;0,2)$

c

Figure 5.5. Examples of overlayer unit meshes. **a**, $(\sqrt{5} \times \sqrt{5})R(26.57)$ on a square substrate, **b**, $c(2 \times 2)$ or $(\sqrt{2} \times \sqrt{2})R(45)$ on a square lattice, **c**, A $(2,1;0,2)$ structure with no Wood equivalent.

is a $c(2 \times 2)$ mesh, where the c indicates a centered structure. The mesh on the right is $(\sqrt{2} \times \sqrt{2})R(45)$. It is left as an exercise to show that the notation for the structure in Figure 5.4b is Ta(100)-(1×3)-O and that for Figure 5.4c is Ti(001)-(1×1)-N.

The overlayer lattice in Figure 5.5c cannot be described in the Wood notation because the angle between $\mathbf{b}_1$ and $\mathbf{b}_2$ is not equal to that between $\mathbf{a}_1$ and $\mathbf{a}_2$. A second, more general notation that includes cases like Figure 5.5c and is more useful in making transformations between real space and reciprocal space representations is the matrix notation originally proposed by Park and Madden.[14] Because

$$\mathbf{b}_1 = m_{11}\mathbf{a}_1 + m_{12}\mathbf{a}_2$$
$$\mathbf{b}_2 = m_{21}\mathbf{a}_1 + m_{22}\mathbf{a}_2 \tag{5.2}$$

every overlayer mesh can be described by the matrix

$$M = \begin{pmatrix} m_{11} & m_{12} \\ m_{21} & m_{22} \end{pmatrix} \tag{5.3}$$

Only the primitive unit meshes are used in the matrix notation and the determinant of M is always a positive number equal to the ratio of the overlayer mesh area to the substrate mesh area. Each of the meshes in Figure 5.4 can be described using the matrix notation. For the $(\sqrt{5} \times \sqrt{5})R(26.57)$ lattice, $M = (2, 1; -1, 2)$ and det $M = 5$. For the $c(2 \times 2)$ lattice, $M = (1, 1; -1, 1)$ and det $M = 2$. Each of these three meshes is an example of a *simple lattice*. Simple lattices are those for which each element of M is an integer and the determinant of M is an integer. Simple lattices have the property that if one lattice site of the overlayer coincides with a substrate lattice site then so do all other overlayer lattice sites. Simple lattices often occur when a single type of adsorption site is occupied by the adsorbed species and when the interaction between the adsorbate and the substrate is stronger than the adsorbate–adsorbate interaction. Many atoms and molecules adsorb at sites on the substrate that have high symmetry; Figure 5.6 shows models of the important high-symmetry adsorption sites on BCC and FCC surfaces. The more densely packed surfaces have fewer types of high symmetry sites, whereas open surfaces (like the BCC (111)) have a wider variety.

Another class of overlayer lattices is the *coincidence lattices* for which some of the elements of M are rational numbers and det M is a simple fraction. At a coverage near 0.67 monolayer, a $(1.5, 0; 0.5, 1)$ coincidence structure occurs for N_2 adsorption on a Ni(110) surface.[15] Figure 5.7 shows the unit mesh of this structure. Coincidence lattices have the property that there is no orientation of the overlayer for which every overlayer lattice point coincides with a substrate lattice point. Coincidence lattices can occur when several types of adsorption sites are simultaneously occupied or when adatom–adatom interactions are of comparable strength to adatom–substrate interactions.

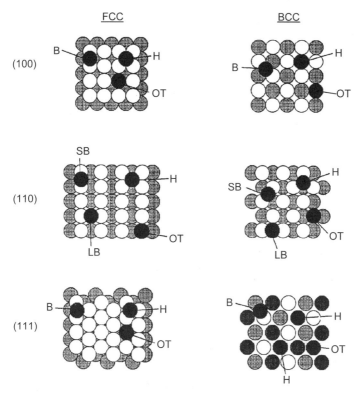

Figure 5.6. Various types of adsorption sites on the low-index FCC and BCC surfaces shown in Figure 5.1. The *solid circles* represent adsorbed atoms. Sites labeled *OT* are on-top adsorption. Hollow sites with high symmetry are indicated by *H*. Bridge sites are indicated by *B*, if all possible sites are the same, otherwise there are long-bridge *LB* and short-bridge *SB* sites, depending on the relative distance between the substrate atoms.

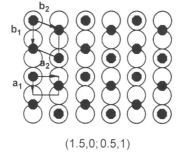

(1.5,0; 0.5,1)

Figure 5.7. A (1.5, 0; 0.5, 1) structure. The adsorbate molecules occupy both on-top and short-bridge sites.

The third class of overlayer structures is *incoherent*, or *incommensurate, lattices*. In this class, some elements of *M* and det *M* are irrational numbers. Incommensurate lattices are typically formed when the interaction between the adsorbate atoms is stronger than their interaction with the substrate and are often observed for submonolayers of inert gases.

It is important to realize that neither the Wood notation nor the matrix notations contain any information about either the number of atoms or their locations within the unit mesh. This information must be obtained from detailed structure determinations. A large fraction of the research in the area of surface crystallography has been devoted to the development of improved diffraction techniques, usually involving low-energy electrons, and their application to more and more complex structures. Atom locations are usually extracted from the data, assuming that the surface is free of defects and imperfections. Current technology can determine interatomic bond lengths to a precision of about 0.01 nm or better. STM has insufficient lateral resolution to be used for structure refinements, but is invaluable for selecting between alternative models. The bond lengths at surfaces typically lie within the range observed for similar bonds in molecular and bulk compounds as demonstrated in Figure 5.8. Several compilations of surface structures are available.[7,16]

5.1.3 Reconstructions

Many clean surfaces take on structures that differ substantially from the ones predicted from the underlying bulk structure. Such surfaces are said to be *reconstructed*. The (7×7) reconstruction of Si(111) is probably the most well known and complicated example of a reconstructed surface. STM images of Si(111)-(7×7) are shown in Figures 4.1 and 4.12. Simpler reconstructions also occur. For the (100) surface of Ir, the surface atoms form a nearly hexagonal (111) layer that is slightly distorted by interactions with the underlying (100) layers.[17] Au(100) and Pt(100) have similar reconstructions. In these three examples, the atomic density of the surface layer is larger than in the bulk. For other reconstructions, like the (1×2) reconstruction of the (110) surfaces of Au, Pt, and Ir, the reconstructed layer has a reduced density.[18-20] The unreconstructed (110) surface of FCC materials consists of rows of close-packed atoms parallel to the $[1, \bar{1}, 0]$ direction (Fig. 5.1). Figure 5.9 shows an end-on view of a ball model of the reconstructed Au(110)-(1×2) surface. This type of (1×2) reconstruction is called a *missing row* structure because every other row of surface atoms has been removed. Reconstruction of the surface layer also causes subsurface atoms to shift from their singular positions, primarily by pairing of $[1, \bar{1}, 0]$ rows of atoms in the second layer, in response to the loss of neighbors in the surface layer.

Theoretical studies[21-23] of metal surface reconstruction show that *d*-electron interactions must be included in a realistic way to account for the experimental result that clean Au, Pt, and Ir (110) surfaces have the (1×2) reconstruction but Ag, Cu, and Ni do not. The models used are sensitive to higher order contributions to the interatomic potential such as three body interactions and have not yet suc-

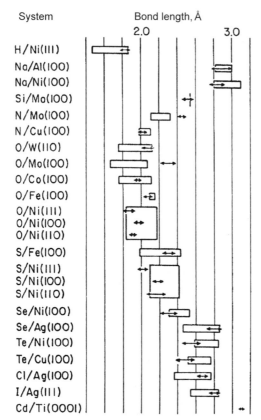

Figure 5.8. Comparison of bond lengths for surface structures (*arrows*) with ranges of bond lengths found in bulk compounds and molecules (*boxes*). (Modified from Ref. 7.)

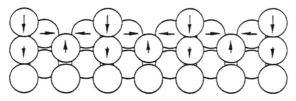

Figure 5.9. A ball model of the Au(110)-(1 × 2) reconstruction. *Arrows*, direction of shifts from singular positions. (Modified from Ref. 20.)

ceeded in calculating every detail of the structure such as the direction of displacements in the second layer.

Most surfaces of elemental and compound semiconductors are also reconstructed. Because bonds in semiconductors are more directed than in metals or ionic crystals, the major contribution made to the surface energy in forming a new piece of semiconductor surface, say by cleavage, is the energy required to break those bonds that cross the interface. These broken bonds are directed out of the surface and are called *dangling bonds*. The free energy of the surface can usually be lowered by reconstruction to a new structure in which the number of broken bonds has been reduced. This process is called *rehybridization*. On the Si(100) surface, rehybridization leads to the formation of symmetric dimers, which lie on a (2×1) reconstructed lattice (Fig. 5.10). Another STM image of this surface is given in Figure 4.5, and an example of the reconstruction of a semiconductor surface, GaAs(110), can be found in Figure 3.5. Some reconstructions such as the Si(111)-(7×7) involve complex rearrangements of the uppermost several layers of atoms. Tight-binding methods have been developed to provide a semiquantitative theoretical basis for many of the experimentally observed reconstructions.[25]

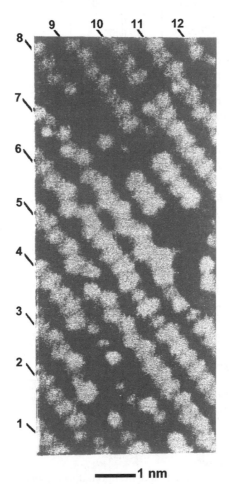

Figure 5.10. An STM image of the Si(100)-(2×1) reconstruction. Each number along the side marks a row of dimers. A unit mesh of the reconstructed surface is also shown. The surface region shown has numerous defects. Reprinted with permission from Ref. 24.

Surface reconstructions can also be induced by adsorption. If CO is adsorbed on the clean Pt(110)-(1 × 2) surface, the missing row structure is destroyed. This process has been studied by Gritsch et al.,[26] and one of their beautiful STM images illustrating an intermediate stage of the deconstruction of the (1 × 2) structure is shown in Figure 5.11. A second example of adsorption-induced changes in the surface structure of the substrate is provided by hydrogen adsorption on Ni(110) and Pd(110) surfaces. H_2 causes adjacent $[1, \bar{1}, 0]$ rows of surface metal atoms to move closer together and form a (1 × 2) 'paired-row' structure.[12,27] Another, more complex, example of adsorption-induced reconstruction is the $(\sqrt{3} \times \sqrt{3})$ structure formed by Al on Si(111), shown in Figures 4.7 and 4.8.

In some cases, extremely small changes in surface composition are sufficient to induce reconstructions. For example, small amounts of alkali metal adsorbates (coverage <0.2 monolayer) cause the (110) surfaces of Cu and Ag to reconstruct into a (1 × 2) missing row structure.[28] For compounds or alloys such as Cu_3Au, the equilibrium surface composition is often different from the bulk composition, which can lead to changes in surface order.[29]

5.1.4 Surface Electronic Structure

The electronic properties of surfaces are discussed in some detail in other portions of this book. In this chapter, I wish to point out that the atomic arraignment and the electronic properties of surfaces are intimately interrelated. If the atomic positions are altered, the electronic properties will also be changed and vice versa. The sensitivity of the surfaces of Cu(110) and Ag(110) to reconstruction in the presence of small amounts of alkali metals mentioned above is probably caused by change in the electronic structure.

10 Å

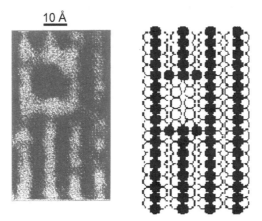

Figure 5.11. a, The STM image shows a hole induced in the reconstructed Pt(110)-(1 × 2) surface by the adsorption of CO. The CO molecule is not visible in the image. (Reprinted with permission from Ref. 26.) **b**, A ball model.

The importance of the coupling between atomic and electronic structures is apparent in the Schrödinger equation for the many electron system:

$$H\Psi_j(\mathbf{r}, \mathbf{R}) = E_j\Psi_j(\mathbf{r}, \mathbf{R}) \tag{5.4}$$

where

$$
H = \sum_{M=1}^{N}\left(\frac{-\hbar}{2M}\right)\nabla_M^2 + \sum_{i=1}^{n}\left(\frac{-\hbar}{2m}\right)\nabla_i^2 + \frac{e^2}{4\pi\varepsilon_o}\sum_{i=1}^{n}\sum_{j>i}^{n}\frac{1}{|\mathbf{r}_i - \mathbf{r}_j|}
$$
$$
+ \frac{e^2}{4\pi\varepsilon_o}\sum_{i=1}^{n}\sum_{M=1}^{N}\frac{Z_M}{|\mathbf{r}_i - \mathbf{R}_M|} + \frac{e^2}{4\pi\varepsilon_o}\sum_{M=1}^{N}\sum_{M'>M}^{N}\frac{Z_M Z_{M'}}{|\mathbf{R}_i - \mathbf{R}_j|}
$$

Both the electronic wavefunctions $\Psi_j(\mathbf{r}, \mathbf{R})$ and the Hamiltonian H depend not only on the positions $\mathbf{r}_i$ of the electrons but also on the positions $\mathbf{R}_j$ and charges Z_M of the nuclei. The $\Psi_j(\mathbf{r}, \mathbf{R})$ for a given set of boundary conditions are determined by a self-consistent solution of Eq. (5.4). For real systems, this problem is most often too complex to be solved exactly but numerous, usually computer intensive, approximate methods have been developed.[30]

This close relationship between the atomic positions and electronic properties of a surface is especially important for techniques like STM that measure electronic quantities and try to deduce the atomic structure from them. For example, certain atoms can appear in or disappear from an STM image, depending on the portions of the electronic density of states that are being sampled. This important point was discussed in detail in Chapters 3 and 4.

5.2   THERMODYNAMIC PROPERTIES OF SURFACES

5.2.1   Surface Energy and Surface Structure

The reversible work required to form the unit area of a new surface of a given orientation $\{hkl\}$ at constant volume, temperature, and chemical potential μ_i of each of the components is called the *surface energy* $\gamma(hkl)$. For solids, $\gamma(hkl)$ lies in the range 0.1 to 2.5 J/m^2.[31,32] The magnitude of $\gamma(hkl)$ for an (hkl) surface is intimately linked to its atomic structure. This linkage is expressed using statistical mechanics.[33]

Interatomic bonds form because they lower the internal energy U of the participating atoms. In the simplest picture, a rough estimate of U can be obtained by considering only nearest-neighbor interactions of strength ε. This bond energy is shared between the atoms so that the internal energy of each atom is lowered by $\varepsilon/2$ for every bond it forms. For example, atoms at an FCC (111) surface have only 9 nearest neighbors compared to 12 in the bulk. Each surface atom has 3 less bonds, which results in an excess internal energy of $3(\varepsilon/2)$ per surface atom. An estimate of ε can be obtained from the latent heat of sublimation L_S because, if 1 mol of bulk solid is vaporized, $12N_A$ bonds are broken. Thus the internal energy per atom for a

(111) surface is on the order of

$$U(111) \sim 0.25L_S/N_A \qquad (5.5)$$

This crude model ignores interactions owing to higher order neighbors, assumes that the value of ε is the same for surface and bulk atoms, and does not include entropic or pressure-volume contributions to $\gamma(hkl)$; i.e.,

$$\int_A \gamma dA = U + PV - TS \qquad (5.6)$$

Nonetheless, it does account for the relative differences in $\gamma(hkl)$ between different materials and for the fact that less dense surfaces of a material have lower $\gamma(hkl)$.

Reliable measurements of $\gamma(hkl)$ are difficult because even fractions of a monolayer of an impurity can change $\gamma(hkl)$ by amounts that are larger than the variation between faces. Thus accurate values are available for only a few materials. The effect of orientation and temperature on $\gamma(hkl)$ is illustrated for the case of Pb in Figure 5.12.

The equilibrium shape of a crystal in contact with its vapor at constant temperature and volume is that shape for which the total energy associated with the surface is minimized and is given by the minimum of the integral $\int_A \gamma dA$. At low temperatures, the equilibrium shapes are regular polyhedra whose facets are made up of low index surfaces such as $\{111\}$ and $\{100\}$. It seems that the edges at the intersection

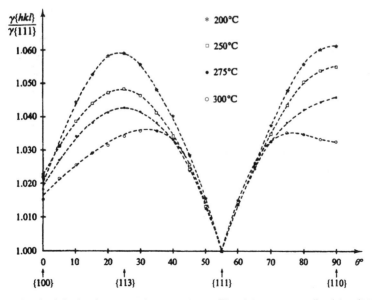

Figure 5.12. $\gamma(hkl)$ for lead at several temperatures. The data are normalized to $\gamma(hkl)$. (Reprinted with permission from Ref. 34.)

of two facets can be either sharp or rounded but only a few reliable measurements are available. Because $\gamma(hkl)$ changes with temperature, the fraction of the surface owing to a particular $\{hkl\}$ will also change with temperature. Nearer the melting point, as can be seen from Figure 5.12, the differences in $\gamma(hkl)$ for nearby $\{hkl\}$ are reduced and large regions of curved surface are possible. In the case of Pb near its melting point (Fig. 5.13), the equilibrium shape is nearly spherical; there are no $\{110\}$ facets and only small $\{111\}$ and $\{100\}$ facets. For liquids, atoms are no

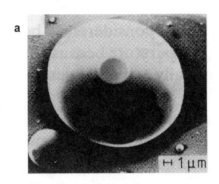

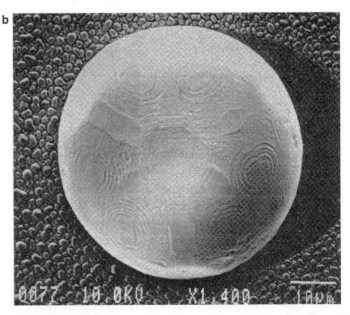

Figure 5.13. Scanning electron micrographs of a small Pb crystal. **a**, Near its melting point, $T = 599\,\mathrm{K}$. The larger flat facets are $\{111\}$ surfaces, and the smaller ones are $\{100\}$ surfaces. (Reprinted with permission from Ref. 35.) **b**, A slightly larger crystal showing the increase in surface topographical structures that occurs at lower temperature; $T = 373\,\mathrm{K}$. (Courtesy of A. Pavlovska.)

longer fixed at crystalline sites so that each point on the surface will have the minimum value of γ; i.e., γ has no angular dependence and the equilibrium shape is a sphere.

The statistical mechanics of vicinal surfaces was reviewed in detail by Williams and Bartelt.[33] If ϕ and θ are the polar and azimuthal angles that describe the orientation of a vicinal surface with respect to the nearest low-index plane, then the variation of the surface energy with orientation is

$$\gamma(\phi, \theta, T) = \gamma^o(T) + \frac{\beta(\theta, T)}{h} |\tan \phi| + g(\theta, T)|\tan \phi|^3 \qquad (5.7)$$

where $\gamma(\phi, \theta, T)$ is the reduced surface energy with respect to $\gamma^o(T)$, the surface energy of the low index surface. The free energy to create unit length of step edge of height h is $\beta(\theta, T)$ and $g(\theta, T)|\tan \phi|^3$ is the free energy per unit area owing to step–step interactions. The specific forms of $\beta(\theta, T)$ and $g(\theta, T)$ must be derived from an atomic model of the stepped surface. STM experiments have been important tools for testing the atomic models.

5.2.2 Phase Diagrams

The ordered overlayer structures and reconstructed surfaces discussed above are examples of discrete surface phases. In addition to crystalline phases, it is also common to observe liquid and gaseous phases. Each surface phase is stable over only a limited range of thermodynamic variables, such as temperature, coverage, or gas pressure. The range of thermodynamic parameters over which any phase can exist in thermodynamic equilibrium is conveniently summarized in a *phase diagram*, such as that for O adsorbed on Ni(111) shown in Figure 5.14.[36] The ranges of T and Θ over which each of the various phases exist are delineated by lines called phase boundaries. The regions labeled Gas, $p(2 \times 2)$, and $(\sqrt{3} \times \sqrt{3})R(30)$ in which only one phase occurs are called *single*, or *pure*, phase regions. If two phases are present simultaneously, the region is called a *coexistence* region. The regions labeled $p(2 \times 2) +$ Gas and $p(2 \times 2)$ Antiphase domains $+$ liquid in Figure 5.14 are examples of coexistence regions. The phase diagram for Ni(100)-Se is shown in Figure 5.15.[37]

The Ni(111)-O system can be considered to be a *closed system*; i.e., the O coverage is independent of temperature because O atoms are so strongly bound to the Ni substrate that thermally activated desorption is insignificant over the temperature range covered by the phase diagram. At higher temperatures, the adsorbed atoms could be desorbed into the gas phase, dissolved into the substrate, or react with the substrate to form new compounds. For the case of Ni(111)-O, the latter occurs. In cases such as N_2 on Ni(110) in which desorption occurs, constant coverage can often still be maintained if a gas phase of the adsorbate is maintained over the sample.[15] This is impractical for other systems such as Ni(100)-Se; in these cases desorption limits the highest temperature, T_m, at which an equilibrium phase diagram can be measured. T_m is determined primarily by $E_D(\Theta)$, the coverage

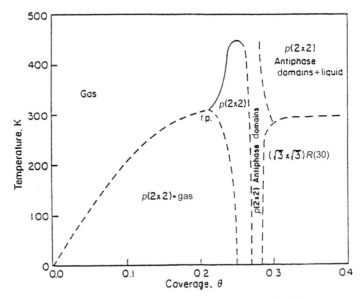

Figure 5.14. Equilibrium phase diagram for Ni(111)-O.

dependent activation energy required to desorb an atom or molecule; E_D can range from a few millielectron volts for weakly bound adsorbates like the inert gasses up to several electron volts for very reactive adsorbates. The rate of desorption can be estimated from

$$\frac{d\Theta}{dt} = \nu \exp\left(-\frac{E_D}{kT}\right) \tag{5.8}$$

where k is the Boltzmann constant and the pre-exponential term ν depends on the detailed desorption mechanism and the attempt frequency. Typically, ν is assumed to be on the order of a vibration frequency ($\sim 10^{13}\,\mathrm{s}^{-1}$).[38] There are also practical lower temperatures for which phase equilibrium can be studied because kinetic processes such as diffusion become too slow for equilibrium to be reached within a practical time interval. Phase diagrams for closed systems are determined by measuring the structure as a function of T and Θ. In the case of weakly bound adsorbates, the chemical potential, and thus the coverage of the system, is controlled by the pressure of the gas phase. Phase diagrams for these *open systems* are usually determined by measuring the surface structure as a function of the substrate temperature and the gas pressure with the gas phase at room temperature.

   The crystallographic structure of each phase and the precise locations of its phase boundaries are determined by the interactions of the adsorbate atoms with each other and with substrate atoms. If the structures of the phases are known, the sequence in which they form as Θ is increased and provides direct information about the relative interaction strengths between pairs of adsorbate atoms on various

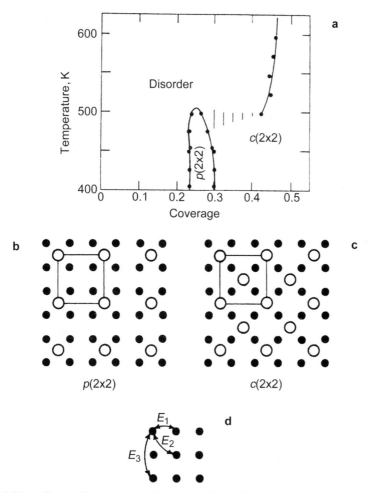

Figure 5.15. a, The equilibrium phase diagram for Ni(100)-Se. **b,** The structure of the $p(2 \times 2)$ phase. **c,** The structure of the $c(2 \times 2)$ phase. The *solid circles* are Ni atoms, and the *open circles* are Se atoms. **d,** Interactions between Se atoms at various separations.

possible adsorption sites. For example, Figure 5.15 shows the phase diagram for Se adsorbed on Ni(100) and the atomic structures of the $p(2 \times 2)$ and $c(2 \times 2)$ phases. The figure also defines the interaction energies between Se adsorbed on first, second, and third nearest-neighbor sites. In the $p(2 \times 2)$ phase, Se does not occupy first and second nearest-neighbor sites. Therefore, the interaction E_3 must be more attractive than E_1 or E_2. Because the $c(2 \times 2)$ phase forms next, E_2 must be more attractive than E_1. Thus, $E_3 < E_2 < E_3$. The numerical values of these energies can be estimated using various statistical mechanical techniques in which the values of the E_i are adjusted until an acceptable fit to the measured phase boundaries is obtained.[39]

A limit to the maximum number of phases p that can coexist on a surface is given by the Gibbs' phase rule.[40] The number of independent variables or the number of degrees of freedom f for a system with c components and p phases is

$$f = c - p + 2 \tag{5.9}$$

In the examples shown in Figures 5.14 and 5.15, $c = 1$. Therefore, in single phase regions ($p = 1$) like the $p(2 \times 2)$ and $c(2 \times 2)$ phases, $f = 2$. This result means that there are two independent variables for each of these phases; in this case they are T and Θ. At the phase boundary between two phases ($p = 2$), $f = 1$. Thus once either T or Θ is fixed, the other is uniquely determined. Three phases can only occur for unique value of T and Θ, because $f = 0$. The Gibbs' phase rule is especially useful in constructing phase diagrams from experimental data, because it is usually impractical to characterize a system at every point in its phase diagram.

5.2.3   Phase Transitions

The structure of a surface is profoundly changed whenever it undergoes a transition between two phases. Furthermore, the structural changes induced by the phase transition cause changes in other important physical and chemical properties, such as mass transport and reactivity. As discussed below, the information needed to characterize a phase transition is statistical and thus requires averages over large numbers of surface atomic positions. Because STM is a local probe, it is at best tedious to acquire sufficient data for meaningful averages. In addition, the time required to obtain an STM image is often long compared to the temporal fluctuations important during phase transitions. Diffraction methods, on the other hand, simultaneously sample the entire surface and directly measure the important statistical quantities.

One useful classification of phase transitions is based on the properties of the Gibbs free energy

$$G = U + PV - TS + \gamma A = H - TS + \gamma A = \sum \mu_i N \tag{5.10}$$

where H is the enthalpy, A is the surface area, N_i is the number of moles of component i, and μ_i is the chemical potential of component i. A transition is said to be *first order* if any of the first derivatives of G—e.g., $(\delta G/\delta T)_{P,N,A}$ and $(\delta G/\delta P)_{T,N,A}$—are discontinuous (Fig. 5.16). From the combined first and second laws of thermodynamics,

$$dG = VdP - SdT + \Sigma \mu_i N_i + \gamma dA \tag{5.11}$$

$(\delta G/\delta T)_{P,N,A} = -S$, $(\delta G/\delta P)_{T,N,A} = V$, etc., first-order transitions have discontinuous changes in V, S, μ_i, and γ. First-order transitions also have a latent heat of transformation ΔH because the enthalpy, $H = U + PV$, is also discontinuous at the transition. The crystal structure also changes abruptly, and every region of the surface is distinctly in one phase or the other. Figure 5.17 shows a reflection electron

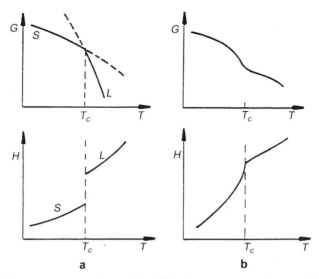

Figure 5.16. Temperature dependence of the Gibbs free energy G and the enthalpy H for (a) a first-order phase transition and (b) a second-order phase transition.

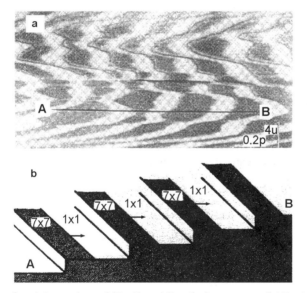

Figure 5.17. a, Coexisting regions of (7×7) and (1×1) phases on a slightly stepped Si(111) surface. The *dark areas* are (7×7) and the *light regions* are (1×1). (Reprinted by permission from Ref. 41.) **b,** The structure along the line *AB* in part **a**. (Courtesy of Y. Yagi.)

microscope image recorded during the first-order (1×1) to (7×7) transition on the Si(111) surface.[41] The dark regions have already transformed to the (7×7) phase, and the light regions are still in the (1×1) phase.

Second order phase transitions have continuous first derivatives of G but discontinuous second derivatives, (Fig. 5.16b). Thus they have no latent heat. The concepts of symmetry and universality are central to the understanding of second-order phase transitions between surface phases.[42,43] To understand them, let's begin with a discussion of how to quantify the degree of order in a surface.

Degree of Order in Surface Phases   Surface order changes during a phase transition. Three measurable quantities that are used to measure this change are the order parameter ϕ, the pair correlation function $\langle n_i n_j \rangle$, and the correlation length ξ. These quantities can be defined in terms of a lattice gas of particles adsorbed on a lattice described by unit mesh vectors $\mathbf{a}_1$ and $\mathbf{a}_2$ and having adsorption sites at positions $\mathbf{r}_i = m_{1i}\mathbf{a}_1 + m_{2i}\mathbf{a}_2$, where m_{1i} and m_{2i} are integers. In this model, n_i, the probability at any instant that a particular site $\mathbf{r}_i$ is occupied, is either 1 or 0; $n_i = 0, 1$. The average occupation probability $\langle n_i \rangle$ is determined by the average of all possible configurations of the particles, each weighted by its statistical probability.[44] As the degree of order changes near a phase transition, $\langle n_i \rangle$ also changes.

The *order parameter* ϕ describes the average degree of order at each site on the lattice as a function of temperature and is, therefore, closely related to the structure of the surface. In simple cases, ϕ can be a single quantity such as the average site occupation $\langle n_i \rangle$ or the magnetization. In more complex cases, ϕ is a set of functions, ϕ_j. Diffraction techniques such as low-energy electron diffraction or grazing incidence X-ray scattering are sensitive probes of the type and degree of surface order and, for this reason, are commonly used to study phase transitions.[7,45] In the case of diffraction, it is most convenient to define the ϕ_j in terms of Fourier coefficients of the average site occupation probability $\langle n_i \rangle$. The number of components j of ϕ_j is equal to the number of terms in this expansion. Consider the example of a (2×1) structure on a square lattice that has a two-component (vector) order parameter, as shown in Figure 5.18. In this case, the (2×1) and (1×2) structures are symmetrically equivalent and can coexist on the surface. Order parameters that describe these possibilities are

$$\phi_1 = \frac{1}{N}\sum_j \langle n_j \rangle e^{i\mathbf{k}_1 \cdot \mathbf{r}_j} \tag{5.12}$$

and

$$\phi_2 = \frac{1}{N}\sum_j \langle n_j \rangle e^{i\mathbf{k}_2 \cdot \mathbf{r}_j} \tag{5.13}$$

where N is the number of lattice sites, $\mathbf{k}_1 = 2\pi\mathbf{a}_1^*/2$ and $\mathbf{k}_2 = 2\pi\mathbf{a}_2^*/2$ are reciprocal lattice vectors characteristic of each possible (2×1) lattice, as shown in Figure 5.18b; and $\mathbf{a}_1^*$ and $\mathbf{a}_2^*$, are the reciprocal lattice unit mesh vectors of the substrate. For a perfect (2×1) phase $\phi_1 = 1/2$ and $\phi_2 = 0$. A $c(2 \times 2)$ phase (Fig. 5.5b) is another possible structure on the same square substrate lattice. The single compo-

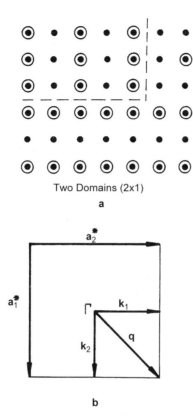

Two Domains (2x1)

a

b

Figure 5.18. a, Two-symmetry equivalent domains of a (2×1) structure on a square lattice. The *dashed line* marks the domain boundary. **b,** The surface Brillouin zone for part **a**. The substrate reciprocal mesh vectors are $\mathbf{a}_1^*$ and $\mathbf{a}_2^*$. The zone center is at G. The reciprocal lattice vectors $\mathbf{k}_1$ and $\mathbf{k}_2$ describe the order parameters of the two (2×1) domains. The wave vector q describes the order parameter of a $c(2 \times 2)$ structure.

nent (scalar) order parameter ϕ_3 for this structure is determined by the reciprocal vector $\mathbf{q}$,

$$\phi_3 = \frac{1}{N} \sum_j \langle n_j \rangle e^{i\mathbf{q} \cdot \mathbf{r}_j}. \tag{5.14}$$

The *pair correlation function* $\langle n_i n_j \rangle$ gives the probability that the lattice sites at $\mathbf{r}_j$ and $\mathbf{r}_i$ are simultaneously occupied and, therefore, $\langle n_i n_j \rangle$ provides more information about the details of the ordering than does ϕ. Because n_i can be written as a sum of its average value $\langle n_i \rangle$ plus its instantaneous fluctuation δn_i from that average, one obtains

$$\langle n_i n_j \rangle = \langle n_i \rangle^2 + \langle \delta n_i \delta n_j \rangle \tag{5.15}$$

The functional form of $\langle \delta n_i \delta n_j \rangle$ for a system that contains some amount of disorder is such that $\langle \delta n_i \delta n_j \rangle$ decreases as $\mathbf{r}_j$ and $\mathbf{r}_i$ become farther apart. The length that characterizes this fall-off is called the *correlation length* ξ. ξ provides an important measure of the distance over which fluctuations from the average coverage $\langle n_i \rangle$ are correlated.

Diffraction from a Disordered Lattice Diffraction is an ideal tool for the study of phase transitions. The large area illuminated by the incident radiation ensures that the diffracted intensity provides a statistical average of the surface. If a perfect 2-D lattice is illuminated by radiation with wave vector $\mathbf{k}_o$ ($|\mathbf{k}_o| = 2\pi/\lambda$, where λ is the wavelength), diffracted beams are observed only for the specific scattered wave vectors $\mathbf{k}$ that satisfy the Laue conditions.[46] If a phase transition is taking place on the lattice, disorder is introduced and intensity is removed from the diffracted beams and redistributed into other directions. In this situation, the measured intensity consists of two components[47]

$$I \propto I_{\text{LRO}} + \chi(\mathbf{S}) \tag{5.16}$$

where $\mathbf{S} \equiv \mathbf{k} - \mathbf{k}_o$ is the scattering vector.[47] Only $\mathbf{S}_\|$, $\mathbf{k}_{o\|}$, and $\mathbf{k}_\|$, the components of $\mathbf{S}$, $\mathbf{k}_o$ and $\mathbf{k}$, which are parallel to the plane of the surface, are important. I_{LRO}, the first term in Eq. (5.16), is the intensity that would have been obtained for the perfectly ordered surface reduced by the factor ϕ^2, i.e.,

$$I_{\text{LRO}} \equiv \phi^2 \delta(\mathbf{S}_\| + \mathbf{G}_{hk}) \tag{5.17}$$

where $\mathbf{G}_{hk}$ are the reciprocal lattice vectors of the 2-D Bravais lattice. $\chi(\mathbf{S})$, the second term in Eq. (5.16), is redistributed intensity owing to the fluctuations from perfect order

$$\chi(\mathbf{S}) = \sum_{ij} \langle \delta n_i \delta n_j \rangle e^{i\mathbf{S} \cdot \mathbf{r}_{ij}} \tag{5.18}$$

In analogy with magnetic systems, $\chi(\mathbf{S})$ is called the *susceptibility*. It is a peaked function; the half-width $\Delta \mathbf{S}_\| (1/2)$ is inversely related to the correlation length ξ by

$$\left| \Delta \mathbf{S}_\| \left(\frac{1}{2} \right) \right| = \frac{2\pi}{\xi} \tag{5.19}$$

For the case of continuous phase transitions, the susceptibility can usually be approximated as a Lorentzian function.[47] $\chi(\mathbf{S}) = 0$ if the surface is perfectly ordered.

STM is useful for the study of fluctuations only if they are very slow or can be frozen in by rapid quenching from a high temperature. A particularly good example is the study of atomic distribution at step edges on Si(100).[48] Information can also be obtained from structures that fluctuate slowly on the time scale of an STM measurement, but the analysis becomes more involved. Examples include step edges on Ag, Cu, and Au.[49,50]

Critical Exponents and Universality The way in which I_{LRO} and $\chi(S)$ change in the vicinity of a second-order phase transition can be described by a set of numbers ω called *critical exponents*. The values of the critical exponents do not depend on the details of the interatomic potentials but are determined entirely by the symmetry properties of the surface. A set of empirical rules relating the ω to the surface symmetry was developed by Landau and Lifshitz.[51] Originally, it was hoped that there would be only one set of critical exponents.[52] Instead, because of the variety of symmetries that can occur, there turn out to be several distinct sets, called *universality classes*. Once the universality class of a particular surface system is known, its behavior near a second-order transition is completely determined; i.e., membership in a universality class is a necessary condition for any surface system to undergo a second-order phase transition. If a system does not belong to a universality class, it is impossible for it to have a second-order phase transition. In addition, even if a system belongs to a universality class, its phase transitions can still turn out to be first-order.

Each critical exponent predicts the variation of specific physical quantities near the critical temperature of the transition.[53,54] For example, if $F(T)$ is a quantity of interest such as the order parameter, its variation near the critical temperature T_c can be described by a power series expansion

$$F(T) \approx F_o|t|^{\omega} + \cdots \qquad (5.20)$$

where $t \equiv (T-T_c)/T_c$ is the reduced temperature and F_o is an amplitude. Higher-order terms in Eq. (5.20) are usually unimportant as long as T is near enough to T_c.

STM has had no role in the study of universality in critical phenomena.

Other Types of Phase Transitions There are several other important types of phase transitions. Here I discuss briefly surface premelting, roughening, and commensurate–incommensurate transitions.

In the *premelting phase transition*, layers of atoms near the surface melt at lower temperatures than the bulk melting temperature.[55] Premelting explains why a solid cannot be superheated above its melting point. There is no nucleation barrier to formation of the liquid phase, because premelting has already created a nucleus from which the liquid phase can grow. Theoretical calculations of premelting, such as the molecular dynamics simulations shown in Figure 5.19, show that surface melting is expected to be significant only at temperatures near the melting point.[56] The thickness of the melted layer exceeds a few monolayers only when T is within a few degrees of the melting temperature. In Figure 5.19**a**, the temperature is $0.96T_m$. In the center of the slab, each atom follows a trajectory that remains near its low temperature equilibrium position. However, at either surface, the atoms are able to wander, and essentially no crystalline order remains. In Figure 5.19**b**, the temperature is $1.01T_m$; the slab has clearly lost both its long-range and short-range order

$$kT/\epsilon = 0.40$$

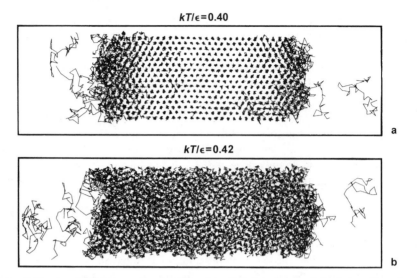

$$kT/\epsilon = 0.42$$

Figure 5.19. A molecular dynamics simulation for a 2-D slab showing trajectories of the atoms. The bulk melting point is $T_m = 0.415\epsilon/k$. **a**, Just below T_m. **b**, Just above T_m. (Courtesy of F. Abraham.)

and has melted. Premelting transitions are poorly understood, and the available experimental studies indicate that there may actually be several types of premelting behaviors.

The *roughening phase transition* is the term given to the spontaneous formation of steps on a surface and is initiated at the temperature T_R at which the free energy change $\Delta F = \Delta U - T\Delta S$ required to form a step becomes zero.[57] Atoms at a step edge have fewer nearest neighbors so that step formation requires that interatomic bonds be broken at an energy cost of ΔU per unit length of step. In a nearest-neighbor model, the energy per bond is $\epsilon/2$, and $\Delta U = \epsilon/2a$, where a is the interatomic spacing along the step. Step formation also increases the entropy of the surface particularly if the step edge is not straight. It is this increase in entropy that drives ΔF to zero. In order for the roughening transition to occur, T_R must be lower than the melting temperature T_m. Thus roughening occurs only for more open surfaces that have low ΔU; e.g., the (113) and (115) surfaces of FCC metals.[58] The roughening transition is distinct from premelting, because the surface retains a high degree of short-range order in the roughened phase. Roughening is important in crystal growth, because the steps provide a high density of sites that provide a template for nucleation and growth.

Frank and van der Merwe[59] originally described *the commensurate–incommensurate phase transition*. This transition has been studied for a large number of systems, ranging from inert gas overlayers on various substrates to metallic layers on metals. It is often observed when the interaction between atoms in the overlayer is of comparable strength to their interaction with the substrate. At low temperatures or pressures the overlayer lattice is commensurate with the substrate, but at a

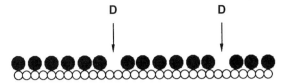

Figure 5.20. Example of an incommensurate phase with a light domain wall.

critical value of T or P, arrays of domain walls are formed and cause the overlayer lattice to lose registry with the substrate.[60] Figure 5.20 shows a 1-D example of a domain wall marked by D. Far from the domain wall, the overlayer atoms remain in registry with the substrate, but near the wall the overlayer atoms may be shifted from registry because of their asymmetric environments. A domain wall, like the one in Figure 5.20, that has lower density than the other parts of the overlayer is called a light wall. Domain walls with higher local density are called heavy walls. The domain walls may be ordered to form a domain wall solid or disordered to form a domain wall fluid. The properties of the commensurate–incommensurate transition can be described completely in terms of the interactions between the domain walls and many variations are possible, including such nonintuitive cases as a commensurate solid-to-fluid-to-incommensurate solid sequence of transitions.[15,61]

5.3 DEFECTS AND THE SURFACES OF REAL MATERIALS

At the atomic level, phase transitions usually involve the creation and motion of defects in the lattice.[62] Furthermore, real surfaces always contain a wide variety of defects that can significantly alter the behavior of the surface. A few of the examples already discussed illustrate this point. In the case of the Si(111)-(7 × 7) to (1 × 1) transition, the (7 × 7) phase nucleates and grows from the top edges of steps (Fig. 5.17). The Au(110)-(1 × 2) surface (Fig. 5.9**a**) is not perfectly ordered and has various types of defects, including regions of (1 × 3) structure. The Si(100)-(1 × 2) surface (Figs. 4.3 and 5.10) has numerous defects, and the CO-induced reconstruction shown in Figure 5.11 is really the image of a point defect. Surface defects are common, and STM has been an important tool for their study.

Surface defects can take on many forms, some of which are illustrated in Figure 5.21. Point defects include self-adatoms and foreign adatoms and vacancies. Steps are examples of extended defects and can have their own point defects, that are called *kinks*. Bulk defects, like edge and screw dislocations, can intersect the surface.

Surface defects occur for many reasons. They can be intrinsic to the surface; in which case they have concentrations that are determined by thermodynamic or kinetic properties of the material. Defects can be created by thermal or mechanical stresses during fabrication of the surface. For example, in the preparation of low-index surfaces, even a slight misalignment of the crystal will introduce steps.

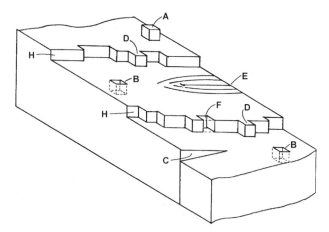

Figure 5.21. Types of surface defects. *A*, Adatoms; *B*, vacancies; *C*, screw dislocations; *D*, step edge adatoms; *E*, edge dislocations; *F*, step edge vacancies; *H*, steps.

Contamination caused by impurities present in the bulk substrate or to adsorption from the gas or liquid environment is another common source of defects.

Many solids such as glasses and polymers have little or no crystallinity. Liquid surfaces also have no long-range order. The concepts of surface crystallography are of little use in describing the surface properties of these highly noncrystalline materials. A more extensive review of surface defects is given by Tringides.[63]

5.3.1 Surface Point Defects and Their Properties

Surface point defects, such as vacancies and impurity atoms, have significant effects on the surface properties of materials. For example, one half monolayer of oxygen adsorbed on a clean W(112) surface below 350 K forms a random structure that orders into a (2×1) structure on heating. The addition of randomly distributed nitrogen impurities drastically slows down the kinetics of ordering, breaks the long-range order of the (2×1) structure, and lowers the critical temperature of the (2×1) order–disorder transition.[64] Another example is given by adsorption of small amounts of alkali metal atoms, which not only can significantly alter the structure, as already discussed, but can also change the reactivity of the surface. For example, small amounts of potassium increase the selectivity of the Fisher-Tropsch reaction toward higher molecular weight products.[65] Figure 4.18 illustrates how the electronic structure of a reconstructed Si(111) surface is altered in the vicinity of a vacancy; Chapter 4 also includes examples.

Oxides are a good example of materials whose surface properties are strongly affected by the presence of defects.[66] The bonding in many oxides is strongly ionic and an important type of point defect is caused by removal of a surface oxygen ion. This removal creates a vacancy, and the cations around it modify their positions to adjust to their changed environment. In addition, to maintain charge neutrality, one or two electrons become bound to the vacancy. The electronic properties in the vicinity of an oxygen vacancy are drastically altered from those of the perfect surface, and this in turn changes the chemical reactivity of the surface. Many oxides are inert to adsorption of common gases like CO unless defects are present.

Free dislocations, provide another way to destroy the long-range order of a layer of adsorbed atoms and are believed to play a central role in 2-D melting of adsorbed layers (Fig. 5.22).[67] A free dislocation results from the distortions created when an extra partial row of atoms is added to an ordered structure. Free dislocations interact with each other through their long-range strain fields and can also be pinned by defects on the substrate. On an infinite surface, free dislocations must always occur in pairs. Surface dislocations are also created when a bulk edge dislocation encounters the surface. Surface electronic and magnetic properties might be expected to change near a dislocation but there are no studies of this.

Dislocations are also important in crystal growth and dissolution. Screw dislocations, for example, act as sources of steps that in turn provide new sites for the incorporation or loss of atoms. Figure 12.5 shows an STM image of steps emerging from a gold surface and illustrates how they change during cycling in an electrochemical cell.

The energy required to form various defects like those in Figure 5.21 can be estimated from the simple nearest-neighbor bond model, introduced in Section 2.1. For example, removal of an atom from the (111) surface of an FCC crystal requires that a total of nine bonds (six between neighbors in the surface layer and three with neighbors in the second layer) be broken. This requires an energy of $u_v = 9\varepsilon/2$. However, if the atom is not completely removed but only placed on top of the (111) surface as a self-adatom, three new bonds are formed. Thus the energy u_s to form a vacancy-self-adatom pair is only 3ε. An even smaller energy is required if the atom is placed at a step edge or at a kink site on the step edge. Because the vacancy-

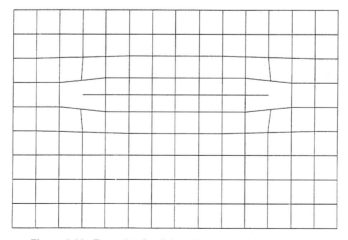

Figure 5.22. Example of a dislocation pair on a square lattice.

adatom pairs increase the entropy of the surface, their equilibrium concentration will increase as the temperature increases. At low concentrations, this fraction of vacancy-adatom pairs is determined by the Boltzmann factor $e^{-3\varepsilon/kT}$. In the case of metals, ε is on the order of an electron volt and the equilibrium concentration of point defects at room temperature will be small. However, the preparation of clean, well-ordered surfaces often involves annealing at temperatures above 1000 K where the vacancy-adatom pair concentration can be on the order of several percent. If the crystal is quenched too rapidly, this higher defect concentration will be frozen in.

If the defect formation energies are known, the surface configuration at any temperature can be determined from statistical thermodynamics using Monte Carlo methods. Figure 5.23 shows a typical result for the case of a simple cubic crystal with only nearest-neighbor interactions for various values of kT/e.[68]

As the densities of various types of defects increase, interactions between them become important and the energies of defect formation become concentration dependent. One consequence is that catastrophic changes in the surface properties can occur. The roughening transition, illustrated by the simulations in Figure 5.23, is an example. At the roughening temperature, the density of steps increases dramatically and average width of the crystal-vacuum interface begins to diverge logarithmically. For open-faced ($11n$) surfaces of FCC metals, roughening is found to occur at temperatures in the range 0.5 to $0.75T_m$.

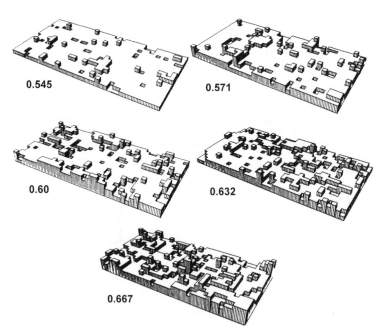

Figure 5.23. Examples of increasing surface defect density as a function of temperature in the solid-on-solid model. (Courtesy of G. H. Gilmer.)

5.3.2 Extended Surface Defects

Examples of extended surface defects include steps, grain boundaries, and screw dislocations. Until the advent of STM, steps were the only extended defects that had been extensively studied experimentally. Ordered arrays of nearly equally spaced and parallel steps can be created if a crystal surface is slightly misoriented from a low-index plane. By changing the angle of misorientation, the step density can be varied. This type of slightly misoriented surface is called a vicinal surface and is specified by the Miller indices of the bulk planes parallel to it. Because the steps on a vicinal surface are in uniform arrays, they form a diffraction grating with translational periodicity different from the nonstepped surface. Thus ordered step arrays are easily studied by diffraction and the average values of step height and step spacing can be directly measured. Stepped surfaces are often modeled by staircases for which the terraces, or treads, consist of strips of low-index surface separated by single or multiple atomic height risers.

Real steps are more complex in their structure, as is easily seen for the steps surrounding the large defects in Figure 4.3. Step edges are not straight and, even on carefully prepared vicinal surfaces, have numerous kinks along them. Kinks increase the surface entropy, and their equilibrium density is determined thermodynamically in analogy to the spontaneous generation of point defects discussed in the preceding section.[69] The local environments of atoms at step edges and at kink sites along step edges are different from the environments of atoms on a perfect surface. These differences alter the electron densities and thus change the local electrical and chemical behavior of the surface. For example, the redistributed charge density at a step edge creates a local dipole moment that in turn changes the surface work function. This effect has been demonstrated on several materials.[70]

Steps also have important effects on the rates of crystal growth from the vapor phase. In the case of epitaxial growth of Si on a Si(001) substrate,[71] adsorbed Si atoms condense into one-monolayer-thick islands whose edges are atomic height steps. Island growth occurs when Si atoms that are diffusing about on the surface encounter the edge of an island and are accommodated to it. Steps running in different directions on the surface have different atomic structures. For Si(001), these differences cause the accommodation coefficient to be much higher on step edges that are perpendicular to the dimer rows of the substrate. (The dimer reconstruction is illustrated in Figure 5.10.) This causes the islands to grow anisotropically and become much longer along the directions of dimer rows. If the growth is stopped and the surface annealed, the islands become more circular; this demonstrates that it is the kinetics of the sticking process that are different rather than differences in the energetics of adsorption at the various types of step edge sites.

In an important series of experiments, Blakely and Somorjai[72] showed that steps can be strongly alter the chemical reactivity of a surface. In the case of Pt surfaces, they found that C—H and C—C bond breaking in adsorbed hydrocarbon molecules occurs preferentially on stepped (111) surfaces but not on the (111) surface itself. Steps can also cause the distribution of adsorbed atoms to be nonuniform. For example, Figure 5.24**a** is a photoelectron microscope image of a 40-μm diameter

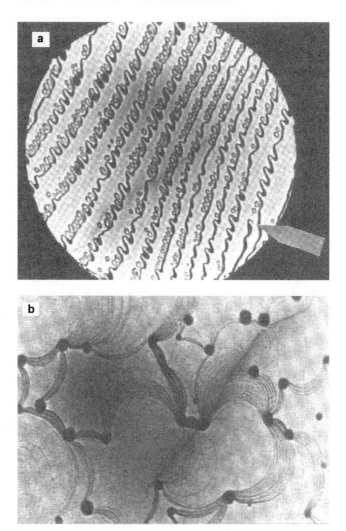

Figure 5.24. a, A 0.2-monolayer of Cu on a slightly misoriented Mo(110) surface. The image diameter is 40 µm. **b**, Step pinning caused by surface impurities on Si(111). The image is 2.6 × 3.75 µm. (Courtesy of E. Bauer.)

region on a Mo(110) surface covered with about 0.2 monolayer of Cu, shown as bright lines. The Cu atoms are preferentially adsorbed at the bases of steps.

Impurities can also greatly alter the properties of steps on surfaces. Eizenberg and Blakely[73] studied the effects of segregation of carbon to vicinal surfaces near Ni(111). At high temperatures, the surface coverage of carbon is low and the vicinal surfaces have the expected uniform staircase structures with monatomic height steps. Graphite precipitation begins to occur at temperatures below the equilibrium precipitation temperature T_p. This causes the vicinal surfaces to become unstable and to break up into {111} facets. However, in the intermediate temperature range between T_p and about 1.1 T_p, a single monolayer of graphite is precipitated; and depending on the original step directions, the monatomic step arrays are replaced by either (111) and (110) facets or by (111) and (311) facets. Such phenomena are complex and poorly understood from a theoretically point of view.

At high temperatures, surface diffusion causes steps to migrate across the surface to relieve localized regions of misorientation or strain. When impurities are present, step migration can be impeded. Figure 5.27**b** shows an example for Si(111). Surface contaminants (probably SiC precipitates, shown by the *black specks*) pin steps while motion far from the impurities is able to continue giving rise to complex bowed step arrays.

5.3.3 Surfaces of Noncrystalline Materials

The surfaces of many materials are not crystalline. Examples include glasses and polymers. Other materials that are crystalline in their equilibrium state can also be prepared in metastable states that are either amorphous or highly polycrystalline. Before the development of STM and atomic force microscopy (AFM), it was not possible to study the atomic and mesoscale structure of such materials. Thus the advent of STM and AFM has opened up entire new areas of research that are only beginning to be explored.

The conformation of adsorbed biomacromolecules and their interactions on surfaces is of interest in immunology for applications such as the development of new medical diagnostic techniques.[74] Such systems provide another example in which the traditional surface structure determination techniques, discussed in the following section, are of little use. Other examples of STM images of biologic samples are given in Chapter 9.

5.4 TRADITIONAL TECHNIQUES FOR SURFACE STRUCTURE DETERMINATION

A wide variety of experimental techniques have been used to study surface crystal structures. It is beyond the scope of this chapter to make a detailed presentation of these techniques, and the reader is referred to the extensive literature.[2,75] Structural techniques can be roughly divided into two basic categories: diffraction and geometrical.

5.4.1 Diffraction methods

The regular rows of atoms at the surface of a crystal form a diffraction grating. If a beam of monochromatic radiation of wavelength λ illuminates the surface, Bragg's law gives the conditions for constructive interference of the reflected radiation

$$n\lambda = 2d \sin \varphi \qquad (5.21)$$

where d is the spacing between equivalent rows or planes of atoms, 2φ is the angle through which the incident radiation is scattered, and n is an integer that determines the order of the diffraction. In general, $\lambda < 2d/n$ so that λ must be somewhat shorter than the d-spacings of interest. Highly collimated, monochromatic beams of atoms, electrons, and X-rays can all be prepared with the appropriate wavelengths and are the most widely used diffraction probes for practical surface crystallography. The wavelength is determined from the de Broglie relationship

$$\lambda = h/p \tag{5.22}$$

where p is the momentum, and h the Planck constant. Electrons and atoms of energy E have $p^2 = 2mE$, where m is the particle mass; and photons have $p = h\nu/c$, where ν is the frequency and c the speed of light.

For an incident beam, Bragg's law defines the directions along which diffracted beams are emitted from the sample. The directions and intensities of these beams provide the basic data needed to determine the structure of the crystal surface. Bragg's law and the spatial pattern of the diffraction beams are sufficient to determine the size, shape, and space group of the surface unit mesh but, to obtain the positions of the basis atoms within the unit mesh, the intensities of the diffraction beams must be analyzed. The procedures for doing this depend on the detailed interaction potential between the incident particles and the surface, i.e., on the elastic and inelastic scattering cross sections.

Atom diffraction is the most surface sensitive of all the diffraction techniques, because the elastic scattering cross sections of atoms are so large that they cannot penetrate the surface at all. Instead, they are scattered from the electronic potential several angstroms outside the ion cores of the surface atoms. Because the atoms are neutral, both conducting and nonconducting surfaces can be studied. An atom must have thermal energy (a few millielectron volts) to have the appropriate wavelength for diffraction, as is easily calculated from Eq. (5.22).

Low-energy electrons (10 to 500 eV) have large inelastic cross sections that restrict their penetration to only a few atomic layers into the material. Their elastic cross sections are also large; this results in high intensities in the diffracted beams so that experiments are relatively easy. For these reasons LEED is by far the most widely used technique for surface crystallography. A complication occurs in the determination of the atomic positions because the large elastic cross sections cause the electrons to have a high probability of being scattered more than once. This *multiple scattering* complicates structure determinations. However, sophisticated, straightforward procedures for LEED structure analysis have been developed[7] and allow atomic positions to be routinely determined to a precision of about 0.1 Å. LEED is not generally useful for studies of nonconducting materials.

Both X-rays and neutrons have small inelastic cross sections so that, in most situations, scattering from the bulk swamps the part of the diffracted signal owing to surface atoms. For this reason, neutron diffraction is used only for materials such as intercalated graphite that have large surface to volume ratios. This would also be the case for X-rays if not for the fact that their high frequencies result in an index of refraction n_x that is slightly smaller inside materials than in vacuum, i.e., $n_x = 1 - \delta_x$ where δ_x is of order 10^{-5}.[76] This causes X-rays to be totally reflected if their grazing angle of incidence is smaller than the critical angle $\theta_c = \sqrt{2\delta_x}$. Typically, θ_c is less than $1°$. For $\theta < \theta_c$, an evanescent wave with decay length of only a few 10s of angstroms penetrates the surface. This shallow penetration makes grazing incident X-ray diffraction highly surface sensitive. However, the small cross sections require incident intensities that are higher than available in laboratory X-ray sources. For this reason, routine X-ray studies of surfaces can be carried out only at synchrotron radiation sources. Because of their high penetration, X-rays are also useful probes of buried interfaces.

5.4.2 Geometric Methods

Several techniques yield direct information about the atomic structure of surfaces.[77] Field ion microscopy (FIM)[78] and transmission electron microscopy (TEM)[79] can obtain direct images of surface or interfacial structures with atomic scale resolution. Other techniques such as Rutherford backscattering (RBS)[80] and medium-energy ion backscattering[81] provide direct information about surface structure but do not yield a direct image. Spectroscopic techniques like infrared absorption, angle-resolved ultraviolet photoelectron spectroscopy (ARUPS), and high-resolution electron energy loss spectroscopy (HREELS) are sensitive to the vibrational modes of surface species and thus contain information about adsorption site symmetries.[82] Because the spectroscopic techniques are sensitive to local structure, they are particularly useful for studies of noncrystalline systems.

Field ion microscopy was invented by Müller[83] and was the first technique to yield atomic resolution images of surfaces. In FIM, the image is obtained from the tip of a sharply pointed fine wire with a radius of a few hundred angstroms or less. A strong electric field is applied between the tip and an imaging screen a few centimeters away. Because of the tip's small radius of curvature, the electric field near it is extremely high—several volts per angstrom. The highest field strengths are near surface protrusions, such as step edges. An imaging gas of He or Ne is introduced into the volume around the tip. Near the tip, the electric field is strong enough to ionize atoms of the imaging gas. These positively charged ions are then accelerated away from the tip by the strong electric field and strike the imaging screen to yield an image like that in Figure 5.25. The bright spots correspond to the positions of atoms on the surface of the tip. Atoms in exposed positions feel stronger electric fields and are, therefore, more visible—atoms near the center of planes may not be visible. The magnification achieved by FIM is on the order of one million.

5.4.3 Evaluation of Traditional Methods

Diffraction measurements yield statistical information about a surface structure. One reason for this is that the incident beam illuminates a large area on the surface—typically $1 mm^2$ or larger—so that many atoms, including atoms at defect

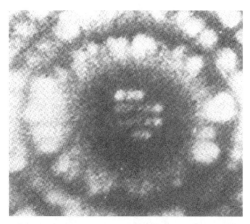

Figure 5.25. A FIM image of a small island of adsorbed Si atoms on a W(110) surface. Every Si atom is visible. (Reprinted with permission from Ref. 84.)

sites, simultaneously contribute to the signal. Thus diffraction techniques are best used for crystallographic studies of surfaces with nearly ideal structures and good long-range structural order. Disorder in the surface can profoundly alter the properties of the diffraction.[85] For example, if the regions of perfect crystalline order are small with dimension L, the diffraction beams are broadened by an amount proportional L^{-1}. On a real surface, L has a distribution of sizes; and if this distribution is known, the variation of intensity within each of the diffraction beams can be calculated. For example, the angular width of the diffracted beam is determined by the mean size of the finite size regions. Unfortunately, there is no unique solution to the inverse problem, i.e., determination of the distribution function of finite size regions directly from the measured beam profile. Thus, in practice, diffraction measurements yield little more than the average size of the ordered regions.

Point defects are even more difficult to study using diffraction, because they have little effect on the diffraction beams themselves. Instead, point defects contribute a weak, nearly uniformly distributed diffuse background to the diffraction pattern. The situation is even more difficult because scattering from surface phonons causes intensity from the diffracted beams to be redistributed throughout the diffraction pattern as *thermal diffuse scattering*.[86] From a practical viewpoint, it can be extremely difficult to distinguish the thermal diffuse scattering from the point defect scattering.

Atom diffraction is limited because it is sensitive only to the outermost layer of atoms, as are STM and AFM, and therefore is unable to provide information about the structure in the second or deeper atomic layers. X-rays, on the other hand, can penetrate well below the surface and thus are an attractive probe to study buried interfaces. No other diffraction method offers this possibility. X-rays and atoms can also have the advantage that they can be used for crystallographic studies of insulating materials, whereas electrons and ions cannot. Because of their high penetration, X-rays can be used with high gas pressures over the sample and make it possible to study surface structures under ambient conditions. High-temperature and time-dependent studies are also possible.

LEED is the most highly developed method for structure determination. The capability for immediate visual display of the diffraction pattern coupled with its long history of successful use, continue to make LEED the technique most widely used to monitor surface structures, including surface phase transitions. The resolution of the best LEED instrumentation rivals that of X-ray diffractometers. Furthermore, LEED is not limited to near grazing angles of incidence.

Several electron microscopy techniques provide direct images of surface structures. TEM uses high-energy electrons ($\sim$100 keV) and can be used to study the atomic structure of solid–solid interfaces, surface structure, and the structure of mesoscopic particles. However, the need for special sample preparation techniques in TEM is a major barrier to wide application. Low-energy electron microscopy (LEEM) uses more surface sensitive electrons with several tens of electron volts of kinetic energy. Examples of LEEM images are given in Figure 5.24. Neither LEEM nor the technique of reflection electron microscopy has achieved atomic resolution, but they provide the capability for in situ measurements at elevated temperature in real time. For these reasons, both are important for the study of surface processes at high temperatures, especially when only mesoscale information is sought. Unfortunately, none of these direct electron based methods is useful for studies of dielectric materials.

Ion scattering is an easily interpreted technique because of the geometrical nature of the scattering but provides only local information averaged over the entire surface. It is also relatively insensitive to defects and not useful for insulators. Ions are strongly interacting particles and in many cases radiation induced damage limits applicability. Quantitative analysis of ion scattering data is also complicated by the sensitivity to lattice vibrations.

Vibrational spectroscopies such as HREELS and infra red (IR) absorption provide powerful information about local site geometries but are relatively insensitive to long range order or defect structures and types. IR does not require a vacuum environment over the sample and is, therefore, one of the only techniques that can probe the liquid–solid interface or the gas–solid interface at high pressures.

Finally, no single method has the capability to study all problems in surface structure; rather each has a range of applications in which it is superior, and often several techniques must be applied to the same system before a complete structural characterization is possible.

5.5 THE ROLE FOR STM AND AFM

The first STM images had an immense impact on surface science research simply because they demonstrated the possibility to directly image the atomic structure of surfaces. This potential to view the structural properties of surface defects such as vacancies, atomic height steps, and domain boundaries meant that new advances in our understanding of the properties of real surfaces were at hand. Previously only FIM and, under special circumstances, TEM had demonstrated similar capabilities.

5.5.1   Strengths of STM

The most important contributions of STM are the result of its capability to image the neighborhood around individual atoms. Thus it yields direct information about point defects, local electronic structure, kinks, and other step morphologies. Furthermore, STM does not require large single crystalline samples and can, therefore, be used to study the local atomic structures on amorphous and polycrystalline materials. No other technique rivals STM for these types of studies.

Because of their limited lateral resolution and lack of subsurface resolution, STM and AFM are unlikely to ever rival the precision of diffraction methods for complete crystal structure determinations on well-ordered surfaces. However, STM and AFM do provide reliable information about the symmetry and size of the unit mesh and some information about vertical distortions in the surface layer. They quickly determine if the surface is well ordered, if more than one type of ordered structure is present, and if finite size effects are significant. This type of complimentary information is invaluable in eliminating incorrect models and thus providing a better-defined starting point for subsequent LEED or X-ray structure determinations.

The ability to apply scanned probe techniques at the liquid–solid interface opens up whole new areas of study of important problems in adhesion and friction, microbiology, and materials processing. For amorphous materials or materials for which large single crystals do not exist, scanned probe techniques may be the best structural techniques available.

5.5.2   Limitations of STM and AFM in Surface Crystallography

A few of the fundamental limitations of scanned probe methods, such as the inability of STM to image insulating materials, were apparent from the beginning. Because of the strong interaction between the tip and the surface, there is also always the potential for the surface properties to be altered by the measurement. The practitioner must, therefore, always be on the lookout for measurement induced artifacts. For example, when does the tip alter the surface structure? Is always possible to uniquely separate the structural and spectroscopic content of the signal? Under what conditions are atoms transferred between the tip and the sample? When can the tip cause adatoms to move about on the surface? How can ghost images caused by tunneling at more than one point on the tip be recognized? Many of the possible artifacts and related topics are discussed in Chapter 6. As the theoretical understanding of STM and AFM improve, these limits will become better understood and appreciated.

In the case of AFM, quantitative understanding of the tip–surface interactions and the imaging mechanisms is poor but improving rapidly. Although atomic scale features are often observed, true atomic resolution is rare.[77] Chapter 7 provides a good overview of the state of the art.

One of the most important limitations is that STM does not probe the subsurface structure. Thus STM is not an effective technique to study topics such as oscillatory relaxations or subsurface reconstructions, although ballistic electron energy microscopy (BEEM) does give limited information (see Chapter 8). The lateral resolution of STM is also limited by the area on the surface that contributes to the current so that atomic positions cannot be determined with the same degree of precision possible with LEED or other diffraction techniques. STM is also unable to obtain images fast enough to study the local surface structure of fluid-like adsorbate phases, because images cannot be obtained in times that are short compared to diffusion times, although improvements in temporal resolution continue to be made.

A final category of limitations is the result of technological limits of existing instrumentation. In many cases, these will be removed or reduced in the future by improved instrumentation. One important example is that STM cannot be used to study surfaces at very high temperatures because of thermal drift and because the piezoelectric drivers cease to operate above several hundred degrees. Another is the limited ability to study kinetic processes owing to the relatively long times required to acquire images.

As with all other techniques used to study surface structure, STM and AFM have specific strengths and weaknesses. They cannot solve all problems in surface structure but must be applied with skill and insight in conjunction with other complementary methods.

REFERENCES

1. N. W. Ashcroft and A. D. Mermin, *Solid State Physics*, Holt, Rinehart and Winston, New York, 1976.

2. Ed. W. N. Unertl, ed. *Physical Structure*, Vol. 1 of *Handbook of Surface Science*, Elsevier Science, Amsterdam, 1996.

3. B. K. Vainshtein, *Modern Crystallography*, Springer-Verlag, Berlin, 1981.

4. H. B. Nielsen, J. N. Andersen, L. Petersen, and D. L. Adams, *J. Phys. C* **15**, L113 (1982); D. L. Adams, H. B. Nielsen, J. N. Andersen, I. Stensgaard, R. Friedenhans'l, and J. E. Sorensen, *Phys. Rev. Lett.* **49**, 669 (1982); H. L. Davis and J. R. Noonan, *J. Vac. Sci. Technol.* **20**, 842 (1982).

5. M. W. Finnis and V. Heine, *J. Phys. F* **4**, L37 (1974).

6. U. Landman, R. N. Hill, and M. Mosteller, *Phys. Rev. B* **21**, 448 (1980).

7. M. A. Van Hove, W. H. Weinberg, and C. M. Chan, *Low-Energy Electron Diffraction* Springer-Verlag, Berlin, 1986.

8. C. J. Davisson and L. H. Germer, *Phys. Rev.* **29**, 908 (1927).

9. K. Christmann, R. J. Behm, G. Ertl, M. A. Van Hove, and W. H. Weinberg, *J. Chem. Phys.* **70**, 4168 (1979).

10. A. V. Titov and H. Jagodzinski, *Surface Sci.* **152–153**, 409 (1985).

11. H. D. Shih, F. Jona, D. W. Jepsen, and P. M. Marcus, *Surface Sci.* **60**, 445 (1976).

12. K. Griffiths, P. R. Norton, J. A. Davies, W. N. Unertl, and T. E. Jackman, *Surface Sci.* **152–153**, 374 (1985).

13. E. A. Wood, *J. Appl. Phys.* **35**, 1306 (1964).

14. R. L. Park and H. H. Madden, *Surface Sci.* **11**, 188 (1968).

15. W. N. Unertl, M. Golze, and M. Grunze, *Prog. Surface Sci.* **22**, 101 (1987).

16. J. M. MacLaren, J. B. Pendry, P. J. Rous, D. K. Saldin, G. A. Somorjai, M. A. Van Hove, and D. D. Vvedensky, eds., *Surface Crystallography Information Service*, Reidel Publishing, Dordrecht, 1987.

17. M. A. Van Hove, R. J. Koestner, P. C. Stair, J. P. Birberian, L. L. Kesmodell, I. Bartos, and G. A. Somorjai, *Surface Sci.* **103**, 189, 218 (1981).

18. C. M. Chan, M. A. Van Hove, and E. D. Williams, *Surface Sci.* **91**, 440 (1980); D. L. Adams, H. B. Nielsen, M. A. Van Hove, and A. Ignatiev, *Surface Sci.* **104**, 47 (1981).

19. L. D. Mark, *Surface Sci.* **139**, 281 (1984).

20. I. K. Robinson, Y. Kuk, and L. C. Feldman, *Phys. Rev. B* **29**, 4762 (1984).

21. M. Guillope and B. Legrand, *Surface Sci.* **215**, 577 (1989).

22. S. M. Foiles, *Surface Sci.* **191**, L779 (1987).

23. K. M. Ho and K. P. Bohnen, *Phys. Rev. Lett.* **59**, 1833 (1987).

24. R. M. Tromp, R. J. Hamers, and J. E. Demuth, *Phys. Rev. B* **34**, 5343 (1986).

25. W. A. Harrison, *Electronic Structure and the Properties of Solids*, W. H. Freeman, San Francisco, 1980.

26. T. Gritsch, D. Coulman, R. J. Behm, and G. Ertl, *Phys. Rev. Lett.* **63**, 1086 (1989).

27. K. H. Rieder, M. Baumberger, and W. Stocker, *Phys. Rev. Lett.* **51**, 1799 (1983).

28. B. E. Hayden, K. C. Prince, P. J. Davie, G. Paolucci, and A. M. Bradshaw, *Solid State Comm.* **48**, 325 (1983); M. Copel, W. R. Graham, T. Gustafsson, and S. Yalisove, *Solid State Comm.* **54**, 695 (1985).

29. H. C. Potter and J. M. Blakely, *J. Vac. Sci. Technol.* **12**, 635 (1975); T. M. Buck, G. H. Wheatley, and L. Marchut, *Phys. Rev. Lett.* **51**, 43 (1983).

30. N. D. Lang, *Solid State Phys.* **28**, 225 (1973); F. Forstmann, B. Feuerbacher, B. Fitton, and R. F. Willis, eds. in *Photoemission and the Electronic Properties of Surfaces*, John Wiley, New York, 1978, J. E. Inglesfield, *Rept. Prog. Phys.* **45**, 223 (1982).

31. G. A. Somorjai, *Principles of Surface Chemistry*, Prentice-Hall, Englewood Cliffs, N.J., 1972.

32. J. M. Blakely, *Introduction to the Properties of Crystal Surfaces*, Pergamon, Oxford, UK, 1973.

33. E. D. Williams and N. C. Bartelt in *Physical Structure*, Vol. 1 of *Handbook of Surface Science*, Elsevier Science, Amsterdam, 1996.

34. J. C Heyraud and J. J. Metois, *Surface Sci.* **128**, 334 (1983).

35. A. Pavlovska, K. Faulian, and E. Bauer, *Surface Sci.* **221**, 233 (1989).

36. A. R. Kortan and R. L. Park, *Phys. Rev.* **B23**, 6340 (1981).

37. J. S. Ochab, G. Akinci, and W. N. Unertl, *Surface Sci.* **181**, 452 (1987).

38. D. Menzel in M. Grunze and H. J. Kreuzer, eds., *Kinetics of Interface Reactions*, Springer-Verlag, Berlin, 1987.

39. T. L. Einstein in R. Vanselow and R. Howe, eds., *Chemistry and Physics of Solid Surfaces IV*, Springer-Verlag, Berlin, 1983.

40. E. A. Guggenheim, *Thermodynamics*, North-Holland, Amsterdam, 1988.

41. N. Osakabe, K. Yagi, and G. Honjo, *Jpn. J. Appl. Phys.* **19**, L309 (1980).

42. M. Schick, *Prog. Surface Sci.* **11**, 245 (1981); C. Rottman, *Phys. Rev. B* **24**, 1482 (1981).

43. P. Kleban in R. Vaneslow and R. Howe, eds., *Chemistry and Physics of Solid Surfaces V*, Springer-Verlag, Berlin, 1984.

44. L. E. Reickl, *A Modern Course in Statistical Physics*, University of Texas Press, Austin, 1980.

45. R. J. Birgeneau and P. M. Horn, *Science* **232**, 329 (1986).

46. C. Kittel, *Introduction to Solid State Physics*, John Wiley, New York, 1986.

47. J. C. Campuzano, M. S. Foster, G. Jennings, R. F. Willis, and W. N. Unertl, *Phys. Rev. Lett.* **54**, 2684 (1985).

48. B. S. Swartzentruber, *Phys. Rev. B* **47**, 13432 (1993).

49. J. S. Ozcomert, W. W. Pai, N. C. Bartelt, and J. E. Reutt-Robey, *Mat. Res. Soc. Symp. Proc.* **355**, 115 (1995).

50. M. Poensgen, J. F. Wolf, J. Frohn, M. Giesen, and H. Ibach, *Surface Sci.* **274**, 430 (1992).

51. L. D. Landau and E. M. Lifshitz, *Statistical Physics, Part I*, Pergamon Press, Oxford, UK, 1980; I. P. Ipatova and Y. E. Kitaev, *Prog. Surface Sci.* **18**, 189 (1985).

52. H. E. Stanley, *Introduction to Phase Transitions and Critical Phenomena*, Oxford, UK, 1971.

53. P. Bak, *Solid State Commun.* **32**, 581 (1979).

54. W. N. Unertl, *Comments Cond. Mat. Phys.* **12**, 289 (1986).

55. J. F. van der Veen and J. W. M. Frenken, *Surface Sci.* **178**, 382 (1986).

56. F. F. Abraham, *Phys. Rev. B* **23**, 6145 (1981).

57. J. D. Weeks in T. Riste, ed., *Ordering in Strongly Fluctuating Condensed Matter Systems*, Plenum, New York, 1980.

58. I. K. Robinson, E. H. Conrad, and D. S. Reed, *J. Phys. France* **51**, 103 (1990).

59. F. C. Frank and J. H. van der Merwe, *Proc. R. Soc. London Ser. A* **198**, 205 (1949).

60. D. R. Nelson and B. I. Halperin, *Phys. Rev.* **19**, 2457 (1979).

61. R. J. Birgeneau, G. S. Brown, P. M. Horn, D. E. Moncton, and P. W. Stephens, *J. Phys. C* **14**, L49 (1981); F. F. Abraham, S. W. Koch, and W. E. Rudge, *Phys. Rev. Lett.* **49**, 1830 (1982).

62. L. D. Roelofs in *Physical Structure*, Vol. 1 of *Handbook of Surface Science*, Elsevier Science, Amsterdam, 1996.

63. M. C. Tringides in *Physical Structure*, Vol. 1 of *Handbook of Surface Science*, Elsevier Science, Amsterdam, 1996.

64. J. K. Zuo, G. C. Wang, and T. M. Lu, *Phys. Rev. B* **40**, 524 (1989).

65. D. J. Dwyer in H. P. Bonzel, A. M. Bradshaw, and G. Ertl, eds., *Physics and Chemistry of Alkali Metal Adsorption*, Elsevier, New York, 1989.

66. V. E. Henrich, *Prog. Surface Sci.* **14**, 175 (1983); R. J. Lad and V. E. Henrich, *J. Vac. Sci. Technol.* **A6**, 781 (1988); M. Tsukada, H. Adachi, and C. Satoko, *Prog. Surface Sci.* **14**, 113 (1983).

67. K. J. Strandburg, *Rev. Mod. Phys.* **60**, 161 (1988).

68. H. J. Leamy, G. H. Gilmer, and K. A. Jackson in J. M. Blakely, ed., *Surface Physics of Materials*, Academic Press, New York, 1975.

69. E. D. Williams and N. C. Bartelt in *Physical Structure*, Vol. 1 of *Handbook of Surface Science*, Elsevier Science, Amsterdam, 1996.

70. K. Besocke, B. Krahl-Urban, and H. Wagener, *Surface Sci.* **68**, 39 (1977).

71. Y. W. Mo, R. Kariotis, B. S. Swartzentruber, M. B. Webb, and M. G. Lagally, *J. Vac. Sci. Technol. B* **8**, 232 (1990).

72. D. W. Blakely and G. A. Somorjai, *J. Catal.* **42**, 181 (1976).

73. M. Eizenberg and J. M. Blakely, *J. Chem. Phys.* **71**, 3467 (1979).

74. J. C. Andle, J. F. Vetelino, R. Lec, and D. McAllister, in *Proceedings of the IEEE Ultrasonics Symposium*, 1989.

75. G. Ertl and J. Kuppers, *Low-Energy Electrons and Surface Chemistry*, Verlag Chemie, Weinheim, 1974; W. N. Unertl, *Appl. Surface Sci.* 11–12, 64 (1982); H. Ibach, ed., *Electron Spectroscopy for Surface Analysis*, Springer-Verlag, Berlin, 1977; G. Margaritondo, P. K. Larsen and P. J. Dobson, eds., *Introduction to Synchrotron Radiation*, Oxford University Press, Oxford, UK, 1988; *Reflection High-Energy Electron Diffraction and Reflection Electron Imaging of Surfaces*, Plenum Press, New York, 1988.

76. A. Guinier, *X-Ray Diffraction*, Freeman, San Francisco, 1963.

77. W. N. Unertl and M. E. Kordesch in *Physical Structure*, Vol. 1 of *Handbook of Surface Science*, Elsevier Science, Amsterdam, 1996.

78. T. T. Tsong, *Atom Probe Field ion Microscopy*, Cambridge University Press, Cambridge, UK, 1990.

79. J. E. Bonevich and L. D. Marks, *Microscopy* **22**, 95 (1992).

80. L. C. Feldman, *Surface Sci.* **299–300**, 233 (1994).

81. J. W. Rablalais, *Surface Sci.* **299–300**, 219 (1994).

82. Evans Jr., and S. Wilson, eds., *Encyclopedia of Materials Characterization*, Butterworth-Heinemann, Stoneham, Mass., 1992.

83. E. W. Müller, *Z. Physik.* **131**, 136 (1951).

84. R. Casanova and T. T. Tsong in P. S. Ho and K. N. Tu, eds., *Thin Films and Interfaces*, Elsevier North-Holland, New York, 1982.

85. M. G. Lagally in R. Vaneslow and R. Howe, eds., *Chemistry and Physics of Solid Surfaces IV*, Springer-Verlag, Berlin, 1983.

86. M. B. Webb and M. G. Lagally, *Solid State Phys.* **28**, 301 (1973).

6

THE PREPARATION OF TIP AND SAMPLE SURFACES FOR SCANNING PROBE EXPERIMENTS

Richard L. Smith and Gregory S. Rohrer

6.1   THE PREPARATION OF SCANNING TUNNELING MICROSCOPY TIPS

6.1.1   Introduction

The chemical identity and arrangement of the tip atoms that are closest to the sample surface determine the stability and resolution of the tunneling tip. These two parameters are, however, extremely difficult to control or quantify, as illustrated in Figure 6.1 with reflection electron microscopy (REM) images of a W tip before and after contacting a PbS surface. Despite this difficulty, a variety of methods have been developed empirically for the fabrication of scanning tunneling microscopy (STM) tips capable of producing atomic resolution images. Before describing these methods, it is useful to consider the "model" STM tip and the manner in which tip composition and morphology can affect an STM image.

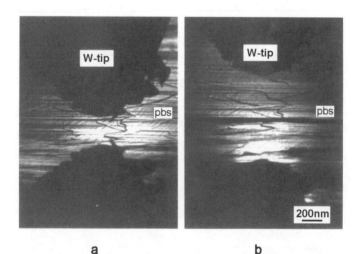

a   b

Figure 6.1. REM images showing the change in tip shape that occurs as a result of a W tunneling tip contacting a PbS surface. The images were recorded in a combined REM/STM. **a,** The tip before contact with the surface. **b,** The tip after contact. (Reprinted with permission from Ref. 1.)

6.1.2   A Description of the STM Tip

The STM tip has been described theoretically as a spherical potential well with no atomic features, as a small cluster of three or four atoms, and as a single atom (Fig. 6.2). Although the spherical potential well model works for most cases, it does not adequately predict the lateral resolution of some STM images. For example, using this model, Tersoff and Hamann[2, 3] estimated that the lateral resolution of a spherical tip with radius R, separated by a distance s from the sample surface, is $[2 \text{ Å } (R+s)]^{1/2}$. Since then, individual copper atoms that are 2.55 Å apart on the Cu(110) surface have been resolved.[10] Assuming that the resolution function given above is accurate, this experimental observation implies that the sum $R+s$ is < 1.6 Å, giving the tip radius a roughly atomic dimension and, therefore,

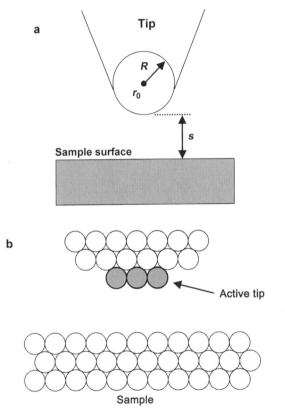

Figure 6.2. Various models used to describe the tunneling tip. **a,** Continuum model in which the tip has s-like wave functions and no atomic features.[2, 3] **b,** Atomic cluster or multiatom tip in which several atoms contribute equally to the tunneling image.[4, 5] **c,** In this case, the single atom at the apex of a crystal is responsible for essentially all of the tunneling signal. This model was used to compute tunneling current in independent electrode (large sample–tip separation) and interacting electrode (small sample–tip separation) regimes.[6] **d,** A single atom (acting as a tip) adsorbed on one of two identical metallic electrode surfaces that are described by the Jellium model. This model was used to perform numerous tunnel current calculations.[7–9]

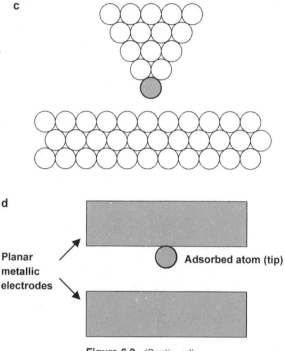

Figure 6.2. *(Continued).*

contradicting the continuum assumption. The breakdown of this model in the high-resolution limit suggests that single atom tip models may be more accurate. In fact, such models have been used to calculate the dependence of the tunnel current on sample–tip separation with considerable accuracy[7–9] and to explain high-resolution images of metal surfaces.[11, 12] The success of the single atom tip models supports the notion that the active part of the tunneling tip is limited to the atom or small atomic cluster that is nearest the surface.

Field ion microscope (FIM) observations have provided the most convincing experimental evidence in support of the single atom and atomic cluster tip models.[4] Small clusters of atoms 6 to 30 Å in diameter were observed at the ends of UHV-prepared tunneling tips and combined FIM-STM observations demonstrated that tips with smaller atomic clusters had greater resolving power on gold surfaces. The surface corrugation of the (1 × 5) reconstruction on the Au(001) surface was measured to be 0.6 Å using a tip with a cluster radius of 6.3 Å, whereas the corrugation was measured to be only 0.13 Å, using a tip with a cluster radius of 20 to 30 Å.

6.1.3 Tip Shape

Because vibration isolation is a critical factor in microscope design, a tunneling tip should be rather thick (0.5 to 1 mm) and come to a point rapidly (Fig. 6.3). Such a tip has a higher resonant frequency than a long thin one; and in practice, tips with

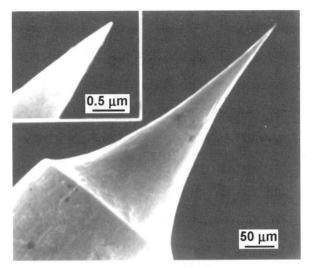

Figure 6.3. SEM micrograph showing a tip that comes rapidly to a point. This is an etched gold tip prepared by Stemmer et al.[13]

exponentially decreasing radii are rather easy to produce.[14] It must, however, be emphasized that the critical factor for tip performance is the tip's microscopic structure, which is much more difficult to control. Experimental evidence suggests that the critical components of high-resolution tunneling tips are actually the mini-tips, with radii <100 Å, that have been observed at the ends of the probes.[15–17] A transmission electron microscopy (TEM) micrograph of a polycrystalline W tip with minitips is shown in Figure 6.4a. TEM micrographs of mechanically cut Pt-Ir tips (the macrostructure of one such tip is shown in Figure 6.4b) have also demonstrated the presence of such minitips.[19] Although minitips are frequently formed

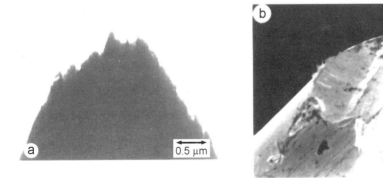

Figure 6.4. **a**, TEM image of an ion milled, polycrystalline W tip. Irregularities and minitips are easily observed. (Reprinted with permission from Ref. 18.) **b**, SEM image at the same magnification as Figure 6.3 showing the rough macrostructure of a mechanically formed Pt tip. (Reprinted with permission from Ref. 13.)

with a variety of tip preparation techniques, the origin of the structures is not clear, and there are no truly reproducible means of creating them. The shape and arrangement of the minitips are important factors that can affect the stability of the tip and limit the observable surface roughness. For example, if a tip consists of several, widely spaced minitips, each of which is close to the sample surface, the minitip from which tunneling occurs may change during imaging. This situation will often lead to disjointed and confusing images and, in special cases, can lead to interesting double images that are easily recognized by the repetition of identically shaped surface features. An example of such an image is shown in Figure 6.5. Image artifacts associated with tip shape also arise when features with sizes comparable to that of the tip are encountered on the surface. This can lead to misrepresentations

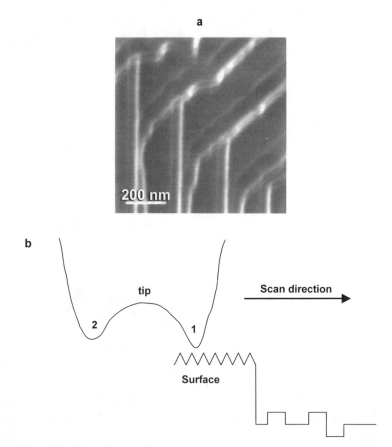

Figure 6.5. a, Double tip image of steps on a molybdenum oxide surface. The probable tip configuration that caused the image is also shown. **b,** Initially, tip 1, which is closest to the surface, traces the topography of the left side of the surface. **c,** Later, when tip 1 extends downward to reach the right-hand side of the surface, tip 2 interrupts by tunneling from the left-hand part of the surface. **d,** The double image is composed of the topography of the right-hand side of the image, traced first by tip 1 and later by tip 2.

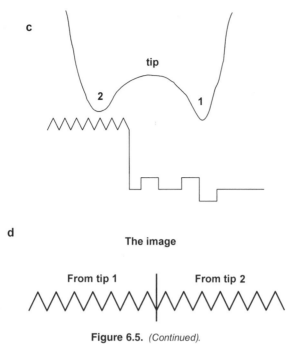

c

tip

2

1

d

The image

From tip 1 | From tip 2

Figure 6.5. *(Continued).*

of the size, shape, or height of such features; some examples of these artifacts are illustrated in Figure 6.6. For more on tip artifacts, see Pelz and Koch.[20]

The atomic structure of the tip, which is often changing and difficult to control, is also critical to STM performance. If several atoms or small clusters of atoms at the end of the tip act independently, the STM image will actually be a superposition of two or more images and, therefore, appear distorted. Images published by Park et al.[21] demonstrated that tips as close together as 3 Å and as far apart as 20 Å can act independently and produce superimposed images of the (7×7) reconstruction of the Si(111) surface. Colton et al.[5] used a model to compute the images of graphite surfaces produced by tips consisting of four or less independent atoms. By positioning the tip atoms in different arrangements and orientations with respect to the graphite lattice, the authors were able to produce a number of the different periodic structures that were observed in STM images of the graphite(0001) surface.

6.1.4 Tip Material

The chemical identity of the tip atom or atoms may also be an important factor in the performance of the STM. Metals such as Na, which have primarily s-band electrons at the Fermi level, are expected to perform differently from metals such as W, which have d-band electrons at the Fermi level. It has been proposed that higher resolution can be attained when tunneling to or from d-orbitals because electrons in these orbitals are more strongly localized than electrons in s-states.[11] The d-band tip is more sensitive to small features, because the tunneling matrix element is enhanced by the greater charge localization. There are, however, no known experimental correlations between tip material and resolution. Although most atomic resolution images have been acquired with d-band metal tips such as W and Pt, atomic resolution has also been achieved with gold tips, a metal that has primarily s-electrons at the Fermi level.[22]

Regardless of the material used, it is nearly impossible to know the chemical identity of the active atom or atoms of the tunneling tip. The atom is just as likely to be an impurity, an atom from the sample surface, or an atom adsorbed from the gas phase as it is to be from the bulk of the tip. Tiedje et al.[17] found carbon on the end of platinum tips after scanning on graphite. Tromp et al.[23] studied the effect that a

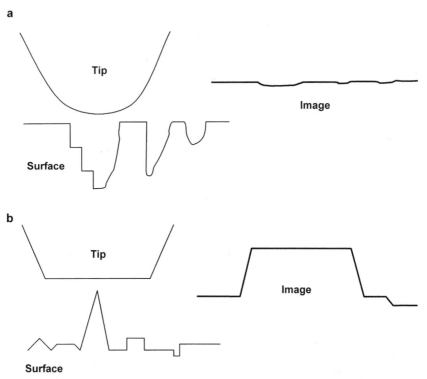

Figure 6.6. A variety of tip artifacts. **a,** A blunt tip cannot accurately image topographic changes on the surface that are smaller than the tip. Because the tip cannot fit in the valleys, the topography is not reproduced in the image. **b,** If the surface is rougher than the tip, the tip may tunnel from the same location on the surface, even though it is scanning. **c,** In this case, it is actually the surface topography of the tip that is traced. When this happens, repeated shapes (from the tip surface) are apparent throughout the image. **d,** In this case, a portion of the surface is not imaged because of a multiple tip effect. As the tip scans from left to right, the point of nearest sample–tip approach changes owing to the ledge. This artifact can usually be detected by discontinuities in the image.

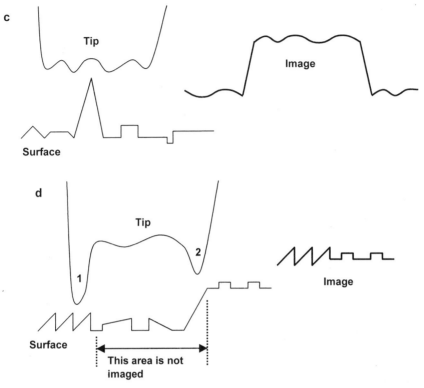

Figure 6.6. *(Continued)*.

single electronegative impurity atom has on imaging. Because an electronegative atom contributes more to the filled density of states of the tip than to the empty, higher resolution is possible at negative tip biases when electrons flow from the occupied states. The reason is that in this bias condition, the tunneling electrons come primarily from the single electronegative adatom at the end of the blunt tip. In the opposite bias, on the other hand, electrons tunnel to a large number of surface metal atoms because the adatom contributes only weakly to the unoccupied density of states. This explanation has been proposed to explain the simultaneous acquisition of atomic resolution images at negative sample biases and nonatomic resolution images at positive sample biases during current imaging tunneling spectroscopy (CITS) scans.[7–9, 23] This effect is shown in Figure 6.7.

It was mentioned earlier that simultaneous tunneling from two different tip atoms could produce a superposition image. What happens if the atoms are chemically different? If we assume that the first tip atom has a relatively symmetric distribution of the density of states around the Fermi level and the second contributes only to the density of states (DOS) below the Fermi level, then the images should be bias dependent. In this case, images acquired at negative tip voltages should appear to be superpositions, because both atoms contribute equally, whereas

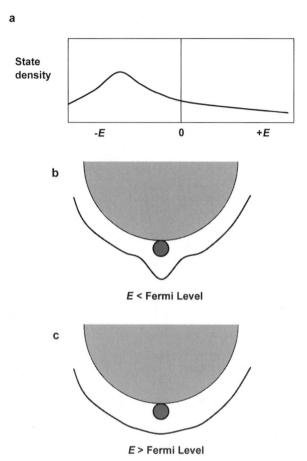

Figure 6.7. Resolving power of a tip can vary with bias if the tip is an impurity atom. These figures illustrate the electronic features of a metallic tip with an adsorbed electronegative impurity such as Na. **a,** The density of states of the electronegative impurity is centered below the Fermi level. **b,** Because of this, there is a protrusion in the state density at the position of the adsorbed atom when the bias is negative. **c,** But no such feature is seen in the positive bias condition. Therefore, at negative biases, the atom alone acts as the tip, and high-resolution images can be produced; whereas at positive biases, the tip is blunt and electronically featureless. (Modified with permission from Refs. 7 and 23.)

images acquired at positive potentials should appear as single images, because the second tip atom should not contribute to the image. This effect was observed by Park et al.[24] during CITS scans on the Si(111)−(7 × 7) surface.

Chemical reactivity and surface diffusivity can also be a basis for choosing a tip material. For example, noble metal tips such as gold or platinum are less susceptible to contamination than tips formed from easily oxidized metals such as tungsten. This can be particularly important when using STM under ambient conditions, in which a tungsten tip is expected to have an oxide coating in addition to an adsorbed layer of atmospheric contaminants. Because image quality depends on the arrangement of a single atom or small number of atoms, displacements of the atoms by surface diffusion could cause abrupt changes in image quality. Therefore, the tip is most stable when the active atoms are tightly bound to the tip surface. For example, the room temperature stability of tips formed on a W(111) surface is ensured by the high activation energy for surface diffusion of tungsten atoms, 1.8 eV.[25]

6.1.5 Tip Fabrication

In this section, methods for the fabrication of STM tips are described. Although it appears difficult to control the atomic structure at the end of the tunneling tip, a number of methods have been developed empirically that allow for the preparation of tips capable of achieving atomic resolution. The wide variety of materials and methods used vary in elegance and complexity; and although some procedures are less reproducible than others, all can be used to fabricate tips capable of producing atomic resolution images.

Mechanical Forming Binnig et al.[26, 27] produced the first tunneling tips from 1-mm-diameter tungsten wire that was sharpened mechanically by grinding it at 90° angles. Tips produced in this way had radii ranging from thousands of angstroms to 1 μm and were decorated with numerous minitips, the small structures mentioned earlier whose radii range in size from tens to hundreds of angstroms. Feenstra and Fein[28] acquired atomic resolution images of GaAs surfaces using tips produced in this way but also reported that such tips had limited stability and often exhibited abrupt changes in resolution and z-position. Mechanically sharpened Pt and Pt-Ir tips have been used with similar results. In the cases of these two materials, which are softer than W, an adequate tip can often be formed simply by clipping the end of the wire at an angle with wire cutters or scissors. A scanning electron microscopy (SEM) image of a Pt tip produced by clipping is shown in Figure 6.4**b**. Fischer and co-workers[29] used Ir tips produced mechanically.

Electrochemical Techniques Electrochemical etching is the most common method used to prepare W probes. Typically, a tungsten wire, 0.1 to 1 mm in diameter, is used as the working electrode of the electrochemical cell. The electrolyte is a 1 to 5 M KOH or NaOH solution and a carbon rod may be used as a counter electrode. The immersed portion of the W wire is etched as an ac or dc current is passed between the electrodes. Typical cell parameters are 5 to 12 V and 10 mA. Because etching occurs preferentially at the meniscus of the solution, a neck forms at the surface and the immersed portion of the tip eventually falls off. Etching is usually stopped at this point by a feedback controller that senses the reduction in current. The tip is then washed with distilled water and can be used immediately or stored in an inert environment. The surface composition and structure of W tips produced in this manner will be addressed at the end of this section.

Gold and Pt-Ir wires can also be etched electrochemically. Stemmer et al.[13] etched gold tips in a 32% HCl solution using the gold wire, immersed only 0.5 mm into the etchant, as the working electrode and a Pt-Ir ring as the counterelectrode. The gold wire was rotated at 950 rpm while a 1 kHz, 5 V peak-to-peak sine wave added to a 2.5 V dc offset was applied across the cell until the current dropped to 0. The tip was then lowered into the electrolyte for 20 s at zero bias, removed, washed with distilled water, and stored in a 0.05% aqueous azide solution. The tips were washed again with distilled water before use. Pt-Ir wires, typically 10% Ir, can be etched in a 20% KCN solution using ac voltages in the range of 3 to 15 V[30] or an acidified aqueous $CdCl_2$ solution using 3 to 10 V ac.[19]

Nicolaides et al.[14] studied the effect of a number of etching parameters on the final shape of tungsten tips. The authors found that if the portion of wire immersed in the electrolyte was too long, then the resulting tip was bent at the end. This is believed to be because the weight of the wire below the surface stretched the neck during the etching process. As a result, when the immersed portion eventually fell off, the stored elastic energy was released as a recoil that bent the tip. It was also found that the overall tip shape was parabolic if more than 1 cm of the wire was exposed to the etchant and an ac current was used. Exponential tip shapes were favored when the reaction rate or current density was high.

To further refine the shape and sharpness of the tip, the single-step polishing procedures described above can be followed by a finer electropolish, such as zone electropolishing.[31, 32] In this technique, a small quantity of electrolyte is held in a loop of metal wire that also serves as the counterelectrode of an electrochemical cell. A micromanipulator is then used to move the loop of electrolyte around the end of the tip while the process is observed with the aid of an optical microscope. Beyond the tip geometry control it affords, another useful aspect of this method is that the ends of blunted tips can be repolished with less effort and material waste then would be necessary for the preparation of a completely new tip.

If the tip is to be used to perform STM in an electrochemical cell, it must be isolated as nearly as possible from the electrolyte so that no appreciable current flows from the tip to either of the electrodes in the cell. This can be accomplished by first insulating the entire tip with a silica coating and then etching away a small portion of the coating near the tip with HF so that tunneling can take place.[33, 34]

Although electrochemical techniques allow one to exercise some control over the macro structure and microstructure of the tip, it must be emphasized that the atomic structure of the tip is far more important and that this structure is both unpredictable and difficult to control. For example, as mentioned earlier, TEM has been used to show that the ends of etched W tips are not smooth round hemispheres, but rather have small protrusions or minitips with radii of <100 Å. The same is true for mechanically formed platinum tips. Furthermore, as one would expect, the surfaces of as-formed tips are typically covered with an oxide or a layer of adsorbates, depending on the tip material. For example, the results of Auger electron spectroscopy (AES) analysis of a tungsten tip after etching have shown that the surface is composed entirely of carbon and oxygen, i.e., no tungsten signal was visible.[16] High-resolution transmission electron microscopy (HRTEM) has been used to demonstrate that W tips produced by electrochemical etching are coated with an oxide layer 3 to 10 nm thick (Fig. 6.8).[18, 19] Because it is impossible to tunnel through such a thick layer and still maintain a reasonable vacuum gap, the tip will contact the surface and be deformed until the layer is broken or removed abrasively. This process is apparently effective, because untreated etched tips have been used to produce atomic resolution images. However, removing the oxide layer by an acid etch, ion milling, or heating in vacuum[18] is likely to produce a more stable tip. Of course, tips will never remain free of contamination when used in air.

UHV Treatments   Considering the impure state of the as-prepared surface of the STM tip, it is not surprising that in situ methods for cleaning or sharpening tips in UHV have been developed. Similar results have been achieved with both elegant, time-consuming methods and with some rather simple and apparently arbitrary methods. The first in situ procedure, used by Binnig et al.,[26] involved contacting the tungsten tip to a platinum plate and then applying a 10 kHz, 20 Å peak-to-peak vertical oscillation to the tip. The authors proposed that this procedure ultra-sonically cleaned the tip. A more common procedure is to apply a short-duration

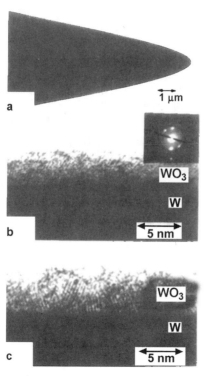

Figure 6.8. HRTEM images showing the WO_3 shell that surrounds an electrochemically etched W tip. **a**, Macrostructure of the tip. **b**, Tip detail showing thin surface oxide layer. **c**, Specimen showing polycrystalline ordering in the oxide layer. (Reprinted with permission from Ref. 18.)

voltage pulse to the tip. Biases reported in the literature for such pulses range from 6 V to 5 kV. Another common technique is to operate the tip in the field emission regime for several minutes. Typical conditions for this procedure are to operate the tip–sample junction in the bias range of 10 to 100 V and the current range of 10 nA to several mA. In either case, the procedure may be performed on a portion of the sample that is not to be analyzed or on a piece of metal reserved for this purpose. Another common cleaning method is heating the tip to $\geq 900°C$ in UHV. The fact that atomic resolution has been obtained on etched tungsten tips with no further treatment reminds us that continuous scanning is another simple, albeit arbitrary, way to prepare a good tip.

A number of SEM, TEM, FIM, and AES studies have been undertaken to determine how the treatments described above affect the structure and chemistry of the tip. Electron microscopy observations have shown that high-voltage pulses, heating, and field emission treatments cause the ends of tungsten tips to melt and that, although the new tip is globally more blunt than the as-etched tip, small minitips are formed when the tip solidifies.[15] The fact that the active region of the tunneling tip was actually a minitip, with a radius of approximately 10 Å, was verified by driving a cleaned tungsten tip several hundred angstroms into a silver surface and then removing it and scanning the indentation left behind. FIM observations of the atomic structure of a tunneling tip after a high-current cleaning procedure showed that a cluster 20 to 30 Å in diameter was formed at the apex of the probe.[4] AES has shown that oxygen is also removed from tungsten tips by high-voltage pulses and that after such treatments, the surfaces are composed of tungsten and carbon.[16] Minitips formed from carbon deposits have also been observed on the end of tunneling probes.[17] Both long fibers and graphitic carbon deposits have been detected on tips after scanning on graphite. It is not clear, however, whether these deposits form from impurities already on the tip or by the transfer of carbon to the sample.

One tip sharpening procedure noteworthy for its ability to reproducibly make a high-resolution tip was described by Wintterlin et al.[35] An Al(111) surface was scanned with a blunt, electrochemically etched tungsten tip cleaned in vacuum by field desorption. The tip bias was increased from -0.5 to -7.5 V for four lines while scanning. The tip displacement indicated that during this period, the tip became approximately 25 Å longer. When the bias was reduced, the resolution of the tip was much higher and individual Al atoms on the surface, separated by 2.86 Å, could be imaged. Although the high-resolution state of the tip is unstable, the process can be repeated. The mechanism by which this procedure enhances tip resolution is not clear.

Ion sputtering is a somewhat more sophisticated method of fabricating tunneling tips. Biegelsen et al.[18] used argon ion sputtering to remove the oxide layer from electrochemically etched, single crystal, and cold-drawn polycrystalline tungsten wires and have found that no detectable oxide (detectability limit of HRTEM $\sim$ 10 Å) grows after being exposed to air for up to 1 day. Although the method apparently works well for single crystal wires, anisotropic sputtering of the polycrystalline wires often leads to rough tips.[36] Micrographs of the ion milled, single crystal tips are shown in Figure 6.9. Tiedje et al.[17] reported that clean tips capable of atomic resolution imaging could be produced by simultaneously ion milling and sputter coating etched tungsten tips with gold, platinum, or nickel. A somewhat simpler approach is to simply coat an etched W tip with a more inert metal, such as gold, with no previous ion milling. This procedure has been used to fabricate tips capable of atomic resolution imaging.[22]

The in situ use of field ion microscopy to directly observe and control the atomic structure of the tunneling tip is certainly one of the most elegant methods applied to tip fabrication and characterization. Kuk and Silverman[4] etched a 1-mm-diameter (100) oriented W single-crystal wire and sharpened it in vacuum by heating in the presence of a high electric field. With this procedure, a pyramidal tip is formed by the intersection of the low-energy {110} crystal faces and the sharpness of the tip can be monitored with FIM. The authors found that although monomer, dimer, and trimer tips were often unstable, a six-atom cluster was stable and capable of resolving the (1 × 5) reconstruction of the Au(001) surface. Fink,[25] on the other hand, produced monoatomic tips with (111) oriented tungsten wire with a more complex procedure. First, a (111) surface capped with a stable trimer of tungsten atoms was formed by field evaporation. Next, a tungsten evaporator was used to deposit a single tungsten atom onto the trimer from the gas phase. Because of the high

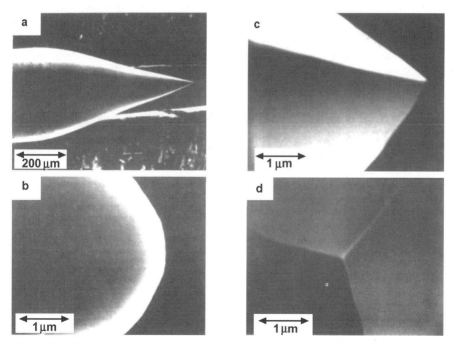

Figure 6.9. SEM images of tungsten ⟨111⟩ wire. **a**, Macrostructure of a 0.5 mm wire after electrochemical etching. **b**, Tip after electrochemical etching. **c**, Side view of the tip after ion milling for 8 h at 4.5 keV at 30° below horizontal. **d**, Axial view. The apparent nonlinearities in the facet edges are owing to an artifact of the SEM video display.[36]

activation energy for diffusion on this surface, this single atom tip should be stable at room temperature.

In some cases, tips formed mechanically, with no additional treatment, perform as well as those carefully formed in situ. There is little evidence at this time to suggest that even the most careful tip preparation will significantly increase the resolution of an STM experiment. However, it should be emphasized that a carefully prepared tip will generally exhibit greater reliability and stability. The ease with which an adequate tip may be formed was demonstrated by Colton et al.[37] who resolved the atomic surface structure of graphite using a piece of broken pencil lead (polycrystalline graphite).

6.1.6 Tip Performance

There are several criteria that can be used to evaluate the performance of the STM tip. The easiest and most common is simply the quality of the images obtained with the tip. If atomic resolution is attained, the tip is obviously adequate. If atoms are not resolved, but single atomic steps are, the resolution can be estimated from the step width, which should be abrupt. Tips with poor stability often exhibit abrupt changes in resolution. In the image in Figure 6.10, tip resolution changes abruptly in the lower portion of the image and subtly in the upper portion. It has already been mentioned that tip morphology (blunt tips, multiple tips) can cause image artifacts such as discontinuities and abrupt changes in vertical position (Fig. 6.6). Pelz and Koch[20] suggested that the simultaneous acquisition of STM and scanning tunneling potentiometry images can help detect such phenomena. A more quantitative standard of tip quality was proposed by Tiedje et al.,[17] who used a spectrum analyzer to characterize the current noise from a variety of tunneling tips. The authors suggested that the noise in the current is a characteristic of the tunnel junction and is caused, in part, by adsorbate interactions at the surface and the tip.

A second aspect of tip quality is the chemical purity and homogeneity of the tip surface, an especially important factor for atomic resolution imaging and spectroscopic measurements in UHV. The greatest vertical sensitivity is achieved when there is a vacuum gap between the tip and sample surface. If the surface of the tip is covered with reaction or contamination layers, and the tip contacts the sample through an insulating layer, the tunnel barrier height is decreased and the sensitivity of the tunnel junction to small displacements is reduced.[40] This possibility is illustrated in Figure 6.11. The only way to be certain that there is a vacuum gap between the sample and the tip is to measure the variation of the tunnel current with tip displacement. From such data, the apparent barrier height ϕ can be calculated using a simple expression for the tunneling current $i = c \exp(-s\phi^{1/2})$. Apparent barrier heights of approximately 4 eV indicate that there is a vacuum gap at the junction, whereas lower barriers indicate that tunneling is occurring through an insulating layer. One reason for the reduced sensitivity of the tunnel junction when the tip is in contact through an insulating layer is that the displacements not only occur in the tunnel gap but are taken up by the sample, tip, and insulating layer as the whole

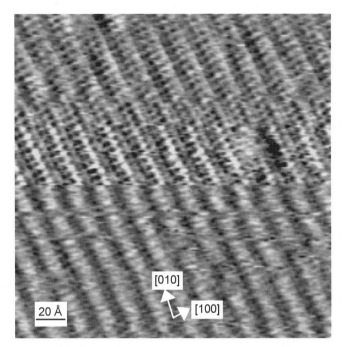

Figure 6.10. Several changes in tip resolution are evident in this constant current STM image of a cleaved $V_2O_5(001)$ surface. Although atomic scale features are visible throughout the image, the highest resolution is achieved in the center of the image. We assume that the abrupt changes in image contrast are the result of changes in the arrangement of the atoms at the end of the Pt-Ir tip, e.g., as a result of surface diffusion or a sample–tip interaction. For an explanation of the image contrast, see Smith et al.[38, 39]

system is compressed. Therefore, the gap distance s changes by only a fraction of the total displacement.[40]

One noteworthy method for determining the cleanliness of a metal sample–tip junction was described by Gimzewski and Moller.[41] They noted that for nanometer scale metal–metal contacts, cohesive bonding should occur between the sample and the tip. They found experimentally that when a clean tip was retracted from a clean sample surface after contact, a protrusion was left behind on the surface from the material that was pulled out under tensile stress as the tip was retracted. If the surfaces were dirty, however, the cohesive force was reduced, so bonding did not occur. In this case, a depression was left behind after retraction of the tip.

This last example hints at the importance of sample surface preparation, which varies greatly depending on the material analyzed and the conditions of interest. Common sample preparation techniques are described and evaluated in Section 6.3 chapter. In the next section, the characteristics and preparation of scanning force microscopy (SFM) tips are described.

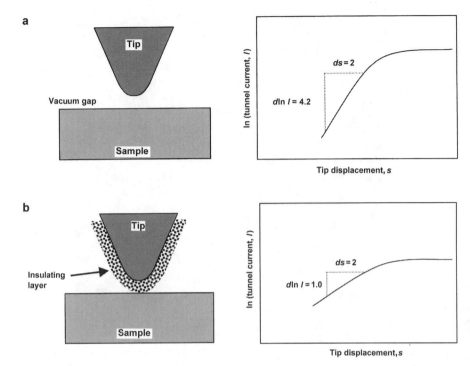

Figure 6.11. The differences between a vacuum gap and a contact gap. **a**, As the tip is ramped toward the sample, the current should increase exponentially as the separation is decreased. The apparent barrier height can be computed by graphing $d(\ln I)$ vs. s. **b**, In this case, the displacement of the piezo actuator does not occur exclusively in the gap. The actual displacement is related to the compressibilities of the constituents. The apparent barrier height is reduced, making the tip less sensitive to small changes.

6.2 THE PREPARATION OF SCANNING FORCE MICROSCOPY TIPS

SFM probes generally consist of a sharp tip mounted on the end of a small, flexible cantilever. When the probe is scanned over a sample within approximately 20 nm of its surface, the cantilever will deflect in response to attractive and repulsive forces acting between the tip and the surface. If, on the other hand, the cantilever is oscillated (near its resonant frequency) during scanning, the apparent resonant frequency of the cantilever will change in response to changes in the gradient of the forces (∇F) acting between the tip and the sample. By detecting, processing, and feeding back on the changes in the cantilever's deflection or resonant frequency, as the probe is rastered over a sample, the topography or properties of the surface can be mapped.

In principal, any force interaction sufficiently large to produce measurable changes in the cantilever's deflection or resonant frequency can be used to image, and a family of force microscopy techniques has evolved since the initial development of contact atomic force microscopy (AFM).[42] These techniques include non-contact (NC) and intermittent contact (IC) AFM, magnetic force microscopy (MFM), and lateral force microscopy (LFM). Related techniques that rely on the ability of SFM to accurately track surface topography while another surface property, such as capacitance or temperature, is measured have also emerged. Finally, the force sensitivity of the SFM has been exploited to measure and quantify a variety of interaction forces, such as those between complimentary DNA strands and similar and dissimilar organic functional groups.

Obviously, the shape and sharpness of the probe tip will determine the ultimate resolution of the SFM. Equally important, however, is the sensitivity of the cantilever to the tip–sample force interactions of interest. The first SFM probes were sharpened metal wires or consisted of a small piece of diamond glued to the end of a cantilever manually shaped from a metal foil or wire.[42] Soon after, microfabrication techniques, based on Si technology, were adapted to the batch fabrication of entire wafers of microscopic Si and Si_3N_4 cantilevers[43] and integrated cantilevered-tips.[44, 45] Microfabricated probes offer a number of advantages over hand-fabricated probes. First, the microfabrication techniques are more reproducible and less labor intensive. More important, the size and mass of the cantilever can be greatly reduced; and as a result, stable probes with both low spring (force) constants and high resonant frequencies can be formed.[45] Today, nearly all SFM probes are produced with microfabrication techniques; and in most cases, the average SFM user purchases standard probes from a commercial vendor rather than fabricating them personally.

In the following sections, the preparation and physical characteristics of probes that can be applied for a variety of SFM techniques are described. First, the microfabricated probes available commercially for the most common modes of SFM operation (i.e., contact, NC, and IC AFM) are described. Next, procedures that can be used to modify standard probes for related SFM techniques, such as MFM, are outlined. Finally, some of the challenges associated with the interpretation of SFM data are addressed. This discussion will focus, in particular, on image artifacts arising as a result of the finite shape and size of the SFM tip.

6.2.1 Commercially Available SFM Probes

In this section, the characteristics of commercially available SFM probes that can be used for contact, NC, and IC AFM imaging are described. Some simple guidelines for selecting probes, based on the characteristics of the cantilevers and tips, are also outlined. The geometric descriptions of the probes provided here are not specific to any single manufacturer or supplier. Instead, they are general guides to the many probes available commercially and have been deduced from the nominal probe specifications reported by commercial suppliers.

For the most part, commercially obtained probes are ready to use when they are received. They may be delivered in the form of a large wafer containing hundreds of probes, a small strip of probes that has been separated from a wafer, or a small case of individual probes. If the probes are supplied in wafer (or strip) form, each probe must, of course, be separated from the wafer before it can be used. This can usually be accomplished by carefully breaking the probe along trenches or scores that were etched into the wafer during the microfabrication process. An optical micrograph of a typical SFM probe is presented in Figure 6.12 along with a schematic representation. The large rectangular section, commonly referred to as the chip, provides a carrier for the probe and is usually either silica glass or silicon. One or more probes may be mounted on the end of the chip. The chip in Figure 6.12 has two probes, and these are the small triangles indicated by arrows in the figure. The backs of the cantilevers, Si_3N_4 in particular, are frequently coated with a reflective,

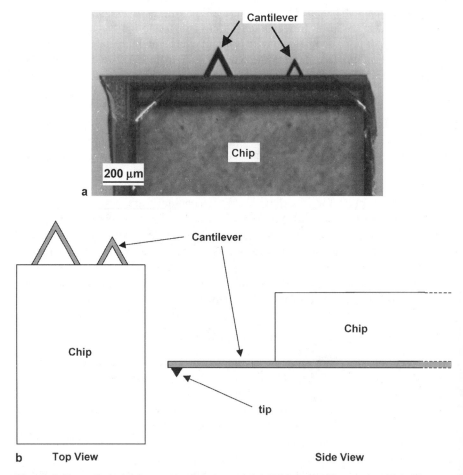

Figure 6.12. a, Optical micrograph of a commercial SFM probe (Nanoprobe, Digital Instruments) with two cantilevered tips. The tips are not visible in the micrograph but are located at the very ends of the V-shaped cantilevers. **b**, Representations of the SFM probe.

metal (Au or Al) film to enhance the intensity of the reflected laser signal in microscopes in which force detection is based on optical beam deflection or interferometry. The tip, which is located at the end of the V-shaped cantilever, is not visible in the micrograph but is shown in Figure 6.12**b**.

SFM probes are fragile and should be handled with care at all times. Even the slight swipe of a finger across the probe will invariably break the cantilever from the chip and render the probe useless. Probes can be stored under ambient conditions indefinitely but should be kept free from dust, which can stick to the cantilever or tip and lead to unpredictable and irreproducible performance. If a probe is suspected to be dirty or contaminated, it can be cleaned gently in an ultrasonic bath. Now that the probe and its care have been introduced, SFM cantilevers and tips are addressed separately and some simple guidelines for selecting cantilevers and tips for SFM experiments are presented.

SFM Cantilevers Commercial SFM probes are fabricated exclusively of Si, SiO_2, and Si_3N_4. Two basic cantilever geometries are available, the V-shaped and the beam cantilever (Fig. 6.13). V-shaped cantilevers are often preferred for contact AFM imaging because they offer greater mechanical stability with respect to torsional forces that can develop as the tip is scanned over the sample in contact with the surface.[46] On the other hand, when these torsional or frictional forces are of interest, for example in LFM, beam cantilevers may be preferred. Both types of cantilevers are readily applicable for NC and IC AFM imaging.

A cantilever's mechanical properties are determined by its geometry, its dimensions, and the material from which it is fabricated. The force (spring) constant C and resonance frequency f_o of a beam cantilever are given by the following equations:[45]

$$C = \frac{E}{4} \times \frac{w \times t^3}{l^3} \tag{6.1}$$

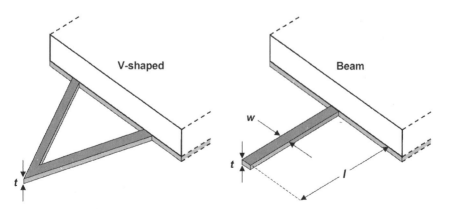

Figure 6.13. V-shaped and beam SFM cantilevers.

$$f_{\mathrm{o}} = 0.162 \sqrt{\frac{E}{\rho}} \times \frac{t}{l^2} \qquad (6.2)$$

where E and ρ are the Young's modulus and density, respectively, of the cantilever fabrication material and w, t and l are the width, thickness, and length of the cantilever, respectively, indicated in Figure 6.13. By varying the geometry, dimensions, and material, cantilevers are fabricated commercially with C ranging from 0.01 to 200 N/m and f_{o} between 1 and 400 kHz. Although it's trivial to calculate the spring constant and resonant frequency with Eqs. (6.1) and (6.2) and the manufacture's nominal dimension specifications, these values should be taken only as a rough estimate because cantilever dimensions, in particular thickness, can vary significantly from probe to probe across a fabricated wafer. If accurate values are needed, e.g., for the quantification of measured forces, the cantilever must be calibrated; a number of procedures have been outlined in the literature.[47–49]

Cantilever selection should be guided by the SFM technique for which it will be applied and the mechanical stability of the sample of interest. Because the tip is in direct contact with the sample during contact AFM imaging, large forces can potentially be applied to the surface. The possibility of inelastically deforming or otherwise modifying the surface during analysis can be minimized by choosing a cantilever with a low spring constant ($C < 1$ N/m). Although surface modification may be unlikely for materials that are harder than the tip, the tip can also be modified during analysis and high, applied forces can lead to abrasion and blunting of the tip. For NC and IC AFM imaging, stiffer cantilevers ($C > 5$ N/m) with high resonant frequencies ($f_{\mathrm{o}} > 100$ kHz) are usually preferred. By selecting a cantilever with a high resonant frequency, the signal-to-noise ratio of the microscope can be maximized. With a stiffer cantilever, the probability of the cantilever bending and then contacting and sticking to the sample's surface under the influence of attractive forces is reduced.[50] This can be a particular problem during ambient imaging because the layer of atmospheric adsorbates that invariably coats the tip–sample interface gives rise to a strong, attractive, capillary force.[51]

SFM Tips The SFM tip is usually fabricated at the same time as the cantilever on which it is mounted and, therefore, is usually made of the same material as the cantilever. The as-fabricated geometry of the tip is dictated by the microfabrication technique employed and the three basic geometries available are pyramidal, conical, and tetrahedral. Si_3N_4 tips are, apparently, exclusively pyramidal, whereas Si and SiO_2 tips have been fabricated in all three geometries, with conical being the most common. Following fabrication, the tips may be subjected to one of a number of additional, often proprietary, treatments to either sharpen them or modify their geometry.

In addition to its outward shape, important geometric parameters of the SFM tip include its radius of curvature r, the sidewall angles θ, and its length l_{tip} (Fig. 6.14). In general, tips do not terminate at an atomically sharp point. Frequently, the very end of the tip is approximated as a hemisphere that can be described by a radius of

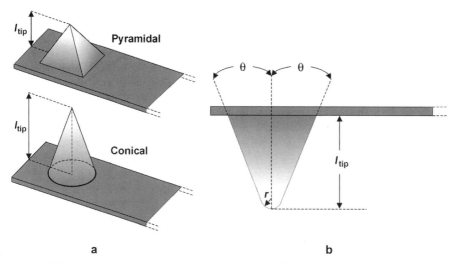

Figure 6.14. **a**, Pyramidal and conical SFM tips. **b**, The SFM tip showing the tip's length l_{tip}, radius of curvature r, and sidewall angles θ.

curvature r. The manufacturer's nominal radii specifications can be taken as a rough estimate of the tip's sharpness. It should be noted, however, that the value is difficult to measure accurately and will vary from tip to tip across a wafer. Furthermore, the size of the area of the tip that interacts with the sample may vary with imaging conditions. Much like STM, the best way to gauge the sharpness of the tip is to image with it. A well-characterized standard specimen, having sharp features of known size and shape, is particularly useful for gauging the sharpness and shape of the SFM tip. Recently, Digital Instruments (Santa Barbara, Calif.) released an automated analysis package that evaluates tip size and shape based on SFM images obtained with the tip of interest.

The tip's sidewall angles and length influence its ability to accurately image steep slopes and measure the depths of trenches and pits on the surface. For example, to accurately measure the depth of a narrow 1-μm-deep pit, the tip must be at least this long. Furthermore, the tip must be sufficiently narrow that it can completely penetrate the pit and reach the bottom. The sidewall angles and geometry of the tip will determine the tip's cross section at a given penetration into a crevice. If this exceeds the cross section of the depression before the probe detects the bottom of the feature, the depression's depth will be underestimated. This possibility was shown in Figure 6.6 for STM tips. Image artifacts associated with tip size and shape are discussed in more detail in Section 6.2.3.

The pyramidal Si_3N_4 tips frequently employed for contact AFM imaging have a standard, and rather large, sidewall angle of 35°. Therefore, these tips are not particularly well suited for quantitative imaging on rough surfaces with deep (>1 μm) and narrow (<100 nm) valleys or holes. On the other hand, they are well suited for relatively flat surfaces, are exceptionally wear resistant compared to Si tips, and can be used for imaging atomic scale contrasts in contact mode.[39, 52] Standard Si_3N_4 tips have a nominal radius of curvature between 20 and 60 nm, whereas sharpened probes may have radii between 5 and 40 nm.

Si tips can be etched or focus ion beam (FIB) milled to modify their sidewall angles, radii of curvature, and aspect ratios. The etched silicon probes used for contact AFM imaging typically have nominal radii of curvature <20 nm and sidewall angles substantially smaller than those of pyramidal Si_3N_4 probes. Identical tips with shorter and stiffer cantilevers are also frequently employed for NC and IC AFM imaging. When surfaces with large sidewall angles (approaching 90°) or deep and narrow gaps are to be probed with NC AFM, the user can turn to Si probes that have been FIB beam machined to have sidewall angles approaching 0°.

6.2.2 Probes for Related SFM Techniques

Magnetic Force Microscopy In MFM, the spatial variation of magnetic fields emanating from a sample's surface are imaged by monitoring the interaction between the sample and a magnetic SFM probe. These long-range magnetic forces are typically monitored while the tip is maintained at a distance (on the order of 5 nm or more) from the sample's surface; and in most cases, detection is based on monitoring shifts in the cantilever's effective resonant frequency. The first probes applied for MFM were electrochemically sharpened wires fabricated from magnetic metals such iron or nickel.[53, 54] A second approach was to sputter coat an etched, nonmagnetic wire with a thin layer of a magnetic alloy.[54] Because this reduces the volume of magnetic material that can interact with the sample, the strength of the stray field emanating from the tip, which may perturb the sample's magnetization, is reduced, and microscope resolution is thought to be enhanced.[54]

Keeping with the advantages that smaller magnetic probes offered, thin magnetic films were then applied to microfabricated Si and Si_3N_4 probes.[55] Because resonant detection is often employed, probes suitable for NC and IC AFM are usually used. Grütter and co-workers[55] prepared both magnetically hard and soft MFM probes by coating Si probes with 15-nm films of Co, $Co_{71}Pt_{12}Cr_{17}$, and $Ni_{80}Fe_{20}$ via evaporation or sputter deposition. Although the authors coated entire wafers of batch fabricated probes, the techniques are equally applicable to the preparation of individual MFM probes. Babcock and co-workers[56] explored the effect of film thickness on the performance of selectively etched Si probes that were RF sputter coated with $Co_{85}Cr_{15}$. The sensitivity, as measured by the effective magnetic moments, of these probes increased linearly with film thickness up to 50 nm, where it saturated, and the tips had a resolution of 50 nm or better. The imaging of low coercivity materials requires special care because the stray field emanating from the MFM probe can potentially induce changes in the magnetization of the sample.[57] Approaches to overcoming these problems include applying a thinner magnetic coating to the probe,[57] employing a soft magnetic coating,[56] weakening the tip's moment by demagnetizing it, and increasing the tip–sample separation during imaging.

Electrostatic Force Microscopy With electrostatic force microscopy (EFM), static charges on a surface can be imaged by scanning a tip that is biased, relative to the sample, across the surface. Because this requires that a voltage be applied across the tip–sample interface, conductive SFM probes are required. In many cases, the conductivity of commercially available Si probes has been found to be sufficient for EFM imaging.[58–60] For example, Nyffenegger and co-workers[60] used Si probes to characterize silver nanocrystals on graphite substrates and Bluhm et al.[59] used Si probes to image domain-inverted gratings in $LiNbO_3$ crystals. Si_3N_4 probes can be coated with a thin metal layer to impart the necessary conductivity. For example, Bluhm and co-workers[58] coated commercially available Si_3N_4 probes with a 15-nm layer of gold and used these tips to study the domain structure of $C(NH_2)_3 Al(SO_4)_2 \cdot 6H_2O$ ferroelectric single crystals.

Scanning Capacitance Microscopy Scanning capacitance microscopy (SCM) can be used to image the spatial variation of a surface's capacitance and has been applied to profile dopant concentrations in semiconducting materials. Like EFM, SCM relies on biasing an electronically conductive tip relative to the sample during scanning; however, in SCM, the tip is maintained in contact with the sample's surface during imaging and the sample's capacitance and topography are mapped simultaneously. The first probes used for SCM imaging were sharpened metal wires.[61, 62] More recently, commercial Si or Si_3N_4 probes (suitable for contact AFM imaging) that have been coated with a thin metallic film have been applied for SCM. Kopanski et al.[63] used Si_3N_4 cantilevers coated with $\sim$20 nm of Cr or Ti as well as commercially available MFM probes—Co/Cr-coated Si probes—to probe dopant profiles in Si substrates and lateral p–n junctions. The authors connected the probes to their capacitance sensor with conductive silver paint and a thin enamel-coated wire. Kang and co-workers[64] also used Ti- and Cr-coated probes to probe dopant profiles in Si. The authors applied their films by electron beam evaporation and annealed the probes before use.

Chemical Force Microscopy—Functionalized SFM Probes The force sensitivity of SFM can be used to probe chemical and molecular interactions such as those between ligand–receptor pairs,[65] organic functional groups,[66–68] and DNA strands.[69] The technique has been used to directly measure the energy of specific binding events and, in some cases, to map out the spatial distribution of functional groups on photolithographically prepared substrates. The first step of most experiments involves the preparation of surfaces and SFM probes that have been functionalized with the molecular species of interest. Once the model probe and sample are formed, cantilever force vs. distance curves are obtained as the tip is approached to and then retracted from the surface. If the mechanical properties of the cantilever are well characterized, the force vs. distance data can be used to quantify the interaction forces of interest. Functional groups have been mapped spatially by relying on the fact that different functional groups have different adhesion forces that give rise to contrast differences in LFM images.[66]

Florin and co-workers[65] measured the adhesion force between Si_3N_4 tips derivatized with avidin and agarose beads functionalized with biotin, desthiobiotin, and iminobiotin. Biotinylated bovine serum albumin (BSA), which nonselectively adsorbs on the Si_3N_4 probes, was used to bind the avidin to the probe by first soaking the sample in BSA and then avidin. Using a similar molecular immobilization technique, Lee et al.[69] measured the interchain forces between complementary strands of DNA. Immobilization was accomplished by covalently attaching thiolated oligonucleotides to self-assembled monolayers (SAMs) of γ-aminopropyla-minoethyltrimethoxysilane on the silica probes and substrates. SAMs have also been used to produce probes and surfaces that are terminated with specific functional groups. For example, $-CH_3$ and -COOH terminated probes have been prepared by simply immersing Au-coated, pyramidal Si_3N_4 probes in 1 mM ethanolic solutions of octadecylmercaptan and 11-thioundecanois acid, respectively, for 2 h.[66] Gold coatings were applied to commercially obtained Si_3N_4 probes by thermal evaporation of a 30 Å Cr adhesion layer followed by a 1000 Å Au layer.

Probing Surface and Colloidal Forces   SFM has also been used to probe the force interactions between similar and dissimilar materials in both the gas and the liquid phase. In order for the force measurements to be quantitative, a reliable knowledge of the geometry over which the force interactions occur is a necessity. Furthermore, one would also like to use a simple geometry that is amendable to calculations or simulations, the results of which can then be compared to the SFM force measurements. These requirements are usually met by modifying the SFM tip geometry. One of the most popular techniques relies on attaching a small sample of known geometry, usually spherical, to the end of the probe and then monitoring the force interactions between the modified probe and a planar substrate.

In light of the availability of microscopic silica spheres, it is not surprising that the silica system was one of the first studied. Ducker and co-workers[70] investigated the colloidal forces between small silica-glass spheres and planar silica substrates immersed in NaCl solutions. The model probes were prepared by gluing a silica sphere ($r \sim 3$ to $4 \mu m$) to the end of a V-shaped Si_3N_4 cantilever with approximately 10^{-15} l of Epikote 1004 resin. Positioning of the sphere was facilitated by the use of a three-dimensional (3-D) microtranslation stage. The planar SiO_2 substrate was formed by oxidizing a Si wafer to a depth of 30 nm. Using a similar approach, Larson and co-workers[71] measured the force interaction between silica spheres (attached to the cantilever) and a rutile TiO_2 crystal of unspecified orientation. Biggs and Mulvaney[72] measured the forces between gold surfaces in water with gold-coated SFM probes and smooth gold substrates. Commercial Si_3N_4 probes and Si_3N_4 probes that had a small glass sphere glued to their tip were sputter coated with a 0.6-μm layer of gold.

6.2.3   SFM Tip Performance

In Section 6.1.3, some of the tip artifacts that can arise as a result of the finite shape and size of the STM tip were introduced. Artifacts are probable when surface features with dimensions comparable to or smaller than that of the tip are encountered and lead to distortions of the shape, size, and vertical scale of both protrusions and depressions on the surface. Artifacts and distortions are perhaps more prevalent in SFM because the techniques can often be applied with ease to surfaces that would be too rough for reliable STM characterization. Although SFM may be easier to apply on such rough surfaces, image interpretation can often be a challenge. Figure 6.15 provides an example of the difficulty one can encounter in the interpretation of SFM data. Both of the images in the figure were acquired in the same area on the same TiO_2 thin film but with two different commercial SFM probes. Figure 6.15a was acquired using a probe with a pyramidal Si_3N_4 tip, and Figure 6.15b was acquired with a conical Si tip. Clearly, the surface morphology appears different in each image and is convoluted with the tip shape.

Even when features with dimensions significantly larger than the tip are imaged, the user must be wary of distortions if quantitative data are to be retrieved from the image. Errors in quantitative measurements are most serious when attempting to measure concave shapes. The central issue is illustrated in Figure 6.16 for the case of a thermal groove at a grain boundary. As the AFM probe descends into the groove, its vertical position is actually determined by a point on the side of the tip, not by the point at the end of the tip. For relatively steep and/or narrow features, this leads to a systematic underestimation of the depth and slope of the bounding wall. The magnitude of this systematic error depends on the aspect ratios (width to depth) of the feature and the tip. Established simulation methods make it possible to determine this systematic error.[73–75] For example, Figure 6.17 shows the results of

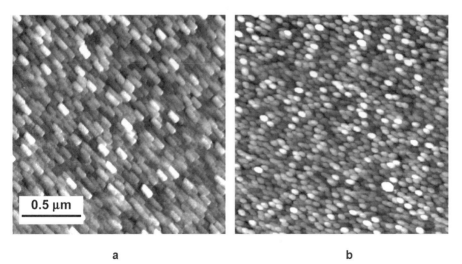

a   b

Figure 6.15. Contact AFM images of the same area of a (001) oriented TiO_2 thin film deposited on an $Al_2O_3(10\bar{1}0)$ substrate by sputtering. **a**, AFM image aquired with a pyramidal Si_3N_4 tip (Digital Instruments). **b**, Image acquired with a conical, etched Si tip (Digital Instruments). Both images are on the same scale and in the same surface orientation. In both cases, when the sample was rotated, the orientation of the surface features did not change. Clearly, the tip shape is convoluted into the surface morphology in both images.

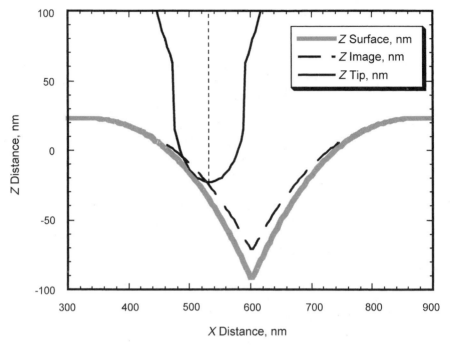

Figure 6.16. The depths of narrow, deep grooves are systematically underestimated in contact AFM topographs because the point at which the tip contacts the groove surface is not on the tip axis; thus the slope of the concave shape and its depth are systematically underestimated. Note that the vertical axis is amplified with respect to the horizontal axis; this aspect ratio distorts the actual shape of the tip, which has a sidewall angle of 35°.

the simulated imaging of an ideal thermal groove by a pyramidal AFM tip. Although the width of the feature is accurately measured, the depths and slopes of the sidewalls are systematically underestimated. In summary, whenever quantitative dimensional information is to be retrieved from contact AFM images, simulations should be used to determine the potential errors associated with the measurement.

6.3 SAMPLE SURFACE PREPARATION

In this final section, methods for preparing sample surfaces for SPM study are reviewed. The primary focus here is methods of producing atomically flat surfaces suitable for high or atomic resolution imaging. As a result, discussion will focus primarily on surface preparations used for STM experiments. We note, however, that the surfaces that have been or could be studied with SFM are essentially limitless; and when gross topography is of interest, special preparations are not generally required.

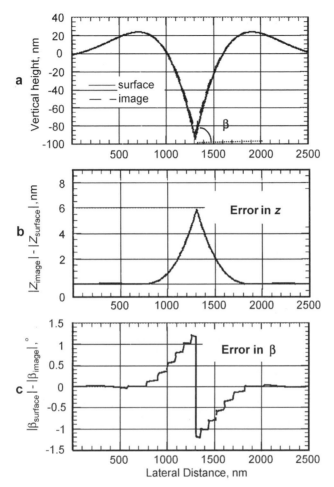

Figure 6.17. a, Simulated contact AFM image (in cross section) of an ideal thermal groove at a grain boundary. The ideal groove has a width of 1.2 μm and $\beta = 25°$. Image simulation is for the case of a pyramidal AFM tip with 35° sidewall angles and a 60-nm radius of curvature. **b,** Error incurred in the measurement of the depth of the groove. **c,** Error incurred in the measurement of the groove's slope, β. Under these conditions, the error in each measurement is approximately 5%.

6.3.1 Metals

In Air Topographic maps of metal surfaces can often be acquired in air with no prior surface preparation, although the surface chemistry of such a specimen is usually questionable. Surprisingly, it is also possible to make atomic resolution images of properly prepared noble metal surfaces in air. Hallmark et al.[76] reported the first such observation on the Au(111) surface. Gold films of 2500 Å thickness were evaporated epitaxially at a rate of 5 Å/s onto cleaved mica substrates maintained at 300°C. Before examination in air, the samples were allowed to cool in the evaporator system at a pressure of approximately 1×10^{-6} Torr. AES examination of these samples after introduction into an UHV chamber indicated a substantial level of surface contamination by carbon and atomic resolution images were not produced until after the surface had been ion sputtered and annealed. Although this seems to suggest that the surface actually imaged in air was an adsorbed layer, the authors identified tip instabilities as the cause of decreased resolution in UHV and emphasized the similarity of the images they obtained in air with those obtained from clean samples in UHV. Kim et al.[77] verified these observations of atomic resolution in air in a later study of similar thin films of copper, silver, and gold evaporated onto graphite. It is interesting, however, to note that atomic resolution images of these metals have not been obtained from bulk samples in air.

In UHV Clean metal surfaces can usually be prepared for UHV SPM analysis in the same manner as they would be for analysis using other surface sensitive techniques. Some examples of these preparations are outlined here. The (100) surface of a gold crystal should be polished mechanically or electrochemically before introduction to the UHV chamber. The surface should then be ion sputtered and heated to 650°C to remove impurities and form the (1×5) reconstruction.[78] The compositional purity of the surface can be checked using AES and the crystallinity can be verified with low energy electron diffraction (LEED). Clean Al(111)–(1×1) surfaces can be prepared by ion sputtering followed by annealing at 400°C,[35] Cu(100)–(1×1) surfaces by ion sputtering followed by annealing at 500°C,[10] and Au(111)–(1×1) surfaces by ion sputtering followed by annealing at 360°C.[76] In each case, the surface preparation procedures involve the same basic steps. The sample is first mechanically polished to a smooth finish. It is then cleaned in the vacuum chamber by sputtering away the contaminated surface layers with inert gas ions. Finally, the surface is recrystallized by annealing. An alternate approach, which eliminates the lattice disorder introduced by ion sputtering, is to use a chemical etch rather than sputtering to clean the surface. For example, de Lozanne et al.[79] cleaned Nb_3Sn thin films by etching them for 15 min in an aqueous 10% HF solution. Such ex situ methods, however, are only suitable for relatively unreactive surfaces.

Although AES is normally used to check the cleanliness of a surface, *I-V* curves acquired with the STM itself can be used to check the cleanliness of both the metal sample and the tip at the tunnel junction. *I-V* curves of clean metal-vacuum-metal tunnel junctions are continuous at low biases because there is no energy gap at the Fermi level. However, if either the tip or the sample is contaminated with a surface oxide or adsorbate layer, the *I-V* curve will typically show a zero conductivity area at low biases. The *I-V* responses of a "clean" and a "dirty" surface are shown in Figure 6.18.

With Adsorbed Layers A clean metal surface prepared by ion sputtering and annealing can be deliberately modified by the adsorption of selected species from

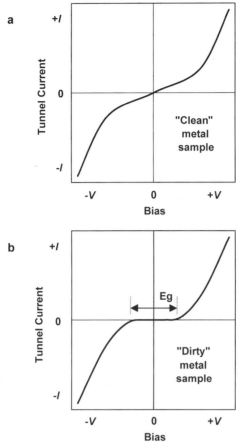

Figure 6.18. The *I-V* curve of a clean, metallic sample–tip junction should not have a zero conductance region, indicating that a continuous density of states exists at the Fermi level. If there is a contamination layer, however, the *I-V* curve should be semiconducting in character.[41]

the gas phase. The first STM study of such a metal surface focused on the (2×1) structure of oxygen on Ni(110). The sample was kept at 150°C and exposed to an oxygen pressure of 5×10^{-9} Torr until the LEED pattern changed from the (1×1) characteristic of Ni(110) to the (2×1) pattern characteristic of the chemisorbed oxygen layer.[80] A similar study was performed on the Ni(771) surface, a vicinal Ni(110) surface.[81] Generally, gases are selectively admitted to an UHV chamber via a precision leak valve in a process known as dosing. The chemisorption of oxygen on a metal surface was also studied dynamically. After a clean copper surface had been prepared, oxygen was admitted to the chamber at a rate of 0.001 monolayers/s via a tube whose orifice was placed 3 cm from the sample surface. With such low flow rates, the surface could be imaged continuously while oxygen accumulated.[82]

There are, of course, a great number of possible surface modifications. For example, other metals can be added to the surface from the gas phase by evaporation. To deposit iron on gold, Kuk and Silverman[83] resistively heated an iron wire to 1050°C. Lower melting point materials, poor conductors, and materials not easily made into a filament can be sublimed from a crucible that is heated independently by a resistive element. This method was used by Lippel et al.[10] to deposit copper phthalocyanine on Cu(100). More complex overlayers can also be prepared in the UHV chamber. For example, Ohtani et al.[84] prepared a Rh(111)$-$(3 × 3) (C$_6$H$_6$ + 2CO) surface by first exposing the clean Rh crystal to 2 × 10^{-7} Torr-s CO and then 3.6 × 10^{-6} Torr-s benzene at room temperature and studied the surface successfully with STM.

Perhaps the most intriguing overlayers are those that passivate an already well-characterized metal surface so that it can be studied later in air. The first example of this was a sulfur-Mo(001) surface that was studied at atomic resolution in air after preparation in an UHV chamber.[85] Similar passivating sulfur layers can be formed on Re(0001) and imaged at atomic resolution after transfer through air to a different chamber. To produce the sulfur overlayer, the crystal is heated to 800°C and given a 6 × 10^{-6} Torr-s exposure to H$_2$S gas. The S overlayer is formed owing to the decomposition of the H$_2$S molecules at the hot Rh surface.[86] Images published by Schardt et al.[87] of iodine on platinum are perhaps the best ambient atmosphere atomic resolution images of an adsorbed layer on a metal. The most interesting thing about these well-ordered layers is that they were not produced in UHV. The polished (111) face of the Pt crystal was cleaned in HClO$_4$, annealed in a hydrogen flame, and then cooled in a flowing I$_2$-N$_2$ atmosphere. The success of this method suggests that it may be possible to form other stable, well-ordered layers on metal surfaces without resorting to UHV techniques.

6.3.2 Semimetals

The basal surfaces of graphite, layered transition metal dichalcogenides, and transition metal trichalcogenides can all be imaged at atomic resolution in air. Surface preparation is usually as simple as cleaving the crystal to produce a fresh surface.[88, 89] In the case of graphite and the layered transition metal dichalcogenides, cleaving is easily accomplished either with a razor blade or by peeling a layer off with a piece of adhesive tape, because the adjacent layers of these structures are held together by only weak van der Waals forces. The newly produced surfaces of these crystals stay clean for minutes to days and most can be imaged at atomic resolution for several hours after cleavage.

The stability of these materials in air makes them excellent STM substrates for the study of adsorbed atoms, layers, and clusters. For example, Ichinokawa et al.[90] used a standard evaporation method to deposit gold onto the basal surface of a MoS$_2$ crystal that had been cleaved in air. The substrate–adsorbate system could later be studied at atomic resolution. It is also possible to perform well-controlled UHV experiments on graphite surfaces that have been cleaved in air. Small numbers of atoms can be deposited on graphite surfaces in UHV so that single atoms, dimers, and trimers can be studied by STM at atomic resolution. Ganz et al.[91] deposited metal clusters on a graphite surface using resistively heated sources of silver, aluminum, and gold placed 0.1 m away from the sample surface. To deposit very small amounts of material, exposure times must be carefully controlled using a shutter and the deposition amount measured with a crystal monitor. For example, the authors estimated that exposure times of less than a second produced Ag coverages of $<1\%$.

Semiconductor clusters have also been studied on graphite substrates. Sarid et al.[92] synthesized BiI_3 clusters ranging from 10 to 1000 atoms in size in a colloidal acetonitrile solution, a drop of which was then placed on a graphite substrate and allowed to dry so that the clusters were left behind. This is a useful general technique that may be used to study a number of microscopic objects, however, there are two associated difficulties. The first is that the density of the microscopic clusters is typically low and, as a result, the surface may have to be searched for some time to find the objects of interest. The second is that dissolved salts and other impurities in the suspension will precipitate out of the solution during evaporation and may make it difficult to unambiguously distinguish these precipitates from the objects of interest.

6.3.3 Semiconductors

In Air Semiconductor surfaces, which have been intensely scrutinized by STM, have been studied primarily under UHV. However, surfaces with oxide overlayers can often be studied at ambient pressures. Typically, the sample is first cleaned with a chemical etch and then capped with an oxide of known thickness grown under controlled conditions.[93] Aqueous HF is commonly used to remove the oxide layer from Si surfaces and the oxide film thickness is usually determined with ellipsometry. Because tunneling through the oxide layer limits the lateral resolution of the STM, atomic resolution imaging is usually only possible at submonolayer oxide coverages. There is one known surface preparation that can be used to prepare oxide- and contamination-free silicon surfaces that are stable in air for reasonable periods of time.[94] First, hydrocarbons are eliminated from the surface by exposure to an UV light source in an oxygen atmosphere. This can be done by warming the silicon up to 200°C in oxygen and exposing it to the illumination from a low-pressure mercury lamp for 10 min. After this, the sample is dipped in a high purity 1% aqueous HF solution and then rinsed with ultrapure water. This treatment creates a primarily hydrogen terminated surface, with some oxygen and fluorine contamination, that is stable in air for several hours and can be imaged at atomic resolution.

Kaiser et al.[95] used an alternate technique to prepare oxide-free silicon surfaces for analysis in nitrogen atmospheres. After the surface was etched, oxidized (1000 Å SiO_2 layer), and re-etched, a 100-Å-thick oxide layer was grown. The sample was then annealed in a dry nitrogen atmosphere for 2 h at 1150°C before it was transferred to a glove box with a dry nitrogen atmosphere. Drops of a 10% HF solution in ethanol were then applied to wafer as it was spun rapidly to remove the oxide, a method referred to as "spin-etching." Surfaces prepared in this way are passivated by hydrogen and remain oxide-free, to less than one monolayer, for up to 20 h in the glove box.

In UHV The methods used to prepare clean, well-defined semiconductor surfaces in UHV are time tested and reliable. Ion sputtering followed by annealing, the same preparation used for metals, is one of the most common methods. For example, well-ordered germanium surfaces have been prepared by ion sputtering followed by a 15-min anneal at 500°C.[96] Cleaving the crystal in UHV is a second method that can be used to obtain a clean semiconductor surface and has been used to prepare silicon, Si(111) and Si(110), and GaAs(110) surfaces.[97-99] Thermal desorption is by far the easiest and most commonly used method for cleaning silicon samples in UHV. A typical heating procedure, which is suitable for most silicon surfaces, is to first degas the sample at 600° to 700°C, "flash" the oxide off by heating the sample to 1100° to 1150°C for 1 to 5 min, and then cool at a rate of 10° to 50°C/min. This treatment is known to produce flat, well-ordered surfaces with regular step distributions. Although the surface oxide can also be removed by 5-min anneals at temperatures as low as 850°C, this procedure is also known to produce irregular step patterns and leave SiC impurities at the surface. It is believed that the SiC precipitates formed by this treatment dissolve into the bulk at higher temperatures.[100]

A number of similar thermal treatments have been used to produce the (7×7) reconstruction of the Si(111) surface. Tokumoto et al.[101] reported flashing the sample at 1200°C several times for only 20 to 30 s while maintaining a pressure $<1 \times 10^{-9}$ Torr. After this treatment, clear (7×7) images could be recorded for several days. Hamers and Demuth[102] reported a procedure that routinely produced (7×7) reconstructions that were atomically flat over 10^6 Å^2. After baking the chamber, the sample was degassed for 8 h at 700°C and then raised to 900°C for 5 min. A final 950°C anneal was used immediately before the experiment to produce the well-ordered surface. The importance of maintaining a pressure $<1 \times 10^{-9}$ Torr during the final anneal was emphasized by the authors.

Both cubic and hexagonal modifications of SiC have been studied at atomic resolution with STM in UHV. β-SiC(001) surfaces have been prepared by epitaxial chemical vapor deposition (CVD) growth on Si(001) substrates.[103] Samples were chemically cleaned before introduction into the UHV chamber where they were further cleaned by heating to 1100° to 1200°C. Surfaces prepared in this temperature range contained regions that exhibited (3×2) and $c(2 \times 2)$ reconstructions of the (001) surface. Kulakov and co-workers[104] studied the $(000\bar{1})$ surface of 6H SiC crystals grown by the Lely[105] method with UHV-STM. Before examination of the as-grown morphology of the $(000\bar{1})$ surface, the samples were degassed at 600°C for 10 h in UHV. To prepare surfaces on which atomic resolution images could be obtained, samples were flashed three times to $\sim$1550°C for 2 min and then degassed at 600°C. This treatment led to a (6×6) surface, which the authors ascribed to the formation of a graphite double-layer on the SiC surface. Owman and Mårtensson[106] studied the 6H-SiC(0001)$-(\sqrt{3} \times \sqrt{3})$ surface at atomic resolution with STM.

Before introduction into the UHV chamber, single crystal surfaces were treated with a $NH_4F:HF$ (7:1) solution for 15 min and then rinsed with deionized water. Treatment of the surfaces for several minutes at 950°C in UHV transformed the previously (1 × 1) surface to the ($\sqrt{3} \times \sqrt{3}$) reconstruction.

Cross-sectional scanning tunneling microscopy was applied to study a number of epitaxial III-V and a few IV semiconductor structures and relies exclusively on cleavage for specimen preparation. Because of this, study is generally limited to (110) surfaces for III-V systems and (111) and (110) surfaces for group IV systems.[99] In most cases, the structures of interest are semiconductor superlattices that were grown by molecular beam epitaxy (MBE). For example, Smith et al.[107] studied the (110) face of dilute $Al_xGa_{1-x}As/GaAs$ heterostructures grown on GaAs(001) substrates with MBE at 580°C and prepared their surface by cleavage in situ. Goldman et al.[108] examined the (110) faces of GaN/GaAs superlattices formed by in situ cleavage and prepared their samples by the nitridation of MBE grown GaAs layers.

Of the in situ methods of semiconductor surface preparation, one of the most elegant is growth in situ during STM imaging. For the most part, this necessitates a variable temperature SPM. Voigtländer and Zinner[109] were responsible for one of the first such studies and examined Ge MBE on the (7 × 7) reconstruction of the Si(111) surface. Images were recorded at substrate temperatures between 600 and 900 K and the Ge beam was formed by the evaporation of Ge from a boron nitride Knudson cell at a rate of 0.07 bilayers/min. Hasegawa and co-workers[110] examined the homoepitaxial growth of Si on Si(111)−(7 × 7) at 350°C and were able to resolve the growth process at the atomic level by using a Si evaporator operated at rates as low as 0.5 bilayers/h.

With Adsorbed Layers As with the materials already discussed, clean semiconductor surfaces can be modified by the addition of adsorbates. The modification methods are typically the same as those used for metal surfaces, atoms being added to the surface from the gas phase. As mentioned earlier, substances that are gases at room temperature can be added to the UHV chamber from a near ambient pressure reservoir through a precision leak valve and substances that are solid at room temperature can be added to surfaces by evaporation. If the dosing rate is sufficiently slow, images can be made as the adsorption reaction proceeds. Tokumoto et al.[101] used this method to study the oxidation and hydrogenation of the Si(111)−(7 × 7) surface.

6.3.4 Metal Oxides

In Air A number of metal oxides have been characterized at or near atomic resolution in the ambient with both STM and AFM. For air studies, one of the most popular methods of surface preparation is cleavage and, therefore, it is not surprising that many of the oxides that have been studied under these conditions are either layered or anisotropically bonded compounds. In one of the first ambient studies, near atomic resolution STM images of the (0001) face of the intergrowth tungsten bronze $Rb_{0.05}WO_3$ were obtained after cleaving the crystal.[111] Several molybdate and vanadate cleavage surfaces have been successfully imaged at or near atomic resolution with STM and contact AFM. Molybdates include ternary alkali molybdenum bronzes, $A_{0.3}MoO_3$[112-115] and $A_{0.9}Mo_6O_{17}$[114] (where A is an alkali metal), as well as the binary molybdenum oxides $MoO_3(010)$, $Mo_{18}O_{52}(100)$, $Mo_8O_{23}(010)$, $Mo_4O_{11}(100)$.[52] Isolated islands of $MoO_3(010)$ have also been prepared for AFM study by oxidizing a cleaved (0001) surface of a MoS_2 crystal.[116] The (001) cleavage surfaces of $Na_{0.003}V_2O_5$ and V_6O_{13} were also examined at the atomic scale in the ambient with STM and AFM.[38, 39]

High critical temperature superconductors have been a popular subject for SPM study and have been imaged in the ambient atmosphere as well as in ambient pressure controlled atmosphere glove boxes. Owing to the layered or highly anisotropic structures of these compounds, some surfaces can often be formed easily by cleavage. Polycrystalline samples of $YBa_2Cu_3O_{7-x}$ have been fractured in air and then imaged in a vacuum of 1×10^{-6} Torr.[117] $Bi_2(Ca,Sr)_3Cu_2O_{8+x}(001)$ was imaged at near atomic resolution in air immediately after cleavage of a single crystal.[118] Wu et al.[119] examined the effects of Pb substitution in $Pb_xBi_{2-x}Sr_2CaCu_2O_8$ single crystals with STM and STS in an Ar-filled glove box and prepared (001) surfaces by in situ cleavage. Hasegawa and Kitazawa[120] studied the (100) surface of $Bi_2Sr_2CaCuO_y$ at atomic resolution with STM and STS in a dry nitrogen atmosphere. To examine this layered compound in cross section, they diamond-filed single crystal samples just before each measurement. The growth structure of c-axis oriented $YBa_2Cu_3O_7$ films grown on $MgO(001)$ and $SrTiO_3(001)$ substrates by sputtering was studied with STM and AFM in the ambient by Hawley and co-workers.[121]

Comparatively few 3-D bonded oxides have been examined in the ambient. Single crystal $TiO_2(110)$ was studied in air after being polished, etched for 2 h in H_2SO_4, washed ultrasonically with distilled water, and dried with nitrogen. After examination of the etched crystal, phenol molecules were adsorbed onto the surface by immersing it in a 0.1 M solution of phenol in methanol.[122] $TiO_2(001)$ was prepared for study in the ambient by mechanical polishing followed by a molten NaOH etch and 1 h reduction at 550°C in H_2.[123, 124] Fan and Bard[125] were able to image the $TiO_2(001)$ surface at atomic resolution in air and prepared their sample by reducing a polished crystal that had been etched ($KHSO_4$ at 620°C for 30 min) for 30 min at 800°C.

In UHV   Both metallic and semiconducting oxides have been studied with STM under UHV conditions, and most surface preparations involve some combination of mechanical polishing, cleaving, heating, ion sputtering, or in situ thermal oxidation treatments.[126] Cleavage is by far the easiest method and can be done in the ambient before sample insertion into the UHV chamber, in a glove-bag attached to a UHV chamber load-lock, or in situ. In some cases, cleavage followed by an in situ anneal in UHV or in the presence of a slight O_2 pressure. Sputtering and annealing procedures, much like those applied on metal and semiconductor surfaces, are frequently practiced on surfaces that have been mechanically polished before insertion into the analysis chamber. These procedures can be complicated by the propensity of metal oxides to give up lattice oxygen (reduce) under the influence of ion beams or heat in vacuum. As a result, they are often followed by reoxidation treatments designed to restore the stoichiometry of the surface. The progress of the oxidation or cleaning (sputtering) treatments can be tracked with analytical techniques such as AES or X-ray photoemission spectroscopy (XPS) and LEED.

Some of the molybdates and vanadates that have been characterized in the ambient have also been studied at atomic resolution in UHV. $Mo_{18}O_{52}(100)$ surfaces were prepared by cleavage with adhesive tape in the ambient air before insertion into the UHV chamber.[127] $Mo_8O_{23}(010)$ surfaces were prepared by cleavage with a razor blade in a N_2-filled glove-bag attached to a load-lock on the UHV analysis chamber.[52] $Na_{0.003}V_2O_5(001)$ surfaces were formed in this manner and also by cleavage in situ.[38] Reduced $V_2O_5(001)$ surfaces were prepared by ex situ cleavage followed by a brief (a few seconds) UHV anneal at 480°C.[128]

Iron oxide surfaces have been prepared using a variety of techniques. Wiesendanger and co-workers[129] prepared $Fe_3O_4(001)$ surfaces by cleavage and also by mechanical polishing followed by a short anneal at 1000 K in UHV. Tarrach et al.[130] examined $Fe_3O_4(001)$ surfaces prepared by a combination of polishing, annealing and, in some cases, Ar ion sputtering and demonstrated that different surface reconstructions could be formed with different procedures. Jansen and co-workers[131] used mechanical polishing, Ar ion sputtering and annealing, $800 < T < 1200$ K, in 10^{-6} mbar O_2 to prepare $Fe_3O_4(110)$ surfaces that exhibited a 1-D reconstruction that ran along the $[\bar{1}10]$ direction. α-$Fe_2O_3(0001)$ surfaces were prepared by cycles of Ar ion sputtering and annealing in 10^{-6} mbar O_2. However, these preparations did not lead to stoichiometric hematite surfaces. After a final oxidizing anneal at 800°C, Condon et al.[132] observed a biphasic surface composed of ordered, mesoscopic islands terminated by $Fe_2O_3(0001)-(1 \times 1)$ and $FeO(111)-(1 \times 1)$. If, on the other hand, the surface was annealed at 730°C for 30 min and cooled to 200°C in the presence of O_2, the surface was terminated by an epitaxial $Fe_3O_4(111)$ layer.[133]

Binary and ternary tungstates have also been studied at atomic resolution in UHV. Jones and co-workers[134] prepared $WO_3(001)$ surfaces that exhibited a (2×2) super structure by annealing surfaces formed by cleavage in the ambient for 15 h at 650°C in 10^{-5} mbar of O_2. Cubic sodium tungsten bronze, Na_xWO_3, (001) surfaces were prepared by both cleavage and combinations of mechanical polishing and UHV annealing. Rohrer et al.[135] observed a (1×2) reconstruction on $Na_{0.82}WO_3(001)$ surfaces that had been either cleaved in the ambient or annealed at 500°C in 6×10^{-8} Torr O_2 for 5 h following cleavage. Jones et al.[136] observed a similar reconstruction on $Na_{0.65}WO_3(001)$ surfaces prepared by annealing a polished surface below 650°C for 15 h in UHV and a $c(2 \times 2)$ reconstruction on surfaces annealed above 650°C.[137] Lu et al.[138] prepared $Rb_{1/3}WO_3(0001)$ surfaces by cleavage in the ambient air before introduction into the UHV chamber.

Perhaps no oxide surface has been studied with STM in UHV more than $TiO_2(110)$.[29, 139–146] Here we review only a few selected surface preparation procedures. One of the first procedures relied on the application of sputtering and annealing cycles to a mechanically polished single crystal sample.[139] The rutile single crystal was first polished by sequentially reducing the size of Al_2O_3 grinding media down to 1 μm. After a 36-h high-temperature (1073 to 1273 K) UHV reduction, the surface was argon ion milled for 5 min at 2 KV, followed by 25 min of ion bombardment at 500 V using an estimated current density of 5 to 10 μamps/cm^2. The surface was then reoxidized using a 30-min anneal at 823 K in 1×10^{-7} Torr of oxygen. Finally, the crystal was exposed to air for several days before being introduced into the UHV STM chamber, where it was briefly heated at approximately 673 K for 2 min and then flashed at approximately 973 K for 5 s. After this treatment, a (1×1) LEED pattern was observed and high-resolution images were obtained of the surface. Zhong and co-workers[140] explored the role of various surface treatments, including UHV and H_2 reduction, Ar ion sputtering, and oxidizing treatments on the morphology of the (110) surface. Sander and Engel[141] also used Ar ion sputtering and annealing cycles to prepare $TiO_2(110)$ surfaces for STM study. Surfaces annealed in UHV at around 1000 K exhibited both (1×1) and (1×2) regions. Exposure of the surface to O_2 at ~1000 K led to oxygen uptake and the formation of a $c(2 \times 1)$ reconstructed surface. Onishi and Iwasawa[142] cleaned the (110) surface of a polished crystal with cycles of Ar ion sputtering and UHV annealing. After several cycles, the crystal took on a deep blue hue. UHV anneals between 600 and 900 K produced $(110)-(1 \times 1)$ terraces bounded by steps parallel to $[1\bar{1}1]$ and $[001]$. Further annealing between 900 and 1100 K led to the formation of strings composed of double ridges along $[001]$. Above 1150 K the strings covered the entire surface to yield a (1×2) superstructure. These authors also observed interactions with acetate ions[147] and between Na clusters and CO_2 molecules[148] on the $(110)-(1 \times 1)$ surface.

A number of high critical temperature superconducting oxides have been studied in UHV. $Bi_2(Ca,Sr)_3Cu_2O_{8+x}(001)$ was imaged at atomic resolution in UHV following cleavage in situ.[149] Shih and co-workers[150] studied the (001) surface of a Bi-Sr-Ca-Cu-O 2:2:1:2 crystal cleaved in situ with STM/S. Ikeda and co-workers[151] studied unsubstituted and Y-substituted $Bi_2Sr_2CaCu_2O_y(001)$ surfaces with STM in UHV and prepared their surfaces by cleavage with adhesive tape in situ. Kazumasa and co-workers[152] examined the morphology and atomic structure of $Ca_{1-x}Sr_{x}$-CuO_2 thin films grown on $SrTiO_3(001)$ substrates with laser MBE. The c-axis oriented films were grown and imaged in a combined growth and UHV STM analysis chamber.

Other notable oxide surfaces that have been characterized in UHV include $TiO_2(100)$, $SrTiO_3(001)$ and the polar ZnO surfaces. Single and polycrystalline ZnO samples, cleaned using a variety of thermal treatments, were studied in UHV, however, atomic resolution images were not obtained.[153] Murray et al.[154] prepared $TiO_2(100)-(1 \times 3)$ surfaces by annealing previously polished and Ar ion sputtered surfaces in UHV at 870 K. This surface could then be transformed to the (1×1) surface by annealing at 870 K for 30 min in 10^{-6} mbar O_2. Matsumoto and co-workers[155] prepared atomically flat $SrTiO_3(001)-(2 \times 2)$ surfaces by annealing Nb-doped crystals for <1 h at 1200°C in UHV. They also found that similar anneals rendered undoped specimens conductive enough for STM study. Liang and

Bonnell[156, 157] examined the reduction of the $SrTiO_3(001)$ surface under a variety of conditions and found that reduction may lead to the formation of intergrowths of $Sr_{n+1}Ti_nO_{3n+1}$ lamellae as well as Sr enrichment at the surface.

6.3.5 Insulators

Although insulators have traditionally been a challenge for STM, the advent of SFM has opened up many new areas of SPM research. Here, we begin by presenting selected techniques for the preparation of insulating surfaces for STM study and then summarize some of the techniques being applied for the preparation of surfaces for SFM study. With the recent demonstrations of true atomic resolution with SFM,[158–160] we anticipate many more high-resolution studies of insulating surfaces in the future.

Insulating surfaces can be prepared for STM study by depositing a thin gold (or other metallic) layer on the sample. Jaklevic et al.[161] imaged gold coatings on fused silica, alkali halide, oxidized metal, and polymer surfaces. Before coating, the substrates were ultrasonically cleaned in a detergent solution, rinsed in deionized water, and dried in hot propanol vapor and air. Organic contaminates were removed from the surface using a 15-min, 600 V dc oxygen discharge treatment at 10^{-2} Torr, and gold was then slowly evaporated onto the surface at a pressure of $<10^{-8}$ Torr. Although this treatment allows for study of the microscopic structure of insulating specimens, the metal coating precludes atomic resolution. Another technique is to deposit thin layers of the insulator of interest onto a conductive substrate for imaging. For example Al_2O_3 layers were grown on NiAl(110) substrates by oxidation and characterized at high, although not atomic, resolution with STM in UHV.[162, 163]

For SFM examination, a conductive layer is, of course, not required. The calcite, $CaCO_3$, $(10\bar{1}4)$ surface has been imaged at the atomic scale with SFM in aqueous environments.[164, 165] In the both of the studies cited, the calcite surface was prepared by cleavage of a single crystal. Bulk polycrystalline and single crystal ceramic surfaces can also be studied. The surface preparations used are typically similar to those used to prepare metal and oxide surfaces, without sputtering. The sample is first polished mechanically to a flat finish and then annealed at high temperature to heal and facet the surface. Mica (001) surfaces are easily formed by cleavage with adhesive tape or a razor blade. Natural crystals of the zeolite Heulandite, $Ca_4Al_8Si_{28}O_{72}$, were cleaved in the ambient air to produce fresh (010) surfaces suitable for atomic scale imaging.[166]

6.3.6 Liquid Crystals

STM can also be used to study the structure of liquid crystals adsorbed on highly oriented pyrolytic graphite. Two approaches have been used to prepare these liquid crystal surfaces with equal success. One method is to simply place a drop of the liquid crystal on the freshly cleaved graphite surface.[167] If the tip is immediately immersed in the liquid and brought into tunneling range, only the graphite substrate is visible. However, several hours after deposition of the liquid, the molecules are visible on the surface. An alternate method is to condense liquid crystal molecules on a freshly cleaved surface of HOP graphite by holding it 2 cm above a warm (100°C) liquid crystal reservoir.[168] It was determined that the images were independent of film thickness, indicating that the tunneling probe simply goes through the uppermost insulating layers and images only the layer directly adsorbed on the graphite. There is some evidence that the structures formed depend on the deposition conditions.

6.3.7 Langmuir-Blodgett Films

Graphite supports have also been used to prepare Langmuir-Blodgett (LB) films for analysis with STM. Films of DL-a-myristoyl-phosphatidic acid (DMPA) and cadmium-arachidate were applied to the substrates using the standard Langmuir-Blodgett technique, described briefly here.[169] Both compounds were dissolved in a 3 to 1 chloroform/methanol mixture, which was then spread on a salt solution that consisted of 10^{-3} M CdCl$_2$ + NaOH. The molecules align themselves with the polar heads toward the salt solution, and single or multiple layers may be added to graphite simply by dipping it into and removing it from the liquid; a new layer is added each time the surface is passed. One complication is that films are stable only when the hydrophobic tails point outward, a situation that normally exists only after the formation of a bilayer. This is because it is the hydrophobic tails of the molecules that adhere to the graphite substrate, the new monolayer surface is formed by the polar groups. The second layer, however, will naturally orient itself with the polar groups inward, leaving an outward, stable layer of hydrophobic tails. To form a stable monolayer on graphite, the polar groups must adhere to the graphite surface, which will happen only if portions of the surface are oxidized. To achieve this condition, the graphite can be shock-oxidized by subjecting it to an 11 Hz, 3 mA peak-to-peak and 0.5 mA offset current pulse lasting 5 s while immersed in a 0.01 M NaOH solution.

6.3.8 Biologic Specimens

Droplet Deposition The first biomaterial to be examined in the STM was bacteriophage ϕ29 particles with lateral dimensions of several hundred angstroms on a side. The phage particles were diluted in 0.1% glycerol solution, deposited on a highly oriented pyrolitic graphite (HOPG) surface and allowed to air dry before analysis in air.[170] Other air-dried tissue specimens were also examined, but were deposited instead on glow discharged carbon films, which are more hydrophilic than a cleaved HOPG surface.[13, 171] For example, porin membranes, which are composed of phospholipid and protein, were deposited on thin carbon films supported by a thicker, fenestrated carbon film, which was in turn supported by a finder grid. Air-dried T4 polyheads were adsorbed onto a glow discharged C film, washed with distilled water, air dried, and examined successfully.

Droplet deposition was also used to prepare samples of molecular proteins, DNA, and RNA. Beebe et al.[172] prepared an aqueous electrolyte solution consisting of 1 mg calf thymus DNA per mL of 10 mM KCl. A drop of this was placed in the middle of a freshly cleaved piece of HOPG and allowed to evaporate. The evaporation of the salt solution leaves a series of concentric rings and a concentrated deposit in the center where the KCl precipitates from the solution. Avoiding the central region, the outlying areas must be searched for DNA molecules, which sometimes occur in unresolved lumps. It is likely that the intercalation of ionic species from solution or the surface forces present during the dehydration alters the DNA structure. Lee et al.[173] attempted to stabilize the molecular structure by depositing bundles of calf thymus DNA and synthetic RNA on freshly cleaved HOPG surfaces rather than individual molecules. A dilute solution of the molecules was dissolved in a pH 7.4 buffer solution of 1 mM sodium cacodylate and 10 nM NaCl. Although the RNA molecules formed bundles spontaneously, DNA bundle formation was encouraged by mixing it for 2 h in either 200 μM spermidine HCl or 100 μM hexaamine cobalt (III). After allowing one drop of the solution to air dry on a HOPG surface, bundles were clearly imaged and the spacings of molecular features could be determined using a 2-D Fourier transform.

Freeze Drying Molecular shapes are more accurately preserved if samples are slowly freeze dried, a process that reduces the effects of dehydration. Amrein et al.[174] adsorbed DNA and recA (a protein involved in recombination) onto 2 nm Pt-C films that had been deposited on mica. After freeze-drying, the samples were shadowed, or coated with a thin, conductive, heavy metal film. The authors reported that the best coating is a 1-nm Pt-Ir-C film evaporated from a 2-mm-diameter hollow carbon rod filled with a 1.5-mm-diameter Pt-Ir (25% Ir) cylinder. The fraction of Pt evaporated decreased with respect to the amount of C and Ir as the coating was applied. These films had several desirable qualities not found in other films: stability in air, reduced granularity, and electrical stability. Similar shadowing techniques were used to coat larger structures, such as TMVs and T4 polyheads.[13] The application of the 1- to 3-nm films occurs at low temperature after freeze-drying, and the evaporation source may be placed directly over the sample (normal incidence) or at an angle as the sample is rotated.

Negative Staining Negative staining is another method used to image biologic materials that was developed from earlier TEM work. In an attempt to apply this method to STM images, Stemmer et al.[13] tested the electrical resistance of several negatively stained air-dried surfaces. Of the eight stained surfaces tested, the sodium phosphotungstate- and uranyl sulfate–coated surfaces had the lowest resistance and were, therefore, likely candidates for STM sample preparation. However, TMV samples stained with these substances did not produce stable images, and the authors suggested that a stain with a lower resistivity in its air-dried state may improve images.

Freeze Fracture Another sample preparation method borrowed from traditional EM work on biologic specimens is the fabrication of freeze-fractured samples. This method replaces the soft, nonconducting biologic specimen with a rigid, conductive replica and was used by Zasadzinski et al.[175] to produce replicas of dimyristoylphosphatidylchlorine (DMPC) bilayers that were imaged by STM. The DMPC was mixed with water; and after the desired bilayers with the appropriate thermal treatment were produced, the liquid was quickly frozen to $-190°C$ and then fractured at $-170°C$ in a vacuum of 10^{-8} Torr. The fracture occurred in a manner that exposed the interiors of the bilayer. A 2.5-nm Pt-C coating, topped off with a 30 nm carbon film to increase stability, was deposited by evaporation immediately after the fracture. These conductive replica films were then easily imaged with the STM.

REFERENCES

1. M. Kuwabara, W. Lo, and L. C. H. Spence, *J. Vac. Sci. Technol.* **A7**, 2745 (1989).

2. J. Tersoff and D. R. Hamann, *Phys. Rev. Lett.* **50**, 1998 (1983).

3. J. Tersoff and D. R. Hamann, *Phys. Rev.* **B31**, 805 (1985).

4. Y. Kuk and P. J. Silverman, *Appl. Phys. Lett.* **48**, 1597 (1986).

5. R. J. Colton, S. M. Baker, R. J. Driscoll, M. G. Youngquist, J. D. Baldeschwieler, and W. J. Kaiser, *J. Vac. Sci. Technol.* **A6**, 349 (1988).

6. S. Ciraci and E. Tekman, *Phys. Rev.* **B40**, 11969 (1989).

7. N. D. Lang, *Phys. Rev.* **B34**, 5947 (1985).

8. N. D. Lang, *Phys. Rev. Lett.* **55**, 230 (1985).

9. N. D. Lang, *Phys. Rev.* **B36**, 8173 (1987).

10. P. H. Lippel, R. J. Wilson, M. D. Miller, C. Woll, and S. Chiang, *Phys. Rev. Lett.* **62**, 171 (1989).

11. C. J. Chen, unpublished manuscript, 1989.

12. E. Tekman and S. Ciraci, *Phys. Rev.* **B40**, 10286 (1989).

13. A. Stemmer, A. Hefti, U. Aebi, and A. Engel, *Ultramicroscopy* **30**, 263 (1989).

14. R. Nicolaides, Y. Liang, W. E. Packard, et al., *J. Vac. Sci. Technol.* **A6**, 445 (1988).

15. G. F. A. van De Walle, H. Van Kempen, and P. Wyder, *Surf. Sci.* **167**, L219 (1986).

16. S. Chiang and R. J. Wilson, *IBM J. Res. Develop.* **30**, 515 (1986).

17. T. Tiedje, J. Varon, H. Deckman, and J. Stokes, *J. Vac. Sci. Technol.* **A6**, 372 (1988).

18. D. K. Biegelson, F. A. Ponce, J. C. Tramontana, and S. M. Koch, *Appl. Phys. Lett.* **50**, 696 (1987).

19. J. Garnaes, F. Kragh, K. A. Mørch, and A. R. Tholen, *J. Vac. Sci. Technol.* **A8**, 441 (1990).

20. J. P. Pelz and R. H. Koch, *Phys. Rev.* **B41**, 1212 (1990).

21. S. Park, J. Nogami, and C. F. Quate, *Phys. Rev.* **B36**, 2863 (1987).

22. D. H. Rich, F. M. Leibsle, A. Samsavar, E. S. Hirschorn, T. Miller, and T. C. Chiang, *Phys. Rev.* **B39**, 12758 (1989).

23. R. M. Tromp, E. J. van Loenen, J. E. Demuth, and N. D. Lang, *Phys. Rev.* **B37**, 9042 (1988).

24. S. Park, J. Nogami, H. A. Mizes, and C. F. Quate, *Phys. Rev.* **B38**, 4269 (1988).

25. H. Fink, *IBM J. Res. Develop.* **30**, 460 (1986).

26. G. Binnig, H. Rohrer, C. Gerber, and E. Weibel, *Appl. Phys. Lett.* **40**, 178 (1982).

27. G. Binnig, H. Rohrer, C. Gerber, and E. Weibel, *Phys. Rev. Lett.* **49**, 57 (1982).

28. R. M. Feenstra and A. P. Fein, *Phys. Rev.* **B32**, 1394 (1985).

29. S. Fischer, A. W. Munz, K.-D. Schierbaum, and W. Göpel, *Surf. Sci.* **337**, 17 (1995).

30. T. Sleator and R. Tycko, *Phys. Rev. Lett.* **60**, 1418 (1988).

31. A. J. Melmed and J. J. Carroll, *J. Vac. Sci. Technol.* **A2**, 1388 (1984).

32. A. J. Melmed, *J. Vac. Sci. Technol.* **B9**, 601 (1991).

33. R. Sonnenfeld and B. C. Schardt, *Appl. Phy. Lett.* **49**, 1172 (1986).

34. D. J. Trevor, C. E. D. Chidsey, and D. N. Loiacono, *Phys. Rev. Lett.* **62**, 929 (1989).

35. J. Wintterlin, J. Wiechers, H. Brune, T. Gritsch, H. Hofer, and R. J. Behm, *Phys. Rev. Lett.* **62**, 59 (1989).

36. D. K. Biegelson, F. A. Ponce, J. C. Tramontana, *Appl. Phys. Lett.* **54**, 1223 (1989).

37. R. J. Colton, S. M. Baker, D. J. Baldeschwieler, and W. J. Kaiser, *Appl. Phys. Lett.* **51**, 305 (1987).

38. R. L. Smith, W. Lu, and G. S. Rohrer, *Surface Sci.* **322**, 293 (1995).

39. R. L. Smith, G. S. Rohrer, K. S. Lee, D.-K. Seo, and M.-H. Whangbo, *Surf. Sci.* **367**, 87 (1996).

40. J. H. Coombs and J. B. Pethica, *IBM J. Res. Develop.* **30**, 455 (1986).

41. J. K. Gimzewski and R. Moller, *Phys. Rev.* **B36**, 1284 (1987).

42. G. Binnig, C. F. Quate, and C. Gerber, *Phys. Rev. Lett.* **56**, 930 (1986).

43. T. R. Albrecht and C. F. Quate, *J. Vac. Sci. Technol.* **A6**, 271 (1988).

44. T. R. Albrecht, S. Akamine, T. E. Carver, and C. F. Quate, *J. Vac. Sci. Technol.* **A8**, 3386 (1990).

45. O. Wolter, Th. Bayer, and J. Greschner, *J. Vac. Sci. Technol. B* **9**, 1353 (1991).

46. R. S. Howland, *How to Buy a Scanning Probe Microscope*, Park Scientific Instruments, Sunnyvale, Calif., 1993.

47. J. P. Cleveland, S. Manne, D. Bocek, and P. K. Hansma, *Rev. Sci. Instr.* **63**, 403 (1993).

48. T. J. Senden and W. Ducker, *Langmuir* **10**, 1003 (1994).

49. H.-J. Butt and M. Jaschke, *Nanotechnology* **6**, 1 (1995).

50. U. D. Schwarz, in S. Amelinckx, D. van Dyk, J. van Landuyt, and G. van Tendeloo, eds., *Handbook of Microscopy—Methods II*, VCH Verlagsgesellschaft mbH, Weinheim, FRG, 1997.

51. S. N. Maganov and M.-H. Whangbo, *Surface Analysis with STM and AFM*, VCH Verlagsgesellschaft mbH, Weinheim, FRG, 1996.

52. R. L. Smith and G. S. Rohrer, *J. Solid State Chem.* **124**, 104 (1996).

53. Y. Martin and H. K. Wickramasinghe, *Appl. Phys. Lett.* **50**, 1455 (1987).

54. D. Rugar, H. J. Mamin, P. Guethner, S. E. Lambert, J. E. Stern, I. McFadyen, and T. Yogi, *J. Appl. Phys.* **68**, 1169 (1990).

55. P. Grütter, D. Rugar, H. J. Mamin, et al., *Appl. Phys. Lett.* **57**, 1820 (1990).

56. K. Babcock, V. Elings, M. Dugas, and S. Loper, *IEEE Trans. Mag.* **30**, 4503 (1994).

57. K. Babcock, M. Dugas, S. Manalis, and V. Elings, *Mater. Res. Soc. Proc.* **355**, 311 (1995).

58. H. Bluhm, A. Wadas, R. Wiesendanger, K.-P. Meyer, and L. Szczesniak, *Phys. Rev.* **B55**, 4 (1997).

59. H. Bluhm, A. Wadas, R. Wiesendanger, A. Roshko, J. A. Aust, and D. Nam, *Appl. Phys. Lett.* **71**, 146 (1997).

60. R. M. Nyffenegger, R. M. Penner, and R. Schierle, *Appl. Phys. Lett.* **71**, 1878 (1997).

61. C. C. Williams, J. Slinkman, W. P. Hough, and H. K. Wickramasinghe, *Appl. Phys. Lett.* **55**, 1662 (1989).

62. C. C. Williams, J. Slinkman, W. P. Hough, and H. K. Wickramasinghe, *J. Vac. Sci. Technol. A* **8**, 895 (1990).

63. J. J. Kopanski, J. F. Marchiando, and J. R. Lowney, *J. Vac. Sci. Technol.* **B14**, 242 (1996).

64. C. J. Kang, C. K. Kim, J. D. Lera, et al., *Appl. Phys. Lett.* **71**, 1546 (1997).

65. E.-L. Florin, V. T. Moy, and H. E. Gaub, *Science* **264**, 415 (1994).

66. C. D. Frisbie, L. F. Rozsnyai, A. Noy, M. S. Wrighton, and C. M. Lieber, *Science* **265**, 2071 (1994).

67. A. Noy, C. D. Frisbie, L. F. Rozsnyai, M. S. Wrighton, and C. M. Lieber, *J. Am. Chem. Soc.* **117**, 7943 (1995).

68. D. V. Vezenov, A. Noy, L. F. Rozsnyai, and C. M. Lieber, *J. Am. Chem. Soc.* **119**, 2006 (1997).

69. G. U. Lee, L. A. Chrisey, and R. J. Colton, *Science* **266**, 771 (1994).

70. W. A. Ducker, T. J. Senden, and R. M. Pashley, *Nature* **353**, 239 (1991).

71. I. Larson, C. J. Drummond, D. Y. C. Chan, and F. Grieser, *J. Phys. Chem.* **99**, 2114 (1995).

72. S. Biggs and P. Mulvaney, *J. Chem. Phys.* **100**, 8501 (1994).

73. D. J. Keller and F. S. Franke, *Surf. Sci.* **294**, 409 (1993).

74. J. S. Villarrubia, *Surf. Sci.* **321**, 287 (1994).

75. J. S. Villarrubia, *J. Vac. Sci. Technol. B* **14**, 1518 (1996).

76. V. M. Hallmark, S. Chiang, J. F. Rabolt, J. D. Swalen, and R. J. Wilson, *Phys. Rev. Lett.* **59**, 2879 (1987).

77. H. S. Kim, Y. C. Zheng, and P. J. Bryant, *J. Vac. Sci. Technol.* **A8**, 314 (1990).

78. G. K. Binnig, H. Rohrer, C. Gerber, and E. Stoll, *Surf. Sci.* **144**, 321 (1984).

79. A. L. de Lozanne, S. A. Elrod, and C. F. Quate, *Phys. Rev. Lett.* **54**, 2433 (1985).

80. A. M. Baro, G. Binnig, H. Rohrer, et al., *Phys. Rev. Lett.* **52**, 1304 (1984).

81. R. Koch, O. Haase, M. Borbonus, and K. H. Rieder, *Surf. Sci.* **272**, 17 (1992).

82. F. M. Chua, Y. Kuk, and P. J. Silverman, *J. Vac. Sci. Technol.* **A8**, 305 (1990).

83. Y. Kuk and P. J. Silverman, *J. Vac. Sci. Technol.* **A8**, 289 (1990).

84. H. Ohtani, R. J. Wilson, S. Chaing, and C. M. Mate, *Phys. Rev. Lett.* **60**, 2398 (1988).

85. B. Marchon, P. Bernhardt, M. E. Bussell, et al., *Phys. Rev. Lett.* **60**, 1166 (1988).

86. D. F. Ogletree, C. Ocal, B. Marchon, G. A. Somorjai, and M. Salmeron, *J. Vac. Sci. Technol.* **A8**, 297 (1990).

87. B. C. Schardt, S. Yau, and F. Rinaldi, *Science* **243**, 1050 (1989).

88. X. L. Wu and C. M. Lieber, *Phys. Rev.* **B41**, 1239 (1990).

89. C. G. Slough, B. Giambattista, W. W. McNairy, and R. V. Coleman, *J. Vac. Sci. Technol.* **A8**, 490 (1990).

90. T. Ichinokawa, T. Ichinose, M. Tohyama, and H. Itoh, *J. Vac. Sci. Technol.* **A8**, 500 (1990).

91. E. Ganz, K. Sattler, and J. Clarke, *J. Vac. Sci. Technol.* **A6**, 419 (1988).

92. D. Sarid, T. Henson, L. S. Bell, and C. J. Sandroff, *J. Vac. Sci. Technol.* **A6**, 424 (1988).

93. J. Jahanmir, P. E. West, A. Young, and T. N. Rhodin, *J. Vac. Sci. Technol.* **A7**, 2741 (1989).

94. Y. Nakagawa, A. Ishitani, T. Takahagi, et al., *J. Vac. Sci. Technol.* **A8**, 262 (1990).

95. J. Kaiser, L. D. Bell, M. H. Hecht, and F. J. Grunthaner, *J. Vac. Sci. Technol.* **A6**, 519 (1988).

96. R. S. Becker, J. A. Golovchenko, and B. S. Swartzentruber, *Phys. Rev. Lett.* **54**, 2678 (1985).

97. R. M. Feenstra, W. A. Thompson, and A. P. Fein, *J. Vac. Sci. Technol.* **A4**, 1315 (1986).

98. R. M. Feenstra and J. A. Stroscio, *J. Vac. Sci. Technol.* **A5**, 923 (1987).

99. R. M. Feenstra, *Semicond. Sci. Technol.* **9**, 2157 (1994).

100. D. Dijkkamp, E. J. van Loenen, A. H. Hoeven, and J. Dieleman, *J. Vac. Sci. Technol.* **A8**, 218 (1990).

101. H. Tokumoto, K. Miki, H. Murakami, H. Bando, M. Ono, and K. Kajimura, *J. Vac. Sci. Technol.* **A8**, 255 (1990).

102. R. J. Hamers and J. E. Demuth, *J. Vac. Sci. Technol.* **A6**, 512 (1988).

103. C.-S. Chang, N.-J. Zheng, I. S. T. Tsong, Y.-C. Wang, and R. F. Davis, *J. Am. Ceram. Soc.* **73**, 3264 (1990).

104. M. A. Kulakov, P. Heuell, V. F. Tsvetkov, and B. Bullemer, *Surf. Sci.* **315**, 248 (1994).

105. J. A. Lely, *Ber. Deut. Keram. Ges.* **32**, 229 (1955).

106. F. Owman and Mårtensson, *Surf. Sci. Lett.* **330**, L639 (1995).

107. A. R. Smith, K.-J. Chao, C. K. Shih, K. A. Anselm, A. Srinivasan, and B. G. Streetman, *Appl. Phys. Lett.* **69**, 1214 (1996).

108. R. S. Goldman, R. M. Feenstra, B. G. Briner, M. L. O'Steen, and R. J. Hauenstein, *Appl. Phys. Lett.* **69**, 3698 (1996).

109. B. Voigtländer and A. Zinner, *Appl. Phys. Lett.* **63**, 3055 (1993).

110. T. Hasegawa, M. Kohno, S. Hosaka, and S. Hosoki, *J. Vac. Sci. Technol.* **B12**, 2078 (1994).

111. M. L. Norton, J. G. Mantovani, and R. J. Warmack, *J. Vac. Sci. Technol.* **A7**, 2898 (1989).

112. J. Heil, J. Wesner, B. Lommel, W. Assmus, and W. Grill, *J. Appl. Phys.* **65**, 5220 (1989).

113. E. Garfunkel, G. Rudd, D. Novak, et al., *Science* **246**, 99 (1989).

114. G. Rudd, D. Novak, D. Saulys, et al., *J. Vac. Sci. Technol.* **B9**, 909 (1991).

115. U. Walter, R. E. Thomson, B. Burk, M. F. Crommie, A. Zettl, and J. Clarke, *Phys. Rev.* **B45**, 11474 (1992).

116. Y. Kim and C. M. Lieber, *Science* **257**, 375 (1992).

117. N. J. Zheng, U. Knipping, I. S. T. Tsong, W. T. Petuskey, and J. C. Barry, *J. Vac. Sci. Technol.* **A6**, 457 (1988).

118. M. D. Kirk, C. B. Eom, B. Oh, et al., *Appl. Phys. Lett.* **52**, 2071 (1988).

119. X. L. Wu, Z. Zhang, Y. L. Wang, and C. M. Lieber, *Science* **248**, 1211 (1990).

120. T. Hasegawa and K. Kitazawa, *Jpn. J. Appl. Phys.* **29**, L434 (1990).

121. M. Hawley, I. D. Raistrick, J. G. Beery, and R. J. Houlton, **251**, 1588 (1991).

122. K. Sakamaki, S. Matsunaga, K. Itoh, A. Fujishima, and Y. Gohshi, *Surf. Sci.* **219**, L531 (1989).

123. S. E. Gilbert and J. H. Kennedy, *J. Electrochem. Soc.* **135**, 2385 (1988).

124. S. E. Gilbert and J. H. Kennedy, *Surface Science Letters* **225**, L1 (1990).

125. F.-R. F. Fan and A. J. Bard, *J. Phys. Chem.* **94**, 3761 (1990).

126. V. E. Henrich, *Rep. Prog. Phys.* **48**, 1481 (1985).

127. G. S. Rohrer, W. Lu, R. L. Smith, and A. Hutchinson, *Surf. Sci.* **292**, 261 (1993).

128. T. Oshio, Y. Sakai, S. Ehara, *J. Vac. Sci. Technol.* **B12**, 2055 (1994).

129. R. Wiesendanger, I. V. Shvets, D. Bürgler, et al., *Science* **255**, 583 (1992).

130. G. Tarrach, D. Bürgler, T. Schaub, R. Wiesendanger, and H.-J. Güntherodt, *Surf. Sci.* **285**, 1 (1993).

131. R. Jansen, V. A. M. Brabers, and H. van Kempen, *Surf. Sci.* **328**, 237 (1995).

132. N. G. Condon, F. M. Leibsle, A. R. Lennie, P. W. Murray, D. J. Vaughan, and G. Thornton, *Phys. Rev. Lett.* **75**, 1961 (1995).

133. N. G. Condon, P. W. Murray, F. M. Leibsle, G. Thornton, A. R. Lennie, and D. J. Vaughan, *Surf. Sci.* **310**, L609 (1994).

134. F. H. Jones, K. Rawlings, J. S. Foord, et al., *Phys. Rev.* **B52**, R14392 (1995).

135. G. S. Rohrer, W. Lu, M. L. Norton, M. A. Blake, C. L. Rohrer, *J. Solid State Chem.* **109**, 359 (1994).

136. F. H. Jones, K. H. Rawlings, J. S. Foord, P. A. Cox, R. G. Egdell, J. B. Pethica, *J. Chem. Soc. Chem. Commun.* **1995**, 2419 (1995).

137. F. H. Jones, K. Rawlings, S. Parker, J. S. Foord, P. A. Cox, R. G. Egdell, J. B. Pethica, *Surf. Sci.* **336**, 181 (1995).

138. W. Lu, N. Nevins, M. L. Norton, and G. S. Rohrer, **291**, 395 (1993).

139. G. S. Rohrer, V. E. Henrich, and D. A. Bonnell, *Science* **250**, 1239 (1990).

140. Q. Zhong, J. M. Vohs, and D. A. Bonnell, *Surf. Sci.* **274**, 35 (1992).

141. M. Sander and T. Engel, *Surf. Sci. Lett.* **302**, L263 (1994).

142. H. Onishi and Y. Iwasawa, *Surf. Sci.* **313**, L783 (1994).

143. P. W. Murray, N. G. Condon, and G. Thornton, *Phys. Rev.* **B51**, 10989 (1995).

144. A. Szabo and T. Engel, *Surf. Sci.* **329**, 241 (1995).

145. U. Diebold, J. F. Anderson, K.-O. Ng, and D. Vanderbilt, *Phys. Rev. Lett.* **77**, 1322 (1996).

146. A. Berkó and F. Solymosi, *Langmuir* **12**, 1257 (1996).

147. H. Onishi and Y. Iwasawa, *Catal. Lett.* **38**, 89 (1996).

148. H. Onishi, Y. Yamaguchi, K.-I. Fukui, and Y. Iwasawa, *J. Phys. Chem.* **100**, 9582 (1996).

149. M. D. Kirk, J. Nogami, A. A. Baski, et al., *Science* **242**, 1674 (1988).

150. C. K. Shih, R. M. Feenstra, and G. V. Chandrashekhar, *Phys. Rev.* **B43**, 7913 (1991).

151. K. Ikeda, K. Takamuku, R. Itti, and N. Koshzuka, *Surf. Sci.* **290**, 207 (1993).

152. K. Kazumasa, T. Matsumoto, and T. Kawai, *Science* **267**, 71 (1995).

153. G. S. Rohrer and D. A. Bonnell, *J. Am. Cer. Soc.* **73**, 3026 (1990).

154. P. W. Murray, F. M. Leibsle, C. A. Maryn, H. J. Fisher, C. F. J. Flipse, and G. Thornton, *Surf. Sci.* **321**, 217 (1994).

155. T. Matsumoto, H. Tanaka, T. Kawai, S. Kawai, *Surf. Sci. Lett.* **278**, L153 (1992).

156. Y. Liang and D. A. Bonnell, *Surf. Sci. Lett.* **285**, L510 (1993).

157. Y. Liang and D. A. Bonnell, *Surf. Sci.* **310**, 128 (1994).

158. F. J. Gissibl, *Science* **267**, 68 (1995).

159. Y. Sugawara, M. Ohta, H. Ueyama, and S. Morita, *Science* **270**, 1646 (1995).

160. N. Nakagiri, M. Suzuki, K. Okiguchi, and H. Sugimura, *Surf. Sci. Lett.* **373**, L329 (1997).

161. R. C. Jaklevic, L. Elie, W. Shen, and J. T. Chen, *J. Vac. Sci. Technol.* **A6**, 448 (1988).

162. R. M. Jaeger, H. Kuhlenbeck, H.-J. Freund, et al., *Surf. Sci.* **259**, 235 (1991).

163. H. J. Freund, H. Kuhlenbeck, and V. Staemmler, *Rep. Prog. Phys.* **59**, 283 (1996).

164. F. Ohnesorge and G. Binnig, *Science* **260**, 1451 (1993).

165. Y. Liang, A. S. Lea, D. R. Baer, and M. H. Engelhard, *Surf. Sci.* **351**, 172 (1996).

166. L. Scandella, N. Kruse, and R. Prins, *Surf. Sci. Lett.* **281**, L331 (1993).

167. J. S. Foster and J. E. Frommer, *Nature* **333**, 542 (1988).

168. D. P. E. Smith, H. Horber, C. Gerber, and G. Binnig, *Science* **245**, 43 (1989).

169. C. A. Lang, J. K. H. Horber, T. W. Hansch, W. M. Henkl, and H. Mohwald, *J. Vac. Sci. Technol.* **A6**, 368 (1988).

170. A. M. Baro, R. Miranda, J. Alaman, et al., *Nature* **315**, 253 (1985).

171. A. Stemmer, R. Reichelt, A. Engel, *Surf. Sci.* **181**, 394 (1987).

172. T. P. Beebe, T. E. Wilson, D. F. Ogletree, et al., *Science* **243**, 370 (1989).

173. G. Lee, P. G. Arscott, V. A. Bloomfield, and D. F. Evans, *Science* **244**, 475 (1989).

174. M. Amrein, A. Stasiak, H. Gross, E. Stoll, G. Travaglini, *Science* **240**, 514 (1988).

175. J. A. N. Zasadzinski, J. Schneir, J. Gurley, V. Elings, and P. K. Hansma, *Science* **239**, 1012 (1988).

PART III

APPLICATIONS OF SCANNING PROBE MICROSCOPY

7

ELECTROSTATIC
AND MAGNETIC
FORCE MICROSCOPY

Sergei V. Kalinin and Dawn A. Bonnell

7.1   INTRODUCTION

The development of scanning probe techniques in the final two decades of twentieth century is one of the major breakthroughs in modern science. Even though profilometry had been used to characterize solid surfaces for a long time, it was not until the seminal work by Binnig, et al.[1] that the concepts of a local probe interacting with the surface and a pliable cantilever used for detection of the probe–surface force was established. Initially, atomic force microscopy (AFM) was developed as a tool sensitive to the strong short-range repulsive forces between a tip and surface. Binnig et al.[1] predicted that scanning probe methods potentially could achieve sensitivity to forces as small as 10^{-18} N, thus allowing them to be used to measure weak long-range magnetic and electrostatic forces. In the 15 years since the invention of AFM the prediction has been fulfilled remarkably well. In this chapter we summarize some of the major developments in the scanning probe imaging of electrostatic and magnetic phenomena.

In lieu of introduction to these techniques we briefly consider the operating concepts of the topographic AFM (see Chapter 2). The original approach to AFM imaging was contact mode imaging. The probe (tip) is brought into contact with the surface (hence the name) and repulsive van der Waals (vdW) forces result in the deflection of the cantilever, which is monitored by an optical, heterodyne, or capacitive system. A feedback loop keeps the deflection constant by adjusting the vertical position of the cantilever while scanning along a surface. The feedback signal provides the topographic profile of the surface. The interested reader is referred to Wiesendanger[2] for further details. Progress in topographic AFM led to the development of intermittent contact mode imaging.[3] This approach uses mechanically driven cantilever-tip system. The mechanical oscillations are usually imposed by the piezoelectric actuator, traditionally referred to as piezo. The oscillation amplitude at a fixed driving voltage on the piezo is detected with a lock-in amplifier. While approaching the surface, the tip eventually comes into intermittent contact with it and the oscillation amplitude decreases. Similar to contact mode imaging, the feedback loop keeps the oscillation amplitude constant while scanning and the feedback signal provides the topographic profile of the surface.

It was almost immediately realized, that these two approaches could be extended to map forces of a different nature, such as attractive vdW forces, magnetic forces, and electrostatic forces.[4–9] This is achieved by varying the separation between the tip and the surface. For small tip–surface separations strong short ranged vdW forces prevail. Magnetic and electrostatic contributions to the signal are negligible and cannot be detected without special techniques. On the other hand, for larger tip–surface separations, the latter interactions will dominate owing to the smaller power in the corresponding force-distance relation. The next major step in long-range force detection was achieved with the invention of an non-contact imaging mode.[10] In this mode, the tip acquires topographic data near the surface, then retraces topographic profile at a predefined height above the surface to measure the force interactions. The general operation of AFM in these modes is described in Chapter 2, along with typical examples of instrumentation. The preparation and characterization of tips are presented in Chapter 6.

This chapter extends the basic concepts of sample–tip interactions to show how electromagnetic interactions can be quantified and to describe variants of scanning probe microscopies (SPMs) that access a variety of properties. SPMs based on electrostatic interactions include surface potentiometry, scanning capacitance microscopy, and piezoelectric and electrostrictive measurements. SPMs based on magnetic forces include MFM, supercurrent imaging, and scanning superconducting quantum interference device (SQUID) microscopy. As with previous chapters, this is not intended to be a comprehensive review, rather it is a description of the various approaches with illustrative examples; the interested reader is referred to the reviews and original papers cited for further details.

7.2 ELECTROSTATIC PROBE IMAGING

The number of electrostatic probe imaging techniques developed since 1990 is immense. Although many are not commonly used, the sheer number of electrostatic force microscopy (EFM) techniques necessitates classification. This is especially true for the techniques commercialized and implemented on major commercial instruments that have long become working horses in physics, chemistry and material science. EFMs can be loosely subdivided into three regimes based on tip–surface separation: long range, intermediate, and short range. Depending on whether the tip is driven mechanically or electrostatically or whether both modulations are combined, three more operational regimes can be defined. Finally, depending on the nature of the contact between the tip and the surface, another three regimes can be established. The tip can be either in constant contact with surface, in intermittent contact (tip touches the surface in the lowest point of the trajectory), or the noncontact mode. These three contact regimes are correlated with tip–surface separation regimes (i.e., the long-range regime clearly indicates a noncontact or intermittent contact mode). The experimental setup for the EFM implies the modulation, contact, and separation regime.

In the long-range regime (tip–surface separation >10 to 50 nm), only electrostatic forces between tip and the surface are significant. No charge injection or band bending takes place. The tip can be either static (deflection detection) or oscillating (dynamic detection). The oscillation amplitude is usually much smaller than the tip–surface separation, which allows relatively easy extraction of force gradient (for mechanically driven cantilever) or force (for electrostatically driven cantilever) data from the experimental signal, provided that the properties of the cantilever are known. Quantification of surface properties from the force gradient data is complicated, because the exact shape of the tip and even the cantilever must be taken into account, especially for large tip–sample separations.

The second operational regime of EFM is characterized by relatively small tip–surface separations (<10 to 50 nm), but the contribution of electrostatic forces still dominates over that of vdW interactions. In this regime the tip can operate both in noncontact static (deflection) mode or can oscillate. Unlike the long-range regime, the oscillation amplitude is comparable to the tip–surface separation and the tip can actually touch the surface, thus rendering extraction of force and force gradient data from the experimental signal more difficult. Also in this case charge transfer between the tip and the surface, tip-induced band bending, electromechanical response of the surface, and electrostatic forces all contribute to the detected force signal. However, the tip can usually be approximated by simple geometrical models (e.g., sphere), because the part closest to the surface provides the major contribution to force signal.

In the last case, the tip is in contact with the surface. Because vdW interactions dominate over electrostatic forces, only contact detection is possible. However, in this case the tip can be used as a local probe of capacitance (scanning capacitance microscopy) or resistivity (scanning spreading resistance microscopy). If biased, the tip can also induce piezoelectric and electrostrictive responses and be simultaneously used for the detection of electromechanical response of the surface. Also in this configuration the tip can be used for electrochemical or physical modification of the surface.

Most SPM techniques use the dynamic response of the tip to a periodic mechanical force (tapping or intermittent contact mode AFM), oscillating bias on the tip or on the sample (voltage modulation techniques), or a oscillating magnetic field. Periodic perturbation can result either in a static (dc) response of the tip, a response at the main frequency of perturbation (first harmonic signal), or a response at twice the main frequency of perturbation (second harmonic signal). Lock-in techniques allow extraction of the amplitude of first or second harmonic of the response for subsequent use as feedback or data signals. Major EFM techniques employ either mechanically or electrostatically driven cantilever, even though some other driving regimes are possible.[11]

During the operation in the mechanically driven mode the voltage on the piezo V_{piezo} driving the cantilever is

$$V_{\text{piezo}} = V_{\text{acp}} \sin (\omega_p t) \qquad (7.1)$$

where the driving frequency, ω_p, is selected equal or close to the resonant frequency of the cantilever. The oscillating voltage on the piezo induces cantilever oscillations and the tip–surface separation is

$$d = d_0 + A(\omega_p) \sin (\omega_p t + \varphi_c) \qquad (7.2)$$

where $A(\omega_p)$ is the frequency-dependent oscillation amplitude and φ_c is the phase shift between the driving voltage on the piezo and the cantilever oscillations. The optimal choice of driving frequency, ω_p, is discussed in Chapter 2. During operation, the conductive tip is either grounded or biased with a dc voltage.* The presence of an electrostatic force gradient near the surface results in a change of the resonance frequency of the cantilever and can be detected as a resonance frequency shift (frequency detection) while the feedback adjusts ω_p to keep $A(\omega_p)$ maximal, as a shift of the oscillation amplitude at constant driving frequency (amplitude

---

*Electric condition of dielectric or semiconductive tips is difficult to control. Contact of such a tip with the surface usually leads to contact electrification and the resulting tip charge and electrostatic forces are generally undefined. However, qualitative imaging with nonconductive tips is possible.

detection), or as a phase shift (phase detection). Details of these detection techniques are provided in Chapter 2. To minimize the influence of surface topography on the dynamic properties of the cantilever, this technique is usually implemented in the noncontact mode. Mechanical contact between the tip and the surface strongly influences the dynamic properties of the cantilever and information on electrostatic interactions can no longer be easily extracted from experimental data. Instead, mechanical tip–surface interactions can be quantified. Phase detection in the intermittent contact regime provides information on the elastic properties of the surface (phase imaging), which are out of the scope of the present chapter. Mechanically driving the cantilever in the contact mode provides the information of the local stiffness of the surface (force modulation technique). Information obtained from SFM operating in the mechanically driven (force gradient) regime is summarized in Table 7.1.

An alternative approach involves voltage modulation techniques. The driving voltage at the piezo is set to zero (the tip is no longer mechanically driven) and a conductive tip is biased by an ac voltage. In this operational regime the tip potential is

$$V_{tip} = V_{dc} + V_{ac} \sin (\omega t) \tag{7.3}$$

Biasing the tip in the large tip–surface separation regime above metallic or linear dielectric ($\varepsilon = $ constant) surfaces results in the static force and forces at the frequency of the tip voltage (first harmonic) and at twice the frequency of the tip voltage (second harmonic). All three force components result in the deflection of the cantilever and tip–surface separation is

$$d = d_0 + A_0 + A_1 \sin (\omega t + \varphi_1) + A_2 \sin (2\omega t + \varphi_2) \tag{7.4}$$

where d_0 is tip–surface separation when $V_{tip} = 0$; A_0 is static response; and A_1, A_2, φ_1, and φ_2 are amplitudes and phase shifts of first and second harmonic responses. Magnitudes of A_0, A_1, and A_2 are relatively small and only the latter two components along with corresponding phase shifts can be determined by lock-in technique. Separation of the first and second harmonic responses allows quantification of

Table 7.1 SPM in a mechanically driven regime with constant tip bias

Mode	First Harmonic Response	Static Response
Noncontact	Phase, amplitude, or frequency shift proportional to the force gradient; both electrostatic and magnetic forces contribute	Force; very difficult to detect
Intermittent contact	Amplitude for topography; phase provides the elastic properties of the surface	
Contact mode	Amplitude provides mechanical properties (stiffness) of the surface	Topography

different components of tip–surface force rather than determination of the total force gradient, as in case of the mechanically driven mode.

Biasing the tip in contact mode results in tip displacement owing to electromechanical effects such as the inverse piezoelectric effect and electrostriction. For linear piezoelectric materials, tip deflection as a function of applied bias is similar to Eq. (7.4). However, this technique is most widely used for the characterization of ferroelectric materials, in which case the electromechanical response of the surface to applied voltage is considerably more complicated (discussed below). Information obtained from SFM operating in voltage modulation regime is summarized in Table 7.2.

These modulation regimes can be combined, i.e., the tip can be driven both mechanically by the piezo at frequency ω_p and electrostatically at frequency ω. The voltage modulation ω, in this case, is selected to be smaller than the cantilever oscillation frequency ω_p (tip voltage is almost constant during single oscillation) but much higher than the frequency corresponding to the lateral motion of the cantilever (information is collected from the single point) and feedback response frequency. Depending on whether these modulations are applied in the intermittent contact or noncontact mode, interpretation of the responses at the main frequency ω and the second harmonic can be done along the lines discussed above. Obviously, quantification of electrostatic, electromechanical, and elastic contributions to the signal is more challenging in this case and the first two modulation modes are far more widespread.

The techniques discussed above are based on the detection of force-induced deflection of the cantilever, referred to as scanning force microscopies. However, the tip can be also used as capacitive sensor to measure tip–sample capacitance, thus giving rise to scanning capacitance microscopy. Alternatively, current through the tip–surface junction can be detected and used to quantify local resistivity of the surface. These experimental setups are easily implemented on conventional SPMs. Different types of field-sensitive probes can also be used, such as SQUID probes or Hall probes. A brief summary of the common operation regimes of the SPM is given in Table 7.3.

Table 7.2   SPM in voltage modulation regime

Tip–Surface Separation	Linear Response (First Harmonic)	Quadratic Response (Second Harmonic)	Static Response
Noncontact mode	Surface charge or potential	Dielectric constant ε	Total force
Contact mode	Inverse piezoelectric effect	Electrostriction	Topography

Table 7.3 Electrostatic and magnetic SPMs

Operating Mode	Technique	Information Obtained
Mechanical modulation in noncontact mode	EFM, MFM	Electrostatic or magnetic force gradient acting on the tip; varying the dc bias on the tip yields surface properties (but multiple scans are required)
Voltage modulation in noncontact mode	scanning surface potential microscopy (SSPM) or Kelvin probe microscopy (KPM)	Surface potential, work function and charge
Voltage modulation in contact mode	piezoresponse imaging (PRI)	Piezoelectric and electrostrictive properties
Capacitance detection	scanning capacitance microscopy (SCM)	Tip–surface capacitance related to the carrier density, etc. in semiconductors
Current detection	scanning spreading resistance microscopy (SSRM) or leakage current microscopy (LCM)	Local resistivity
Other field probes	Scanning SQUID microscopy and scanning Hall probe microscopy	Magnetic field (rather than field gradient as in MFM) near the surface

7.2.1 Tip–Surface Interactions in the EFM

The tip–surface interaction depends both on the electric and geometric properties of the tip (conductive/dielectric, tip shape) and on the surface (conductor with a well-defined potential or an insulator with given volume or surface charge density, surface topography). For all practical purposes, however, conductive probes provide much better control over the tip properties, because the voltage on the tip can be easily controlled. For these probes, the tip shape and tip potential are usually known. To clarify the origin of the different frequency responses in the EFM, the most general case is considered here, in which the tip is supplied with an ac bias and a dc offset relative to the surface. Thus we simultaneously describe both constant-bias and voltage modulation techniques as discussed above.

For a conductive surface with a constant potential, the force between the tip and surface is given by Eq. (2.7) rewritten here for convenience:

$$F(z) = \frac{1}{2}(\Delta V)^2 \frac{\partial C(z)}{\partial z} \tag{7.5}$$

where $F(z)$ is the force, $\Delta V = V_{\text{tip}} - V_{\text{surf}}$ is the potential difference between the tip and the surface, z is vertical tip–surface separation, and $C(z)$ is a tip–surface capacitance. Eq. (7.5) implies that the force is a function of the tip and surface geometry through the $C(z)$ term. It can be easily shown that $C(z)$, and consequently $F(z)$, are rapidly decaying functions of tip–surface separation, thus the measurable signal can be obtained only at relatively small tip–surface separations. The exact functional form of tip–surface capacitance $C(z)$ is complicated even for flat surfaces and can be obtained only by numerical methods, such as finite element analysis (FEA). For the typical probe geometry (Chapter 6), the total capacitance $C(z)$ can be conveniently approximated as a sum of the contributions to the tip apex, tip bulk, and cantilever with the spherical, conical and plane geometry correspondingly:

$$C(z) = C_{\text{apex}}(z) + C_b(z) + C_c(z) \tag{7.6}$$

A number of approximate models have been suggested to quantify capacitive force between the tip and the surface. Some of these models use approximate geometric descriptions of the tip as a plate capacitor, sphere,[12] hyperboloid[13,14] cone,[15] or cone with spherical apex.[17] Alternatively, a simplified image charge distribution is used so that a corresponding isopotential surface represents the tip.[17,18] Examples of such models are point charge or line charge models.[19] Sphere and point charge models represent the $C_{\text{apex}}(z)$ term in Eq. (7.6) and are used for smaller tip–surface separations ($z \ll$ radius of curvature of tip apex), in which case the contribution of tip apex dominates. For larger tip–surface separations the contribution from the conical part of the tip $C_b(z)$ is significant, and hyperboloid, cone, or line charge models are used. The cantilever contribution $C_c(z)$ can usually be modeled by a plate-plate model. Cantilever capacitance can provide a nonnegligible contribution at large tip–surface separations because the effective area of the cantilever is much larger than that of the tip. Because the characteristic cantilever–surface separation ($\sim$10 μm) is much larger than tip–surface separation (1 to 500 nm), $C_c(z)$ can usually be approximated as constant. As seen from Eq. (7.5), for all tip geometries the voltage dependence of the force is parabolic.

Figure 7.1 provides an experimental example of force gradient bias and force gradient distance dependencies for the field emitter determined by EFM.[20] The force gradient bias dependence is well described by a parabola, in complete agreement with Eq. (7.5). The force gradient distance dependence shows rapid decay of the force with tip–surface separation. This dependence can be linearized in the log-log coordinates; however the slope of the curve is noninteger and its absolute value -1.5 is between that for plane-sphere model (slope $= -2$) and plane-cone model (slope $= -1$).

Quantification of the electrostatic tip–surface interaction is significantly more complicated for dielectric materials. In this case both surface charges and volume trapped charges, as well as the tip-induced polarization image charges contribute to the total force. Analytical treatment of this problem is possible only for the simplest cases;[21,22] however, the important features of the tip–surface interaction can be understood even within the framework of these simplified models.

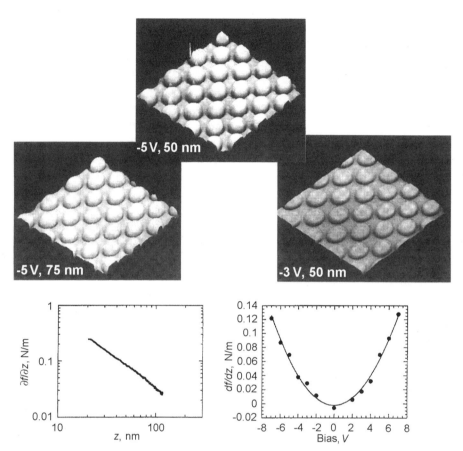

Figure 7.1. A 2×2-μm array of conical field emitters. Electrostatic force gradient images showing height and voltage dependence in image and graphic form. Full scale is 0.08 N/m. The applied voltage and sample–tip separation are noted. (Reprinted with permission from Ref. 20.)

Following Sarid,[11] for a charge Q_s on a dielectric surface the electrostatic interaction between the biased tip and the sample includes three distinct components: effective surface charge on the sample Q_s, constant dc bias applied between the tip and the sample V_{dc}, and periodic ac bias applied between the tip and the sample $V_{ac} \sin(\omega t)$. The charge on the tip can be approximated as

$$Q_t = Q'_s + Q_{dc} + Q_{ac} \tag{7.7}$$

where Q_s is the image charge on the tip and

$$Q_{dc} = V_{dc} C \tag{7.8}$$

$$Q_{ac} = C V_{ac} \sin \omega t \tag{7.9}$$

and C is the effective capacitance of the tip–sample system. Here the interaction between the tip and the surface is approximated by a corresponding capacitance. The induced charge on the tip can be found only for simple geometries. For localized charge Q_s close to spherical tip apex ($z \ll R$, R is the apex radius of curvature) induced charge $Q'_s \approx -Q_s$.

The force between the tip and the sample is a sum of two contributions; the first owing to the charge-induced charge interaction, the other owing to the capacitance. The total expression for the force can be qualitatively written as:

$$F = \frac{Q_s Q_t}{4\pi\varepsilon_0 z^2} + \frac{1}{2}C'(V_{ac} + V_{dc})^2 \tag{7.10}$$

Substituting the expression for Q_t, Eq. (7.10) becomes the sum of three components:

$$F = F_{dc} + F_{1\omega} + F_{2\omega} \tag{7.11}$$

where

$$F_{dc} = \frac{Q_s^2}{4\pi\varepsilon_0 z^2} + \frac{Q_s V_{dc}}{4\pi\varepsilon_0}\frac{C}{z^2} + \frac{1}{2}C'\left(V_{dc}^2 + \frac{1}{2}V_{ac}^2\right) \tag{7.12}$$

is a static component of the force,

$$F_{1\omega} = \left(\frac{Q_s}{4\pi\varepsilon_0}\frac{C}{z^2} + C'V_{dc}\right)V_{ac}\sin\omega t \tag{7.13}$$

is the first harmonic component, and

$$F_{2\omega} = -\frac{1}{4}C'V_{ac}^2\cos 2\omega t \tag{7.14}$$

is the second harmonic component. The expression for the corresponding force derivatives can be easily obtained from Eqs. (7.12–7.14).

In summary, the total electrostatic force on the tip due to the oscillating voltage at frequency ω applied between the tip and linear dielectric surface can be subdivided into three categories:

- A harmonic response with the same frequency as external bias owing to surface and volume trapped charges and dipole charges of remanent (tip bias independent) polarization.
- A second harmonic response owing to the interaction of the tip charges and bias-induced polarization image charges on the surface.
- A static response owing the surface charge–tip image charge interaction, tip charge–surface image charge interaction, and static component of second harmonic response.

7.2.2   Electrostatic Force Microscopy

The different variants of EFM are accomplished by monitoring different terms in Eq. (7.11), as summarized in Tables 7.1 and 7.2. In the following section we present examples of force gradient imaging, scanning surface potential imaging, and a combined approach.

Force Gradient Imaging   Force gradient imaging with conductive cantilevers is one of the most common EFM techniques and is available on most commercial instruments. This technique is usually implemented in the non-contact mode. The grounded tip first acquires the surface topography using standard intermittent contact AFM. Electrostatic data are collected in the second scan, in which the tip retraces the topographic profile separated from the surface 50 to 100 nm, thereby maintaining a constant tip–sample separation. During operation in this mode, the cantilever is mechanically driven by the piezo. Force gradients owing to electrostatic and magnetic forces result in a frequency, phase, or amplitude shift relative to the tip oscillating in the absence of the force gradient, as discussed in Chapter 2. Measurements can be performed with dielectric, semiconductive, or conductive tips. In the former two cases the tip usually acquires a charge owing to contact electrification, the magnitude and sign of which are usually unknown. In the latter case, the electrostatic state of the tip can be controlled by external bias. For a zero-biased (grounded) tip, surface charges induce image charges of the opposite sign on the probe and result in the attractive interaction that will deflect the cantilever. This deflection is proportional to the square of the charge and thus charges of the opposite sign cannot be distinguished. For ac-biased tips, the force between the tip and the surface depends on the tip bias and charges of the opposite sign can be distinguished.

It is interesting to consider the development of various EFM techniques with the useful example of a surface with uniform surface charge density as exemplified by a ferroelectric surface. Surface charge density in this case is

$$\sigma = \mathbf{P} \cdot \mathbf{n} \qquad (7.15)$$

where $\mathbf{P}$ is polarization vector and $\mathbf{n}$ is the normal to the surface. The normal component of polarization and the surface charge density are constant within a domain and abruptly change at the domain boundary.[23]

In the EFM experiment with a nonbiased tip, the resulting force gradient is proportional to the product of the electric field owing to polarization and the induced charge and thus depends only on magnitude of polarization but not on the sign. Regions with equal but opposite surface charge densities (e.g., c^+ and c^- domains[†]) cannot be distinguished. At a domain wall, however, $\mathbf{P}$ changes sign and passes through zero and the tip experiences a minimum in force gradient. In their

---

[†]Polarization in ferroelectrics is generally screened by free carriers and surface adsorbates, and EFM contrast depends on the screening mechanism.

early work, Saurenbach and Terris[24] used the feedback loop to maintain a constant force gradient by adjusting the tip–surface separation. The minimum on the images corresponded to domain wall position. This feedback scheme, however, relies on the force gradient to maintain tip–surface separation and in the vicinity of domain wall, the tip can crash into the surface. As mentioned above, this drawback is now avoided by using the non-contact scheme.

Advantages of the frequency-detection EFM can be illustrated on materials with built-in nonuniform charge distribution, such as charged grain boundaries in ceramics. Dopant or vacancy segregation or intrinsic grain boundary (GB) states result in the formation of interface charge on grain boundary. This charge is screened by free carriers, resulting in formation of Schottky depletion regions. The characteristic width of these depletion region depends on carrier concentration and can be as high as a few micrometers. Figure 7.2 compares EFM images of a grain boundary

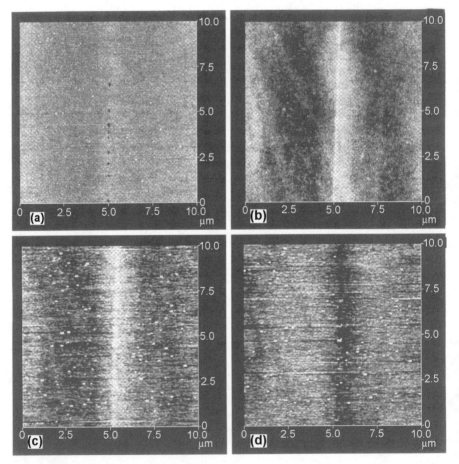

Figure 7.2. a, Topography of Nb-doped 36.8° $SrTiO_3$ bicrystal in the vicinity of grain boundary (a). **b**, Surface potential (SSPM) image of the same region. **c**, EFM (force gradient) images at tip bias $V_{tip} = 5V$ and (**d**) $V_{tip} = -5V$. The range is (**a**) 5 nm, (**b**) 20 mV and (**c, d**) 2 Hz.

in a Nb-doped SrTiO$_3$ bicrystal at positive and negative tip biases with the surface topography and a surface potential image. For positive tip bias, the grain boundary appears as a protrusion corresponding to a positive frequency shift, i.e., repulsion between the tip and GB charges. For negative tip bias, GB contrast is dark, i.e., the force is attractive. This observation implies that the charge on GB is positive. Dependence of the difference in frequency shifts above GB and far from GB, on bias voltage is shown in Figure 7.3. This dependence is linear; and for zero tip bias, the GB contrast vanishes. At the same time, the absolute frequency shift is parabolic in tip bias. The quadratic dependence indicates induced charge interactions. The linear dependence results from the electrostatic attraction–repulsion of the GB charge with the tip charge. Note that the magnitude of the tip–surface interaction determined from the quadratic term of the parabolas in Figure 7.3 and the slope of the lines rapidly decreases with tip–surface separation. The GB-induced effect is also much smaller than total tip–surface interaction. Analysis of the force gradient distance dependence provides the depletion width and interface charge density on the GB.[25]

Scanning Surface Potential Microscopy (Kelvin Probe Microscopy) As shown in the previous section, a tip can be made sensitive to electrostatic forces by biasing it with respect to the surface. However, for a dc-biased tip the force gradient signal results from the combination of several possible interactions and individual components can be resolved only if the measurements are performed at different scan heights (thus varying $C(z)$) or different tip biases. In many cases, the sensitivity of the tip to the electrostatic forces and topographic artifacts doesn't allow high potential resolution (Fig. 7.2). These disadvantages of force gradient EFM led to the development of alternative voltage modulation EFM techniques, one of the most well known examples of which is scanning surface potential microscopy (SSPM), also known as Kelvin probe microscopy (KPM).[26–28]

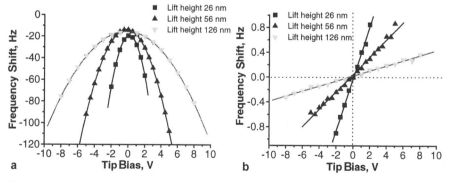

Figure 7.3. a, Total frequency shift (image average) of Nb-doped 36.8° SrTiO$_3$ bicrystal. **b**, The difference in frequency shifts above GB and far from the GB as a function of tip bias for different tip–surface separations.

As in EFM, this technique is usually implemented in the non-contact mode. In the second scan, the piezo is disengaged and oscillating bias is applied to the tip. Assuming that the surface is characterized by uniform (or, rather, slow varying) surface potential, V_{surf}, the capacitive force between the tip and the surface, is given by Eq. (7.5). Substitution of V_{tip} (Eq. (7.3)) yields

$$F(z) = \frac{1}{2} \frac{\partial C(z)}{\partial z} \left[(V_{dc} - V_{surf})^2 + \frac{1}{2} V_{ac}^2 [1 - \cos(2\omega t)] \right.$$
$$\left. + 2(V_{dc} - V_{surf})V_{ac}\sin(\omega t) \right] \tag{7.16}$$

Eq. (7.16) shows that the force between the tip and the surface owing to the application of an ac bias on the tip has static component, F_{dc}, first harmonic component, $F_{1\omega}$, and second harmonic component, $F_{2\omega}$, explicit expressions for which are given in Eq. (7.17) to (7.19). Note that the dc component of tip bias results in static and first harmonic components, whereas the ac component contributes to all three components.

$$F_{dc}(z) = \frac{1}{2} \frac{\partial C(z)}{\partial z} \left[(V_{dc} - V_{surf})^2 + \frac{1}{2} V_{ac}^2 \right] \tag{7.17}$$

$$F_{1\omega}(z) = \frac{\partial C(z)}{\partial z} (V_{dc} - V_{surf})V_{ac}\sin(\omega t) \tag{7.18}$$

$$F_{2\omega}(z) = \frac{1}{4} \frac{\partial C(z)}{\partial z} V_{ac}^2 \cos(2\omega t) \tag{7.19}$$

The lock-in technique allows extraction of the first harmonic signal in the form of first harmonic of tip deflection proportional to $F_{1\omega}$. A feedback loop is employed to keep it equal to zero (hence the term nulling force approach) by adjusting V_{dc} on the tip. Obviously, the condition $F_{1\omega} = 0$ is achieved when V_{dc} is equal to V_{surf} (Eq. (7.18)). Thus the surface potential is directly measured by adjusting the potential offset on the tip and keeping the first harmonic response zeroed. It is noteworthy that the signal is independent of the geometric properties of tip–surface system (i.e., $C(z)$) and the modulation voltage. This technique allows high ($\sim$mV) potential resolution. Comparisons in Figure 7.2 illustrate that it is also much less sensitive to topographic artifacts.

For a realistic surface with a nonuniform surface potential distribution and nonuniform topography, Eq. (7.5) can be rewritten as the sum of partial capacitive interactions of the tip with the different regions on the surface (Fig. 7.4).[29] The force in this approximation is

$$F_{1\omega} = V_{ac} \sum_{i=1}^{n} \frac{\partial C_{eff,i}}{\partial z} (V_{dc} - V_{S,i}) \tag{7.20}$$

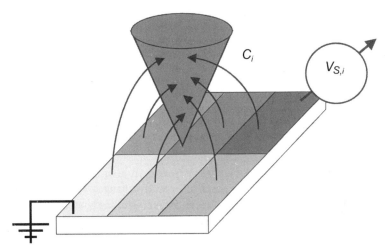

Figure 7.4. How a heterogeneous potential distribution on a sample surface contributes to the capacitance measured by a tip. Regions directly below the tip affect a smaller tip area than do surrounding regions, resulting in a nonlinear interaction.

and effective surface potential determined by SSPM is

$$V_{\mathrm{dc}} = \sum_{i=1}^{n} \frac{\partial C_{\mathit{eff},i}}{\partial z} V_{S,i} \Bigg/ \sum_{i=1}^{n} \frac{\partial C_{\mathit{eff},i}}{\partial z} \tag{7.21}$$

where $C_{\mathit{eff},i}(z)$ is partial capacitance between the tip and ith region on the surface. For inhomogeneous surfaces, local potential $V_{S,i}$ also reflects the difference in the surface work functions of dissimilar materials.[30]

The high spatial and voltage resolution of SSPM makes it a prominent tool for the characterization of current carrying devices, especially integrated circuit analysis.[31, 32] However, simple interpretation of SSPM in terms of surface potential is applicable only for conductive surfaces. Even though SSPM has been successfully applied to the characterization of semiconductor and dielectric surfaces, the image formation mechanism is not yet clearly understood. Such factors as tip-induced band bending, charge injection, a contact electrification, and redistribution of surface adsorbates all can contribute to the observed potential image, rendering its interpretation and quantification complicated tasks. Nevertheless, sensitivity of SSPM to the local variations of work function and Fermi level allows its application for dopant profiling in semiconductor devices.[33] The presence of local stresses,[34] impurities, and/or damage can also be measured. This technique is also sensitive to the local charge on the surface and, therefore, is used to study contact electrification and charge retention phenomena.[35] Presence of the trapped charges, e.g., at the surface–electroactive interface junctions can also be detected by the SSPM.[25, 36]

The versatility of SSPM allows its application with in situ external perturbations, including applied biases,[37] stresses, light illumination, or variable temperatures. An example is shown for a varistor in Figure 7.5.[38] Although the topography image exhibits no differences associated with multiple phases or grain boundaries, they can be detected in the surface potential images. In the potential image of a sample with no external perturbation, a depression of approximately 60 mV is observed that is associated with the difference in work functions of the ZnO surface and a pyrochlore phase. A difference of only 2 mV is measured between the left and right sides of Figure 7.5 which is ascribed to the differences owing to a variation in the orientations of adjacent ZnO grains. The surface potential map with the sample under applied lateral bias shows the abrupt drop at the grain boundary, i.e., the conductivity of the boundary is much lower than that of the bulk. A second example of SSPM measurement under lateral bias is shown in Figure 7.6. Again, the potential is almost constant within the grains but rapidly changes at the interfaces. The variation in local potential variation in this polycrystalline material is significant, ranging from about 100 to 700 mV. The voltage dependence of these grain boundary voltage drops can be used to determine local interface trap state densities and related parameters. The variation in GB properties in a single device certainly affects its performance and reliability. Such interface phenomena are the basis of

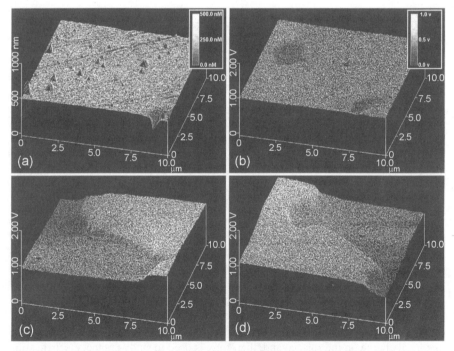

Figure 7.5. Polycrystalline ZnO. **a**, Topographic structure in an AFM image. **b**, Surface potential image identifying secondary phases with a 100 mV difference at the surface. Surface potential with (**c**) positive and (**d**) negative lateral voltage applied on the sample showing a 0.3 V potential drop at the grain boundary. (Reprinted with permission from Ref. 36.) Figure also appears in Color Plate section.

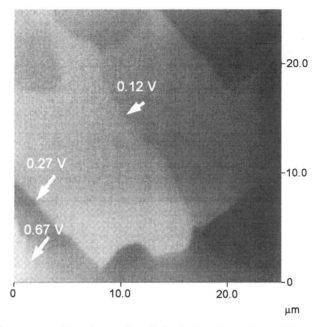

Figure 7.6. Large area of a polycrystalline ZnO showing the variation of potential drops at individual grain boundaries. (Reprinted with permission from Ref. 36.)

operation of positive temperature coefficient of resistivity (PTCR) thermistors, varistors, etc., and SSPM is a technique that allows easy study of in situ operation.

The nulling force technique used in the SSPM also provides significant advantages for variable-temperature measurements.[39] Elastic constants of the cantilever strongly depend on the temperature; therefore, the dynamic properties of the tip oscillations are influenced not only by tip–surface interaction but also by temperature variations. The application of a nulling force technique, which zeroes the force on the cantilever rather than measures the dynamic properties of cantilever oscillations, is much less sensitive to temperature fluctuations. Ferroelectric materials have surface potential dependent on polarization near the surface. The polarization magnitude decreases with temperature and disappears above Curie temperature T_c. Figure 7.7 illustrates the evolution of surface topography and surface potential with temperature through ferroelectric phase transition.[40] Topographic surface corrugations are related to 90° a-c domain walls in the y-direction, whereas the potential images also reveal 180° c^+-c^- domain walls. Upon transition to the paraelectric state, surface corrugations disappear, evidencing a first-order phase transition. Simultaneously, the surface potential exhibits large spurious potential amplitudes, possibly related to the release of charges that screened polarization in ferroelectric state. After annealing above T_c, charged species desorb or diffuse and the surface potential amplitudes disappear. On the reverse transition, surface corrugations form

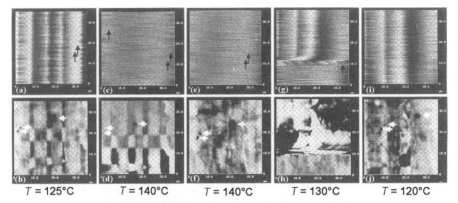

$T = 125°C$ $T = 140°C$ $T = 140°C$ $T = 130°C$ $T = 120°C$

Figure 7.7. Surface topography (*top*) and surface potential distribution (*bottom*) on a Ba-TiO$_3$(100) surface (**a, b**) before ferroelectric phase transition 125°C (**c, d**), 4 min after transition, (**e, f**) after 2.5 h annealing at 140°C (**g, h**), during the reverse transition, and (**i, j**) 1 h after the reverse transition. Note that potential features not related to the domain structure are unchanged. Scale is 30 nm for topographic images. Note the difference in scales for potential images: 0.1 V for (**b, h, j**), 0.5 V for (**d**), and 0.05 V for (**f**). *Arrows* indicate topographic and potential features unrelated to domain structure.

instantly, and the potential exhibits large transient variations. After equilibration below T_c, new surface domain structure and new surface potential are established.

Combined Techniques   The examples considered above are EFM techniques that use either a mechanically driven (force gradient techniques) or an electrostatically driven (voltage modulation) cantilever. A combined approach simultaneously imposing both modulations is also possible. One such approach was developed by Terris et al.[12] and successfully applied to studies of contact electrification. In this technique, the voltage modulation frequency is higher than the feedback loop response frequency but much lower than the oscillation frequency of the tip ω_t. This range of frequencies is chosen so that the surface performs many oscillations while the cantilever has almost constant lateral position (which contributes to resolution), but there is no resonance between oscillations of the cantilever and the sample. The force equations in this case can be obtained from Eq. (7.10) for $V_{dc} = 0$. Thus

$$F = \frac{Q_s Q_t}{4\pi\varepsilon_0 z^2} + \frac{1}{2}\frac{\partial C}{\partial z} V_{ac}^2 \sin^2 \omega t \qquad (7.22)$$

Because $Q_t = -(Q_s + Q_{ac})$, $Q_{ac} = C(z)V_{ac}\sin(\omega t)$, the equation for the force gradient can be obtained from this expression as:

$$F' = \frac{1}{2}V_{ac}^2 \sin^2(\omega t)\frac{\partial^2 C}{\partial z^2} + \frac{Q_s V_{ac}\sin \omega t}{2\pi\varepsilon_0 z^2}\left(\frac{C}{z} - \frac{1}{2}\frac{\partial C}{\partial z}\right) + \frac{Q_s^2}{2\pi\varepsilon_0 z^3} \qquad (7.23)$$

For the uncharged surface, $Q_s = 0$, only the second harmonic term is nonzero, and the force gradient oscillates at the frequency 2ω. This second harmonic term causes the tip oscillations at ω_t to be modulated at 2ω. For the charged surfaces, the harmonic term is also present, and tip oscillations are modulated at frequency ω. This harmonic signal is detected as the output of feedback lock-in amplifier (reference frequency ω_t) with a second lock-in (reference frequency ω). The phase of the signal reflects the sign of the surface charge. The combined modulation technique was further developed by Ohgami et al.[41] who considered the effect of additional constant bias applied to the tip, $V_{dc} \neq 0$, on the first harmonic signal. The general property of these techniques is that topographic and electrostatic information are collected in a single scan. The observed signal is the sum of electrostatic and electromechanical contributions (Tables 7.1 and 7.2). Quantification of images in terms of surface charge, dielectric constant, and electromechanical properties can be complicated.[42]

7.2.3 Scanning Capacitance Microscopy

The electrostatic scanning probe techniques discussed above were based on the detection of the force between the tip and the surface. However, the tip can also be used as a probe of local tip–surface capacitance. As discussed in Section 7.2.1, capacitance is a function of tip–surface separation, topography, tip shape, etc. For semiconductor materials, capacitance depends on applied voltage, and capacitance-voltage (C-V) curves obtained in macroscopic measurements provide a convenient way to study the properties of the semiconductor.[43] This approach can be extended to the SPM, which is the basis of scanning capacitance microscopy. Correspondingly, this technique is particularly useful for the spatially resolved characterization of semiconductor devices.[44]

Consider the example of a n-type semiconductor under an applied ac field in the classical metal-oxide-semiconductor configuration. The presence of the oxide (or, more generally, any insulating material) layer prevents charge transfer between the metal and semiconductor. Application of a positive bias to the semiconductor surface results in downward band bending; carriers are effectively attracted to the surface (accumulation), and the capacitance of the system is large (Fig. 7.8a). In contrast, the application of a negative bias results in upward band bending; carriers are "pushed" from the surface (depletion), and the capacitance decreases. Formation of accumulation and depletion regions can be represented as the movement of equivalent capacitor plates. Because capacitance is inversely proportional to the distance between the plates, the dependence of the total capacitance on applied bias follows the curve shown in Figure 7.8a. The width of the depletion region and, therefore, the capacitance are inversely proportional to the donor concentration. Hence the difference in the C-V curves for high and low donor concentrations. For p-doped material, the C-V curve is reversed.

The exact shape of C-V curve depends on the applied bias, thickness of the oxide layer and doping level of the semiconductor. Provided the oxide thickness is known, a C-V curve can be used to determine the doping level of a semiconductor.

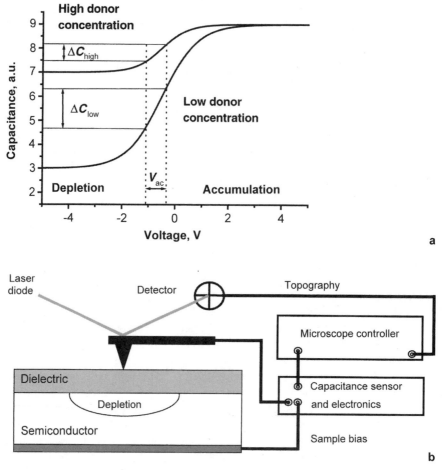

Figure 7.8. a, high-frequency *C-V* curve for an *n*-doped semiconductor with high and low donor concentrations. **b**, Experimental setup for scanning capacitance microscopy.

Obviously, the application of a small capacitor (SPM probe) provides spatial localization of *C-V* curves and thus is used to map a two-dimensional (2-D) dopant profile of the semiconductor device.[45] In many cases, information on doping can be obtained even without the detailed knowledge of the *C-V* curve. Also, the *C-V* curve for a plane-plane arrangement necessarily differs from that for the SPM geometry, because in the latter case fringe effects are significant and for large tip–semiconductor separation will dominate the capacitance. Despite these difficulties, both quantitative and qualitative SCM are proven to be some of the most versatile tools for dopant profiling in semiconductor devices.

Image acquisition in SCM is performed on the semiconductor surface covered with an oxide film of known thickness in contact mode with conventional deflection feedback.[46,47] The advantage of this approach is that contact mode imaging is relatively insensitive to local dielectric constant, surface charge density, or conductivity of the surface and thus provides reliable topographic data. For the SCM signal, an ac bias in the kiloHertz range is applied between the tip and the surface, and the capacitance is detected with the ultrahigh frequency (UHF) capacitance sensor (Fig. 7.8**b**). The variation of the capacitance ΔC owing to depletion–accumulation below the tip is usually small compared to overall tip–surface capacitance and is measured with the lock-in technique. As seen from Figure 7.8**a**, the measured quantity is essentially the derivative of capacitance with respect to bias voltage $\Delta C/\Delta V$, or the slope of the C-V curve. In regions with higher carrier concentration, the depletion depth is smaller, consequently the slope is small. In regions with lower carrier concentration the depletion depth and the slope are large. Alternatively, for a uniformly doped sample the thickness of the surface oxide film is inversely related to the slope of the C-V curve. This imaging technique (ΔC mode) is easily realized; however, measurable change in the capacitance can often be obtained only if the driving amplitude is sufficiently high (several volts). This effectively smoothes C-V curve. For the samples with low carrier concentration, high driving voltages result in big depletion volumes and the loss of lateral resolution. An alternative SCM imaging technique uses additional feedback loops that keep ΔC constant by adjusting the driving voltage ΔV. This technique has an advantage of constant depletion geometry, but the experimental setup is more complicated.[48]

Because the shape of the C-V curve depends strongly on the oxide thickness, SCM data can be obtained only from semiconductor surfaces with uniform oxide thickness. Also, the surfaces should be flat to avoid the geometric effects in the capacitance. Interpretation of the SCM data obtained even under the ideal imaging conditions represents an extremely complicated problem owing to the three-dimensional (3-D) geometry of the problem, which precludes the analytical solution and requires application of numerical techniques. Nevertheless, even semiquantitative SCM provides a remarkable example of technique for the imaging of semiconductor devices. Figure 7.9 shows an SCM image of a 0.18 µm field effect transistor (FET) device. The p-doped source and drain are exhibit by different in the image. The bright line is the compensated region between the source–drain and the well (black), which prevents the leakage. The local capacitance distribution for this image is more clearly seen in color (see color plates). Application of external potential between the gate and n-doped well induces depletion–accumulation in the gate region, thus rendering it conductive or nonconductive for currents between the source and the drain.

7.2.4 Scanning Spreading Resistance Microscopy

The techniques considered so far detect the force acting on the tip or the capacitance in the tip–surface system. Possible charge transfer between the tip and the sample in these techniques is avoided either by maintaining finite a tip–surface separation or by ensuring the existence of the dielectric layer on the tip (e.g., oxidized silicon tip) or on the surface. Detection of dc or ac current through

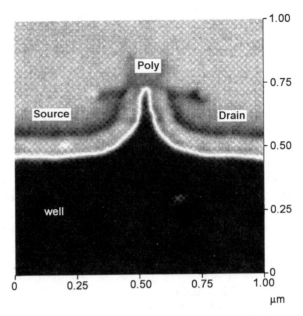

Figure 7.9. SCM image of a cross-sectioned 0.18-μm pFET device. (Courtesy of J. S. McMurray and C. C. Williams, University of Utah.) Figure also appears in Color Plate section.

tip–surface junction under applied bias could provide information on the resistivity of the sample. It can be easily shown that owing to the local nature of the probe, current is limited by the resistivity of the surface directly below the tip.[49–51] This technique—referred to as scanning spreading resistance microscopy (SSRM)—is another example of a local probe technique applicable to semiconductor materials. Measuring the current for constant bias voltage allows local resistivity mapping that can be related to the chemical composition, providing complimentary information to the SCM. A similar approach is used for probing leakage currents in thin dielectric films and is known as leakage current microscopy (LCM).[52,53]

7.2.5   Piezoresponse Imaging

Some of the EFM techniques discussed above are based on the detection of the electrostatic field far from the surface, where the contribution of strong short-range vdW forces is negligible. Alternatively, the tip is in contact and the electric response of the surface to the applied bias is measured. However, for a broad class of materials, i.e., piezoelectrics, the application of bias results in a significant mechanical response that can be detected by SPM. The set of techniques based on the detection of mechanical response to the applied bias[54,55] is generally referred to as piezoresponse imaging (PRI). These techniques are valuable tools for the characterization of piezoelectric and ferroelectric materials.

Piezoelectric properties are described by the tensor of piezoelectric components d_{ijk}.[56] The number of independent components depends on the symmetry of the system. Strain χ_{ij} induced by the applied electric field is

$$\chi_{ij} = d_{kij}E_k = (d_{ijk})^t E_k \qquad (7.24)$$

and is a linear function of applied field. The corresponding second-order effect is electrostriction, in which case

$$\chi_{ij} = M_{ijkl}E_k E_l \qquad (7.25)$$

where M_{ijkl} are the components of the tensor of electrostrictive coefficients.

In the general case, the calculation of the response of a piezoelectric surface to an applied electric field requires the solution of a coupled electroelastic problem.[57–59] With an analytical solution to the point-plane configuration, the orientation of polarization vectors at each position on the sample as well as the piezoelectric and electrostrictive coefficients of the material can be obtained from piezoresponse measurements. Unfortunately, the general solution is complicated,[‡] and the usual approach is to reduce the problem to a simple form under special assumptions. For example, assuming a homogeneous field in the xy plane and a polarization vector oriented to the normal of the surface, the problem is reduced to one dimension and the measurement of sample deformation as a function of applied field gives information on the piezoelectric constant.

A typical value of the piezoelectric constant for conventional piezoelectric and ferroelectric materials is on the order of a few angstroms per volt. Therefore, application of 10 V between the tip and the surface will cause surface deformation on the order of several nanometers. The actual value may be smaller, because the piezoelectric constant for clamped crystal is smaller than that for free body, and the field distribution under the tip is not uniform. The deformation is relatively small and is hard to detect directly. As in SSPM and SCM, the breakthrough in the detection of local electromechanical properties of materials was achieved by the application of voltage modulation techniques. The oscillating electric field applied between the tip and the surface results in the deformation of sample surface. As in other voltage-modulation techniques, the applied field results in a static response as well as first and second harmonic terms. The static signal is superimposed on the height data for the nonbiased surface and usually cannot be detected. First and second harmonic signals can be extracted with the lock-in technique and used to map local piezoelectric and electrostrictive properties of the surface. The amplitude of the first harmonic signal provides information on the absolute value of the z-component of the polarization vector. The phase of the piezoresponse signal depends on the sign of piezoelectric coefficient and allows determination of the polarization direction. This approach is limited to materials with sufficiently high

---

[‡]For the simplest case of opposite c^+-c^- domains in perovskites, the displacement under applied tip bias depends on the contact force, tip radius of curvature, and 10 electroelastics constants. Stresses, point, and extended defects as well as surface charges can also contribute to piezoresponse contrast.

piezoelectric coefficients; otherwise, the electric field necessary to produce the observable deformations of the sample will be greater than the coercive field, and ferroelectric switching below the tip will complicate image interpretation.[60]

Superposition of a dc potential offset to the tip allows measurement of the local hysteresis loop for the ferroelectric material and investigation of fatigue effects in ferroelectric structures.[61–63] Quantification of piezoresponse images in terms of the local piezoelectric constant is simplified for poled, columnar, and epitaxial films and heterostructures with well-defined orientations of domains within the film. For such systems, interpretation of piezoresponse images is relatively straightforward owing to the well-defined orientation of polar axis. Also the electrostatic field distribution can be well approximated without electromechanical coupling. On the other hand, for thick films or bulk solids the coupled problem must be solved. If the grain size of material is smaller than the piezoresponse region, the image contains the sum of responses from several adjacent grains. Figure 7.10 illustrates the experimental setup for PRI and the contours of the E_z field distribution under the point tip, delineating the shape of piezoresponse region. Further development of this technique based on simultaneous measurement of all components of displacement (rather than only the vertical component) allows determination of all components of polarization vector in each point of the surface and yields a reconstruction of surface crystallography.[64] A better estimate of the piezoelectric coefficient can be obtained by using a slightly different piezoresponse scheme. A macroscopic disk

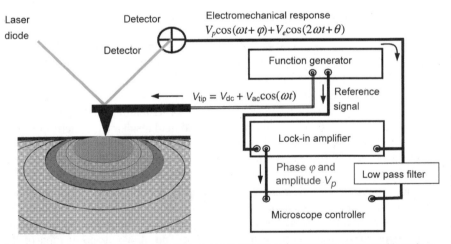

Figure 7.10. Experimental setup for PRI. A conductive tip is brought into contact with surface. The ac bias results in an electromechanical response of the surface at the main frequency owing to the inverse piezoeffect and twice the main frequency owing to electrostriction. Phase shift and amplitude of the first harmonic signal are detected by the lock-in amplifier and stored by the computer. The same signal is used after low pass filtering for determining the surface topography, as in conventional contact mode AFM. The distribution of the z-component of the electric field is shown in a logarithmic scale, which gives some indication of the region characterized by PRI.

electrode is deposited on the surface, displacement of which is detected by the AFM tip.[65] Lateral resolution, however, is lost with this technique.

The advantage of PRI is that images are related to the subsurface structure of the material and thus are not subject to deterioration owing to adsorbed charges (which can effectively cancel the surface polarization charge and thus render such techniques as SSPM inapplicable). Furthermore, imaging (and straightforward interpretation of the observed data) is possible even for rough surfaces and surfaces covered by contamination layers (which can form, for example, because of the loss of oxygen or other volatile components).[66] Figure 7.11 compares surface topography and piezoresponse image for micropatterned $PbZr_xTi_{1-x}O_3$ (PZT) lines on Pt-coated silica substrate. Individual grains are resolved within the lines in the piezoresponse image, reflecting the nonuniform nature of the material and the existence of grains with different polarization orientations. Dark areas indicate the polarization component oriented out of the surface plane, and light regions correspond to the polarization oriented inward. The regions with intermediate contrast correspond to polarization with a large in-plane component. Note that no piezoresponse contrast is seen between the lines on the Pt-coated regions.

7.2.6   Electrostatic Sample Modification

In the last few years significant attention has turned toward modifying surfaces with tips. Surface modification can be performed on small length scales compared to the effective tip size and thus, in principle, can surpass the capabilities of

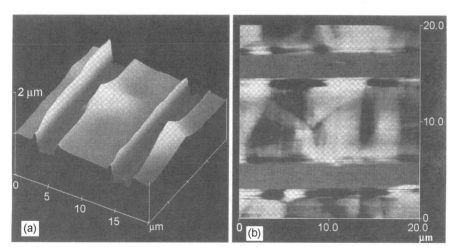

Figure 7.11. Surface topography in (**a**) a 3-D rendering and (**b**) a piezoresponse image of micropatterned PZT lines on Pt-coated silicon. The piezoresponse image shows grains with different polarization within the lines. At the same time, platinum between the lines do not contribute to piezoresponse image. (Sample courtesy of M. Ozenbas and I. Aksay, Princeton University.)

photolithography. Of course, this case is serial rather than parallel and may not be sufficiently fast to be commercially viable. Mechanical modification of the surface, i.e., nanoindentation has become a paradigm in the AFM and is supported by most commercial instruments. It is widely discussed in a number of reviews[67, 68] and in Chapter 10. Application of electric potential through the tip in contact mode can also be used to modify the surface. Depending on the nature of the surface, the material of the tip, the surrounding media, and applied bias, a wide spectrum of electrochemical or physical processes can occur. Most attention has been focused on the electrochemical oxidation of semiconductor surfaces and polarization reversal on ferroelectric surfaces and heterostructures. An example of the former is the electrochemical oxidation of silicon, which allows formation of extremely thin SiO_2 lines.[69] Proposed mechanism includes formation of a water droplet between the surface and the tip; the latter serves as an electrode and causes the oxidation of the substrate. This method allows patterning of silicon surface with characteristic feature size of few 10s of nanometers.[70]

Another spectacular example of potential-induced surface modification is ferroelectric switching in epitaxial films. One of the most prominent features of ferroelectrics is their ability to change the direction of polarization under the influence of electric field or mechanical stress and to retain polarization direction after the field is switched off. It is this property that makes ferroelectric materials and ferroelectric heterostructures perspective materials for memory devices. Owing to the small size of the probe tip in SFM, even a small applied field or force can cause local domain reversal and thus can be used for storing information on a ferroelectric substrate.[71] It has been shown that the minimum achievable size of the written domain is governed by the microstructure of the material and that the highest possible storage density is expected in thin epitaxial films. The stability of induced domain structures depends both on the intensity and on the length of the applied voltage impulse and the structure of ferroelectric material. Domains larger than a critical size were found to be stable for a long periods of time. Feature resolution of 50 nm was achieved for epitaxial PZT(001)/Pt(111)/sapphire(0001) films by Hidaka et al.[72] Comparable resolution of <100 nm was achieved for PZT films on insulating $SrTiO_3$ or metallic Nb-doped $SrTiO_3$ substrates.[73]

Similar experiments have been done recently on semiconductor and superconductor ferroelectric heterostructures. In this case, ferroelectric switching in the ferroelectric layers results in electron or hole injection to the second component and can result in the metal-insulator or superconductor-dielectric phase transitions, as demonstrated by Ahn and co-workers.[74,75] Combined with the possibility to write features as small as a few 10s of nanometers, this provide multiple possibilities for the development of nanoelectronic devices.

7.3 MAGNETIC PROBE IMAGING

It was realized early that use of a ferromagnetic tip can make the AFM sensitive to magnetic forces. High resolution, pertinent to all scanning force microscopies, makes MFM a unique tool for the characterization of micromagnetic structures in materials and their dynamic behavior under applied stress, bias, etc. Examples of prominent applications of MFM include analysis of magnetic recording media, especially studies on the stability of recorded data under mechanical or magnetic influences. Understanding of domain wall motion and switching in the magnetic media is benefited by real-space imaging of magnetic walls behavior under applied fields, e.g., domain wall dynamics and pinning. The high resolution of the MFM allows imaging of the internal structure of domains walls as well, even though limitations exist in this case. MFM has been further extended to magnetic field imaging in current carrying devices and thus for the failure analysis in electrical circuits. Finally, MFM provides a tool for in situ studies of superconducting currents in high-temperature superconductors. Analysis of magnetic field distribution provides information about the flux vortex pinning and behavior of the current in the vicinity of the grain boundaries.

It should be noted that there exist numerous techniques for the micromagnetic studies, including Lorentz electron microscopy,[76] Kerr optical microscopy,[77] spin-polarized electron microscopy (SEMPA), and scanning probe microscopies.[78–80] All techniques provide different options from the point of view of spatial resolution, magnetic contrast, influence of measurement technique on the surface, and image acquisition time. Even among the SPM techniques, different kinds of local probes can be used. In this section we consider only conventional MFM that uses a ferromagnetic tip as field sensor. As an example of alternative SPM techniques that use other types of field probes, we briefly present scanning SQUID microscopy. All other local probes—such as Hall probe—are not discussed and the reader is referred to the original literature.

7.3.1 Tip–Surface Interactions in MFM

As discussed in Chapter 2, the most wide spread MFM is based on the use of a ferromagnetic tip as local field sensor. Magnetic interaction between the ferromagnetic tip and the surface results in a force acting on the tip, which can be detected from the deflection of the cantilever. Further development of the MFM technique uses the dynamic properties of the tip oscillation, i.e. amplitude, phase or frequency detection. In this case, however, the observed signal is proportional to the derivative of the force. Unlike the EFM, the magnetic state of the tip can hardly be controlled during the experiment, thus currently there are no analogs to the voltage modulation techniques, even though the possibility of applying external dc or ac magnetic fields to modify the tip–surface system and achieve maximal resolution and force sensitivity is being studied.[81–83]

As for other scanning force microscopies, the ultimate goal of MFM is to infer the surface magnetization distribution (e.g., domain structure) from the force or force gradient data collected in static or dynamic mode. In addition to magnetostatic interaction between the tip and the surface, such effects as tip-induced reorientation of surface magnetization (and surface-induced reorientation of the tip magnetization) are also possible and can complicate the image interpretation.

Even under the assumption of rigid magnetic structure of a tip and surface (magnetostatic approximation) the force depends both on the state of the tip (thickness of ferroelectric coating, domain orientation, especially near the tip end) and on the magnetic structure of the surface. For arbitrary sample magnetization, the magnetostatic potential, $\phi_s(\mathbf{r})$, is[21]

$$\phi_s(\mathbf{r}) = \frac{1}{4\pi} \int d^2s' \cdot \frac{\mathbf{M}_s(\mathbf{r}')}{|\mathbf{r} - \mathbf{r}'|} - \frac{1}{4\pi} \int d^3r' \frac{\nabla \cdot \mathbf{M}_s(\mathbf{r}')}{|\mathbf{r} - \mathbf{r}'|} \qquad (7.26)$$

where $\mathbf{M}_s(\mathbf{r}')$ is the magnetization vector field of the sample, $\mathbf{s}'$ is an outward normal from sample surface and $\mathbf{r}$ and $\mathbf{r}'$ are radius vectors. The stray field produced by the sample is

$$\mathbf{H}_s(\mathbf{r}) = -\nabla \phi_s(\mathbf{r}) \qquad (7.27)$$

The magnetostatic energy $\psi(\mathbf{r})$ of a microprobe exposed to this field is

$$\psi(\mathbf{r}) = \mu_0 \int d^2s' \cdot \mathbf{M}_p(\mathbf{r}')\phi_s(\mathbf{r}') + \mu_0 \int d^3r' \nabla_{\mathbf{r}'}[\mathbf{M}_p(\mathbf{r}')\phi_s(\mathbf{r}')] \qquad (7.28)$$

where $\mathbf{M}_p(\mathbf{r}')$ is the magnetization vector field of the probe, μ_0 is permeability of vacuum. The force is

$$\mathbf{F}(\mathbf{r}) = -\nabla \psi(\mathbf{r}) \qquad (7.29)$$

Despite the fact that Eqs. (7.28) and (7.29) are rigorously correct, implementation for the quantification of MFM images is virtually impossible because the magnetic state of the tip is generally unknown. A number of simple models have been suggested to describe the magnetic state of the tip $\mathbf{M}_p(\mathbf{r}')$. Because only the parts of the tip close to the surface effectively interact with local stray fields, the tip can be conveniently modeled as prolate spheroid of uniform magnetization.[84] The polarization of the tip can be altered to a desired configuration by magnetizing it in a strong magnetic field. In the pristine (nonmagnetized) state, shape asymmetry of the tip usually results in axial orientation of magnetization.

An even more simplified description of tip–surface forces is achieved by the introduction of the point-probe approximation.[85] In this case, the magnetic state of the tip is described by its effective first- and second-order multipole terms, and the force acting on the probe is

$$\mathbf{F} = \mu_0(q + \mathbf{m} \cdot \nabla)\mathbf{H} \qquad (7.30)$$

where q and $\mathbf{m}$ are effective probe monopole and dipole moments. Relative contributions of monopole and dipole contributions depend on the domain structure in the vicinity of the end of the tip, as illustrated in Figure 7.12. If the length of the effective domain at the tip apex is small compared to the characteristic decay length

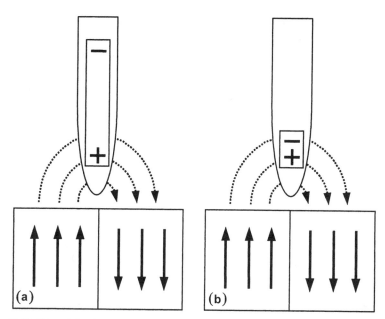

Figure 7.12. The origin of monopole and dipole terms in MFM imaging. **a**, the characteristic size of magnetic domain in the tip is much larger than the spatial extent of the surface stray field, the monopole term dominates. **b**, If the characteristic domain size in the tip is small, the dipolar term dominates. For a realistic tip shape both terms are present.

of the sample stray field, the tip can be approximated as point dipole, and the force is

$$\mathbf{F} = \mu_0(\mathbf{m} \cdot \nabla)\mathbf{H} \qquad (7.31)$$

In this case, the gradient of the stray magnetic field in the vicinity of the surface rather than the field itself contributes to the force signal. The force gradient signal detected in dynamic SPM has an even more complicated structure. For a large effective domain size, only the front part of domain effectively interacts with surface stray field, and the force in this case is

$$\mathbf{F} = \mu_0 q \mathbf{H} \qquad (7.32)$$

i.e., the field (or field gradient for dynamic MFM) is detected.

Quantification of MFM data requires both monopole and dipole moments. Corresponding moments, along with the effective probe–surface separation, can be inferred from the fit of the experimental data to Eq. (7.30), provided that the stray field distribution is known. Conveniently, these parameters can be obtained from calibration of the tip against the standard system, e.g., microfabricated coils or lines carrying known current, and can often be used for the quantification of the MFM data from the surface under investigation.[86–88]

As mentioned, stray fields produced by the tip can significantly alter the surface magnetic structure of the sample. Alternatively, the magnetic structure of the tip can be altered by the surface stray fields. The former effect is expected when the magnetic stray field by the tip H_s(tip) exceeds the anisotropy field at the surface H_k(sample), H_s(tip) $> H_k$(surface). Alternatively, tip magnetization will be influenced by the surface when H_s(surface) $> H_k$(tip). Because the stray field is proportional to the saturation magnetization M_s in the vicinity of a ferromagnetic surface, the conditions for non-destructive MFM operation are[89]

$$\mu_0 H_k(\text{sample})/M_s(\text{tip}) > 1 \qquad (7.33)$$

$$\mu_0 H_k(\text{tip})/M_s(\text{sample}) > 1 \qquad (7.34)$$

These stability conditions can be achieved by the optimal choice of tip material (high coercive field and large magnetic anisotropy) and imaging at appropriate tip–surface separations. The latter, however, results in the loss of spatial resolution. An alternative approach minimizing tip-induced effects is the application of soft magnetic tips or superparamagnetic tips with large susceptibility. The magnetic structure of such a tip aligns parallel to the surface stray field and, therefore, the tips are sensitive to the absolute value of magnetic field. For imaging with soft magnetic tips the force is

$$\mathbf{F} = \frac{1}{2} \chi V \nabla \mathbf{B}^2 \qquad (7.35)$$

where χ is the magnetic susceptibility of material and V is effective tip volume.

Interpretation of the MFM data from nonhomogeneous materials and current-carrying devices can suffer from electrostatic contributions to force or force gradient signal contrast, especially for conventional metal or metal-coated tips. Electrostatic and magnetic forces have different distance dependencies and, in principle, can be distinguished by variable lift height measurements.[90] Alternatively, the appropriate choice of active coating for the tip makes it preferentially sensitive either to electrostatic or to magnetic forces or to both. Several examples of magnetic/nonmagnetic conductive/nonconductive tips are listed in Table 7.4. Eq. (7.26) also implies that magnetic force or force gradient detected in the MFM is determined by the subsurface magnetization distribution of the sample. Thus MFM is not sensitive to the surface contamination that often poses a problem in electrostatic force microscopy.

Among the simplest and most spectacular applications of MFM is visualization of complex magnetic domain structures in ferromagnetic materials. In such systems, magnetization is almost uniform within domains and rapidly changes at domain boundaries. The high spatial resolution of MFM not only allows real-space imaging of magnetic domains but also gives insight into the structure of domain boundaries. Interpretation of MFM images of domain structures and examples of tip-induced effects on imaging are given in next section.

Table 7.4   Coatings for SPM tips

	Conductive	Nonconductive
Magnetic	Metal-coated tips (ferromagnetic or soft magnetic metal)[a]	Ferrite or other nonconductive magnetic particles attached to the tip
Nonmagnetic	W_2C or TiN coated tips,[b] conductive diamond tips[c]	Conventional Si or Si_3N_4 tips[d]

[a]For example, MESP (hard) and MESP-LC (soft) magnetic probes from Digital Instruments.
[b]For example, NSC15 or NSC16 probes with W_2C or TiN coating from Silicon-MDT/NT-MDT.
[c]For example, conductive diamond probes from ThermoMicroscopes.
[d]Available from all major SPM producers.

7.3.2 Micromagnetic Studies by MFM

A variety of materials have built-in nonuniform magnetization, e.g., form magnetic domain structures. The magnetization vector within a domain is oriented along one of the easy axes of magnetization. The magnetization field is uniform within the domain and rapidly changes at domain boundaries. Depending on the behavior of the magnetization vector in the vicinity of the boundary, Bloch and Neel domain walls can be discerned.[91] In the former, the magnetization vector rotates in the plane of the domain wall. In the latter, the magnetization vector rotates in the plane perpendicular to the domain wall. The width of magnetic domain walls is usually on the order of $\sim 100\,nm$.

For single-crystal materials, the magnetic domain structures are related to the magnetocrystalline anisotropy of a material, and possible magnetic configurations can be obtained from the following considerations. Consider magnetite Fe_3O_4, for which the easy magnetization axis is (111), i.e., magnetization is aligned along one of the eight equivalent (111) directions. In the vicinity of the surface, the most stable domain configurations correspond to the domains with purely in-plane magnetization or with minimal out-plane magnetization if the (111) direction does not belong to the surface plane. For the (110) surface, the most favorable domain arrangements are (1-11), (-1 1 1), (1-1-1) and (-1 1 1), all of which have in-plane magnetization. This combination of magnetization vectors gives rise to 71°, 109° and 180° domain walls, which are observed on magnetite surface by MFM (Fig. 7.13).[92] Minimization of free energy requires the normal to the domain wall component of magnetization to be constant (i.e., there is no magnetization charges on the wall), which indeed is the case for the domain structure shown. Note that the 180° domain wall on the top of the image is seen as a dark line, whereas the 180° wall on the bottom of the image is bright. The 71° and 109° walls are composed of dark and bright parts. A further example of contrast variation at domain wall is presented in Figure 7.14.

These observations indicate that high spatial resolution makes it possible to investigate not only the overall domain distribution within the sample but also

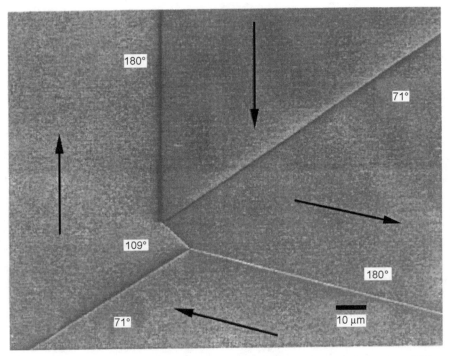

Figure 7.13. MFM image of domain structure on (110) magnetite. *Arrows* indicate magnetization orientation. Types of domain walls are also indicated. (Reprinted with permission from Ref. 92.)

the fine details of domain wall structure. Even though single spins cannot be resolved by MFM because magnetic interactions are collective, resolutions well below 100 nm can be achieved. This suffices for the symmetry of domain walls to be determined and for characterizing the type of domain wall. For example, a 180° Bloch wall between two antiparallel domains with in-plane magnetization (Fig. 7.14)[93] is characterized by chirality, i.e., whether the spins rotate clockwise or counterclockwise in the wall plane. The sign of the force between MFM tip polarized in the z-direction and 180° Bloch walls depends on spin rotation direction. If the spin rotation is such that positive ends point upward in the wall region, the force will be repulsive for a tip with the magnetization vector directed downward. As discussed in the section on EFM, repulsive forces lead to a positive frequency shift, and the walls are seen as bright lines on the MFM image. Conversely, if the positive ends point downward, the force is attractive and the MFM contrast is negative (dark lines). For a tip magnetized in upward direction, the MFM contrast inverts.

Near the surface, minimization of depolarization energy requires the magnetization vector to rotate in the surface plane, i.e., the formation of a so-called Neel cap.[94,95] The MFM contrast will have contributions both from the bulk wall and from the surface Neel region. This behavior was predicted and observed in micro-

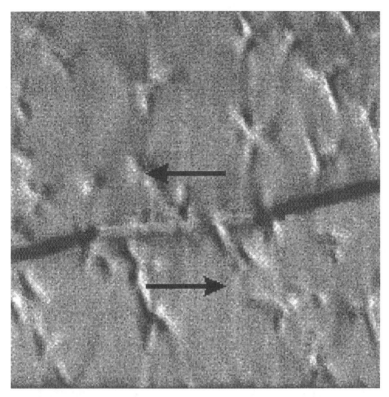

Figure 7.14. A 20 × 20-μm MFM image of a subdivided 180° Bloch wall in an Fe single crystal. Adjacent and mutually titled wall segments exhibit different wall chirality. (Reprinted with permission from Ref. 93.)

magnetic simulations long ago, but it was not until recently that it was observed experimentally. Figure 7.15 shows the development of the magnetic wall structure with film thickness for Fe films. Bloch walls are more stable in the bulk. However, at small film thickness, the formation of a Bloch wall is associated with energetically unfavorable stray fields. On the other hand, a Neel wall with in-plane magnetization doesn't produce large fields and, therefore, is more stable. For larger film thickness, the bulk contribution to free energy dominates, and a Bloch wall is stabilized. Figure 7.15 compares experimentally observed MFM profiles of domain walls with the results of micromagnetic simulation.[96] The domain wall types can be easily identified.

Conventional ferromagnetic tips can exert relatively strong magnetic fields that can alter magnetic structure of the sample. Significant difficulties in the MFM can arise from tip-induced effects during the imaging. The magnetic induction at the tip axis of typical tips with CoCr coating is 40 mT at 40 nm above tip apex[97] and can distort the surface-domain structure for soft magnetic materials. The tip can

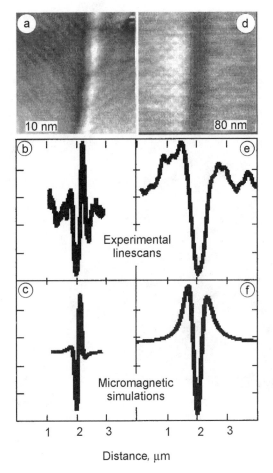

Figure 7.15. MFM results and model calculations for domain boundaries in Fe films for two film thicknesses. The experimental results were obtained under UHV conditions. Comparison between experimental and theoretical results clearly confirms that a 90° Neel wall is present for the 10-nm film, whereas a 90° Bloch wall is present for the 80-nm film. (Reprinted with permission from Ref. 93.)

effectively "drag" the domain walls. This effect is more pronounced for small tip–surface separations and can be minimized by increasing the separation. Figure 7.16 shows the influence of lift height on the MFM image of a garnet film. For large tip–sample separations, the influence of the tip on the magnetic structure of the sample is relatively small, hence the observation of magnetic structure is possible. Lateral resolution in this case is limited by the large tip–surface separations. For smaller tip–surface separation, magnetic force exerted by the tip scanning across the surface results in "dragging" domains across the surface, resulting in a distortion. When the tip is close to the surface, the magnetic structure of the sample is completely destroyed and can no longer be observed. Images obtained after the tip is retracted to the initial height indicate that the magnetic domain structure has indeed changed during imaging at small separations (Fig. 7.16).

Magnetic interaction between the tip and the surface can also modify the structure of the domain walls. For example, micromagnetic simulations predict that a 180° Bloch wall between antiparallel domains at the surface will terminate with a Neel cap. Figure 7.17 presents the MFM profile across the wall obtained for MFM tips polarized "up" and "down".[92, 98] The Neel cap at the Bloch wall–surface

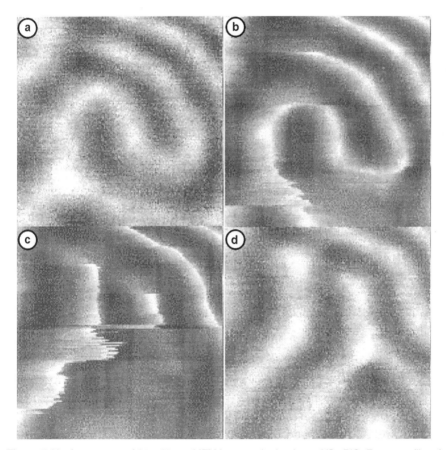

Figure 7.16. A sequence of 25 × 25-μm MFM images obtained on a YSmBiGaFe garnet film of 4.5 μm thickness at probe–sample separations of (**a**) 910 nm, (**b**) 520 nm, (**c**) 390 nm, and (**d**) 910 nm. It is obvious that the domain configuration of the film with perpendicular anisotropy is perturbed if the probe–sample separation becomes too small. If, after such a destructive probe–sample interaction, the probe is retracted to the original working distance, the sample magnetization has completely changed to a new remanent state. The images were taken with a standard MFM probe. (Reprinted with permission from Ref. 93.)

junction is approximated as a dipole. The MFM tip field results in the tilting of the dipole and an additional attractive interaction. Hence the difference in MFM profiles.

7.3.3 MFM of Superconductors

The properties of polycrystalline high temperature superconductors (HTSC) deviate from those of single crystals of the same material by orders of magnitude. It was realized early that the origin of the behavior is the relatively small correlation

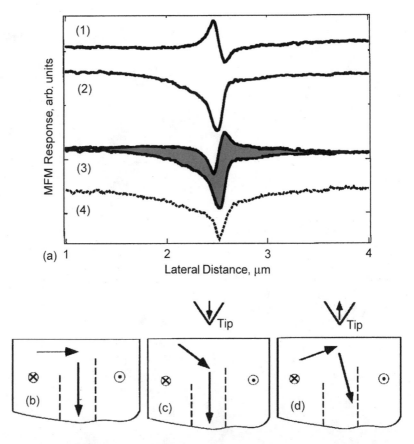

Figure 7.17. a, MFM response profiles measured across a 180° Bloch wall on (110) magnetite for the repulsive *1* and attractive *2* cases. In case *3*, profile *1* was inverted and superimposed on profile *2*. Profile *4* is the difference profile, which shows the effect of an additional attractive MFM response owing to the tip-induced effect on the domain wall structure. **b,** Representation of the Neel cap dipole at the Bloch wall–surface junction and its reorientation owing to the interaction with the MFM tip in **(c)** attractive and **(d)** repulsive cases. (Reprinted with permission from Ref. 92.)

length in these materials, which renders transport properties extremely sensitive to the imperfections, such as GBs and impurities. Grain boundaries act as Josephson junctions and limit critical current. Impurities, on the other hand, are often deliberately introduced to act as pinning centers. The GB structure in HTSC ceramics strongly depends on the preparation route, and significant work has gone into understanding the influence of processing conditions on the morphology of HTSC bulk ceramics, thin films, and filaments as well as related transport properties. This is an obvious case in which the direct visualization of current and particularly the behavior near grain boundaries and impurities is invaluable. Numerous magnetic imaging techniques have been applied to the characterization of micromagnetic structures, such as single Abricosov vortices and vortex lattices in HTSC single crystals and ceramics. Some of the results are summarized by Bending,[99] and the reader is referred to the next section for further examples. In this section, the use of MFM at low temperature is described to characterize current paths in a HTSC material under operational conditions, i.e., during current flow.

Figure 7.18 shows magnetic fields emanating from a textured Bi-2212 ($Bi_2Sr_2Ca_1Cu_2O_x$) film owing to current flow.[100] The physical microstructure consists of lathe-like grains oriented in the plane of the film, and the measurement is made on a cross section of the film. Current is obstructed at various grain boundaries and makes a tortuous path through the film. This is reflected in the complex pattern of the magnetic fields induced by the current. A more detailed view is shown in Figure 7.19. The superconducting transition of this film occurs between 93 and 96 K. The topographic structure measured at low temperature is not directly related

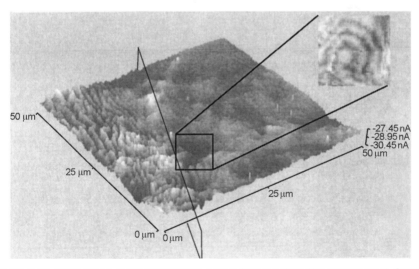

Figure 7.18. MFM image of superconducting current flow in a thick film at 60 to 65 K. Current flows from top to bottom. (Reprinted with permission from Ref. 100.) Figure also appears in Color Plate section.

to the underlying microstructure; rather it represents the surface polishing. The MFM image acquired between 60 and 65 K with no current (Fig. 7.19a) exhibits extremely low contrast with isolated bright spots owing to field focusing at sharp microstructural features. These contributions are subtracted from subsequent images. This image (Fig. 7.19a) can be compared to that obtained with $100 \, A/cm^2$ current flowing from top to bottom (Fig. 7.19b). At this resolution, the contrast variations are large and oriented with respect to the underlying microstructure. The details of the magnetic field variations are more clearly seen in a higher resolution image of a subsection of this region (Fig. 7.19c). The contrast variations are bright and dark lines with widths ranging from 0.8 to 4 µm and extending in length to >10 to 15 µm. The fractional area of bright contrast in Figure 7.19b amounts to 10 to 20%, indicating that only a fraction of 10 to 20% of the cross-section of the film carries current in the superconducting state.

These observations suggest that in Bi-2212, the superconducting current flows preferentially in thin layers close to the grain boundaries. This is unlike other well-known HTSC compounds, such as 123 and 2223, in which the critical current density is large inside the grains. Inhomogeneous current flow through this com-

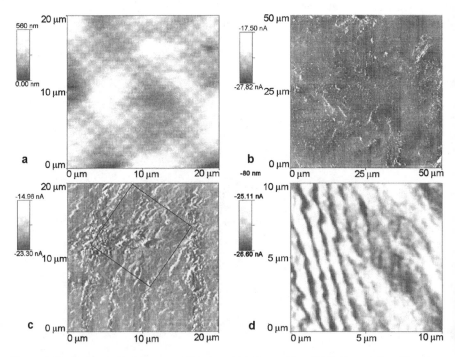

Figure 7.19. Superconducting current flow in a thick film at 60 to 65 K. Current flows from top to bottom. **a**, Topographic AFM image. **b**, MFM image with no current. **c**, MFM image, $J = 100 \, A/cm^2$. **d**, Higher magnification MFM image from the region marked by a box, $J = 100 \, A/cm^2$. (Reprinted with permission from Ref. 100.)

pound was verified with similar measurements on a single crystal. These observations are in line with theoretical predictions for the temperature regime below the transition temperature but above the temperature at which the flux lattice freezes. At these temperatures, the flux vortices interact with the surface (or a grain boundary) of a crystal in a manner that focuses more current in that vicinity. The single-crystal MFM results are similar to Hall-sensor measurements[101] on Bi-2212 superconducting single crystals but have substantially improved spatial resolution.

7.3.4 Scanning SQUID Microscopy

A ferromagnetic tip is not the only potential magnetic field sensor. Another well-known example of a magnetic field sensitive device is a superconductive quantum interference device—SQUID. Since their invention in the 1970s, SQUIDs have shown superb performance for measuring extremely weak magnetic fields and have been used for numerous applications. The pickup coils for SQUID can be as small as a few microns. Recently, significant progress was achieved in the development of scanning SQUID microscopes.[102] Unlike MFM, a SQUID sensor is sensitive to the absolute value of the field rather than to its derivative. It also provides much faster response time and, therefore, can be employed for the investigation of fast processes inaccessible by conventional MFM.

As in other SPM techniques, the typical experimental setup consists of a probe, the sample and motion system that allows 2-D scanning. Because SQUIDs are sensitive to magnetic noise, the sample is usually moved while the probe remains at place. Operation of a SQUID device requires low temperatures, therefore, the sensor is always placed in a cryogenic system. Depending on the experimental setup, the sample either can be cooled or can reside at ambient temperature. In the latter case, thermal insulation between the sensor and the sample is important. The SQUID can be used directly as a probe or an inductively coupled pickup loop or ferromagnetic tip can be used to improve the spatial resolution (Fig. 7.20). Operation at low temperatures usually requires mechanical motors, because the range of piezoelectric positioners is rather limited.

The extreme sensitivity of SQUID sensors renders them effective for studies of systems with weak magnetic fields. One of the most straightforward applications of SQUID microscopy is studies of superconductive materials. Kirtley et al.[103,104] use this technique to visualize flux vortices and directly determine the field in such a vortex. Figure 7.21**a** shows SSM images of four micropatterned, 50-μm-diameter, thin-film rings of the high-T_c superconductor $YBa_2Cu_3O_{7-\delta}$, epitaxially grown on a tricrystal substrate with a geometry chosen to be frustrated at the tricrystal meeting point for a d-wave superconductor. A d-wave superconductor in this geometry will relax the frustration by generating supercurrents that circle the central point. The magnitude of these supercurrents are such as to produce exactly half of a magnetic flux quantum of flux, i.e., the half-flux quantum effect. The central ring circles the tricrystal point and should, therefore, show half-integer flux quantization. The outer three control rings should show integer flux quantization. The sample was cooled in

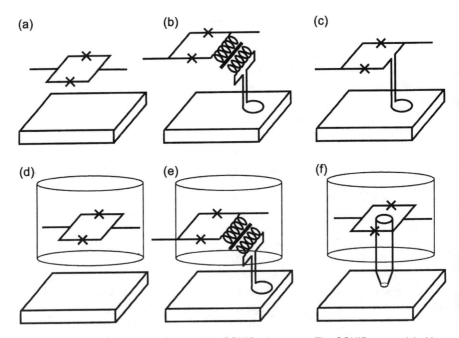

Figure 7.20. Experimental setup for scanning SQUID microscopy. The SQUID sensor (**a**) either can be directly used for imaging under cryogenic conditions or (**b**) can be inductively coupled or directly integrated into the SQUID sensor pick-up loop to improve the resolution. For ambient measurements, the cooled sensor is directly used for imaging or the pick-up loop or ferromagnetic tip is used to couple the flux from the sample at room temperature to the SQUID sensor. (Modified from Ref. 102.)

a zero field and imaged at 4.2 K in liquid helium. The outer rings are visible by a small change in the inductance of the SQUID sensor when it passes over the superconducting material in the ring. They contain no trapped flux. The central ring contains a spontaneously generated half-flux quantum. Superimposed on the image are lines showing the boundaries between the three grains making up the tricrystal, and polar plots with the orientation of the assumed d-wave pairing wavefunctions in the three grains. The images were taken with a low-Tc SQUID with an octagonal pickup loop 10 μm in diameter.

Figure 7.21**b** shows an Scanning SQUID microscopy (SSM) image of a similar sample; but in this case, the film was not patterned into rings after fabrication. The frustration in this case is relaxed by circulating supercurrents around the tricrystal point in the form of a Josephson vortex. This image was of the sample cooled in a field of a few mG and again imaged at 4.2 K. The image has three types of vortices in it. At the tricrystal point is the half-flux quantum Josephson vortex. There are four integer Josephson vortices trapped in the grain boundaries, and seven integer Abrikosov vortices trapped in the grains. This image was taken with a low-Tc SQUID with an octagonal pickup loop 4 μm in diameter. The results from this type

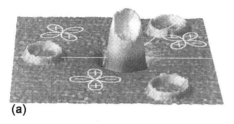 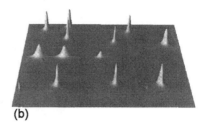

(a)   **(b)**

Figure 7.21. A scanning SQUID microscopy image of four micropatterned thin film rings of HTSC $YBa_2Cu_3O_{7-x}$ on $SrTiO_3$ tricrystal. **a,** The central ring spontaneously generates half of the superconducting flux quantum upon cooling through phase transition, whereas the other rings have no trapped flux **b,** SSM image of a field-cooled $YBa_2Cu_3O_{7-x}$ film. The Half-flux quantum Josephson vortex is at the tricrystall point. There are also four integer Josephson vortices trapped in the grain boundaries and seven integer Abrikosov vortices trapped in the grains. (Reprinted with permission from Refs. 103 and 104.) Figure also appears in Color Plate section.

of phase sensitive experiment are the best evidence to date for unconventional pairing in the high-Tc cuprate superconductors.

This example illustrates the extremely high field sensitivity of SSM, which can exploited in numerous other applications. SSM can be used for investigation of the processes accompanied by development of small magnetic fields, such as electrochemical corrosion or biologic studies.

7.4 RECENT ADVANCES IN SPM

As clearly seen from the preceding sections, SPM has become a universal tool for the study of electrostatic and magnetic phenomena on micron and nanometer length scales. The well-established SPM techniques are accessible with commercially available microscopes. In this section, however, some of the most recent advances in EFM and MFM, which are likely to become more mature techniques in the next few years, are presented. As in other microscopies, progress in the SPM can be associated with the improvement of existing techniques, development of new measurement techniques and mathematical tools for the quantification of experimental results.

As mentioned in Section 7.3, the magnetic field produced by the ferromagnetic tip can alter the magnetic structure of the sample and render MFM studies impossible. Tip-induced effects can be minimized and resolution can be improved by using magnetic supertips prepared by deposition of iron whiskers on a conventional MFM probe.[105] This produces a magnetic probe with a large aspect ratio. Alternatively, improvement of MFM imaging can be achieved through the development of localized probes, such as tips with ferromagnetic coating localized at the tip apex.[106] Such tips can be prepared by electron beam deposition of a protective coating on the end of the conventional metal-coated tip and subsequent etching of the unprotected metal coating. High localization of the magnetic region provides superior spatial resolution. The magnetic state of such a tip is much better defined because small particles exist in the monodomain state. The small volume of ferromagnetic material implies that the magnetic field produced by the tip is also small and, therefore, the problem of the magnetic structure disruption is alleviated. However, small particle size also implies that the magnetic force between the tip and the sample is small, which in turn means that thermal noise is a bigger problem. The magnetic state of such probe is also sensitive to the magnetic field from the sample and thermal fluctuations (superparamagnetic behavior).

This approach is inapplicable for EFM (one cannot bias the tip apex only), and improved EFM probes are created by increasing the tip aspect ratio. The most illustrative example is nanotube tips.[107] Carbon nanotubes with a diameter of $\sim$10 nm and a length of $\sim$2 μm are attached to the apex of conventional EFM tips. Such superprobes both minimize the topographic imaging artifacts and allow superb lateral resolution for topographic and EFM measurements.

A second direction in the development of the SPM is related to the development of new measurement techniques. Operation of the SPM in different tip–surface separation regimes under ac and dc bias and the separation of tip response harmonics provide great versatility in EFM measurements. Many new combinations have been explored; however, a complete survey is impossible in the context of this chapter. One illustrative example presented here is magnetic dissipation force microscopy.

The magnetic field produced by a MFM tip can change the magnetic structure of the sample, e.g., displace the domain walls. For a relatively small magnetic particle, the field produced by the tip can even change the polarization direction in the particle. This process is associated with energy transfer from the tip to the sample,[108–110] resulting in additional damping mechanism for tip oscillations. Analogous damping mechanisms in conductive materials is eddy losses.[111] Additional damping results in a decrease of tip oscillation amplitude at the resonant frequency. This amplitude change can be detected, along with the frequency, and provide information on the losses in the sample, constituting the basis of the magnetic dissipation force microscopy. Alternatively, feedback can be used to keep the oscillation amplitude constant by adjusting the driving voltage at the piezo. The feedback signal provides an alternative magnetic dissipation force microscopy (MDM) signal. In this case, damping does not influence the MFM signal. During the operation of MDM, three images are acquired simultaneously: topography in the tapping mode, the conventional MFM frequency signal, and the MDM signal. This technique can be used to investigate spatial localization of lossy regions during the movement of magnetic domain walls and to visualize magnetization reversal in inhomogeneous materials.

A similar approach was recently employed for the EFM measurements on semiconductor surfaces. In this case, the depletion region formed under the tip results in effective damping of lateral tip oscillations.[112] Again, feedback can be used to keep oscillation amplitude constant by adjusting the driving voltage. Dissipation force is related to the size of depletion region and carrier mobility in the sample and thus electrostatic DFM can be used to obtain 2-D doping profiles.

It is should also probably be noted that the vast majority of EFM and MFM techniques use single-frequency perturbation and detect either static response or first and second harmonic responses. Operation at several modulation frequencies will certainly be used in the future to study nonlinear wave-mixing phenomena. This approach has indeed been successfully applied in the SCM measurements.[113]

The third direction of progress will invariably be related to the development of mathematical tools for SPM image quantification. As repeatedly mentioned throughout this chapter, both quantification of tip–surface interactions in typical SPM geometries and calculation of the associated responses are extremely complicated problems. In the scanning force microscopies such as EFM and MFM the image is strongly influenced by the tip and surface geometries. Significant progress has been achieved in introducing topographic and tip-shape corrections to SFM images, e.g., in MFM[114,115] and SSPM[116] imaging. Application of transfer function based image reconstruction algorithms allows significant improvement of image contrast.[117] For SPMs based on nonlinear phenomena, such as PRI and SCM, this problem is largely unresolved. Therefore, the development of efficient numerical and analytical tools for SPM image quantification is a matter of paramount importance.

REFERENCES

1. G. Binnig, C. F. Quate, and C. Gerber, *Phys. Rev. Lett.* **56**, 930 (1986).

2. R. Wiesendanger, *Scanning Probe Microscopy and Spectroscopy—Methods and Applications*, Cambridge University Press, Cambridge, UK, 1994.

3. Q. Zhong, D. Innis, K. Kjoller, and V. Elings, *Surf. Sci.* **290**, 688 (1993).

4. P. Grutter et al., *J. Vac. Sci. Technol.* **A 6**, 279 (1988).

5. Y. Martin, D. Rugar, and H. K. Wikramasinghe, *Appl. Phys. Lett.* **52**, 244 (1988).

6. J. E. Stern, B. D. Terris, H. J. Mamin, and D. Rugar, *Appl. Phys. Lett.* **53**, 2717 (1988).

7. Y. Martin and H. K. Wikramasinghe, *Appl. Phys. Lett.* **50**, 1445 (1987).

8. D. Rugar, H. J. Mamin, P. Guethner, S. B. Lambert, J. E. Stern, I. McFaduen, and T. Yogi, *J. Appl. Phys.* **68**, 1169 (1990).

9. J. J. Saenz, N. Garcia, and J. C. Slonczewski, *Appl. Phys. Lett.* **53**, 1449 (1988).

10. V. Elings and J. Guzley, US Patent Nos. 5, 266, 801 and 5, 308, 974, Digital Instruments Santa Barbara, CA.

11. D. Sarid, *Scanning Force Microscopy*, Oxford University Press, New York, 1991.

12. B. Terris, D. Stern, J. E. Rugar, and H. J. Mamin, *Phys. Rev. Lett.* **63**, 2669 (1989).

13. V. J. Peridier, L. H. Pan, and T. E. Sullivan, *J. Appl. Phys.* **78**, 4888 (1995).

14. L. H. Pan, T. E. Sullivan, V. J. Peridier, P. H. Cutler, and N. M. Miskovsky, *Appl. Phys. Lett.* **65**, 2151 (1994).

15. H. Yokoyama, T. Inoue, and J. Itoh, *Appl. Phys. Lett.* **65**, 3143 (1994).

16. S. Hudlet, M. Saint Jean, C. Guthmann, and J. Berger, *Eur. Phys. J.* **B 2**, 5 (1998).

17. G. Mesa, E. Dobado-Fuentes, and J. J. Saenz, *J. Appl. Phys.* **79**, 39 (1996).

18. S. Belaidi, P. Girard, and G. Leveque, *J. Appl. Phys.* **81**, 1023 (1997).

19. H. W. Hao, A. M. Baro, and J. J. Saenz, *J. Vac. Sci. Technol.* **B 9**, 1323 (1991).

20. Y. Liang, D. A. Bonnell, W. D. Goodhue, D. D. Rathman, and C. O. Bozler, *Appl. Phys. Lett.* **66**, 1147 (1995).

21. J. D. Jackson, *Classical Electrodynamics*, John Wiley & Sons, New York, 1998.

22. N. N. Lebedev, I. P. Skal'skaya, and Y. S. Uflyand, *Problems in Mathematical Physics*, Pergamon Press, Oxford 1966.

23. M. E. Lines and A. M. Glass, *Principles and Applications of Ferroelectric and Related Materials*, Clarendon Press, Oxford, UK, 1977.

24. F. Saurenbach and B. D. Terris, *Appl. Phys. Lett.* **56**, 1703 (1990).

25. S. V. Kalinin and D. A. Bonnell, unpublished manuscript.

26. J. M. R. Weaver and D. W. Abraham, *J. Vac. Sci. Technol.* **B 9**, 1559 (1991).

27. M. Nonnenmacher, M. P. O'Boyle, and H. K. Wickramasinghe, *Appl. Phys. Lett.* **58**, 2921 (1991).

28. J. Janata, *AVS Newslett.* **May/June** 4 (1997).

29. A. K. Henning and T Hochwitz, *Mater. Sci. Eng. B-Solid State* **42**, 88 (1996).

30. M. P. O'Boyle, T. T. Hwang, and H. K. Wickramasinghe, *Appl. Phys. Lett.* **74**, 2641 (1999).

31. P. Girard, G. Cohen Solal, and S. Belaidi, *Microelec. Eng.* **31**, 215 (1996).

32. R. Said, G. Bridges, and D. Thomson, *Appl. Phys. Lett.* **64**, 1442 (1994).

33. M. Tanimoto and O. Vaterl, *J. Vac. Sci. Technol.* **B 14**, 1547 (1996).

34. J. M. R. Weaver, and H. K. Wickramasinghe, *J. Vac. Sci. Technol.* **B 9**, 1562 (1991).

35. S. Cunningham and J. M. R. Weaver, *Inst. Phys. Conf. Ser.* **143**, 215 (1995).

36. B. D. Huey, D. Lisjak and D. A. Bonnell, *J. Am. Ceram. Soc.* **82**, 1941 (1999).

37. B. A. Haskell, S. J. Souri, and M. A. Helfand, *J. Am. Ceram. Soc.* **82**, 2106 (1999).

38. B. D. Huey and D. A. Bonnell, *Appl. Phys. Lett.* **76**, 1012 (2000).

39. S. V. Kalinin and D. A. Bonnell, *Z. Metallkunde*, **90**, 983 (1999).

40. S. V. Kalinin and D. A. Bonnell, *J. Appl. Phys.* **87**, 3950 (2000).

41. J. Ohgami, Y. Sugawara, S. Morita, E. Nakamura, and T. Ozaki, *Jpn. J. Appl. Phys.* **35**, 2734 (1996).

42. K. Franke, H. Huelz, and M. Weihnacht, *Surf. Sci.* **41**, 178 (1998).

43. S. M. Sze, *Physics of Semiconductor Devices*, John Wiley & Sons, New York, 1981.

44. C. C. Williams, *Annu. Rev. Mater. Sci.* **29**, 471 (1999).

45. A. C. Diebold, M. R. Kump, J. J. Kopanski, and D. G. Seiler, *J. Vac. Sci. Technol.* **B 14**, 196 (1996).

46. C. C. Williams, W. P. Hough, and S. A. Rishton, *Appl. Phys. Lett.* **55**, 203 (1989).

47. C. C. Williams, J. Slinkman, W. P. Hough, and H. K. Wikramasinghe, *Appl. Phys. Lett.* **55**, 1662 (1989).

48. Y. Huang, C. C. Williams, and J. Slinkman, *Appl. Phys. Lett.* **66**, 344 (1995).

49. P. De Wolf, J. Snauwaert, L. Hellemans, T. Clarysse, W. Vandervorst, M. D'Olieslaeger, and D. Quaeyhaegens, *J. Vac. Sci. Technol.* **A 13**, 1699 (1995).

50. A. C. Diebold, M. Kump, J. J. Kopanski, and D. G. Seiler, *Proc. Electrochem. Soc.* **94**, 78 (1994).

51. P. De Wolf, M. Geva, T. Hantschel, W. Vandervorst, and R. B. Bylsma, *Appl. Phys. Lett.* **73**, 2155 (1998).

52. S. Landau, B. O. Kolbesen, R. Tillmann, et al., *Proc. Electrochem. Soc.* **98**, 789 (1998).

53. Z. Xie, E. Z. Luo, H. B. Peng, et al., *J. Non-Cryst. Solids* **254**, 112 (1999).

54. A. Gruveman, O. Auciello, and H. Tokumoto, *J. Vac. Sci. Technol.* **B 14**, 602 (1996).

55. K. Takata, K. Kushida, K. Torii, and H. Miki, *Jpn. J. Appl. Phys.* **33**, 3193 (1994).

56. D. Damjanovic, *Rep. Prog. Phys.* **61**, 1267 (1998).

57. V. Z. Parton and B. A. Kudryavtsev, *Electromagnetoelasticity*, Gordon & Breach, New York, 1988.

58. A. E. Giannakopoulos and S. Suresh, *Acta Mater.* **47**, 2153 (1999).

59. S. A. Melkumyan and A. F. Ulitko, *Prikladnaya Mech.* **23**, 44 (1987).

60. A. Gruverman, O. Auciello, and H. Tokumoto, *Annu. Rev. Mater. Sci.* **28**, 101 (1998).

61. G. Zavala, J. H. Fendler, and S. Trolier-McKinstry, *J. Appl. Phys.* **81**, 7480 (1997).

62. A. Gruverman, O. Auciello, and H. Tokumoto, *Appl. Phys. Lett.* **69**, 3191 (1996).

63. E. L. Colla, S. Hong, D. V. Taylor, A. K. Tagantsev, N. Setter, and K. No, *Appl. Phys. Lett.* **72**, 2763 (1998).

64. L. M. Eng, H.-J. Guntherodt, G. A. Schneider, U. Kopke, and J. Munoz Saldana, *Appl. Phys. Lett.* **74**, 233 (1999).

65. J. A. Christman, R. R. Woolcott, A. I. Kingon, and R. J. Nemanich, *Appl. Phys. Lett.* **73**, 3851 (1998).

66. A. Gruverman, O. Auciello, and H. Tokumoto, *Integr. Ferroelectr.* **19**, 49 (1998).

67. B. Bhushan, *Nanomechanical Properties of Solid Surfaces and Thin films*, 2nd ed., Handbook Micro/Nano Tribology, 1999, CRC, Boca Raton, Fla.

68. F. M. Serry, Y. E. Strausser, J. Elings, S. Magonov, J. Thornton, and L. Ge, *Surf. Eng.* **15**, 285 (1999).

69. J. A. Dagata, *Science* **270**, 1625 (1995).

70. J. A. Dagata, *NATO ASI Ser.* **E 264**, 189 (1994).

71. A. Gruverman, O. Auciello, R. Ramesh, and H. Tokumoto, *Nanotechnology* **8**, A38 (1997).

72. T. Hidaka, T. Maruyama, M. Saitoh, et al., A*ppl. Phys. Lett.* **68**, 2358 (1996).

73. T. Tybell, C. H. Ahn, and J.-M. Triscone, *Appl. Phys. Lett.* **72**, 1454 (1998).

74. C. H. Ahn, S. Gariglio, P. Paruch, T. Tybell, L. Antognazza, and J.-M. Triscone, *Science* **284**, 1152 (1999).

75. C. H. Ahn, T. Tybell, L. Antognazza, K. Char, R. H. Hammond, M. R. Beasley, O. Fischer, and J.-M. Triscone, *Science* **276**, 1100 (1997).

76. K. Tsuno, *Rev. Solid State Sci.* **2**, 623 (1988).

77. Z. Q. Qiu and S. D. Bader, *Int. Ser. Monogr. Phys.* **98**, 1 (1998).

78. E. Dan Dahlberg and R. Proksch, *J. Magn. Magn. Mater.* **200**, 720 (1999).

79. J. Unguris, M. R. Scheinfein, R. J. Celotta, and D. T. Pierce, *Springer Ser. Surf. Sci.* **22**, (*Chem. Phys. Solid Surf.* **8**), 239 (1990).

80. M. Bode, M. Dreyer, M. Getzlaff, M. Kleiber, A. Wadas, and R. Wiesendanger, *J. Phys. Condens. Matter* **11**, 9387 (1999).

81. R. Proksch, G. D. Skidmore, E. Dan Dahlberg, et al., *Appl. Phys. Lett.* **69**, 2599 (1996).
82. K. L. Babcock, L. Folks, R. Street, R. C. Woodward, and D. L. Bradbury, *J. Appl. Phys.* **81**, 4438 (1997).
83. R. D. Gomez, E. R. Burke, and I. D. Mayergoyz, *J. Appl. Phys.* **79**, 6441 (1996).
84. U. Hartmann, *Adv. Electr. Electron Phys.* **87**, 49 (1994).
85. U. Hartmann, *Phys. Lett.* **A 137**, 475 (1989).
86. L. Kong and S. Y. Chou, *Appl. Phys. Lett.* **70**, 2043 (1997).
87. U. Hartmann, *J. Vac. Sci. Technol.* **A 8**, 411 (1990).
88. J. Lohau, S. Kirsch, A. Carl, G. Dumpich, and E. F. Wassermann, *J. Appl. Phys.* **86**, 3410 (1999).
89. U. Hartmann, *J. Appl. Phys.* **64**, 1561 (1988).
90. R. D. Gomez, A. O. Pak, A. J. Anderson, E. R. Burke, A. J. Leyendecker, and I. D. Mayergoyz, *J. Appl. Phys.* **83**, 6226 (1998).
91. A. Hubert and R. Schafer, *Magnetic Domains*, Springer Verlag, Berlin, 1998.
92. S. Foss, R. Proksch, E. Dan Dahlberg, B. Moskowitz, and B. Walsh, *Appl. Phys. Lett.* **69**, 3426 (1996).
93. U. Hartmann, *Annu. Rev. Mater. Sci.* **29**, 53 (1999).
94. M. R. Scheinfein, J. Unguris, J. L. Blue, K. J. Coakley, D. T. Pierce, R. J. Celotta, and P. J. Ryan, *Phys. Rev.* **B 43**, 3395 (1991).
95. M. R. Scheinfein, J. Unguris, R. J. Celotta, and D. T. Pierce, *Phys. Rev. Lett.* **63**, 668 (1989).
96. U. Memmert, P. Leinenbach, J. Losch, and U. Hartmann, *J. Magn. Magn. Mater.* **190**, 124 (1998).
97. S. McVitie, R. P. Perrier, and W. A. P. Nicholson, *Int. Phys. Conf. Ser.* **153**: Section 6, 201 (1997).
98. S. Foss, E. Dan Dahlberg, R. Proksch, and B. M. Moskowitz, *J. Appl. Phys.* **81**, 5032 (1997).
99. S. J. Bending, *Adv. Phys.* **48**, 449 (1999).
100. F. Kral, D. Perednis, B. Huey, D. A. Bonnell, G. Kostorz, and L. J. Gauckler, *Adv. Mater.* **10**, 1442 (1998).
101. D. Fuchs, E. Zeldov, M. Rappaport, T. Tamegai, S. Ooi, and H. Shtrikman, *Nature* **391**, 373 (1998).
102. J. R. Kirtley and J. P. Wikswo, *Annu. Rev. Mater. Sci.* **29**, 117 (1999).
103. C. C. Tsuei, J. R. Kirtley, C. C. Chi, L. S. Yu-Jahnes, A. Gupta, T. Shaw, J. Z. Sun, and M. B. Ketchen, *Phys. Rev. Lett.* **73**, 593 (1994).
104. J. R. Kirtley, C. C. Tsuei, M. Rupp, et al., *Phys. Rev. Lett.* **76**, 1336 (1996).
105. G. D. Skidmore and E. Dan Dahlberg, *Appl. Phys. Lett.* **71**, 3293 (1997).
106. P. Leinenbach, U. Memmert, J. Schelten, and U. Hartmann, *Appl. Surf. Sci.* **144–145**, 492 (1999).
107. S. B. Arnason, A. G. Rinzler, Q. Hudspeth, and A. F. Hebard, *Appl. Phys. Lett.* **75**, 2842 (1999).
108. Y. Liu and P. Grutter, *J. Appl. Phys.* **83**, 7333 (1998).
109. Y. Liu, B. Ellman, and P. Grutter, *Appl. Phys. Lett.* **71**, 1418 (1997).
110. R. Proksch, K. Babcock, and J. Cleveland, *Appl. Phys. Lett.* **74**, 419 (1999).

111. B. Hoffmann, R. Houbertz, and U. Hartmann, *Appl. Phys. A Mater. Sci. Process.* **A 66** (Suppl., Pt. 1), S409 (1998).

112. T. D. Stowe, T. W. Kenny, D. J. Thomson, and D. Rugar, *Appl. Phys. Lett.* **75**, 2785 (1999).

113. J. Schmidt, D. H. Rapoport, G. Behme, and H.-J. Frohlich, *J. Appl. Phys.* **86**, 7094 (1999).

114. H. Saito, J. Chen, and S. Ishio, *J. Magn. Magn. Mater.* **191**, 153 (1999).

115. J.-G. Zhu, X. Lin, R. Shi, and Y. Luo, *J. Appl. Phys.* **83**, 6223 (1998).

116. S. Cohen and A. Efimov, in Y. Kuk, I. W. Lyo, D. Jeon, and S.-I. Park, eds., *Preliminary Proceedings of STM '99*, Seoul (1999).

117. H. J. Hug, B. Stiefel, P. J. A. van Schendel, et al., *J. Appl. Phys.* **83**, 5609 (1998).

8

BEEM AND THE CHARACTERIZATION OF BURIED INTERFACES

W. J. Kaiser, L. D. Bell, M. H. Hecht, and L. C. Davis

8.1 INTRODUCTION

The physics of carrier transport through thin films and across interfaces is fundamental to the understanding of microelectronic devices. Ballistic electron emission microscopy (BEEM)[1,2] comprises a family of spectroscopies that addresses the transport and scattering of electrons and holes in multilayer structures. BEEM is based on scanning tunneling microscopy (STM) technology and is, therefore, highly spatially resolved—the typical resolution is about 2 nm. The spectroscopic information that can be extracted from a BEEM measurement is also extraordinarily rich. BEEM, as a result, offers a fundamentally new and powerful way to investigate problems such as Schottky barrier formation, heterojunction band lineups, and quantum confinement.

The final properties of an interface often develop only when the interface is fully formed and beyond the range of surface-sensitive techniques. Fully formed interfaces have been best investigated by the conventional tools of current-voltage (*I-V*), capacitance-voltage (*C-V*), and internal photoemission or photoresponse measurements.[3,4] The principal new technology applied to the problem has been the ability to make ever smaller and more uniform structures to minimize statistical averaging of spectral information. BEEM, in a sense, reverses that trend, allowing near-atomic spatial resolution from macroscopic structures.

Since the development of BEEM in 1988, the technique has evolved beyond spatially resolved barrier height measurement to include quantitative interface transport characterization as well as analysis of band structure, heterojunction band offsets, scattering processes, and interface abruptness. In many respects, these are the ideal characteristics for an interface probe. The importance of high-energy resolution (principally limited by thermal broadening) and of the ability to perform a true energy spectroscopy cannot be overemphasized. For example, subtle variations in the energies of critical band structure points with temperature, strain and composition may ultimately distinguish different models of interface formation. Such critical points correspond to distinct spectral features in BEEM. With conventional electrical techniques, in contrast, these properties must be extracted from amplitude analysis and extrapolation. Properties of potential steps at interfaces may be dominated by lateral inhomogeneity, a characteristic that cannot be evaluated at all using averaging techniques.

BEEM employs an STM tip under conventional feedback control to inject ballistic electrons (or holes) into a sample heterostructure. The sample consists of at least two layers forming an interface that presents a potential barrier to the carriers. The top layer and (at least) one other layer are electrically contacted. Figure 8.1 illustrates the simplest possible configuration, with the sample being a metal/semiconductor Schottky barrier (SB) system. The STM tip is biased with respect to the metal layer on the top surface, which serves as system ground. Current flow I_c between the metal base and the semiconductor collector is measured through an ohmic contact to the collector.

As a tip-base bias voltage is applied, electrons tunnel across the vacuum gap and enter the sample as hot carriers. Because characteristic attenuation lengths in metals

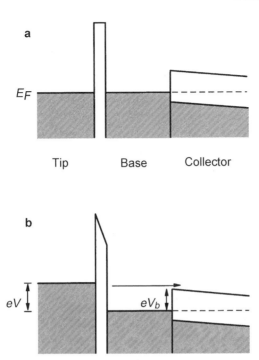

a

E_F

Tip   Base   Collector

b

eV eV_b

Figure 8.1. Energy diagram for BEEM of a metal/semiconductor SB system. **a,** Tunnel voltage $V = 0$. **b,** Applied tunnel voltage $V > V_b$. For this case, some of the injected electrons have sufficient energy to enter the semiconductor.

and semiconductors may be 10s of nanometers,[5] many of these hot electrons may propagate through the sample for a long distance and reach the interface before scattering. If conservation laws restricting energy and wave vector are satisfied, these electrons may cross the interface and be measured as a current in the collector layer. An n-type semiconductor is used for electron collection, because the band bending accelerates the collected carriers away from the interface and prevents their leakage back into the base. By varying the voltage between tip and base, the energy distribution of the hot carriers can be controlled, and a spectroscopy of interface transport may be performed.

Figure 8.2 shows a model BEEM spectrum calculated for $T = 0$. The theory that describes this spectrum will be discussed later, but the qualitative features may be mentioned here. For tunnel voltages less than the interface barrier height, none of the injected electrons has sufficient energy to surmount the barrier and enter the collector, and the measured collector current is zero. As the voltage is increased to values in excess of the barrier, some of the hot electrons cross the interface into the semiconductor conduction band, and an increase in collector current is observed. The location of the threshold in the spectrum defines the interface barrier height.

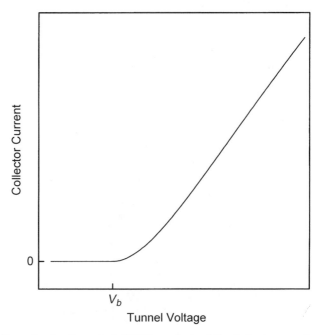

Figure 8.2. Example of a theoretical BEEM spectrum at $T = 0$. Such a spectrum displays a threshold at a voltage equal to the interfacial barrier height V_b.

The magnitude of the current above threshold and the threshold spectrum shape also yield important information on interface and bulk transport.

It is also important to probe valence band structure at an interface. A complete description of an interface requires separate knowledge of both conduction and valence band SB heights and band alignments. An example of this requirement is provided by strained interfaces, where the semiconductor band gap is perturbed. A spectroscopy of valence band structure may also be performed with BEEM techniques, but using ballistic *holes* as a probe.[6] Since most *p*-type barriers are relatively low, ballistic hole spectroscopy has required the development of a low-temperature BEEM apparatus. This apparatus has in turn enabled a study of interface properties with temperature, which provides fundamental insights into interface formation mechanisms.

The details of carrier transport are also of great importance, because carrier scattering in materials dominates transport properties. BEEM is a valuable technique for the study of ballistic transport and interface structure, but it does not directly provide a means for an analysis of the carriers that scatter in the base electrode and are not collected. However, a related technique provides a direct spectroscopy of electron and hole scattering. This method is sensitive to carrier-carrier scattering and allows the first direct probe of the hot secondary carriers created by the scattering process.[7] As a spectroscopy of scattering, it is complementary to BEEM, because it provides a way of analyzing *only* the carriers that scatter and are not observed by the ballistic spectroscopies. Scattering phenomena in metal/semiconductor systems were investigated using this new scattering spectroscopy, and a theoretical treatment for the collected current was developed that yields excellent agreement with experiment. The observed magnitudes of the currents indicate that the carrier-carrier scattering process is dominant in ballistic carrier transport.

BEEM has been applied to many interface systems, and important examples will be given in this chapter. A theoretical framework for this technique, and for the corresponding ballistic hole process, will be presented; this theory provides an excellent description of experimental results. The theory that was developed for carrier-carrier scattering spectroscopy will also be described. Experimental methods particular to BEEM and related techniques will be discussed. Both spectroscopy and imaging of interfaces is possible with ballistic electron and hole probes, and examples of each will be reviewed. Finally, the initial application of the new scattering spectroscopy will be presented.

8.2 THEORY

8.2.1 Electron Spectroscopy

BEEM may be understood by using a simple theoretical model, which may be built on to include more complicated processes. The essence of the model is the description of the phase space available for interface transport. The simplest case to consider is that of a smooth interface, which dictates conservation of the component of the electron wave vector parallel to the interface (transverse to the interface normal) k_t. Conservation of total energy across the interface provides a second constraint on transport. The process is visualized in Figure 8.3. This diagram may be taken to represent a simplified metal/semiconductor interface system.

A particle incident on a simple potential step, with total energy in excess of the step height, loses a portion of its kinetic energy as it crosses the step, owing to a reduction in k_x, the component of **k** normal to the interface. If the particle is normally incident, the direction of propagation $\hat{\mathbf{k}}$ is unchanged. For the case of incidence at a nonzero angle to the normal, however, conservation of k_t demands that $\hat{\mathbf{k}}$ changes across the interface; i.e., the particle is "refracted." If the angle of incidence is greater than the "critical angle," the particle is not able to cross the interface and is reflected back. This critical angle may be expressed as[8]

$$\sin^2 \theta_c = \frac{E - E_F - eV_b}{E} \tag{8.1}$$

where the particle total energy is E and the step height is $E_F + eV_b$. The quantities eV_b and eV are defined as positive for both electrons and holes.

The situation becomes somewhat more complicated for real band structures if the two regions have dispersion relations that are dissimilar. In this case, the normal

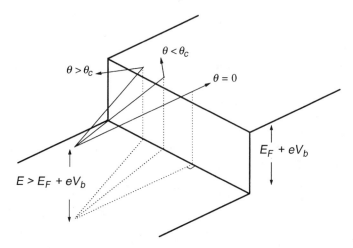

Figure 8.3. Schematic diagram representing a particle incident on a potential step of height $E_F + eV_b$, with an initial energy E in excess of this step height. There exists a critical angle θ_c such that a particle incident at an angle in excess of this value is reflected. For an angle of incidence $< \theta_c$, the particle is transmitted.

and parallel components of the problem are no longer trivially separable. In particular, if E_x is the energy associated with k_x, $E_x > E_F + eV_b$ is no longer the condition for collection. A change in effective mass, or in the location of the conduction band minimum within the Brillouin zone, alters the critical angle conditions.

For zone-centered conduction band minima in both the metal and semiconductor, we can define this critical angle with respect to the interface normal, in terms of electron energy in the base and the potential step $E_F + eV_b$, as

$$\sin^2 \theta_c = \frac{m_{tf}}{m_{xi}} \frac{E - E_F - eV_b}{E + [(m_{tf}/m_{xi}) - (m_{tf}/m_{ti})](E - E_F - eV_b)} \tag{8.2}$$

where m_{xi} and m_{ti} are the components of effective mass in the metal normal to and parallel to the interface, respectively, and m_{xf} and m_{tf} are the corresponding masses in the semiconductor. For evaporated metal base layers which are polycrystalline, the approximation of an isotropic free-electron mass for the base is made, and the expression reduces to

$$\sin^2 \theta_c = \frac{m_t}{m} \frac{E - E_F - eV_b}{E} = \frac{m_t}{m} \frac{e(V - V_b)}{E_F + eV} \tag{8.3}$$

where the second equality is for an incoming electron with $E = E_F + eV$ and m is the free-electron mass. This expression predicts that for Si(100) or GaAs, with a small component of effective mass parallel to the interface and for $e(V-V_b) \leq 0.3$ eV, this critical angle is $<6°$. Critical angle reflection has important implications for the spatial resolution of interface characterization, because only electrons incident on the interface at small angles can be collected. Thus single bulk scattering events may decrease collected current but should not degrade spatial resolution. For a 10-nm base layer, about 2 nm spatial resolution is expected.

The s-wave approximation for STM tunneling was described in Chapter 3. For simplicity, here the electron injection into the sample by tunneling is treated using a planar tunneling formalism.[9] This description provides simple analytic expressions for the $(E, \mathbf{k})$ distribution of the tunneling electrons and for the total tunnel current. Current across the metal/semiconductor interface is then calculated based on this initial distribution, by considering the fraction of the total tunnel current that is within the critical angle. The vacuum barrier is taken to be square at $V=0$; the WKB form of the tunneling probability $D(E_x)$ is thus written as[10]

$$D(E_x) = \left[-\alpha s \left(E_F + \Phi - \frac{eV}{2} - E_x \right)^{1/2} \right] \tag{8.4}$$

where $\alpha = (8m/\hbar^2)^{1/2} = 1.024\,\mathrm{eV}^{-1/2}\,\mathrm{\AA}^{-1}$. Tunnel current is given by the standard expression:

$$I_t = 2ea \iiint \frac{d^3\mathbf{k}}{(2\pi)^3} D(E_x) v_x [f(E) - f(E+eV)] \tag{8.5a}$$

where a is the effective tunneling area and $v_x = \hbar k_x/m$. This expression may conveniently be expressed in terms of integrals over E_x and E_t:

$$I_t = C \int_0^\infty dE_x D(E_x) \int_0^\infty dE_t [f(E) - f(E+eV)] \tag{8.5b}$$

where $C = 4\pi ame/h^3$, and the integration is over all tip states with $E_x > 0$. E_x is the energy associated with k_x, and E_t is that associated with k_t. (It will be convenient in the remainder of this section to express energies with reference to the bottom of the tip conduction band, unless otherwise noted.) The Fermi function $f(E)$ is given by the standard expression (Eq. (4.15)). For simplicity, the tip and base are taken to be identical free-electron metals, which will be the assumption for the remainder of this chapter. At $T=0$, therefore, the appropriate states occupy a half shell within the tip Fermi sphere[11] between $E = E_F - eV$ and $E = E_F$ (Fig. 8.4).

A similar expression may be written for the collector current, with the allowed phase space within the tip determined by the critical angle conditions. For the case of $m_t < m$, these restrictions on tip states are

$$E_t \leq \frac{m_t}{m - m_t} [E_x - E_F - e(V - V_b)] \tag{8.6}$$

$$E_x \geq E_F - e(V - V_b) \tag{8.7}$$

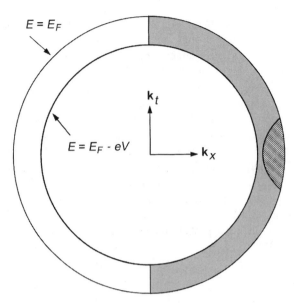

Figure 8.4. k-space diagram representing the free-electron Fermi sphere of the STM tip, for the case of electron tunneling from tip to sample. For a tip–sample voltage V, states within the shell delimited by E_F and $E_F - eV$, and with $\mathbf{k}_x > 0$, are eligible to tunnel. If $eV > eV_b$, a subset of these tunneling electrons, defined by the intersection of a hyperboloid and a spherical shell in k-space, satisfy phase space conditions for collection. These volumes are shown.

where E_F is the Fermi energy of the tip. Taking the equalities as limits $E_t^{\max}$ and $E_x^{\min}$, collector current can be expressed as

$$I_c = RC \int_{E_x^{\min}}^{\infty} dE_x D(E_x) \int_0^{E_t^{\max}} dE_t [f(E) - f(E + eV)] \qquad (8.8)$$

R is a measure of attenuation owing to scattering in the base layer, which is taken to be energy independent for these energies.[12] If $eV_b \gg kT$, as with all examples discussed in this chapter, the second Fermi function $f(E + eV)$ in Eq. (8.8) may be neglected. In Eqs. (8.5b) and (8.8), the integrals over E_t may be performed analytically; for $T = 0$, this is true also of the integrals over E_x.

Eqs. (8.6) and (8.7) define a hyperboloid in tip k-space (Fig. 8.4). The integration is performed over this hyperboloidal volume, with the cutoff at higher energies provided by the Fermi function $f(E)$ centered at the tip Fermi level. The threshold shape of the BEEM I_c-V spectrum is determined by the behavior of this k-space volume with voltage, which in turn is determined by the dispersion relation of the collector conduction band. A parabolic conduction band minimum and the assumption of k_t conservation across the interface, therefore, result in a parabolic threshold shape to the I_c-V spectrum. A similar treatment for I_c can be derived for the case of $m_t > m$.

Unlike STM spectroscopy as described in Chapter 4, BEEM spectra are conventionally obtained with the STM operating at constant tunnel current, which normalizes the collected current to the tunnel current and linearizes the BEEM spectrum. In addition, this normalization removes the lowest order effects owing to structure in the tunneling density of states, which may obscure interface structure. Eq. (8.8) for I_c contains factors of s, the tip–sample spacing, which changes with tunnel voltage at constant I_t. Therefore, Eq. (8.8) will not describe accurately an entire I_c-V spectrum. This effect may be included within the theory in two ways. Eq. (8.5b) for $I_t(s, V)$ may be inverted to give $s(I_t, V)$, which is then inserted into Eq. (8.8). In practice, this inversion must be done numerically; in addition, the prefactor C, which includes effective tunneling area, must be known. The second method is to treat s as a constant s_0, but to normalize $I_c(s_0, V)$ by $I_t(s_0, V)$ for each voltage. This requires only that the tunnel *distribution* be relatively insensitive to small changes in s. For the barriers considered here, this assumption is valid, producing errors only on the order of a percent. The expression for I_c then takes the form

$$I_c = RI_{t0} \frac{\int_{E_x^{\min}}^{\infty} dE_x D(E_x) \int_0^{E_t^{\max}} dE_t [f(E) - f(E + eV)]}{\int_0^{\infty} dE_x D(E_x) \int_0^{\infty} dE_t [f(E) - f(E + eV)]} \qquad (8.9)$$

where I_{t0} is the constant tunnel current at which the BEEM spectrum is measured.

BEEM is sensitive to higher minima in the collector conduction band structure, rather than just the lowest minimum that determines the Schottky barrier height. This results from the opening of additional phase space for electron transport as electron energy exceeds each minimum in turn. This capability is enabled by the control over injected electron energy provided by BEEM.

The conduction band minima of a particular semiconductor are not in general zone-centered; however, the critical angle restrictions are unchanged, provided the minimum is "on-axis", i.e., located at $k_t = 0$. The above phase space requirements may, therefore, be used for a metal on Si(100) or Ge(111), although the wave function coupling into these minima may be different. For GaAs of any orientation, off-axis conduction band minima must be included for energies only several hundred meV above the conduction-band minimum. Similar critical angle requirements exist for these off-axis minima, with the center of the critical cone located at an angle to the interface normal (defined in the base) given by

$$\sin^2 \theta_0 = \frac{E_{0t}}{E_F + eV_b}, \qquad E_{0t} = \frac{\hbar^2 k_{0t}^2}{2m} \qquad (8.10)$$

and where k_{0t} is the component parallel to the interface of $\mathbf{k}_0$, the location of the minimum. Off-axis minima may be included in the treatment for BEEM, although they do not provide an analytic expression for I_c within the theory discussed here.

This case is of great importance, however, because collection via these minima provides a powerful probe of fundamental quantities such as the $(E, \mathbf{k})$ tunneling distribution, parallel wave vector conservation at the buried interface, and carrier scattering.

The foregoing discussion has assumed that all electrons incident on the interface within the critical angle are collected. This classical assumption may not be appropriate for abrupt interfaces, where quantum-mechanical reflection (QMR) must be considered. In this case the integrand of Eq. (8.8) must be multiplied by the quantum-mechanical transmission factor $T(E, \mathbf{k})$ appropriate to the potential profile of the interface. Using the approximation of a sharp step potential, and for normally incident electrons, this factor may be written as[13]

$$T = \frac{4\dfrac{k_{xi}\,k_{xf}}{m_{xi}m_{xf}}}{\left(\dfrac{k_{xi}}{m_{xi}} + \dfrac{k_{xf}}{m_{xf}}\right)^2} \tag{8.11}$$

where k_{xi} and k_{xf} are the components of $\mathbf{k}$ normal to the interface in the base and collector, respectively. In terms of energies referred to the tip conduction band minimum,

$$T = \frac{4\sqrt{\dfrac{(E_x + eV)(E_x + eV - E_F - eV_b)}{m_{xi}m_{xf}}}}{\left(\sqrt{\dfrac{E_x + eV}{m_{xi}}}\sqrt{\dfrac{E_x + eV - E_F - eV_b}{m_{xf}}}\right)^2} \tag{8.12}$$

which increases as $(|V| - V_b)^{1/2}$ for $|V|$ close to the threshold V_b. In the limit of a gradual potential transition from metal to semiconductor, T approaches a step function, and the regime of $(|V| - V_b)^{1/2}$ behavior becomes small. Alternative expressions for other potentials may also be used. The analyses of data in this chapter do not include QMR terms, although later discussions of ballistic hole spectra will address this point further.

8.2.2 Hole Spectroscopy

The previous section dealt with the investigation of ballistic electron transport and its use as a probe of interface conduction band structure. It was mentioned in Section 8.1 that n-type semiconductors are necessary for electron collection to repel the carriers away from the interface into the collector and prevent leakage back into the base. It is possible to use ballistic *holes*, however, as a probe of valence band structure. The unique aspects of ballistic hole spectroscopy will be discussed in this section.

The implementation of ballistic hole spectroscopy using BEEM techniques is shown in the energy diagram in Figure 8.5 for the case of a metal/semiconductor sample structure. Here, a *p*-type semiconductor serves as a collector of ballistic holes injected by the STM tip. The tip electrode is biased positively; because injection is through a vacuum tunnel barrier, the injection process must be treated in terms of electron tunneling from the base to the tip. This tunneling process deposits ballistic electrons in the tip and creates a ballistic hole distribution in the base. These distributions are illustrated in Figure 8.5. Note that, because the tunneling is strictly by electrons, the energy and angular distributions of the ballistic holes are determined by the vacuum level. The distribution is, therefore, peaked toward the base Fermi level.

As a positive voltage is applied to the tip, ballistic holes are injected into the sample structure. As in the ballistic electron case, collector current is zero until the voltage exceeds the barrier height; at higher voltages, collector current increases. It is apparent, however, that the peaking of the hole distribution toward the base Fermi

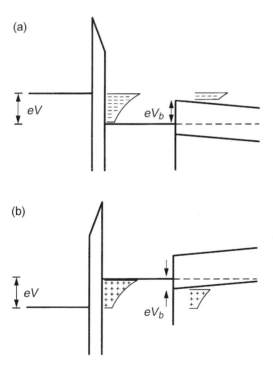

Figure 8.5. Energy diagrams of ballistic electron and ballistic hole spectroscopies. **a,** In ballistic electron spectroscopy of interface conduction band structure, a hot electron distribution is created, of which the most energetic electrons are available for collection. **b,** With ballistic hole spectroscopy of interface valence band structure, a hot hole distribution is created in the base by electron vacuum tunneling. The *least* energetic holes are eligible for collection by the semiconductor valence band.

level introduces an asymmetry between the ballistic electron and hole spectroscopies (Fig. 8.5). In BEEM, the portion of the hot electron distribution that is eligible for collection is toward the higher energies, where the distribution is maximum. For the case of holes, the tail of the distribution is collected. This asymmetry introduces a corresponding asymmetry into a ballistic hole spectrum.

The threshold behavior of the ballistic hole I_c-V spectrum, however, is the same as for the case of BEEM. For a hole barrier at $E_F - eV_b$, the critical angle condition is

$$\sin^2 \theta_c = \frac{m_{tf}}{m_{xi}} \frac{E_F - eV_b - E}{E + [(m_{tf}/m_{xi}) - (m_{tf}/m_{ti})](E_F - eV_b - E)} \tag{8.13}$$

which may be compared with Eq. (8.2). For the case of an isotropic free-electron mass for the base, the expression reduces to

$$\sin^2 \theta_c = \frac{m_t}{m} \frac{E_F - eV_b - E}{E} = \frac{m_t}{m} \frac{e(V - V_b)}{E_F - eV} \tag{8.14}$$

the second expression being for an incoming hole with $E = E_F - eV$. The ballistic hole I_c-V spectrum threshold has a $(V - V_b)^2$ dependence, in agreement with the ballistic electron case, although the appropriate $\mathbf{k}$-space volume of integration is most conveniently taken over states in the base and is quite different from the electron case, as illustrated in Figure 8.6.

8.2.3 Scattering Spectroscopy

In this section, we present a theory for the generation, in reverse bias, of electron-hole pairs within the base and their subsequent collection by the collector. The collector is assumed to be a p-type semiconductor; analogous processes occur with n-type collectors. For simplicity, the theory will treat only the $T = 0$ case. In reverse bias, that is, opposite in sign to that used in ballistic hole spectroscopy, a negative voltage, $-V$, is applied to the tip so that electrons are injected into the base. For $V < V_{bn}$, the position of the semiconductor conduction band edge at the interface, the injected electrons propagate to the interface and are reflected back into the base. Consequently, one might expect that no current will be collected by the semiconductor. However, the injected electrons can scatter off the Fermi sea in the base, exciting electrons below E_F to states above the Fermi energy. The holes that are thereby created may act in the same manner as holes injected by the tip in forward bias, although with a different energy and angular distribution. Those holes that have energy $E < E_F - eV_{bp}$ (V must exceed V_{bp}) and are directed toward the interface within the critical angle θ_c of the normal will be collected, resulting in a collector current I_c. This process is shown in Figure 8.7. The corresponding process for hole injection and secondary electron collection with an n-type collector is also shown.

We can construct a theory for these processes along the lines of the theory already developed for forward bias with the addition of a simple but reasonably

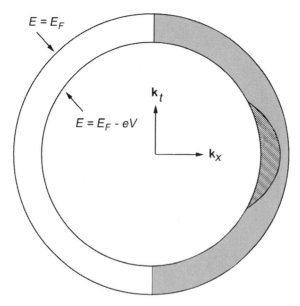

Figure 8.6. k-space diagram representing the free-electron Fermi sphere of the sample base, for the case of electron tunneling from sample to tip. Shaded areas correspond to those in Figure 8.4. For $eV > eV_b$, a subset of the tunneling electrons, defined by the intersection of an ellipsoid and a spherical shell in **k**-space, create holes that satisfy phase space conditions for collection.

accurate treatment of the electron-electron (ee) scattering. As in previous sections, the free-electron picture is used, where $E_n = \hbar^2 k_n^2 / 2m$. An electron injected into the base with energy E_0 can lose energy and scatter to a new state with energy E_1 above E_F. Conservation of energy requires that an electron excited from E to E_2, which must also be above E_F, satisfy the following relationship:

$$E_0 - E_1 = E_2 - E \tag{8.15}$$

Likewise, in a free-electron metal, wave vector is conserved so that

$$\mathbf{k}_0 - \mathbf{k}_1 = \mathbf{k}_2 - \mathbf{k} \tag{8.16}$$

Summing over all states $\mathbf{k}_1$ and $\mathbf{k}_2$ that are above the Fermi energy and satisfy the conservation laws, we obtain the total rate $R(\mathbf{k})$ at which holes with wave vector $\mathbf{k}$ and energy E are generated:

$$R(\mathbf{k}) = 2\pi/\hbar \sum_{\substack{\mathbf{k}_1, \mathbf{k}_2 \\ E_1, E_2 > E_F}} |M|^2 \delta(E_0 + E - E_1 - E_2) \delta_{\mathbf{k}_0 + \mathbf{k}, \mathbf{k}_1 + \mathbf{k}_2} \tag{8.17}$$

(a)

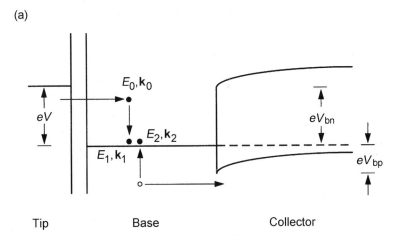

(b)

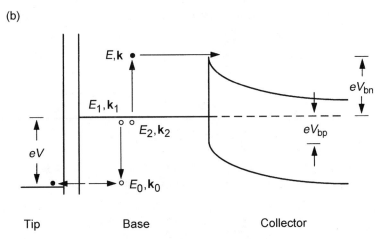

Figure 8.7. Energy diagram of reverse-bias scattering spectroscopy. **a**, The process is shown for electron injection and collection, by a **p**-type collector, of the scattered holes. **b**, The comparable process for the case of hot hole injection and secondary electron collection by an n-type collector. In both cases, the incoming carrier with $(E_0, \mathbf{k}_0)$ scatters to final state $(E_1, \mathbf{k}_1)$, exciting the secondary carrier (of opposite type) from $(E_2, \mathbf{k}_2)$ to $(E, \mathbf{k})$.

In addition to energy and momentum conservation, this expression contains a matrix element M, which in general depends on momentum transfer $\mathbf{q} = \mathbf{k}_0 - \mathbf{k}_1$. For an unscreened Coulomb interaction, $M \propto 1/q$. However, metallic screening removes this singularity, making $M \propto 1/(q^2 + q_{\mathrm{FT}}^2)^{1/2}$. Because q_{FT} is comparable to the Fermi wave vector k_F, and typically the momentum transfer is small compared with k_F, it is reasonable to treat M as a constant. If we make this assumption, Eq. (8.17) can be evaluated. We find, for $2E_F - E_0 < E < E_F$,

$$R(\mathbf{k}) = \text{constant} \cdot \frac{E_0 + E - 2E_F}{[E_0 + E + 2\sqrt{E_0 E}\cos\theta_{\mathbf{k},\mathbf{k}_0}]^{1/2}} \tag{8.18}$$

where $\theta_{\mathbf{k},\mathbf{k}_0}$ is the angle between $\mathbf{k}$ and $\mathbf{k}_0$.

The rate $R(\mathbf{k})$ can be related to the lifetime of the injected electron as follows. The inverse lifetime or width Γ consists primarily of two terms. The first, Γ_{ph}, is due to phonon scattering. The second, Γ_{ee}, is due to electron-electron scattering and is expected to be dominant except near the Fermi level. Γ_{ee} can be found by summing $R(\mathbf{k})$ over all $\mathbf{k}$,

$$\Gamma_{ee} = \hbar \sum_{\mathbf{k}} R(\mathbf{k}) \tag{8.19}$$

or, transforming to an integral,

$$\Gamma_{ee} = \hbar \int \frac{d^3\mathbf{k}}{(2\pi)^3} R(\mathbf{k}) \tag{8.20}$$

where the integration is over all $\mathbf{k}$ such that $2E_F - E_0 < E < E_F$. It is straightforward to show that

$$\Gamma_{ee} \propto (E_0 - E_F)^2 \tag{8.21}$$

for E_0 near E_F, a result obtained from a proper treatment of screening.[14] This gives us confidence that taking the matrix element M to be constant gives reliable results.

The distribution of holes generated can be written in terms of a branching ratio, $\hbar R(\mathbf{k})/\Gamma_{ee}$. This has the advantage that $|M|^2$ cancels out and we do not need to know its magnitude. The physical basis is that every electron injected by the tip decays by the creation of an electron-hole pair. That is, for every electron injected a hole is created. The probability that the hole has momentum $\mathbf{k}$ is the branching ratio. Hence, if we know the distribution of injected electrons, we can find the resulting distribution of holes.

Note that multiple scattering is neglected. An electron that loses only a small amount of energy may have enough energy remaining to excite an electron (create a hole) from a low-energy state by a subsequent scattering. However, near threshold, such multiple scattering will mainly produce holes too close to E_F to be collected. At higher voltages, this assumption breaks down and multiple scattering should be included.

To proceed with the calculation, let $P(\mathbf{k}_0)$ be the rate at which electrons with momentum $\mathbf{k}_0$ are injected into the base. $P(\mathbf{k}_0)$ contains the tunneling probability $D(E_x)$ and other factors discussed previously in connection with forward bias. All electrons with $\mathbf{k}_0$ such that $E_F + eV_{bp} < E_0 < E_F + eV$ can produce holes that may be collected, provided that the hole energy satisfies $E < E_F - eV_{bp}$ and the angle of incidence of the hole, θ, is within the critical angle θ_c, which is given in Eq. (8.14). Thus the collector current is

$$I_c = \sum_{\mathbf{k}_0} P(\mathbf{k}_0) \sum_{\mathbf{k}} \frac{\hbar R(\mathbf{k})}{\Gamma_{ee}} \quad (8.22)$$

or, writing the expression as an integration,

$$I_c = 2ea \int \frac{d^3\mathbf{k}_0}{(2\pi)^3} v_{0x} D(E_x) \int \frac{d^3\mathbf{k}}{(2\pi)^3} \frac{\hbar R(\mathbf{k})}{\Gamma_{ee}} \quad (8.23)$$

with the constraints

$$E_F + eV_{bp} < E_0 < E_F + eV$$
$$\theta < \theta_c$$
$$2E_F - E_0 < E < E_F - eV_{bp}$$

We include the possibility that the injected electrons may be specularly reflected at the metal-semiconductor interface. Owing to the $\cos(\theta_{\mathbf{k},\mathbf{k}_0})$ factor in the denominator of Eq. (8.18), electrons are more effective in producing holes with $\mathbf{k}$ pointing opposite to $\mathbf{k}_0$. Therefore, the reflected electrons make a significant contribution to I_c.

For the results discussed in this chapter, scattering spectra were taken at constant tunnel current, as in the case of the ballistic carrier spectroscopies. This is accounted for by a normalization of I_c by I_t, as previously discussed. Near threshold, $I_c \propto (V - V_{bp})^4$; in contrast, the collector current in forward bias varies as $(V - V_{bp})^2$. An expression similar to Eq. (8.23) may be written for the analogous process of hole injection, secondary hot electron creation, and collection with an n-type semiconductor collector. The same $(V - V_{bn})^4$ threshold dependence is found for this case.

8.3 EXPERIMENTAL CONSIDERATIONS

The experiments described in this chapter have been performed using a standard STM that has been modified for the particular requirements of BEEM. The most important of these requirements will be discussed in this section.

BEEM is implemented here as a three-terminal experiment; in addition to maintaining tip bias and a single sample bias, individual control of two sample bias voltages is required. This entails controlling base and collector voltages while measuring currents into each of these electrodes. The sample stage on the STM provides contact to both base and collector. The base electrode is contacted by a Au wire mounted on a soft spring. Contact to the collector is by spring to a back ohmic contact on the semiconductor. The arrangement requires only minor modifications to the existing STM design. The STM design has been discussed in detail.[15] Au tips were used for all measurements.

The I_c-V spectra were obtained at constant tunnel current (normally 1 nA) using standard STM feedback techniques. This method has the advantage of linearization of the acquired spectra, as mentioned in the theoretical discussion. All imaging was also performed at constant tunnel current. This maintenance of gap spacing during imaging avoids artificial variations in collector current that would result from changes in tunnel current as the tip is scanned across the surface.

Owing to the necessity for measuring small collector currents, a high-gain, low-noise current preamplifier is used. The amplifier provides a gain of 10^{11} V/A in four stages. The reference input of the amplifier is attached to the base electrode and is maintained at ground potential.

Collector current is measured with zero applied bias between base and collector. The effective input impedance of the amplifier is about $10\,\Omega$, which is much smaller than the zero bias resistance R_0 of the diode, which, for reasons discussed below, is always greater than about $100\,k\Omega$. This low input impedance prevents leakage back into the base of electrons which enter the collector. Note that a measurement of collector current by the use of a series resistor would require such a resistor to be at least $10^8\,\Omega$ for adequate sensitivity; this large resistance would cause difficulties owing to this same leakage back across the interface.

The current amplifier has an inherent input noise, necessitating a large sample source impedance across its terminals. An amplifier input noise of $100\,nV/\sqrt{Hz}$ across a source impedance of $100\,k\Omega$ produces a noise current of $1\,pA/\sqrt{Hz}$, which is on the order of the signal to be measured. R_0 must exceed this impedance value for on adequate signal-to-noise ratio. R_0 may be increased either by reducing the interface area or by a reduction in temperature. The former method is required if R_0 is low because of ohmic regions at the interface. Low temperature measurement is more effective for samples in which R_0 is small because of a low interfacial barrier height and a consequent thermionic current. In the thermionic emission approximation,[16] the differential resistance at zero bias can be written

$$R_0 = \left(\frac{dV}{dI}\right)_{V=0} = \left(\frac{eA^*Ta}{k_B}\right)^{-1}\exp\left(\frac{eV_b}{k_BT}\right) \qquad (8.24)$$

where A^* is the Richardson constant and a is the diode junction area. Diode areas are approximately $0.01\,cm^2$; this requires, at room temperature, a SB height of at least 0.7 eV. A reduction in temperature from 293 to 77 K lowers this value to about 0.2 eV. This dependence of lowest measurable barrier height on temperature is plotted in Figure 8.8. To obtain an equivalent capability for low barrier height measurements, a reduction in diode area by more that a factor of 10^{10} would be required. In addition to the increase in resistance, low-temperature operation provides increased energy resolution for interface spectroscopy, resulting from a

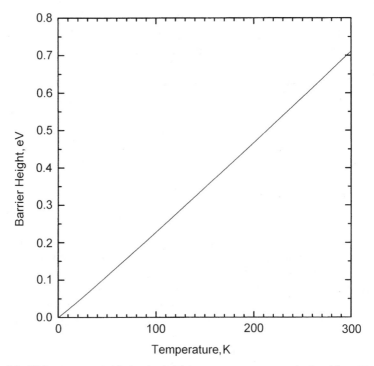

Figure 8.8. Minimum measurable barrier height versus temperature calculated from Eq. (8.24) for n-Si(l00). The criterion for measurement by BEEM is $R_0 = 100\,\text{k}\Omega$, as discussed in the text. For this calculation, $A^* = 252\,\text{A}/\text{K}^2/\text{cm}^2$ and the diode area is $0.01\,\text{cm}^2$.

narrowing of the Fermi edge of the tip. A decrease in Johnson noise of the large tunneling gap resistance also results.

A low-temperature BEEM apparatus, designed for operation at 77 K, was developed for use with low-barrier-height interface systems.[6] This includes most important p-type Schottky barriers; 77 K operation was, therefore, required for ballistic hole spectroscopy. All p-type SB characterization discussed in this chapter was performed at 77 K. This was accomplished by direct immersion of the STM head in liquid nitrogen, with the entire BEEM apparatus enclosed in a nitrogen-purged glove box.

The results presented here serve as examples of the varied capabilities of BEEM. For this work, lightly doped sample substrates were used. Final substrate preparation was performed in a flowing nitrogen environment, using a nonaqueous spin-etch for removal of oxides before metal base layer deposition. Samples were transferred directly into ultra-high vacuum without air exposure. Unless otherwise stated, Au base layer thickness was 10 nm.

Light shielding is important during data acquisition, because photocurrents generated by normal laboratory lighting can be orders of magnitude larger than currents resulting from collection of tunneling electrons. Checkout of the collector current circuitry and sample contacts is conveniently accomplished by exposure to light.

8.4 RESULTS

BEEM has been applied to many different interface systems. Examples will be given that illustrate important aspects of BEEM capabilities.

The first application of BEEM was to the Au/n-Si(100) SB interface system. Au/Si is important from a device standpoint, and it is known to provide high-quality interfaces and reproducible barrier heights. Figure 8.9 illustrates a representative I_c-V spectrum. A theoretical spectrum that was fit to the data is also shown. The agreement between theory and experiment is excellent; fitting was performed by varying only V_b and R.

As previously mentioned, planar tunneling[9] is used for the initial electron injection from the STM tip. A more realistic treatment has been introduced for the nonplanar STM geometry[17] and is discussed in Chapter 3. For a 3-eV work

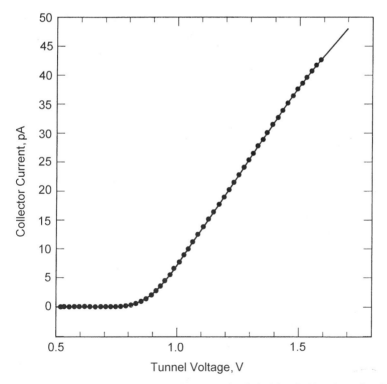

Figure 8.9. BEEM I_c-V spectrum (*solid circles*) obtained for Au/n-Si(100). Also shown is a fit (*line*) to the data. A threshold $V_b = 0.82$ V was derived from the fit to this spectrum.

function, this yields a tunneling gap of $s \approx 6$ Å, and Lang et al.[18] pointed out that an energy width of ~ 0.2 eV is expected for the tunneling distribution. In the planar tunneling model a value of $s = 15$ Å, which was previously used to fit BEEM spectra,[2] reproduces this energy width. This value is used here.

These considerations, however, do not change the quality of the fit to the threshold shape or location. Only the on-axis X minimum is considered when calculating the theoretical spectrum; the angles of the critical cones for the off-axis minima are so large (about 43°) that the tunneling formalism that is used[9] does not provide appreciable current into the large angles. If scattering is included in the transport model, the off-axis minima should be included.

It can be seen that the threshold shape is especially well fit by theory and that the predicted quadratic behavior is present. A much more sensitive test of this agreement may be performed by comparing the derivatives dI_c/dV of the experimental and theoretical spectra. These are shown in Figure 8.10 for the spectra given in Figure 8.9. The agreement provides a good test of the assumptions made in the BEEM theory.

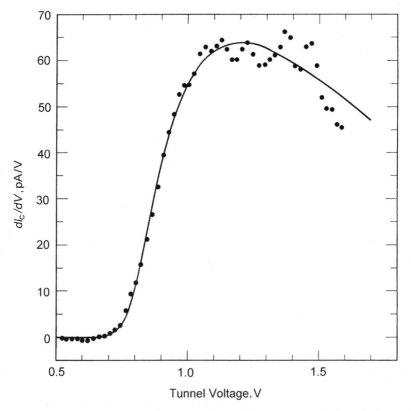

Figure 8.10. Derivatives dI_c/dV of the experimental and theoretical spectra shown in Figure 8.9.

Several observations may be made concerning the spatial variation of Au/ n-Si(100) spectra. Many individual spectra were obtained for many samples, reflecting a small spread of Schottky barrier heights, from 0.75 to 0.82 eV. The variation in R values was also small; differences between spectra were no more than a factor of three. Some level of variation in the magnitude of I_c (of which R is a measure) is expected from variations in base electrode thickness from point to point. This homogeneity has been probed directly by BEEM interface imaging and will be discussed in the next section.

One capability that BEEM provides is the ability to determine attenuation lengths of hot carriers in metals and semiconductors. This method provides a straightforward way of determining attenuation lengths that are not weighted toward the thinnest base regions, as would be the case for conventional measurements on large area interfaces. The simplest experiment consists of measuring collector current as a function of base thickness. The effective attenuation length is then given by

$$I_c = I_0 e^{-t/\lambda} \tag{8.25}$$

where t is the base thickness and λ is the effective attenuation length. Complete I_c-V spectra of Au/n-Si(l00) were obtained and fits performed to obtain R values, which provide a measure of the magnitude of I_c. A semilogarithmic plot of R versus t is shown in Figure 8.11. The expected exponential relationship is apparent, and the derived attenuation length is 13 nm. To obtain the data points that are shown, the R values of many spectra were logarithmically averaged at each thickness and different samples of each thickness were fabricated to account for spatial variations of base thickness about its average value. Note that this yields a true average, which does not weight thin base regions more heavily than others.

Although Au/Si provides an interface that is relatively well characterized, the Au/n-GaAs(100) interface system is considerably more complex. In addition to the direct minimum at Γ, the lowest conduction band has two satellite minima, at the L and X points of the zone.[19] In addition to the more complicated conduction band structure, reproducibility of interface characteristics is known to be difficult. A representative BEEM spectrum and fit to theory are shown in Figure 8.12. In this case, the data are fit with a three-threshold model, in which three separate R values are also allowed to vary. Vacuum barrier parameters $\Phi = 3$ eV and $s = 15$ Å are also used for this case.

The multiple-threshold nature of the Au/n-GaAs spectrum is clear; the derivative of data and theory, given in Figure 8.13, makes the three thresholds even more apparent. A free fit of the threshold locations yields $E_\Gamma - E_F = 0.89$ eV, $E_L - E_F = 1.18$ eV, and $E_X - E_F = 1.36$ eV. These values are in agreement with the accepted relative locations $E_L - E_\Gamma = 0.29$ eV and $E_X - E_\Gamma = 0.48$ eV.[19] The relative R values for the three minima indicate a somewhat wider angular distribution than planar tunneling produces, because the critical cones for the L minima are at about $36°$. This may be the result of elastic scattering of the electrons in the base,

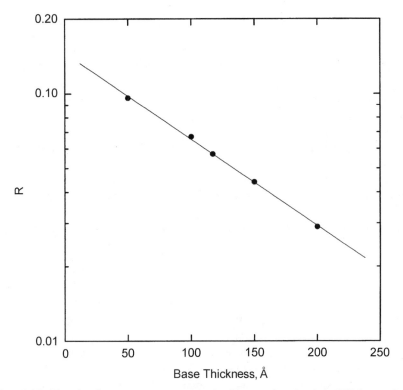

Figure 8.11. Plot of collector current, parameterized by R value, for Au/n-Si(l00) samples of different Au thicknesses. An effective attenuation length $\lambda = 13$ nm is derived from the slope of this line.

although the expected R values are complicated to calculate without knowing details of the scattering. For this fit only the on-axis X minimum is included.

The spatial variation of spectra for the Au/n-GaAs(100) interface is much greater than for Au/n-Si(100). Measured SB heights for Au/GaAs range from 0.8 to 1.0 eV. In addition, the range of R values is as much as two orders of magnitude, much greater than can be explained by a variation in Au thickness. This heterogeneity was also probed by interface imaging.

Interface imaging is made possible by the capability of the STM to scan a tip across the sample, and by the highly localized electron beam that is injected by the tip. The correspondingly high resolution at the interface has already been discussed theoretically. An interface image is acquired simultaneously with an STM topograph by measuring I_c at each point on the surface during the scan, while at a tip bias voltage in excess of threshold. Examples of the experimental results will be presented here.

It was mentioned previously that I_c-V spectra for Au/n-Si(100) show only a small variation from point to point. This variation was probed directly by BEEM

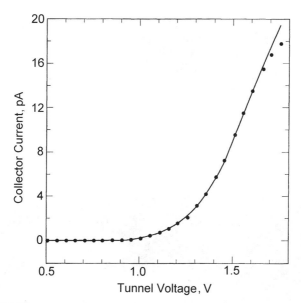

Figure 8.12. BEEM I_c-V spectrum (*solid circles*) for Au/n-GaAs(100). Also shown is a fit of a three-threshold model to the data. Three different V_b parameters and three separate R values are allowed to vary during the fit. The thresholds derived from the fit are at 0.89 V, 1.18 V, and 1.36 V.

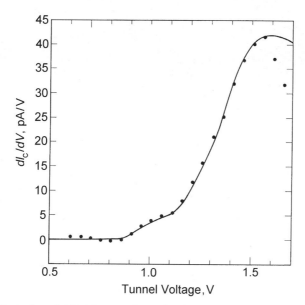

Figure 8.13. Derivatives dI_c/dV of the experimental and theoretical spectra shown in Figure 8.12. The multiple-threshold nature of the data can clearly be seen.

imaging, and a representative surface topograph and interface image for this system are shown in Figure 8.14. The average value of I_c over the image is 18 pA, whereas the RMS variation is only 0.7 pA, which is on the order of the noise level in the acquisition apparatus. This image, therefore, represents an extremely uniform interface. It is notable that the dependence of the collected current on the surface Au topography is quite weak.

In marked contrast to this uniformity is the heterogeneity observed for the Au/ n-GaAs(100) interface, for which a typical image pair is shown in Figure 8.15. The variation from dark to light in the interface image represents about two orders of

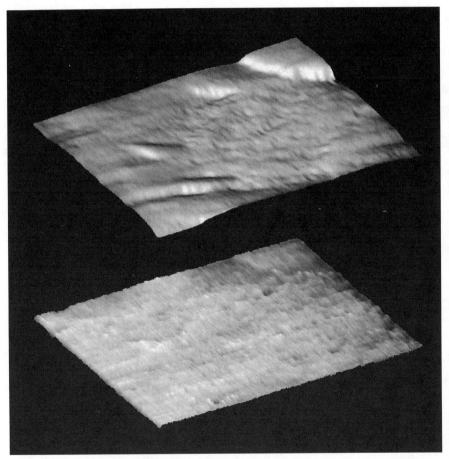

Figure 8.14. Images of Au/n-Si(l00). **a**, Conventional STM topographic image of the Au surface of the sample structure. The image is presented in a light-source rendering. Imaged area is 51 × 31 nm. Surface height range from minimum to maximum is 8 nm. **b**, Corresponding BEEM collector current image of the same sample area, obtained at a tunnel bias of 1.5 V and $I_t = 1.0$ nA. Collector current is shown in gray scale, with the largest currents in white. Average current across the image is 18 pA.

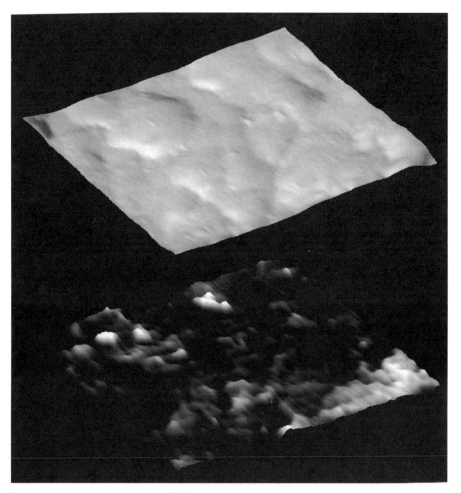

Figure 8.15. Surface topograph/BEEM image pair for the Au/n-GaAs(100) structure. The images are presented as in Figure 8.14. Image area is 51 × 39 nm. **a**, Topograph of the Au surface. The range of height across this image is 6.3 nm. **b**, BEEM gray scale interface image, obtained at $V = 1.5$ V and $I_t = 1.0$ nA. Collector current ranges from <0.1 pA (black) to 14 pA (white).

magnitude in collected current. This large range of intensity values is typical of this interface system and is too great to be explained by a simple thickness variation of the base electrode. Moreover, the variation in intensity is primarily the result of a variation in R rather than to a change in threshold. Spectra taken within lower current areas where there is enough signal to determine thresholds do not indicate a systematic relationship to threshold position.

It has been demonstrated[20] that interfaces formed both on melt-grown and MBE-grown GaAs substrates exhibit this heterogeneity, indicating that bulk defect density does not play a large part in the presence of this interface disorder. The heterogeneity also persists over a wide range of surface preparation conditions. Chemically cleaned GaAs substrates that were exposed to air before Au deposition exhibit this behavior; a careful chemical cleaning of the GaAs substrate in flowing nitrogen gas followed by direct transfer to the Au deposition chamber also produces such interfaces, even though x-ray photoelectron spectroscopy (XPS) shows them to be oxide-free.[20] This is an indication that the interface defect structure is not simply the result of surface contamination.

GaAs is known to dissociate at the interface of a Au/GaAs contact;[21] the Ga is soluble in Au and tends to migrate to the Au surface, whereas As is insoluble and remains at the interface. The low-current areas actually dominate the Au/GaAs interface and are interpreted in terms of interfacial islands of As created by GaAs dissociation and Ga migration.[22] The experimental results indicate that this diffusion process dominates the interface formation process in the Au/GaAs(100) system.

Ballistic hole spectroscopy was first performed on the SB systems that were previously probed by BEEM, allowing a characterization of the interface valence band structure of these structures. As mentioned in the theoretical discussion, a p-type semiconductor collector is used for ballistic hole spectroscopy, and the typically low barrier heights for p-type semiconductors necessitate low-temperature measurements. A ballistic hole spectrum for Au/p-Si(100) obtained at 77 K is shown in Figure 8.16. Also shown is a fit of the ballistic hole theory to the data. The sign of the collected current is not arbitrary; a negative sign for the current indicates the collection of holes. The SB height V_b for this system is measured to be 0.35 eV, with only a small spread in this value from point to point. It should be mentioned that the n and p barrier heights measured at 77 K add to 1.17 eV, which agrees well with the Si 77 K band gap value of 1.16 eV.

Derivatives of both the electron and hole spectra, both obtained at 77 K, are shown in Figure 8.17. The derivative of the ballistic hole spectrum maximizes and turns over more quickly than that of the electron spectrum. This can be interpreted in terms of the asymmetry between the collected distributions for electron and hole injection. For the hole case, the additional current per unit voltage is decreasing, causing the I_c-V spectrum to inflect more quickly. A second feature to notice is the extremely sharp thresholds in the derivatives produced by data acquisition at low temperature.

For this and other ballistic hole spectra, R values are larger than for the corresponding electron spectra. In addition, the best fits to the data are obtained for the vacuum gap parameters $\Phi = 3$ eV and $s = 8$ Å, whereas the values $\Phi = 3$ eV, $s = 15$ Å were used for electron spectra. This may be an indication that the tunneling formalism being used does not accurately describe the energy and angular distributions of tunneling electrons, at least for tunneling from sample to tip. Ballistic hole spectroscopy is more sensitive to the details of the tunneling distribution than is ballistic electron spectroscopy, because the collected carriers come from the tail of the tunneling energy distribution. However, as with BEEM spectroscopy, these considerations do not affect the quality of the fit near threshold or the determination of the interface barrier height.

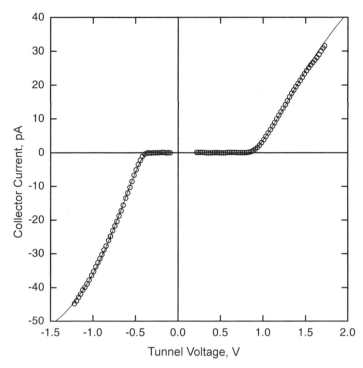

Figure 8.16. Ballistic hole spectrum of Au/p-Si(100) (*open circles*) with fit to theory (*line*), is shown in the lower left quadrant of the figure. Tunnel voltage refers to $V_{sample} - V_{tip}$. These data were obtained at 77 K. The fit yields a threshold $V_b = 0.35$ V for the SB. Also shown for comparison in the upper right quadrant is a ballistic electron spectrum for Au/n-Si(l00). It should be emphasized that the two spectra were obtained from two different samples.

Although the conduction bands of Si and GaAs are quite different, the valence band structures are similar.[19, 23] For both semiconductors, the valence band system is composed of three separate bands: the light-hole, heavy-hole, and split-off bands. The light- and heavy-hole bands are degenerate at the center of the Brillouin zone and define the p-type Schottky barrier; away from the zone center, the mass of the light hole band increases until the two bands are of essentially equal mass, but split by a small energy difference. The third band is nondegenerate at the zone center due to spin-orbit splitting.

In the case of Au/p-Si(100), this light-hole/heavy-hole splitting away from $\mathbf{k} = 0$ is only about 30 meV[23] and is not resolved with the present apparatus; however, this is not the case for Au/p-GaAs(100). A ballistic hole spectrum of Au/p-GaAs(100) is shown in Figure 8.18. Also plotted are two theoretical curves. The first considers only a single threshold at the valence band edge; the second includes the light-hole band, with the approximation that this band changes abruptly from its light-hole mass to the heavy-hole mass away from the zone center. (The point of transition was obtained from Figure 47 in Ref. 20.) It is apparent that this second fit agrees well with the threshold shape of the experimental spectrum, whereas the first

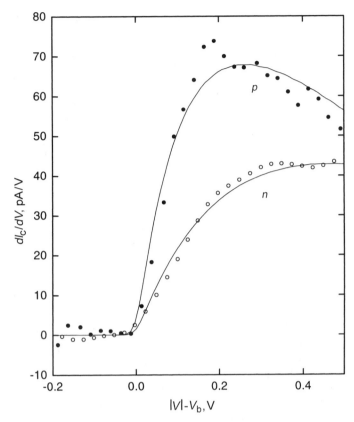

Figure 8.17. Derivatives dI_C/dV of the ballistic hole experimental and theoretical spectra shown in Figure 8.16. Also shown for comparison are equivalent derivatives of ballistic electron spectra for Au/n-Si(100), also from Figure 8.16. The electron and hole spectra are plotted with a horizontal displacement so that the thresholds coincide. Both experimental spectra were obtained at 77 K.

fit is poor at threshold. The measured barrier height is 0.70 eV, and the effective splitting of the light- and heavy-hole band away from the zone center is about 100 meV, in good agreement with current values of this splitting obtained by other methods.[19]

It should be noted that the simple phase space model that has been used to analyze electron and hole spectra thus far would not predict the observation of the light hole band in the Au/p-GaAs data. The phase space for collection by the light-hole band is completely within that for the heavy-hole band; therefore, there is no additional phase space for collection when the onset of the light-hole band is reached. The assumption of the phase-space model, that all carriers within the critical cone are collected, is thus inappropriate here; this would not allow additional current from the light-hole band. The additional current is expected, however, if there is quantum-mechanical reflection at the interface that allows only a portion

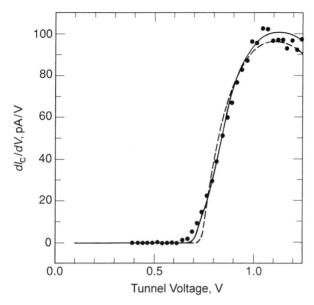

Figure 8.18. Derivative dI_c/dV of a ballistic hole spectrum for Au/p-GaAs(100) (*solid circles*). Also shown are derivatives of two theoretical fits to the data. The *dashed line* represents the best fit obtainable using a single-threshold model; the *solid line* illustrates a two-threshold fit, incorporating a heavy-hole band, with constant mass 0.51 m, and a light hole band, with mass 0.082 m near the zone center, and mass 0.51 m away from zone center. The transition between these masses is taken as abrupt, and the point of transition produces the second threshold.

of the carriers within the critical cone to be collected. A second channel for collection would then produce an increase in current.

It should be mentioned that the split-off bands for both Si and GaAs should in principle be observable. However, the masses of these bands are smaller than those of the heavy-hole bands, and the increase in current as a result of them is difficult to detect.

The first application of the carrier scattering spectroscopy discussed previously was to the interface systems that have been investigated by BEEM: Au/Si(100) and Au/GaAs(100).[7] The importance of these interfaces has already been emphasized; a characterization of transport within the electrodes themselves is also of great importance to a complete understanding of the entire transport process. In addition, analysis of the results of this scattering spectroscopy is aided by the previous characterization of interface transport and the collection process, which is provided by the ballistic spectroscopies that have already been discussed.

Figure 8.19 displays two spectra obtained for Au/p-Si(100). The spectrum at positive tip voltage is a ballistic hole spectrum similar to that in Figure 8.16, with a threshold that yields a SB height of 0.35 eV. Shown at negative tip bias is a spectrum of holes created by carrier scattering. In this case, the negative tip voltage injects ballistic electrons into the sample, some of which scatter with equilibrium

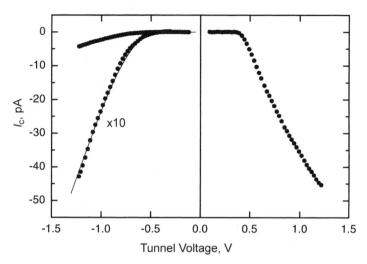

Figure 8.19. Carrier-carrier scattering spectrum obtained for the Au/p-Si(l00) system. Also shown at the positive tip bias is a ballistic hole spectrum for the same sample. For clarity, the scattering spectrum is also plotted on an expanded scale, with a fit (*line*) to the theory described in the text. The value of $V_b = 0.35$ V, obtained from the ballistic hole spectrum, was used as a fixed parameter for the fit; Φ and s were fixed at 3 eV and 15 Å.

electrons in the base and result in the creation of secondary electron-hole pairs. The secondary electrons eventually decay to the Fermi level, but some of the secondary holes may be collected by the Si valence band before they thermalize. The plotted spectrum shows collection of these holes. Also shown is an expanded version of this scattering spectrum. It is clear that the spectral shapes and the magnitudes of the currents are quite different for the ballistic and scattering spectra. A theoretical curve is also shown, which was fit to the data by adjustment of only an overall scaling factor reflecting the magnitude of the current. Agreement with experiment is excellent. The scattering spectrum exhibits a threshold at approximately 0.35 eV, the value of the SB height for this system. This observation of collected current well below the Si conduction band edge ($eV_b = 0.82$ eV) rules out any processes involving transport in the Si conduction band.

A similar experiment may also performed for Au on p-GaAs(100). Because the barrier height for this interface is roughly twice that of Au/p-Si, it provides a good test of the theoretical description developed for the process. As shown in Figure 8.20, the spectra are qualitatively similar to those for Si. This is to be expected, because this spectroscopy is primarily a probe of processes with in base layer, independent of the collector. However, the measured spectra do reflect aspects of the particular collector electrode used. Most noticeably, the threshold is determined by the Au/p-GaAs Schottky barrier height, 0.70 eV. In addition, the current of scattered carriers for this system is smaller than that for Au/p-Si, for equivalent voltages in excess of threshold. This is because of the smaller fraction of phase space available for scattering and collection, and the ratio of the currents for the

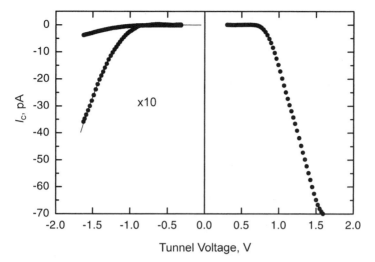

Figure 8.20. Scattering spectrum and theoretical fit obtained for the Au/p-GaAs(l00) system. Also shown at the positive tip bias is a ballistic hole spectrum for the same sample. V_b was fixed at 0.70 V, the value obtained from the ballistic hole spectrum. Φ and s were fixed at 3 eV and 15 Å.

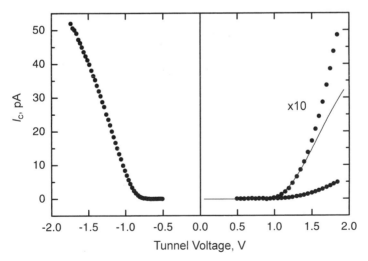

Figure 8.21. Scattering spectrum and theoretical fit obtained for Au/n-Si(100). Also shown at the positive tip bias is a ballistic electron spectrum for the same sample. V_b was fixed at 0.82 V, the value obtained from the ballistic electron spectrum. In this case, only data at tip voltage <1.4 V were used for the fit. Φ and s were fixed at 3 eV and 8 Å.

two systems agrees well with that predicted by theory. This agreement indicates that there is no large interface-specific contribution to the scattering process itself. This scattering spectroscopy has also been performed for n-type collectors. Here, injected ballistic holes may scatter with equilibrium electrons in the base and create electron-hole pairs, and the secondary electrons are collected in the conduction band of the n-type semiconductor. Figure 8.21 shows ballistic and scattering spectra for Au/n-Si(100). The threshold reflects the n-type SB height of 0.82 eV. In this case, there is good agreement of the theory with the data near threshold; however, the measured collector current increases more rapidly at higher voltages than the theory predicts. Interestingly, the onset of this deviation is at about $eV = E_{band\ gap} + eV_{bp}$, raising the possibility that excess current is resulting from pair creation by holes that enter the valence band of the semiconductor. The overall magnitude of the measured current is also high, about an order of magnitude higher than expected from the theory. Fitting of the reverse-bias spectrum for n-Si required inclusion of all conduction-band X minima, even those that are off-axis. This discrepancy is reduced by inclusion of multiple electron scattering in a Monte Carlo calculation.

8.5 RECENT APPLICATIONS OF BEEM

Since the first edition of this book, BEEM has been applied to a large number of interface and transport problems. A few of these may be mentioned here. Reviews that cover this work in more detail have also appeared. [24, 25]

BEEM measurements have been performed at temperatures <77 K to probe details of spectrum threshold behavior[26] and details of transport.[27] The question of parallel wave vector conservation at a metal/semiconductor interface has also recently been addressed.[28] An attempt to investigate the fundamental question of BEEM imaging resolution has yielded surprisingly high resolution of interface structures.[29] Intentional interface modification by high-energy electron injection has also been observed in localized areas.[30, 31]

Examples of epitaxial metal/semiconductor systems investigated by BEEM include NiSi$_2$/n-Si(111) and CoSi$_2$/n-Si(111).[32, 33] Theoretical BEEM calculations for these interfaces were also presented.[34, 35] Furthermore, thorough measurements on CoSi$_2$/n-Si(111) were performed in UHV and at low temperature, at which individual interface dislocations[36] and individual point defects[37] were observed. An energy-dependent attenuation length was measured, which was found to be orientation dependent.[38] Quantized changes in BEEM current owing to the monolayer variation of silicide thickness were also mapped.[39] Finally, corrugation in BEEM images was observed that reflected the atomic periodicity of the surface,[40] resulting from a modulation of the tunneling distribution by the atomic corrugation of the surface.

Other energy-dependent attenuation lengths were measured for other metals,[41] and transport in semiconductors was probed using pn structures[42] and later at higher voltage using a thin NiSi$_2$ base layer with pinholes.[43] These high-voltage measurements revealed features of semiconductor density of states and affect ionization.[44]

BEEM spectroscopy of transmission through oxide layers[45,46] yielded a measurement of the SiO_2 conduction-band edge and oxide transmission probability. Suppression of BEEM transmission was observed in areas that had been charged by the STM tip.[47] Additional work focused on oxide breakdown[48] and extraction of the image-force dielectric constant.[49]

Transport in semiconductor heterostructures was studied by BEEM techniques. BEEM was applied to transport resonances in double-barrier structures,[50] imaging of misfit dislocations,[51] and electronic properties of quantum dots.[52] Band-offset was measured and observed to decrease with increasing alloy ordering.[53] A thorough study of the AlGaAs/GaAs system was also performed.[54]

8.6 SUMMARY

This chapter covered the basic concepts of ballistic carrier spectroscopies using STM and an extension to the study of nonballistic processes. BEEM and its related techniques are extremely powerful probes of subsurface interfaces and of carrier transport through materials and across these interfaces. BEEM provides the most precise method available for the determination of Schottky barrier heights and is sensitive to other characteristics of the interface band structure, such as effective mass and satellite conduction band minima. The implementation of BEEM using STM is a vital aspect of the method, because the highly localized electron injection enables a corresponding spatial resolution of interface properties. This implementation, therefore, provides an interface imaging capability, which has been used to reveal hitherto unknown heterogeneity at semiconductor interfaces. The theoretical description of BEEM spectroscopy was discussed as well as the implications of the theory for interface imaging resolution.

The precision of barrier height measurements allows the determination of small changes in height resulting from such factors as strain or temperature change. Such measurements are one important means for evaluating theories of interface formation.

BEEM also provides information on fundamental questions regarding carrier transport. For instance, spectroscopy of interfaces formed on semiconductors with off-axis minima should yield insight concerning the tunneling distribution, in E and **k**, of electrons in a tip-plane geometry as well as the much-argued notion of conservation of k_t across a real interface. Two of the methods for probing carrier transport through materials were presented in this chapter. The first, using conventional BEEM, involves the measurement of collector current as a function of base thickness and a determination of carrier attenuation lengths. This measurement may also be performed for the case of off-axis collector band minima, which should elucidate the contributions of elastic and inelastic scattering to the measured attenuation length.

The second method for the investigation carrier transport involves the use of the scattering spectroscopy, or reverse-bias BEEM experiment, discussed in this chapter. This spectroscopy allows the first direct probe of carrier-carrier scattering in materials and can be performed as a function of base thickness and temperature and with off-axis collectors to provide a wealth of data on carrier scattering mechanisms. The theoretical treatment for this process yields good agreement with experiment and, as in the case of BEEM theory, may be built on to incorporate more complicated processes.

ACKNOWLEDGMENT

The work described in this paper was performed by the Center for Space Microelectronics Technology, Jet Propulsion Laboratory, California Institute of Technology and was sponsored by the Office of Naval Research and the Strategic Defense Initiative Organization/Innovative Science and Technology Office through an agreement with the National Aeronautics and Space Administration (NASA).

REFERENCES

1. W. J. Kaiser and L. D. Bell, *Phys. Rev. Lett.* **60**, 1406 (1988).

2. L. D. Bell and W. J. Kaiser, *Phys. Rev. Lett.* **61**, 2368 (1988).

3. W. G. Spitzer, C. R. Crowell, and M. M. Atalla, *Phys. Rev. Lett.* **8**, 57 (1962); C. R. Crowell, W. G. Spitzer, and H. G. White, *Appl. Phys. Lett.* **1**, 3 (1962).

4. M. P. Seah and W. A. Dench, *Surf. Interface Anal.* **1**, 2 (1979).

5. C. R. Crowell, W. G. Spitzer, L. E. Howarth, and E. E. LaBate, *Phys. Rev.* **127**, 2006 (1962).

6. M. H. Hecht, L. D. Bell, W. J. Kaiser, and L. C. Davis, *Phys. Rev. B* **42**, 7663 (1990).

7. L. D. Bell, M. H. Hecht, W. J. Kaiser, and L. C. Davis, *Phys. Rev. Lett.* **64**, 2679 (1990).

8. C. R. Crowell and S. M. Sze, *Solid State Electron.* **8**, 673 (1965).

9. J. G. Simmons, *J. Appl. Phys.* **34**, 1793 (1963).

10. C. B. Duke, *Tunneling in Solids*, Academic Press, New York, 1969.

11. E. L. Wolf, *Principles of Electron Tunneling Spectroscopy*, Oxford University Press, New York, 1985.

12. S. M. Sze, C. R. Crowell, G. P. Carey, and E. E. LaBate, *J. Appl. Phys.* **37**, 2690 (1966).

13. S. Gasiorowicz, *Quantum Physics*, John Wiley & Sons, New York, 1974.

14. J. J. Quinn and R. A. Ferrell, *Phys. Rev.* **112**, 812 (1958).

15. W. J. Kaiser and R. C. Jaklevic, *Rev. Sci. Instrum.* **59**, 537 (1988).

16. S. M. Sze, *Physics of Semiconductor Devices*, 2nd ed, John Wiley & Sons, New York, 1981.

17. J. Tersoff and D. R. Hamann, *Phys. Rev. Lett.* **50**, 1998 (1983).

18. N. D. Lang, A. Yacoby, and Y. Imry, *Phys. Rev. Lett.* **63**, 1499 (1988).

19. J. S. Blakemore, *J. Appl. Phys.* **53**, R123 (1982).

20. M. H. Hecht, L. D. Bell, W. J. Kaiser, and F. J. Grunthaner, *Appl. Phys. Lett.* **55**, 780 (1989).

21. P. W. Chye, I. Lindau, P. Pianetta, C. M. Garner, C. Y. Su, and W. E. Spicer, *Phys. Rev. B* **18**, 5545 (1978).

22. J. L. Freeouf and J. M. Woodall, *Appl. Phys. Lett.* **39**, 727 (1981).

23. J. S. Blakemore, *Semiconductor Statistics*, rev. ed., Dover, New York, 1987.

24. M. Prietsch, *Phys. Reports* **253**, 164 (1995).

25. L. D. Bell and W. J. Kaiser, *Annu. Rev. Mater. Sci.* **26**, 189 (1996).

26. G. N. Henderson, P. N. First, T. K. Gaylord, and E. N. Glytsis, *Phys. Rev. Lett.* **71**, 2999 (1993).

27. D. K. Guthrie, L. E. Harrell, G. N. Henderson, P. N. First, and T. K. Gaylord, *Phys. Rev. B* **54**, 16972 (1996).

28. L. D. Bell, *Phys. Rev. Lett.* **77**, 3893 (1996).

29. A. M. Milliken, S. J. Manion, W. J. Kaiser, L. D. Bell, and M. H. Hecht, *Phys. Rev. B* **46**, 12826 (1992).

30. A. Fernandez, H. D. Hallen, T. Huang, R. A. Buhrman, and J. Silcox, *Appl. Phys. Lett.* **57**, 2826 (1990).

31. H. D. Hallen, T. Huang, A. Fernandez, J. Silcox, and R. A. Buhrman, *Phys. Rev. Lett.* **69**, 2931 (1992).

32. A. Fernandez, H. D. Hallen, T. Huang, R. A. Buhrman, and J. Silcox, *J. Vac. Sci. Technol. B* **9**, 590 (1991).

33. W. J. Kaiser, M. H. Hecht, R. W. Fathauer, L. D. Bell, E. Y. Lee, and L. C. Davis, *Phys. Rev. B* **44**, 6546 (1991).

34. M. D. Stiles and D. R. Hamann, *Phys. Rev. Lett.* **66**, 3179 (1991).

35. M. D. Stiles and D. R. Hamann, *J. Vac. Sci. Technol. B* **9**, 2394 (1991).

36. H. Sirringhaus, E. Y. Lee, and H. von Känel, *Phys. Rev. Lett.* **73**, 577 (1994).

37. T. Meyer and H. von Känel, *Phys. Rev. Lett.* **78**, 3133 (1997).

38. E. Y. Lee, H. Sirringhaus, U. Kafader, and H. von Känel, *Phys. Rev. B* **52**, 1816 (1995).

39. E. Y. Lee, H. Sirringhaus, and H. von Känel, *Phys. Rev. B* **50**, 14714 (1994).

40. H. Sirringhaus, E. Y. Lee, and H. von Känel, *Phys. Rev. Lett.* **74**, 3999 (1995).

41. R. Ludeke and A. Bauer, *Phys. Rev. Lett.* **71**, 1760 (1993).

42. L. D. Bell, S. J. Manion, M. H. Hecht, W. J. Kaiser, R. W. Fathauer, and A. M. Milliken, *Phys. Rev. B* **48**, 5712 (1993).

43. A. Bauer and R. Ludeke, *Phys. Rev. Lett.* **72**, 928 (1994).

44. R. Ludeke, *Phys. Rev. Lett.* **70**, 214 (1993).

45. R. Ludeke, A. Bauer, and E. Cartier, *Appl. Phys. Lett.* **66**, 730 (1995).

46. R. Ludeke, A. Bauer, and E. Cartier, *J. Vac. Sci. Technol. B* **13**, 1830 (1995).

47. B. Kaczer, A. Meng, and J. P. Pelz, *Phys. Rev. Lett.* **77**, 91 (1996).

48. R. Ludeke, H. J. Wen, and E. Cartier, *J. Vac. Sci. Technol. B* **14**, 2855 (1996).

49. H. J. Wen, R. Ludeke, D. M. Newns, and S. H. Lo, *J. Vac. Sci. Technol. A* **15**, 784 (1997).

50. T. Sajoto, J. J. O'Shea, S. Bhargava, D. Leonard, M. A. Chin, and V. Narayanamurti, *Phys. Rev. Lett.* **74**, 3427 (1995).

51. E. Y. Lee, S. Bhargava, M. A. Chin, V. Narayanamurti, K. J. Pond, and K. Luo, *Appl. Phys. Lett.* **69**, 940 (1996).

52. M. E. Rubin, G. Medeiros-Ribeiro, J. J. O'Shea, M. A. Chin, E. Y. Lee, P. M. Petroff, and V. Narayanamurti, *Phys. Rev. Lett.* **77**, 5268 (1996).

53. J. J. O'Shea, C. M. Reaves, S. P. DenBaars, M. A. Chin, and V. Narayanamurti, *Appl. Phys. Lett.* **69**, 3022 (1996).

54. J. J. O'Shea, E. G. Brazel, M. E. Rubin, S. Bhargava, M. A. Chin, and V. Narayanamurti, *Phys. Rev. B* **56**, 2026 (1997).

9

THE SCANNING PROBE MICROSCOPE IN BIOLOGY

Stuart M. Lindsay

9.1   INTRODUCTION

Microscopy has laid the foundation for many revolutions in biology since Leeu-wenhoek first glimpsed "animicules" through a glass lens. Feynman saw micro-scopy of single *molecules* as the key to the problems of modern molecular biology. His famous 1959 lecture "There's Plenty of Room at the Bottom"[1] put it this way:

> It is very easy to answer many of these fundamental biological questions; you just look at the thing! You will see the order of the bases in the chain; you will see the structure of the microsome. Unfortunately the present microscope sees at a level which is just a bit too crude. Make the microscope a hundred times more powerful, and many problems of biology would be made very much easier.

The powerful microscope of Feynman's dreams has not been built yet. But progress in scanning probe microscopy (SPM) since the first edition of this book[2] has been remarkable. The atomic force microscope (AFM) is now used routinely as a diag-nostic probe of biomaterials. Fundamental new structural information is being revealed as AFM imaging becomes routine as a complement to transmission elec-tron microscopy (TEM). New types of measurements on single molecules use the scanning probe microscope as a local probe. These developments have been ac-companied by similar advances in optical methods for studying single molecules, and we are beginning to learn that the world is a different place when look at one molecule at a time.[3]

Advances in instrumentation have been rapid. The AFM cam now reproduce the atomic detail of clean surfaces once thought to be the exclusive domain of the scanning tunneling microscope (STM).[4, 5] This level of performance is being ap-proached in liquids.[6, 7] The STM, which featured so heavily in the first edition of this book, has almost retired from the scene, although it merits a short postscript here.

Some quite remarkable resolution has been achieved,[8-11] even in solution[12] where movies of enzyme activity have been recorded.[13, 14] Cryo-AFM is beginning to produce spectacular results.[15, 16] A new field is emerging in which the probe is used as a chemical or force sensor or manipulator rather than as an imaging device.[17, 18]

This chapter is not intended to be a comprehensive review. A number of recent review articles are available, and they are useful starting points for searching the current literature.[8, 10-12, 19-27] The laboratories of Hansma and Bustamante have been particularly active in developing techniques for imaging important processes as they occur, for example, the real-time binding of protein to DNA,[28] the nuclease digestion of DNA,[29] and the action of RNA polymerase.[13] News of further exciting developments will undoubtedly appear in the literature in the near future.[14] This chapter focuses on the physical and instrumental aspects of AFM in fluids, with an emphasis on biological applications. The field is hardly a mature one, and it is expected that instrumental developments will continue apace.

To make this chapter self-contained, I begin with a brief introduction to the physics of the AFM. I then discuss the factors that control resolution and interaction forces. I then turn to a rather extensive discussion of dynamic force microscopy (DFM) in fluids. In the dynamic force microscope, interactions are sensed via the change in amplitude or phase of an oscillated tip. There have been important developments in the field not pulled together in another review, and this section forms the bulk of this chapter. My goals are to explain the physics of the imaging mechanism and operating criteria and to describe the instrumentation. I then consider methods of sample preparation. Many new approaches have become possible as a result of the flexibility of the SPM, and the effect of different preparation methods can be explored in a systematic manner. I then show some examples of images (several of which are included separately as color plates). I turn finally to measurements on individual molecules, such as elastic properties, unfolding, chemical bonding, and charge transport. I end with discussion of STM and conducting AFM.

9.2   FUNDAMENTALS OF AFM: SENSITIVITY AND RESOLUTION

9.2.1   Intrinsic Sensitivity of the Optical Lever

The intrinsic sensitivity of even the simple optical lever is more than adequate for atomic resolution. A well-designed instrument, free of thermal, electronic, and vibrational noise, is limited only by laser shot noise. This determines the smallest change in signal detectable on the segments of the photodiode position sensitive detector, assuming that adequate gain is available. The shot noise can be made vanishingly small with adequate laser power. A 1 mW optical laser emits about

10^{15} photons/s and the corresponding shot noise is $\approx \sqrt{10^{15}}$ or about 10^7 photons/s, so that the signal-to-noise ratio is about 1 part in 10^7. Thus changes in the cantilever bending as small as this may be detected if the signal is accumulated for 1 s. For small deflections, the cantilever bending angle is proportional to the tip height, so this angular sensitivity corresponds to a height sensitivity. With a $100\,\mu$m lever, a motion on the order of 10 pm is detectable. On more typical imaging time scales (1 ms/pixel) the shot-noise-limited sensitivity is on the order of 300 pm. It is other factors (the profile of the tip, thermal fluctuations, deformation of the sample, and intervening contamination and fluid layers) that limit the resolution.

9.2.2 Limits on Interaction Forces—Thermal Noise

Subangstrom resolution is realized only at low temperatures. At finite temperature, the tip undergoes thermal motion. At its simplest, this requires that the sample does more work on the tip than thermal energy. Thus, with a tip of spring constant k N/m, the smallest resolvable vertical displacement, Δz is given approximately by

$$\left\langle \frac{1}{2}k(\Delta z)^2 \right\rangle \approx \frac{1}{2}k_B T \tag{9.1}$$

With $k_B T$ on the order of 4×10^{-21} J at room temperature, the best vertical resolution obtainable with a cantilever with $k = 1\,$N/m is on the order of 0.05 nm. For operation at finite bandwidth, it is if some interest to consider the spectral distribution of noise. An approximate description of the displacement of the cantilever is given in terms of a harmonic oscillator model[30]

$$\left\langle \Delta z^2 \right\rangle = \frac{4m\omega_0 k_B T}{Q} \int_0^\infty \frac{(1/m^2)}{(\omega_0^2 - \omega^2)^2 + \left(\dfrac{\omega_0 \omega}{Q}\right)^2}\, d\omega \tag{9.2}$$

where the cantilever is represented by a harmonic oscillator of mass m, resonant frequency ω_0, and mechanical Q factor Q. The Q parameterizes the sharpness of the resonance, being the ratio of the resonant frequency to the full width at half height of the peak. It is also the ratio of the amplitude at resonance to that at a frequency well below resonance. In air or vacuum, high Q (on the order of thousands) can be achieved and Eq. (9.2) shows that there is a large reduction of the thermal noise away from the resonant peak because of the factor $1/Q$ in front of the integral. Contact-mode operation senses the static deflection of the tip and, therefore, a signal must be passed from zero frequency up to an upper limit $\omega_{\max}$ set by the scan speed and resolution of the microscope. Clearly, an enhanced signal-to-noise ratio can be obtained at the expense of operating speed by reducing $\omega_{\max}$ and scanning more slowly. In practice, contact mode is limited by $1/f$ noise (discussed below), and this can overwhelm any advantage of a slower scan. The effect is marked in dynamic force microscopy where $1/f$ noise is much reduced.

The noise spectrum as given by Eqs. (9.1) and (9.2) forms the basis of a nondestructive method for calibrating cantilever spring constants, a difficult experimental problem.[31-33] If the detector signal is first accurately calibrated in terms of absolute cantilever displacement, then the profile given under the integral in Eq. (9.2) can be fitted over a range of frequencies for which amplitude data are available. This is conveniently obtained by Fourier transforming thermal noise recorded as a function of time. The integral is then carried out on the fitted function and Eq. (9.1) used to extract the spring constant from the ratio of $k_B T$ to $\langle \Delta z^2 \rangle$.

We are concerned almost exclusively with operation in water or aqueous buffer, in which case the effective Q of the cantilever is reduced by the viscous action of the surrounding liquid to a small value (typically three[34]). In this case, according to Eq. (9.2), noise is almost equally distributed as a function of frequency because the resonant peak is so broad. However, inspection of measured amplitude vs. frequency curves[35] shows that Eq. (9.2) is only an approximation. Other sources of mechanical motion and mechanical pointing instabilities of the laser become progressively more important at low frequencies. Lower resonant frequency is equivalent to a smaller spring constant, so that the amplitude of fluctuation increases hyperbolically as ω_0 (and hence k) go to zero (see Eq. (9.1)). This is referred to as $1/f$ noise, and its reduction by operation at a finite frequency is one of the main reasons for the use of DFM as opposed to the conventional contact mode.

9.2.3   Limits to Resolution and Imaging in Fluid

There are several working definitions of resolution used in AFM. One of the more rigorous approaches is based on the spectrum of Fourier components visible in images of periodic arrays.[36] This is often not useful in biologic imaging, particularly for processes involving complexes of molecules. It is often more useful to compare the full width of an image with the known crystallographic dimensions of a structure (although this, in turn, suffers some obvious drawbacks). If such a comparison is used, it is important that some measure is given for a representative field of molecules, for "best" regions of images can be entirely spurious. For example, parts of a molecule may vanish in the noise, resulting in a "width" that appears to be *less* than the crystallographic dimension. The molecular packing is also important. A densely packed monolayer will sample the tip in quite a different manner from the way an isolated molecule does.

The resolution of an AFM is primarily a function of the radius of curvature of the end of the tip, as discussed by many authors.[37-39] Features on the surface of smaller radius of curvature than the tip itself will be mapped out as a replica of the tip, and all information about the surface in the region of high curvature is lost. One approach to correcting for tip broadening (to the extent that this is possible) is based on detection of the regions of steepest slope, using these to reconstruct the tip geometry.[39,40]

The physics of the tip–sample interaction is also important in determining resolution. In fact, even the bluntest of tips has a rough surface, being covered with fine asperities,[41] so that, on a relatively flat sample the apex can appear atomically sharp. Thus, providing one is willing to select tips, no special preparation is needed to find a reasonable fraction (on the order of 10%) that gives particularly high resolution. Ideal and reproducible tips may not be far off. Nanotubes have shown promise as durable, sharp tips of extremely high aspect ratio,[42] and methods for quantity production of single-walled nanotips are expected to evolve rapidly.

Bustamante and Keller[22] considered an ideal "single line of atoms" tip, showing how the resolution falls off with distance as the tip is scanned at increasing height above a sample. Giessibl[5] showed how local interactions can give rise to atomic

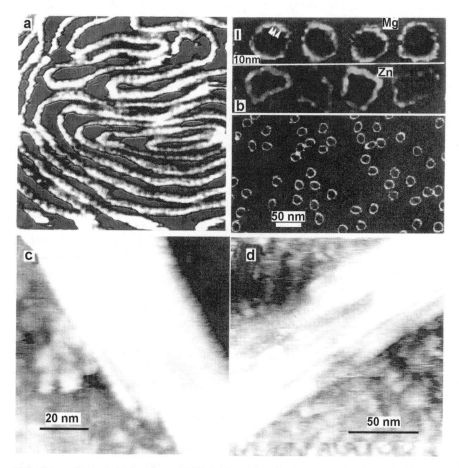

Plate 9.1. **a**, Contact-mode image of DNA tightly adsorbed onto a cationic bilayer in buffer. The double helical repeat is clearly visible. (Courtesy of Z. Shao.) **b**, Magnetic DFM of DNA microcircles on mica in Mg-containing solution. *Lower region*, a field of molecules, *upper portion*, magnified gallery showing helix repeat (*arrows*, 3.4 nm spacing). When Zn (lower part, upper portion) replaces Mg (upper part, upper portion), the circles kink. **c**, Microtubule imaged in buffer by magnetic DFM; 13 protofilaments are visible, as is the tubulin substructure. **d**, When katinin is added, the microtubules are digested. (Courtesy of J. Zhu and R. Vale.) Figure also appears in Color Plate section.

resolution even in the presence of long-range interactions. Yang and co-workers[43] proposed that short-range interactions with a local asperity give rise to high-resolution contrast, whereas longer-range interactions with blunter parts of the tip help support the load of the tip in contact with the sample. Muller and Engel[44] argued that, in solution, the long-range force required for repulsion of the body of the tip is electrostatic. They optimize resolution by adjusting the supporting electrolyte so that the asperity just touches the sample lightly. Thus, provided that the sample can sustain the load of a tip in contact and provided that it is reasonably flat (so that a single asperity can dominate contrast) subnanometers scale resolution can be obtained.[36, 45]

Contact mode imaging requires densely packed samples, both to provide stability against the strong interaction with the tip and to present a flat surface to the tip. It has provided the best resolution to date. Examples of such images are given in plates 9.1 and 9.2. Densely packed samples are not always suitable for experiments that probe biochemical function (although processes in dense arrays of membrane proteins have been studied.[27] It is often desirable to image well-separated molecules bound to an underlying surface rather weakly. In this case, the sample will not survive the intrusion of contact mode scanning and DFM is required. This technique is gentle, in part because it can be operated in a noncontact mode in solution (see below). However, this is also a disadvantage, because layers of semimobile fluid between the tip and sample degrade resolution. It has proved possible to image with about 1 nm broadening (averaged over all the molecules in a field of view) of well-separated molecules (plate 9.1).[46, 47]

There is some evidence that DFM can produce atomic scale resolution on flat enough surfaces when imaged under fluid.[6] Thus, with increasingly

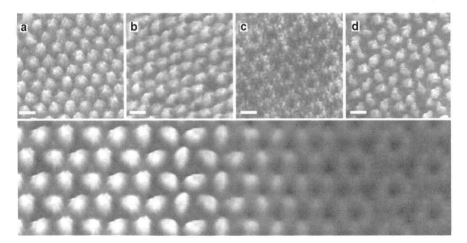

Plate 9.2. A conformational change of *E. coli porin OmpF* is induced by a pH drop, by a voltage across the sample and by a change of the imaging buffer. Here, the extracellular surface is imaged at (**a**) pH 7, (**b**) pH 6, (**c**) pH 2.5, and (**d**) pH 4. **e**, To produce this montage, the different conformations were averaged and merged (from pH 7, left, to pH 2.5, right). *Scale bars*, 5 nm. (Courtesy of A. Engel and D. Muller.) Figure also appears in Color Plate section.

reliable production of sharp tips,[42] subnanometer resolution on soft samples may become routine.

We are primarily concerned with imaging in biological buffers, and this raises the further important question of the nature of the tip–sample interaction in the presence of a fluid. I will deal with aspects of this in some detail below, but it is important to realize that the tip can (and does) image the sample via the medium of an intervening fluid layer. Rapid motion of the molecules in this layer average out any fluid structure, so the layer is not visualized during imaging, but it does affect the overall interaction between tip and sample. I will show that it is possible for the tip to sense a surface many nanometers from it. Resolution is reduced substantially when the tip is far from the surface. In practice, the overall oscillation amplitude is set to the smallest value that gives stable imaging (this minimizes the potential damage to the sample) and then the set-point amplitude change is increased (increasing the resolution) to the largest value that does not damage the sample.

9.3 INTERACTIONS AT A LIQUID–SOLID INTERFACE

The interaction between particles in a fluid is a complex many-body problem. I give the briefest of sketches here, leaving details to expert treatments.[48, 49] One goal is to give the reader a listing of factors to be considered in choosing tip materials and sample preparation methods. Another is to dispel the naive view that height in an AFM image is the true height of an object above the background substrate. Assuming that the sample is not surrounded by a co-adosorbate and assuming that the tip does not compress the sample, the height must still reflect changes in the interaction forces, $F_{1,2}$ (or stiffnesses, $S_{1,2}$) as the tip passes from the substrate (F_1, S_1) to the sample (F_2, S_2). This is illustrated in Figure 9.1, which shows a case in which smaller interaction forces (or interfacial stiffness in the case of DFM) over the sample result in a smaller apparent height ($h_2 < h_1$). Muller and Engel[44] illustrate this distortion using data and calculations for electrostatic interactions.

It is convenient to group the interactions between tip and sample into the following classes: hydrodynamic, van der Waals, ionic, chemical (e.g., hydrophilic and

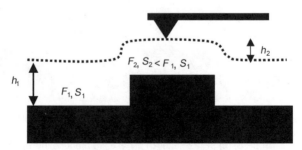

Figure 9.1. Interaction forces (F) and interfacial stiffnesses (S) are a function of the composition of the surface so that the AFM trace will not, in general, faithfully record topography.

hydrophobic), and hydration forces that arise from the changes in the structure of the intervening fluid as the tip is approached.

9.3.1 Hydrodynamic Effects

Hydrodynamic effects cause interactions between the cantilever and substrate over large distances (comparable to the cantilever dimensions). This is because the free flow of fluid around the cantilever is disrupted by the presence of a nearby surface. A simple model for this damping is a sphere (radius R) a distance D above a planar surface for which the damping is

$$\gamma_s = 6\pi\eta\frac{R^2}{D} \tag{9.3}$$

where γ_s is the squeeze damping coefficient and η is the viscosity of the fluid. The expression is valid for $R \gg D$. I used this simple form for comparison with the well known Stokes' law for a free sphere, but more accurate expressions are given by O'Shea and Welland.[50] This term can dominate the overall damping, but it contributes little to image contrast because the cantilever itself is held at an almost fixed height (the distance between the apex and base of the tip) above the surface. This distance is on the order of $10\,\mu$m so that the nanometer scale fluctuations encountered on scanning a surface cause negligible changes in this damping.

9.3.2 Ionic and van der Waals Interactions

Ionic and van der Waals interactions are described semi-quantitatively by the Derjaguin, Landau, Verwey, and Overbeek (DLVO) theory.[49] This ascribes the total force on a sphere of radius R and surface charge density σ_t a distance D ($D \ll R$) from a plane surface of charge density σ_s to the sum of coulomb and van der Waals interactions[44]

$$F_{\text{DLVO}}(D) = \frac{4\pi\sigma_s\sigma_t R\lambda_D}{\varepsilon\varepsilon_0}\exp\left(\frac{-D}{\lambda_D}\right) - \frac{H_a R}{6D^2} \tag{9.4}$$

Here λ_D is the Debye screening length and H_a is the Hamaker constant (which can be taken to have a value on the order of 10^{-20} J). ε and ε_0 are the relative dielectric constant of the medium and the dielectric constant of free space, respectively. The Debye theory is valid only at low salt concentrations and yields

$$\lambda_D = \frac{0.304}{\sqrt{C}}\,\text{nm (for 1:1 electrolytes, e.g., NaCl)}$$

$$\lambda_D = \frac{0.174}{\sqrt{C}}\,\text{nm (for 1:2 electrolytes, e.g., MgCl}_2) \tag{9.5}$$

where C is the molar concentration of the electrolyte. Plots of interaction forces for various conditions are given by Muller and Engel.[44] Here I focus on DFM, so I am

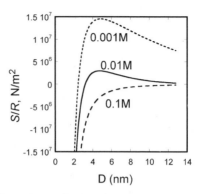

Figure 9.2. Interfacial stiffness (normalized to tip radius) as a function of distance for charged surfaces in various 1:1 salt concentrations as a function of tip–sample distance according to the DLVO theory. This theory accounts for the order of magnitude of the observed interfacial stiffness but fails to account for the fact that negative stiffness (attractive interactions) is rarely observed in water.

more interested in the interfacial stiffness ($S = dF/dD$). It is convenient to express this in terms of the stiffness per unit tip radius, which, from Eq. (9.4) is

$$\frac{S}{R}(D) = -\frac{4\pi\sigma_s\sigma_t}{\varepsilon\varepsilon_0}\exp\left(\frac{-D}{\lambda_D}\right) - \frac{H_a}{3D^3}. \tag{9.6}$$

A highly charged biomembrane may have $\sigma_s = -0.05\,\mathrm{C/m^2}$, whereas a reasonable value for a silicon-nitride tip in neutral (pH 7) solutions is $-0.032\,\mathrm{C/m^2}$.[44] Plots of values of S/R for these parameters are given in Figure 9.2. A tip radius of 100 nm results in an effective stiffness of on the order of $-1\,\mathrm{N/m}$ at nanometer distances from the surface. Similar positive stiffnesses can result from coulomb repulsion at low salt concentrations. In practice, values similar to this are observed, but they often remain positive (i.e., interactions are repulsive) all the way into the surface. This implies that other forces play a role. These forces may be associated the structure of the liquid itself.

The salt dependence of the electrostatic interaction is the basis of one method of using AFM to map the surface charge distribution on biopolymers, as described by Heinz and Hoh[51] and Czajkowsky et al.[52]

9.3.3 Chemical Bonds

Giessibl[5] has ascribed atomic resolution in DFM images to the rapid variations in the atomic scale interaction force between atoms on the surface and a single atom asperity on a tip. He estimated the local fluctuations owing to such forces using a Lennard-Jones potential. By taking the derivative of this potential, we can estimate the interfacial stiffness as follows:

$$S_{ij}(D) = \frac{12E_b}{\sigma^2}\left[13\left(\frac{\sigma}{D}\right)^{14} - 7\left(\frac{\sigma}{D}\right)^{8}\right] \tag{9.7}$$

where E_b is a binding energy that we will take to be characteristic of a 1 eV bond $(1.6 \times 10^{-19}\,\text{J})$ and σ is a an atomic diameter that we will take to be 0.2 nm. Here the subscripts refer to forces between the ith atom in the surface and the jth in the tip. The assumption here is that one dangling atom can dominate fluctuations in force (or stiffness) at small distances. The stiffness calculated with the parameters given above is shown in Figure 9.3. It is positive (repulsive force) only at small (angstrom) distances, suggesting that the observation of positive stiffness at much larger distances is a consequence of the structure of the intervening liquid.

Eq. (9.7) describes the formation of a chemical bond between tip and sample; and at small distances (e.g., when true contact is established), the tip chemistry will play an important role in adhesion. This is the basis of one method of chemical identification using AFM, which will be discussed later in this chapter.

9.3.4 Hydration Forces

One important chemical force in fluids is the hydrophobic–hydrophilic interaction. It is not described by simple pair-wise potentials, arising from the degree to which various groups are miscible or soluble in water and the degree to which they are excluded from favorable interactions. Oily (hydrophobic) moieties are not wetted, so water tends to push them together, giving rise to an apparent (and strong) attractive interaction. Charged or polar species tend to dissolve well and are effectively pulled out of the interior of a structure so that they may be more fully hydrated. Hydrophilic groups tend to interact strongly with each other (although the effect can be repulsive if they are similarly charged). Hydrophobic groups are nearly always repelled from hydrophilic groups. Biomolecules are almost invariably

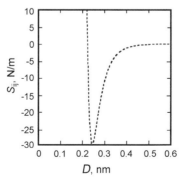

Figure 9.3. Interaction stiffness for an atomic bond described by a Lennard-Jones potential. The interaction is significant only at subnanometer distances.

hydrated (the exception being the intramembrane region of membrane proteins) and thus expose hydrophilic groups on their outer surface. It would clearly be advantageous to use hydrophobic tips for imaging them, because such tips would not stick to the surface on contact. Hydrophobic tips may be made using electron beam deposition of carbon[53a] or chemical vapor deposition of fluorocarbons,[54] but they are not, as yet, commercially available.

Measurements of the effects of solvation forces are shown in Figure 9.4. This shows how the oscillation amplitude of a DFM changes as a graphite surface is approached in an organic solvent (octamethylcyclotetrasiloxane, OMCTS).[53] The large fluctuations in amplitude are spaced by 0.82 nm, the diameter of the minor axis of this molecule, and they are clearly influenced by structuring of the liquid near the graphite surface (Fig. 9.4a). Figure 9.4b shows values for Young's modulus derived from these data (along with data for another fluid, mesitylene). At nanometer distances from the surface, Young's modulus becomes significant (it is, of course, zero in the bulk). The stiffness of the nth layer (which is what is actually calculated from the amplitude decrease; see below) is related to the effective Young's modulus of the nth layer[54a] by[55]

$$E_n^* = \sqrt{\frac{S_n^3}{6RL_n}} \tag{9.8}$$

where R is the radius of the tip and L_n is the load on the nth layer. L_n is calculated from $L_n = \Delta x_n k$, where Δx_n is the amplitude decrease for the nth layer. With R taken as 30 nm, the interfacial stiffness is on the order of 1 N/m at a distance of about 3 nm from the surface, purely as a consequence of the change in fluid properties near the surface. This discussion is based on the low-frequency data of Han and Lindsay.[53] In general, a more complicated approach is needed to separate the effects of stiffening from those of viscosity.[50] Increased viscous losses owing to

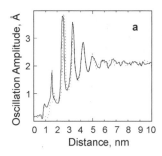
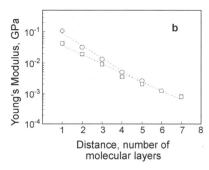

Figure 9.4. Direct experimental observation of solvation forces in an organic solvent at a graphite surface. **a,** Amplitude oscillations as a DFM tip approaches the graphite. The period is equal to the minor axis of the molecule. **b,** Young's modulus deduced from such measurements for two molecular fluids of differing size. (Data from Han et al.[54])

the structuring of the fluid near an interface can also play an important role in DFM.[56]

9.4 MECHANICS OF THE AFM CANTILEVER IN A FLUID

The AFM cantilever operates quite differently in fluid compared to its operation in air or other gasses. In addition to the obvious viscous damping, the cantilever must move fluid with it, leading to a greatly increased effective mass and a substantial reduction of its resonant frequency.[34]

O'Shea and co-workers[56] estimate the Reynolds number for low amplitude DFM from

$$\text{Re} = \frac{\rho_L \omega}{\eta} X \tag{9.9}$$

where ρ_L and η are the fluid density and viscosity, respectively, and X is a characteristic cantilever dimension (e.g., width). Taking values for water ($\rho_L = 1000 \, \text{kg/m}^3$, $\eta = 10^{-3} \, \text{Pa s}$), a lever width of 50 μm and frequencies in the hundreds of hertz to tens of kilohertz gives $\text{Re} > 1$, so that inertial motion is important. This is experimentally evident in the lowering of the resonant frequency upon immersion. O'Shea and co-workers[56] also give the following approximate expression for this effect:

$$\frac{\omega_{0-air}^2}{\omega_{0-liquid}^2} \approx 1 + 0.6 \frac{\rho_L \sqrt{Lb}}{\rho_C t} \tag{9.10}$$

where L is the cantilever length, b its width, t its thickness, and ρ_C its density (the expression was derived for rectangular cantilevers). For a silicon cantilever in water, $\rho_L/\rho_C = 1/2.3$ and taking $L = 250 \, \mu\text{m}$, $b = 50 \, \mu\text{m}$, and $t = 2 \, \mu\text{m}$, gives $\omega_{0-air}^2/\omega_{0-liquid}^2 \approx 14.6$. This is about a factor 4 reduction in resonance frequency in water, which is in reasonable agreement with observations.

The cantilever damping γ scales roughly linearly with the viscosity of the medium and the dimensions of the cantilever. The damping is often parameterized by the mechanical Q-factor. This is the ratio of the amplitude of motion at resonance to that at low frequency (i.e., the driving amplitude if the cantilever is driven by direct displacement). It is also (approximately) the ratio of the resonant frequency to the full width at half height of the resonance peak (plotted as amplitude vs. frequency). For a simple harmonic oscillator, the damping force F_d and Q are given by

$$F_d = \gamma \frac{dD}{dt}$$
$$Q = \frac{m\omega_0}{\gamma} \tag{9.11}$$

Where dD/dt is the cantilever velocity and m is the effective mass of the cantilever (it includes the mass of displaced fluid, as discussed above). The Q-factor, therefore, decreases in water and decreases for larger (i.e., softer) cantilevers.[56] This is because larger cantilevers have a lower resonant frequency and bigger dimensions.

For cantilevers of the dimensions discussed above (i.e., stiffness) k of about 1 N/m), the Q factor in water is about 3 or 4. In air, the same cantilevers would have a Q of >100.

9.5   MECHANISM OF DYNAMIC FORCE MICROSCOPY IN FLUID

9.5.1   Introduction

This section is devoted to a discussion of the mechanism of DFM in liquid, specifically water. The mechanism of the DFM in air has received considerable attention[57-59] as has the operation of DFM on clean surfaces in vacuum.[5] The DFM in fluid is less well understood. The interaction between an oscillating cantilever and a solid surface in fluid has been studied with the cantilever driven solely by thermal fluctuations[60a] or at a low amplitude of oscillation.[50, 53, 57] O'Shea et al.[56] demonstrated that low-amplitude DFM senses changes in the elastic and viscous properties of the surface at nanometer distances, and can, therefore, be operated in a noncontact mode. In this section, I discuss operation of the DFM in water at the larger amplitudes typical of many imaging applications.[61, 62]

The discussion is organized as follows: (1) The criteria for choosing operating amplitude are presented and a procedure for amplitude calibration is outlined. (2) There are two current methods for exciting cantilever motion in DFM; one using acoustic excitation and the other direct magnetic drive of a magnetic cantilever. The difference between these approaches is analyzed and illustrated with experimental data. (3) A formula for interpreting low-frequency, high-amplitude approach curves is derived and applied to surfaces under water. (4) Measurements of the cantilever response as a function of frequency and distance from a mica surface are presented and interpreted with the aid of numerical simulations. These confirm the interpretation of the low-frequency approach curves, which indicate that on initial contact the tip is sensing the surface via compression of an interfacial fluid layer. Thus it appears to be possible to operate the DFM in a noncontact mode in fluid even at high amplitude. Readers not concerned with experimental details can skip to the summary (Section 9.5.7).

9.5.2   Operating Amplitude and Spring Constant for DFM in Liquid

It is desirable to operate the DFM at as small an amplitude and with as small a setpoint change as possible to minimize disturbance of the sample. This is illustrated by considering the energy dissipated in the sample on each cycle in the low-frequency limit:

$$\Delta E(z) = \frac{1}{2}k(A_0^2 - [A_0 - A(z)]^2) \qquad (9.12)$$

where k is the cantilever spring constant, A_0 the amplitude immediately before approach to the surface, and $A(z)$ the amplitude at some distance z from a nominal surface plane (see below). In contrast to DFM in vacuum or air, resonance effects are small, so the microscope is largely sensitive to changes in amplitude (as opposed to changes in resonance frequency). Its sensitivity is thus limited by thermal fluctuations and is not significantly enhanced by operating at resonance. However, stable operation requires an amplitude sufficient to pull the tip out of attractive interactions (such as adhesion). In addition to the usual requirement that $(\partial F_{ts}/\partial z)^{max} < k$ (this ensures that the cantilever will not snap into contact), we need

$$kA_- + F_M + F_{ts}^{max} < 0 \qquad (9.13)$$

where A_- is the amplitude on the downswing (Fig. 9.5), F_M is the maximum magnetic force (for the case of magnetic drive), and F_{ts}^{max} is the tip–sample maximal interaction force. This second condition guarantees that a cantilever that becomes stuck to the surface will be pulled off by the combined action of the bent cantilever and the applied magnetic field. In noncontact operation tip–sample adhesion can be avoided, but in normal imaging conditions on rough surfaces (where accidental contact occurs), a safe value of A_0 is between about 2 and 5 nm with k around 1 N/m.

Soft cantilevers ($k \leq 0.1$ N/m) can give a significant residual signal on contact because the amplitude of higher bending modes (which do not displace the tip but bend the cantilever) can become significant. However, soft cantilevers are probably essential for imaging delicate samples like the surface of a cell (Dirk Badt, personal communication). This residual signal must be measured and taken into account

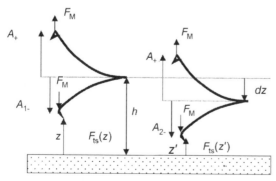

Figure 9.5. Definition of amplitudes of oscillation and cantilever displacements as a DFM cantilever approaches a surface. (Reprinted with permission from Ref. 63.)

when choosing the operating set point. The softer cantilevers also have a much lower resonant frequency in water and so lower scanning speeds are required.

It is common practice to calibrate the oscillation amplitude by pushing the cantilever far into the surface and assuming that, in hard contact, the amplitude falls 1 nm for every nanometer the tip advances (i.e., $dA/dz = 1$; see Fig. 9.5). This procedure assumes that the bending profiles in oscillation and contact are the same. Although this is not true in general,[64] Lantz et al.[63] found that, for small triangular cantilevers, the decay of the oscillation amplitude follows the deflection trace in contact at low driving frequencies. This is illustrated in Figure 9.6a which shows the bending signal when the tip is driven at 1 kHz (well below the resonant frequency of 14.8 kHz in water). The lower part of the swing lies on the line drawn through the contact part of the signal for which $dA/dz = 1$. This is not the case at higher frequencies. dA/dz for the same cantilever, driven at resonance, is 1.64 (Fig. 9.6b). The increased gradient comes about from the dissipation of stored energy, as is clear from the decay of both the top and bottom portions of the swing. The Q for this cantilever ($k = 2\,\mathrm{N/m}$) was about 3. When a higher Q cantilever ($k = 20\,\mathrm{N/m}$,

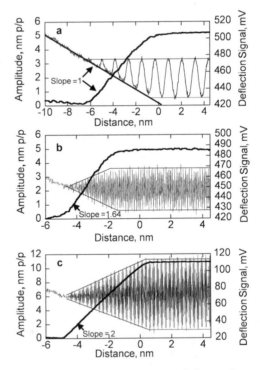

Figure 9.6. Amplitude decay (*heavy curves*) as a function of distance from a surface (the zero is arbitrary) together with a trace of the ac bending signal as the tip is swept toward a surface. **a**, At low frequency, the decay of amplitude follows the advance of the tip ($dA/dZ = 1$) in hard contact with the surface. **b**, At resonance, the slope is increased owing to damping. **c**, With a high Q cantilever at resonance, dA/dz approaches 2 as damping becomes symmetrical.

$Q = 12$ in water) was driven at resonance, the gradient was nearly 2, the limit appropriate for symmetrical decay of the amplitude (Fig. 9.6c). Thus amplitude calibration must first be carried out at low frequency to avoid these complications.

9.5.3 Acoustic Versus Magnetic Drive

Two methods for exciting the cantilever are in common use (Fig. 9.7). Acoustic drive[65, 66] is illustrated in Fig. 9.7**a**. An oscillator (frequency ω) supplies a voltage drive to a piezoelectric actuator (PZT), which generates soundwaves in the cantilever holder. The frequencies are typically from 10s of kilohertz to megahertz, corresponding to long wavelength modes that serve to displace the base of the cantilever. When the driving frequency is near a bending mode resonance of the cantilever, the cantilever is driven into a bending motion that causes an ac signal to be detected by the segmented photodiode detector. Figure 9.7**b** illustrates magnetic drive. In this case an alternating current (frequency ω) is supplied to a solenoid in the vicinity of the cantilever. The cantilever is coated with a magnetic film, and its base (and the sample) remains stationary. Direct bending of the cantilever is induced either as a consequence of magnetostriction or the magnetic interaction of the moment of the film with the applied field, which generates a bending torque.[61, 62] Another approach is to glue a magnetic particle onto the end of the cantilever and apply a field gradient.[67]

Direct comparison of the two methods as used for imaging in fluid[61] has shown that magnetic drive is less noisy. Results with biological samples suggest that higher resolution is often obtained.[46, 47] The origin of these differences is not completely clear, but extra noise might be generated by nonlinear excitation of

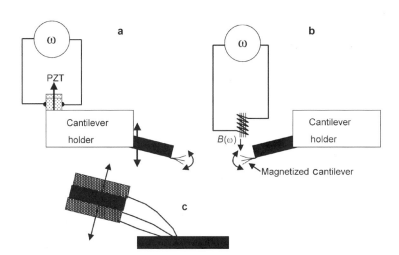

Figure 9.7. (a) Acoustic drive for DFM, (**b**) magnetic drive for DFM, and (**c**) the origin of the residual bending signal when the tip contacts the surface with the base acoustically driven. The optical detection system (not shown) responds to bending rather than to absolute displacement.

vibrational modes of the microscope by the acoustic field. Magnetic drive has some obvious advantages. One lies in its ability to drive the cantilever at any frequency. In the case of acoustic drive, significant displacement of the cantilever base occurs only at frequencies for which there is a mechanical resonance of this assembly. Second (as I will show), significant bending is obtained only near a cantilever bending resonance. Thus acoustic drive is limited to mechanical resonances of the microscope that lie near a bending resonance of the cantilever. The result is a forest of peaks, only some of which yield a usable signal.[66, 68] In contrast, magnetic drive yields a broad response that matches the expected thermal response of the cantilever.[62] Comparative plots of amplitude as a function of frequency are given in Figure 8a (acoustic drive) and Figure 8b (magnetic drive).

A second difference lies with mechanics of the excitation. In acoustic excitation, the long-wavelength sound waves cause displacement of the cantilever holder (Fig. 9.7a) so that the net bending signal is the difference between the displacement of the tip and the displacement of the holder. Writing the complex amplitude in this way and solving for the real part of the signal yields the following result:

$$A(\omega) = A_0 \sqrt{\frac{\omega^4 + \left(\dfrac{\omega\omega_0}{Q}\right)^2}{(\omega^2 - \omega_0^2)^2 + \left(\dfrac{\omega\omega_0}{Q}\right)^2}} \qquad (9.14)$$

Here A_0 is the amplitude of the drive applied to the holder and ω_0 and Q the resonant frequency and Q of the cantilever. Note that the response $A(\omega)$ drops to zero at zero frequency, in contrast to magnetic drive, in which the response is F_M/k [62] (Fig. 9.8). The response of the acoustically driven cantilever falls to zero at low frequency but is finite at low frequency in the case of magnetic drive.

Acoustic drive results in a substantial background signal when the cantilever contacts the sample. This is because of the low Q of the cantilever in fluid. To achieve a displacement A_0 at resonance, the base of the cantilever must be moved by an amount A_0/Q. For example, 3 nm amplitude would require 1 nm drive with

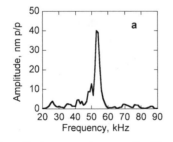
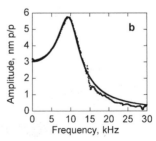

Figure 9.8. a, Tuning curve for acoustic drive: The sharp peaks correspond to mechanical resonances of the microscope. **b**, These are largely eliminated with magnetic drive. (Reprinted with permission from Ref. 63.)

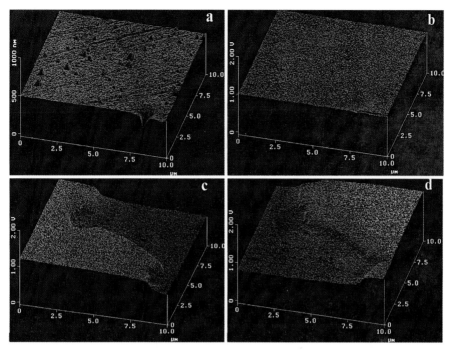

Figure 7.5. Polycrystalline ZnO. **a**, Topographic structure in an AFM image. **b**, Surface potential image identifying secondary phases with a 100 mV difference at the surface. Surface potential with (**c**) negative and (**d**) positive lateral voltage applied on the sample showing a 0.3 V potential drop at the grain boundary. (Reprinted with permission from Ref. 36.)

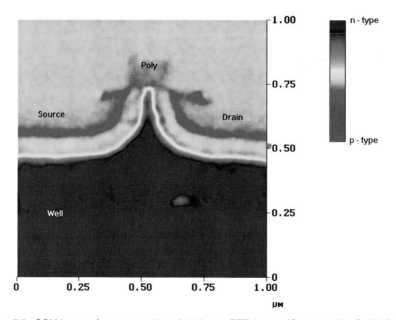

Figure 7.9. SCM image of a cross-sectioned 0.18-μm pFET device. (Courtesy of J. S. McMurray and C. C. Williams, University of Utah.)

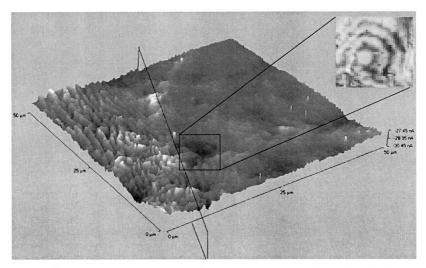

Figure 7.18. MFM image of superconducting current flow in a thick film at 60 to 65 K. Current flows from top to bottom. (Reprinted with permission from Ref. 100.)

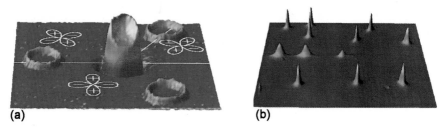

(a)   (b)

Figure 7.21. A scanning SQUID microscopy image of four micropatterned thin film rings of HTSC $YBa_2Cu_3O_{7-x}$ on $SrTiO_3$ tricrystal. **a**, The central ring spontaneously generates half of the superconducting flux quantum upon cooling through phase transition, whereas the other rings have no trapped flux **b**, SSM image of a field-cooled $YBa_2Cu_3O_{7-x}$ film. The Half-flux quantum Josephson vortex is at the tricrystall point. There are also four integer Josephson vortices trapped in the grain boundaries and seven integer Abrikosov vortices trapped in the grains. (Reprinted with permission from Refs. 103 and 104.)

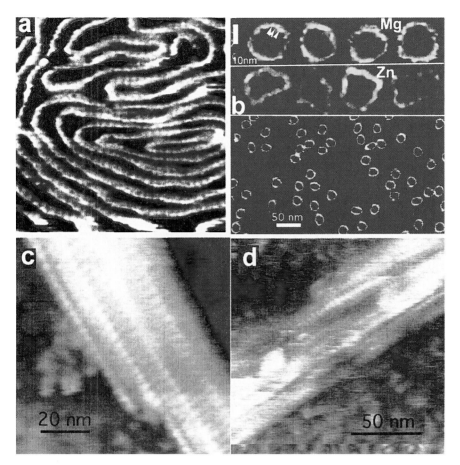

Plate 9.1. a, Contact-mode image of DNA tightly adsorbed onto a cationic bilayer in buffer. The double helical repeat is clearly visible. (Courtesy of Z. Shao.) **b,** Magnetic DFM of DNA microcircles on mica in Mg-containing solution. *Lower region,* a field of molecules, *upper portion,* magnified gallery showing helix repeat (*arrows*, 3.4 nm spacing). When Zn (lower part, upper portion) replaces Mg (upper part, upper portion), the circles kink. **c,** Microtubule imaged in buffer by magnetic DFM; 13 protofilaments are visible, as is the tubulin substructure. **d,** When katinin is added, the microtubules are digested. (Courtesy of J. Zhu and R. Vale.)

COLOR PLATES

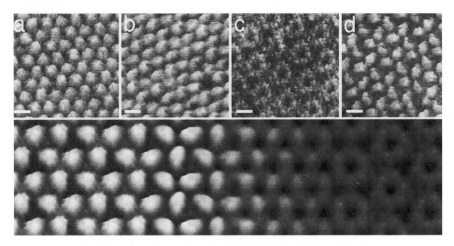

Plate 9.2. A conformational change of *E. coli porin OmpF* is induced by a pH drop, by a voltage across the sample and by a change of the imaging buffer. Here, the extracellular surface is imaged at (**a**) pH 7, (**b**) pH 6, (**c**) pH 2.5, and (**d**) pH 4. **e**, To produce this montage, the different conformations were averaged and merged (from pH 7, left, to pH 2.5, right). *Scale bars*, 5 nm. (Courtesy of A. Engel and D. Muller.)

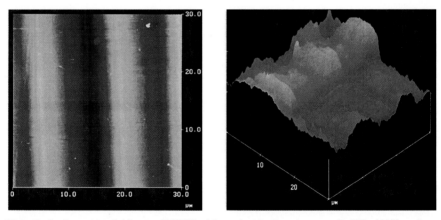

Plate 3. Surface potential image (SSPM) of ferroelectric domains on a $BaTiO_3$ (100) surface. Regions with high local potential correspond to c-domains, while regions with low surface potential correspond to c+domains. This is the figure from the book cover. (Reference: S. V. Kalinin and D. A. Bonnell, *J. Appl. Phys.* **87**, 3950 (2000)).

$Q = 3$. Thus the base of the cantilever must be moved by a significant fraction of the desired amplitude, and this results in a substantial background signal when the tip of the cantilever is in contact with the surface (Fig. 9.7**c**). The exact amount of the signal depends on the bending profile and the location of the laser spot. This residual signal is demonstrated in Figure 9.9**a**, which shows a plot of the oscillation amplitude for an acoustically driven cantilever as a surface is approached in water. Driven near resonance (25 kHz), the signal falls to a little less than 2 nm p/p on contact. The origin of the background is demonstrated by operating well below resonance (5 kHz). In this case, the bending signal falls dramatically, approaching zero at zero frequency (Eq. (9.14) and Fig. 9.8). The residual signal on contact is still present, with the result that the control signal is *reversed* at low frequency (Fig. 9.9**a**). In this case, the bending signal is smaller than the residual signal when the tip does not contact the surface.

This problem is not present with magnetic drive, in which the cantilever bending signal falls to zero as the surface is contacted (Fig. 9.9**b**). Some residual signal can be observed with the softest cantilevers because the first bending mode with both ends of the cantilever pinned might have significant amplitude. With the cantilever used in this case ($k = 0.5$ N/m) this motion was below the level of thermal and other background noise in the instrument. The data for Figure 9.9**b** were obtained well below resonance, so the region labeled "1," where $dA/dz < 1$ corresponds to a region of soft contact in which the motion of the tip is not stopped abruptly at the interface. I show below that the tip probably does not contact the surface at all in this region, sensing the interface by changes in the fluid that extend nanometers away from the surface. The region where $dA/dz = 1$ is labeled "2." The static deflection signal (obtained by filtering the output of the detector) is shown as the *thin solid line*, and this does not begin to rise, indicating contact, until close to the point (3), where the DFM signal falls to zero. In this experiment, the tip was pushed hard into the surface by continuing the advance many hundreds of nanometers (not

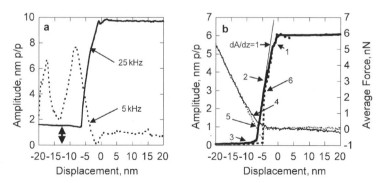

Figure 9.9. a, Approach curves for acoustic drive near resonance (*solid line*) and well below resonance (*dashed line*). Note the inversion of the control signal. **b,** The approach (*heavy solid line*) and retraction (*heavy dashed line*) for magnetic excitation. Average force is shown as the *light lines* (*solid* is approach; *dashed* is retraction). (Reprinted with permission from Ref. 63.)

shown) so that, on retraction, the average deflection signal (*thin dased line*) showed the hysteresis characteristic of adhesion (5). The corresponding DFM signal remained at zero (*heavy dashed line*) until the tip jumped off the surface (6). At this point, the DFM signal was restored to about half its maximum value, indicating that much (if not all—see below) of the swing is out of contact with the surface in the region labeled "1."

The tuning curve shown in Figure 9.8**b** was obtained from an early instrument, and it shows signs of some spurious mechanical resonances (sharp peak near 15 kHz). I found that noise increases significantly with such resonances, suggesting that mechanical excitation of the microscope contributes to the noise background. With care, all spurious resonances can be eliminated, and it is useful to obtain a tuning curve before imaging to ensure that the cantilever really is magnetically excited with no spurious peaks.

9.5.4 Harmonic and Anharmonic Analyses of Low-Frequency Data

O'Shea and co-workers[56, 60] analyzed data for low-amplitude operation using a damped harmonic oscillator model of the cantilever motion in which the displacement amplitude $A(z, \omega)$, is given by

$$A(z,\omega) = \frac{F_M}{\sqrt{(k + \bar{S}(z) - \bar{m}^*(z)\omega^2)^2 + \dfrac{(k + \bar{S}(z))\bar{m}^*(z)\omega^2}{\bar{Q}(z)^2}}} \qquad (9.15)$$

where F_M is the magnitude of the applied magnetic force, k the cantilever spring constant, $\bar{S}(z)$ an effective interface stiffness, $\bar{m}^*(z)$ an effective cantilever mass and $\bar{Q}(z)$ the mechanical Q factor. Bars over the quantities S, m^*, and Q indicate that they are averaged over the motion of the cantilever, which is accurate for small amplitudes. This approximation breaks down at high amplitude. For example, the low frequency limit of Eq. (9.15) yields the well-known result $A_0/A(z) = (S(z) + k)/k$,[54a] implying that the *fractional* change in amplitude is constant at a given z. This is obviously not true at high amplitude where all of the motion that lies beyond the range of surface forces is unaffected by the approach.

It is straightforward to generalize the result of Pethica and Oliver[54a] to the high-amplitude regime. Consider a cantilever initially at a height h above a surface (Fig. 9.5) with a maximum upward swing $A_+ = F_M/k$ and a maximum downward swing A_{1-}. Moved a distance dz toward the surface (Fig. 9.5), the new downward swing is A_{2-}. With $dA = A_{1-} - A_{2-}$ and $z' = z - (dz - dA)$, the force equilibration on the downward stroke is given by

$$F_M = kA_{1-} + F_{ts}(z) = kA_{2-} + F_{ts}[z - (dz - dA)] \qquad (9.16)$$

where we have assumed a symmetric magnetic drive so that the magnitude of the magnetic force on the downward swing is equal to that on the upward swing. If we define a differential force "constant," $S(z)$ by

$$S(z) = \frac{F_{ts}[z - (dz - dA)] - F_{ts}[z]}{(dA - dz)}, \quad (dA - dz) \to 0 \tag{9.17}$$

then Eq. (9.16) yields

$$S(z) = \frac{k}{\left(\dfrac{dz}{dA}(z) - 1\right)} \tag{9.18}$$

with repulsive forces defined as negative. Thus the inverse of the derivative of the approach curve taken at low frequency may be used to estimate $S(z)$. The approach curves of Figures 9.6a and 9.9b were used to generate the values for $S(z)$ displayed in Figure 9.10 as described by Lantz et al.[63] Far from the surface, the stiffness is zero as expected, whereas close to it, the stiffness rises to the $100\,\text{N/m}$ values typical of compression of a solid surface.[54a, 69] There is, however, a remarkably large region in which the stiffness in on the order of $1\,\text{N/m}$, characteristic of compression of a fluid layer.[50, 53, 70] The data taken over mica show an extended region of exponential decay, indicated by the *straight line*.

9.5.5 Amplitude Decay over a Wide Frequency Range

The situation for high amplitude oscillation over a wide frequency range is more complicated. It was investigated for a silicon cantilever and a mica substrate under water by Lantz et al.[63] A small white noise signal was added to a larger sinusoidal (5 nm p/p) driving signal, the cantilever held at a fixed distance from the surface by

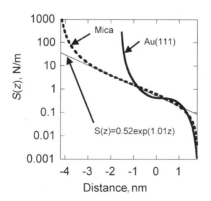

Figure 9.10. Interfacial stiffness deduced from the approach curves shown in Figures 9.6a (mica) and 9.9b (gold) using Eq. (9.18). The *straight line* is the fit used for numerical simulation of the operation of the DFM at high amplitude. (Reprinted with permission from Ref. 63.)

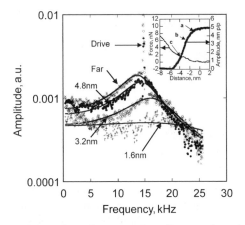

Figure 9.11. Dynamic response of cantilever at various distances from the surface (a, b, and c on approach curve in inset show set points for lower three curves). Data taken in water on mica. *Solid lines* are fits to Eq. (9.15). (Reprinted with permission from Ref. 63.)

feeding back on the sinusoidal signal, and the displacement signal was acquired as a function of time. The Fourier transforms of these signals (Fig. 9.11) gave a sharp peak at the driving signal (off scale in Fig. 9.11) whereas the white noise drive sampled the response of the cantilever over a broad range of frequencies. These are shown fitted with the damped harmonic oscillator response given by Eq. (9.15). The fits suggest that, for the three distances farther from the surface (nominally "far," 4.8 nm, and 3.2 nm), the main effect on oscillation amplitude is owing to the stiffening of the medium. Stiffening accounts for the reduction in amplitude at low frequency (i.e., the data in Figure 9.10 as deduced using the approach curves and Eq. (9.18). The same stiffening also accounts for the observed increase in resonant frequency. Agreement with Eq. (9.15) is somewhat surprising because the motion is highly anharmonic, the stiffness changing substantially over a cycle of cantilever cycle. Yet the form of the curves is in agreement with data taken at low amplitude as a tip approaches a mica surface.[56] I discuss this further with the use of a numerical model in the next section. Close to the surface (1.6 nm, in Fig. 9.11) damping also increases substantially. The dominance of stiffening (as opposed to damping) may be an unusual property of water. In an organic solvent (OMCTS), increased damping contributes most to reduction of amplitude at higher frequencies (Fig. 9.12). This displays the response of the cantilever with a low amplitude drive (0.1 to 0.5 nm) with added white noise as a graphite surface is approached.[56] Significant reduction of amplitude is first observed at about 3 nm from the surface, but Figure 9.12 shows that the cantilever resonant frequency is little affected over the entire approach. Thus, in this case, the dominant mechanism is enhanced dissipation as the surface is approached. Note that the effect decays on a molecular-length scale in contrast to the hydrodynamic damping described by Eq. (9.3).

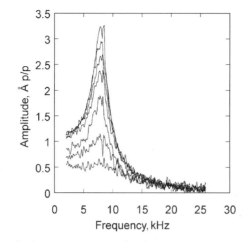

Figure 9.12. Low amplitude response curves for the tip approaching graphite under OMCTS. Damping appears to be the dominant interaction in this case. Noticeable damping begins with the tip about 3 nm from the surface. (Reprinted with permission from Ref. 56.)

9.5.6 A Numerical Model of the High-Amplitude DFM

The cantilever can be modeled as damped harmonic oscillator colliding with a massive surface held in place with a nonlinear spring as described by Lantz et al.[63] The cantilever is described by a spring constant k, mass m, and a Q-factor Q_0. The surface is described by a mass M and a Q factor Q_s. When the surface is indented a distance δ, the stiffness is taken to be $S(\delta) = S_0 \exp(\delta/\lambda)$, where the parameters are fitted to measured stiffness data in the region of liquid–layer compression (Fig. 9.10). Thus the surface corresponds not to the underlying hard substrate, but rather to the interfacial layer of fluid that is compressed in the region where S is on the order of 1 N/m. When the tip is a distance h' above this nominal surface, the equation of motion is

$$m^* \ddot{Z} + \gamma^* \dot{Z} + k^* Z = F_0 \exp i\omega t \qquad (9.19)$$

where Z is measured from the equilibrium position of the tip and for Z_+ and

$$|Z_-| < h' : \quad m^* = m, \quad k^* = k, \quad \gamma^* = \gamma_c$$

For
$$|Z_-| \geq h' : m^* = M, \quad k^* = k + S_0 \exp(\delta/\lambda), \quad \gamma^* = \gamma_c + \gamma_s$$

Z_- is the displacement on the downward stroke, and Z_+ is the displacement on the upward stroke. The damping parameters are given by $\gamma_c = m\omega_0/Q_0$ and $\gamma_s = M\omega_s/Q_S$. Eq. (9.19) can be integrated numerically to obtain the tip trajectory when a stable state is reached after the initial transients.[63] The trajectory was Fourier transformed to yield the time-averaged amplitude as a function of frequency.

With an appropriate choice of parameters, the response looks remarkably like that of a damped harmonic oscillator (Fig. 9.13). The results are quite sensitive to the choice of surface Q factor. If the damping is increased further, the resonance peak appears to be cut off when the tip is quite far from the interface. This occurs when the low-frequency amplitude is too small for the tip to reach the surface. However, as amplitude increases on approaching resonance, the tip eventually touches the surface. When the damping associated with the surface is large, further increase of amplitude as resonance is approached more closely is suppressed, giving rise to a flat-topped response. Thus the harmonic-like amplitude frequency response is an accidental consequence of the particular properties of the interfaces studied. Even small increases in surface damping (Fig. 9.13**b**) produce results that do not agree with experiment. The increased damping with increased frequency results in a peak response that moves to lower frequency as the surface is approached. The data are also quite sensitive to the function used to describe the interfacial stiffness. The curves shown in Figure 9.13**a** were obtained with a stiffness decay length λ of about 1 nm, in agreement with the data for mica–water in Figure 9.10. If this parameter is changed substantially, the amplitude decay no longer matches the experimentally observed rate (Fig. 9.13**c**).

The numerical simulations confirm the speculative interpretation obtained using the (inappropriate) damped harmonic oscillator. (1) For the mica–water interface, much of the amplitude decay on approaching the surface is owing to stiffening of the interfacial water layer at a distance some nanometers from the mica surface. (2) Much of the approach is in the noncontact regime. Significant damping of the cantilever amplitude occurs when the interfacial stiffness *at the lowest point of the swing* becomes equal to the cantilever spring constant.

9.5.7 Summary and Future Developments

In contrast to the contact mode, which senses a *force*, DFM is a derivative technique that senses an *interfacial stiffness*. As frequency is increased, DFM also becomes sensitive to increased interfacial damping when this changes on a nanometer scale.

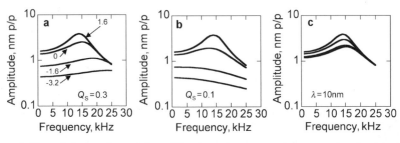

Figure 9.13. **a,** Anharmonic modeling of the dynamic response for approach using interfacial stiffness measured for mica (Fig. 9.11). **(b)** Changing the surface damping or **(c)** the stiffness model yields curves that do not match the experimental results shown in Figure 9.11. (Reprinted with permission from Ref. 63.)

The elastic and viscous properties of a fluid change significantly at distances on the order of several molecular diameters away from a surface. In the case of water, these changes appear to be predominantly elastic up to kilohertz frequencies, and they occur over a distance of some 10 molecular diameters. At this point, we cannot exclude the possibility that contamination may play a role. In this regard, it is worth noting that the surface energy of mica freshly cleaved in air is an order of magnitude less that of mica cleaved in ultra-high vacuum,[49] while gold changes from hydrophilic to hydrophobic on exposure to air.[71]

Direct excitation of the cantilever by magnetic means gives a lower noise, wider frequency range of operation, and less background signal than indirect acoustic excitation. Specially coated cantilevers are required for this mode of operation.

While the long amplitude decay lengths facilitate noncontact operation, they also make atomic resolution difficult. Experience indicates that resolution is increased as the tip is brought closer to the underlying surface by increased the amplitude damping set point. On the other hand, this also results in increased disturbance of the sample (Eq. (9.12)). The discussion in Section 9.3 indicated that extremely close approach would be required for atomic resolution. Recent work[6] does suggest that atomic resolution might be possible in fluid. There is, as yet, no comprehensive analysis of noise sources and the optimal choice of cantilever along the lines of that carried out for ultra-high vacuum DFM.[5] Perhaps adequate sensitivity will not be available with low-Q cantilevers.

Nonetheless, progress has been substantial, and much work can be done at the 1 nm level of resolution that is routinely obtained for DFM imaging in fluid.

9.6   PRACTICAL INSTRUMENTATION

Dynamic force microscopes are available from several manufactures, as are liquid cells. Acoustic-drive DFM is sold as Tapping Mode (Digital Instruments, Santa Barbara, Calif.) or Intermittent Contact (ThermoMicroscopes, Sunnyvale, Calif.). Magnetic-drive DFM is available as MACMode (Molecular Imaging, Phoenix, Ariz.). Systems are available that integrate AFM with an inverted optical microscope, important for large samples such as whole cells. The performance of these systems suffers somewhat from vibrations when compared to the normal compact design of single-purpose AFMs.

The overall layout of a DFM workstation suitable for high-resolution biomolecular imaging is shown in Figure 9.14. The microscope sits on a vibration isolation stage with a resonant frequency of ca. 1 Hz, enclosed in a heavy wooden box designed to attenuate acoustic noise. This low-resonant frequency is adequate to permit atomic resolution with a reasonably well-made SPM, even in quite noisy environments.[72] The microscopes in my busy sample-preparation laboratory are close to noisy fume hoods. Given adequate vibration isolation, the most severe remaining problem is thermal drift, and a well-controlled environment is of considerable benefit. The problem is acute if reagents are flowed into a liquid cell from outside the microscope chamber, because temperature changes of even a tiny fraction

Figure 9.14. A typical setup for fluid DFM. The controller workstation is on the right and the microscope is shown sitting on a bungee-cord vibration isolation system inside an acoustic isolation box on the left. The hand-held terminal on the table controls the DFM operating parameters.

of a degree can cause large bending of the cantilever. An isothermal environment is essential.

A liquid cell for a scanning probe microscope is shown in Figure 9.15. It is designed for a top-down scanner that dips into the open cell from above. Reagents may be flowed into, and out of the cell using the two fluid tubes. An arrangement for changing reagent flow during imaging is described by Thomson et al.[73] Cleanliness, always a problem in high-resolution imaging, is a particularly problematic

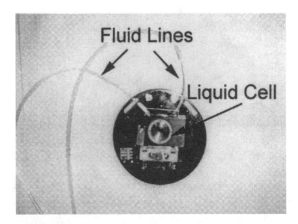

Figure 9.15. Top view of the liquid flow-through cell used with the microscope shown in Figure 9.15. It sits underneath the top-down scanning and sample support system, which forms the main body of the microscope. The cell is an open Teflon ring into which fluid is gravity fed. The level is kept constant by a drain line attached to a peristaltic pump. The terminal block below the cell is for holding electrodes for electrochemical control.

when a flow-through system is used. Vigorous cleaning of components is essential between runs. Samples should be prepared in laminar flow hoods using the highest available grade of freshly deionized, filtered, and distilled water. Buffer salts should be purified by recrystallization wherever possible. One approach is to use low concentrations of the highest available grade of reagent—the concentration of contaminants will be correspondingly lower than in highly concentrated solutions. Repeated column filtration of biopolymer samples is required, particularly if they have been prepared by elution from gels.

9.7 SAMPLE PREPARATION

9.7.1 Introduction

The SPM has permitted an unprecedented level of exploration of the effects of sample preparation on the conformation of biomolecules. This is because of the ease with which samples are prepared compared to electron microscopy and because of the freedom to image under solution. This aspect of biological SPM was recently reviewed by Amrein and Muller.[74] It is worth re-emphasizing the importance of cleanliness. The AFM will see contamination layers not visible in an electron microscope, be they on the sample or the tip. Further details of methods for cleaning of samples, water, and tips is available in Ref. 74. Development will undoubtedly continue apace, but I illustrate some of the issues with studies of the deposition of DNA on mica.

9.7.2 Adsorption from Solution

Rivettii et al.[75] undertook a careful study of the end-to-end length of DNA molecules as a function of the sample preparation conditions. They showed that, in the absence of stirring, the number of molecules adsorbed onto the surface, $n_F(t)$ at a time t varies with the initial concentration n_0 and time, as expected for translational diffusion to the surface followed by irreversible adsorption, for which[76]

$$n_F(t) = n_0 \sqrt{\frac{4Dt}{\pi}} \qquad (9.20)$$

where D is the translational diffusion coefficient for the molecular weight of molecule being deposited.

The chemistry of the substrate is also important. This is because adsorption onto the surface is a replacement reaction: Surface water must be replaced, and the free-energy cost of doing so must be favorable. Thus hydrophobic molecules will rapidly displace water at a hydrophobic surface. Adsorption onto a highly hydrophilic surface is more difficult. Mica is commonly used as a substrate, and it is (nominally) quite hydrophilic. Perhaps contamination plays a role in reducing this hydrophilicity when mica is cleaved in laboratory air.

9.7.3 Equilibration on the Surface

Once trapped by the surface, two-dimensional (2-D) translation of the molecule might still occur at the interface. In the study by Rivetti et al.,[75] molecules were immobilized by drying before imaging. If dried slowly on surfaces that bind them weakly, molecules may equilibrate in two dimensions (2-D equilibration) before becoming fixed. Molecules that become completely immobilized immediately upon adsorption are 3-D trapped in as much as their conformation represents the solution conformation projected onto a surface. These two adsorption processes lead to quite different conformations with different end-to-end lengths (Figure 9.16). 3-D trapping results in a projection that has an end-to-end length that is $\sqrt{2/3}$ of the free value for a worm-like chain of $\sqrt{2PL}$ (in the $L \to \infty$ limit), where P is the persistence length and L is the contour length of the polymer. This follows from the collapse of the dimension perpendicular to the surface, taken to be z, so that $\langle R^2 \rangle = \langle R_x^2 \rangle + \langle R_y^2 \rangle$. For 2-D equilibration, Rivetti et al.[75] showed that $\langle R^2 \rangle = 4PL$. Their study used DNA of 389 nm length, for which the difference in end to end length for the two mechanisms is large. Table 9.1 summarizes the results of some mica pretreatments. In general, the more chemically active surfaces result in 3-D trapping (although the end result must depend on the rate of drying to some extent).

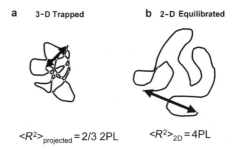

| a   3-D Trapped | b   2-D Equilibrated |

$\langle R^2 \rangle_{\text{projected}} = 2/3\ 2PL$ $\langle R^2 \rangle_{2D} = 4PL$

Figure 9.16. The difference in polymer conformation for (**a**) trapping from solution resulting in a 2-D projection of the solution structure versus (**b**) equilibration on the surface resulting in a more extended structure.

Table 9.1. zSome mica preparation methods with the consequent DNA conformation on deposition[a]

Mica Treatment	Sample Conformation
Bare mica	2-D equilibrated
Glow discharged	3-D trapped
Water rinsed	3-D trapped
$MgCl_2$ or HEPES	2-D equilibrated
Any treatment + rinse	3-D trapped

[a]For details, see Ref.[75]

Thorough rinsing before DNA deposition appears to produce an active surface, presumably because of protonation of the mica.

Ideal and reproducible 3-D trapping would be preferable in as much as the imaged conformation would most closely reproduce a projection of the solution conformation. The processes involved are far from understood at present, but some of the effects of different surface preparations can be illustrated with supercoiled DNA plasmids. These molecules are sensitive to their environment (salt, pH, temperature) and serve to illustrate the difference between bare mica (presumably 2-D equilibration) and spermine-treated mica (presumably 3-D trapping).[77] Results of this study are reproduced in Figure 9.17. The samples were a pUC19 supercoiled plasmid on bare mica (Fig. 9.17a) and spermine-treated mica (Fig. 9.17b). The supercoiling is evident in the trapped samples but absent in the samples that have equilibrated on the mica surface. The equilibrated sample may have substituted writhe for twist to maximize contacts with the surface.

It is important to note that no similar study (trapping vs. equilibration) has been carried out for the more interesting case of sample preparation carried out entirely without drying. Of course, chemically active surfaces are required for this process in the first place. I have observed strong differences even among such surfaces (unpublished data) and more study is needed.

9.7.4   Salt Effects on Adsorption and Conditions for Weak Adsorption

Muscovite mica consists of tetrahedral double sheets of $(Si/Al)_2O_5$ electrostatically linked by potassium ions. Dissolution of these ions at a hydrated surface gives rise to a surface charge of $-0.0025\,C/m^2$ at neutral pH, equivalent to 0.015 electronic charges per square nanometer.[78] Thus electrostatic interactions play an important role in the adsorption and binding of charged biomolecules. In the case of negatively charged molecules like DNA and purple membrane, it is necessary to carry the adsorption out at high salt concentration to screen the electrostatic repulsion and permit adsorption by attractive van der Waals interactions. Muller et al.[79]

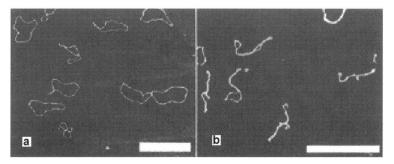

Figure 9.17. **a**, Supercoiled plasmid imaged on untreated mica (2-D trapped. **b**, and Spermine-treated mica (3-D trapping). The supercoiling is retained in (**b**) but is removed in (**a**). *Scale bars*, 500 nm. (Reprinted with permission from Ref. 77.)

investigated the adsorption of purple membrane (surface charge $= -0.05\,C/m^2$) as a function of salt concentration. Optimal adsorption occurred for concentrations of 1:1 electrolytes (LiCl, NaCl, KCl) above 40 mM and for 1:2 electrolytes (MgCl$_2$, CaCl$_2$, NiCl$_2$) above 1 mM. Similar considerations apply to (undesired) adsorption onto the SPM tip. Silicon nitride and silicon tips have a considerable amount of surface oxide, which ionizes by dissolution of surface ions to produce a surface charge in the region of $-0.032\,C/m^2$ at pH 7.[80] Muller et al.[79] suggest adsorbing the target molecules onto the substrate at high salt and then diluting the electrolyte before placing the sample into the microscope to reduce subsequent adsorption onto the scanning probe. Hansma et al.[81] carried out a systematic study of the effect of cation radius on DNA adsorption onto mica, finding stronger adsorption in the presence of highly charged transition metal ions.

Although strong adsorption is desirable for high-resolution imaging, weak attachment is important if processes are to be studied. Thompson et al.[73] described methods for modulating the ionic environment to produce a loose enough attachment for DNA to be visible in DFM images yet still be mobile on the surface. Zuccheri et al.[82] used similar methods to make a movie of DNA undergoing conformational changes as the ionic strength is raised and lowered during imaging.

9.7.5 Silane Treatments of Mica

One of the first methods for reliable AFM imaging of DNA was based on the attachment of positive amine groups to the mica surface using aminopropyl-triethoxysilane (APTES).[83–86] This approach binds DNA very strongly, too much so for studies of processes in situ.[82] As originally described,[83] it is capable of producing unacceptable roughening of the mica surface.[77] However, I have found it to be one of the most reproducible methods for adsorbing negatively charged molecules onto mica; the treatment of the surface is more uniform (in my hands) when carried out in solution. A fresh aqueous suspension of APTES (10,000:1 water:APTES) is contacted to the mica for a few minutes and then rinsed, after which a drop of buffered solution of the biomolecules is placed onto the treated mica surface.

9.7.6 Adsorption of Heterogeneously Charged Material

Chromatin consists of an assembly of highly positively charged histone protein together with negatively charged DNA, and I have found in situ imaging difficult on mica. Allen et al.[87] used glass a substrate for subsequent imaging in air. With a suitable choice of buffer, glass is also an excellent substrate for imaging in situ (S. Leuba, personal communication).

9.7.7 Future Developments

The range of substrate materials explored to date is quite limited. Metals are known to denature many proteins on adsorption,[88] and graphite is so hydrophobic that even the use of electrochemical potential control to induce a large positive surface charge does not appear to permit adsorption of negative molecules like DNA. Attractive though the notion of electrochemical control of an interface is,[89, 90] it is limited by the chemical complexity of the sample. Chemical interactions often overwhelm electrostatic interactions. For example, few salt solutions absorb weakly enough to permit the study of the (relatively dilute) biopolymers in the salt solutions. Tris buffer is easily oxidized, hiding adsorption of DNA at positive   electrodes.

Semiconductors deserve further study and are promising as substrates for further functionalization.[91] Passivated metal surfaces (e.g., using alkane-thiol monolayers[92] with template stripped gold[93] might permit fabrication of flat, appropriately functionalized surfaces. Chemical attachment of the DNA to an electrode surface overcomes some of the problems of adsorption, and the electrical behavior of DNA oligomers chemically attached to an electrode has been investigated.[94] Chemically inert surfaces might facilitate electrochemical potential control of the surface charge over a substantial range.

9.8 IMAGING

This section illustrates the methods outlined earlier with some examples (particularly color plates). Plate 9.1**a** shows a high-resolution image of DNA tightly adsorbed onto a cationic bilayer and imaged by contact mode AFM.[45] The helical repeat of the major groove of the double helix is observed at many places in the image. This is representative of the highest-resolution images of DNA obtained at the time of writing.

Crystal surfaces provide stable packing, and proteins have been imaged in these surfaces at high resolution by contact mode AFM.[95] Insulin crystals have been imaged at somewhat lower resolution using acoustic DFM.[96, 97] Another system in which close packing permits extremely high resolution contact mode imaging is a 2-D crystalline array of proteins.[11, 36] Such films are often made using densely packed arrays of trans-membrane proteins, and these can retain functionality despite immobilization in a 2-D film. They are thus ideal candidates for the study of important conformational changes in situ with the high resolution offered by contact mode AFM and an immobilized sample. Engel's group studied changes in *E. coli porin OmpF* as a function of electric potential applied across the membrane, pH, and buffer changes. Images at low applied potential show an open channel that closes at higher potential. Images of the open structure fail to resolve a soluble domain that presumably floats about in solution, invisible to the scanning probe. At higher potential, this domain is presumably pulled back into the channel to close it. The process can also be induced by a pH drop or a change of the imaging buffer. It is illustrated in Plate 9.2, which shows the extracellular surface imaged at various pH levels. To produce the montage at the bottom, the different conformations were averaged and merged.[98] This work is probably the first study of channel-protein mechanism by direct imaging.

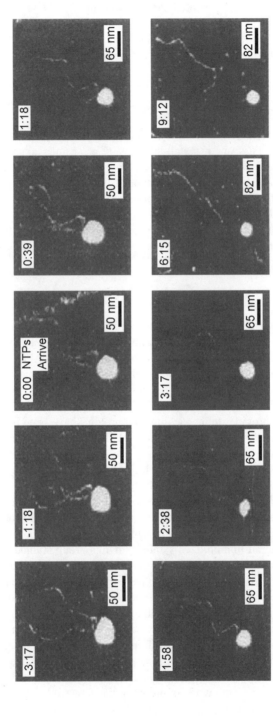

Figure 9.18. Time-lapse series of images showing transcription of a 1047 dsDNA template by an RNAP molecule. The first two images show that before the NTPs enter, the DNA is mobile on the surface. The six images after NTP addition from time 0:00 onward are sequential and show that one arm of the DNA template becomes progressively shorter until it is released (2:38). The 1047 base pair template has a terminating sequence, but it appears that the RNAP read through the terminator to the end of the DNA. (Reprinted with permission from Ref. 13.)

Plate 9.1**b** illustrates DFM imaging of isolated molecules at high resolution. It shows images of DNA microcircles imaged using magnetic DFM.[46, 47] These molecules are made by ligating intrinsically bent DNA oligomers to produce a closed circle of just 168 base pairs (about 18 nm in diameter). The lower region shows a scan over many molecules imaged in a buffer containing $MgCl_2$. The average width of the DNA in the images is a little over 3 nm, implying about 1 nm of broadening. This is just adequate for resolution of the major groove, as illustrated in the magnified gallery of images at the top. The circles change dramatically when imaged in the presence of Zn, becoming straight DNA connected by four kinks.

High-resolution magnetic DFM imaging of protein molecules is shown in Plates 9.1**c** and 9.1**d**. Plate 9.1**c** shows an isolated microtubule imaged in microtubule buffer on ATES-treated mica. The image was stable over a period of hours. The strings of protofilaments are clearly visible (13 in all, corresponding to flattened sheet formed from a whole microtubule), and the tubulin α and tubulin β subunits (spaced by 4 nm) are also clearly visible. When katatin, an enzyme that digests microtubules, is added, the image immediately disintegrates (Plate 9.1**d**).

I end this section with a series of acoustic-DFM images of the polymerization of cRNA from a double-stranded DNA template by RNA polymerase (RNAP).[13] This work relies on pinning the DNA loosely to the mica substrate using a buffer containing zinc. The zinc is then washed away and the reagents needed to bind the RNAP and initiate transcription added. This results in dissociation of Zn from the mica surface with consequent loosening of the DNA, giving rise to a time window in which the DNA is bound loosely enough to permit sliding of the polymerase yet still bound strongly enough to enable SPM imaging. Stalled complexes were generated by omission of one the NTPs (NTP = ATP, CTP, GTP or UTP) so that the RNAP stopped at the appropriate codon. The complex was imaged repeatedly to ensure that the RNA was stationary. The missing triphosphate was added to initiate transcription, and the subsequent motion of the RNAP was observed. Figure 9.18 shows a movie of the process. The first two images show the stalled RNAP. The next six images show progress of the RNAP along the DNA. The RNA transcript is not observed, presumably because it is suspended in solution. The last two images show the discarded RNAP and DNA template, frozen by the addition of Zn.

9.9 NONLINEAR ELASTICITY OF INDIVIDUAL MOLECULES

AFM has been used to measure the linear elastic properties of biomolecules by selective indentation.[99,100] The study of the nonlinear elastic properties is a more recent development. Reif et al.[17,101] demonstrated that remarkable structure appeared in the force versus distance curve when the AFM probe was pulled back after being pushed into a sticky biopolymer attached to a substrate. I use the unfolding of the giant muscle protein, titin, as an example here.[17] Titin consists of a series of interconnected globular domains and may play a role as shock absorber in muscle tissue. It adsorbs readily onto a gold substrate. Attached to an AFM tip by means of nonspecific interactions (the tip is pushed hard into the adsorbed protein), it adheres well enough so that single molecules can be unfolded by pulling the tip away from the surface (Fig. 9.19a). The corresponding force-distance curve has a sawtooth structure, with peaks of about 100 pN, separated by about 30 nm distance. An example is shown in Figure 9.19b. Rief et al.[17] attributed each rapid drop to the unfolding of a single globular domain and proved this to be the case by examining a series of constructs of a known small number of domains. The peaks increase in force from left to right, the weaker domains unwinding first. The same group used this method to pull on polysaccharide molecules, observing a bump in the force curve that they attributed to a conformational change within the interconnected sugar rings.[101]

None of the early work showed images of the substrate for the good reason that the sample sticks to the tip. Images can be important in, for example, ensuring that a single molecule (and not a bundle) sticks to the tip. It may also be important to check the conformation of the molecule on the substrate before pulling. The molecule may be denatured with multiple attachment points to the substrate, so that peaks in the force curve reflect the process of pulling off the substrate rather than unfolding of internal domains. DFM can be useful here, because it permits non-contact imaging. Thus the sample can be surveyed before pulling, the tip can then be pushed into a molecule hard, and the pulling can be done in the normal way. Figure 9.19c shows an image of the area used to obtain the force curve shown in Figure 9.19b. It was taken with magnetically excited DFM.

The exquisite sensitivity of magnetic DFM for interfacial forces[50, 53, 60] leads one to wonder if additional information might be obtained in nonlinear experiments

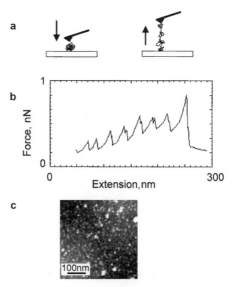

Figure 9.19. a, A molecular unfolding experiment. **b,** Force vs. distance pulling on a titin molecule. (Data courtesy of W. Han; samples courtesy of J. Trinick and L. T. Skhovvebova.) **c,** The titin layer on the surface before pulling. (Courtesy of W. Han.)

such as this. For small oscillation-amplitude and strictly elastic interactions (see Section 5.4)

$$A_0/A(z) = (S(z) + k)/k \qquad (9.21)$$

so, for a conservative system for which the force is given by the integral of stiffness with respect to extension distance z,

$$F(z) = -k \int \left(\frac{A_0}{A(z)} - 1 \right) dz + C \qquad (9.22)$$

Figure 9.20 shows a plot of the dynamic amplitude signal with simultaneously acquired force data for pulling on a chromatin sample.[102] The stiffness is approximately inversely proportional to the amplitude signal according to Eq. (9.21). However Eq. (9.22) cannot apply because the corresponding integral would rise continuously. In regions where the stiffness falls with extension, Eq. (9.22) would predict negative forces, and these are clearly not observed. The reason is that the interactions in the regions marked *NC* are not conservative and thus not elastic. Neither Eq. (9.21) or (9.22) applies. Within the regions of monotonically increasing force, Eq. (9.22) can be applied, as demonstrated in Figure 9.20**b**. There, the constant of integration was set to match the experimentally measured force at the start of each integration, and the stiffness was integrated to produce the solid lines. The dots are experimental force data. There are no fitting parameters beyond the known spring constant of the cantilever and the initial force. Excellent agreement was obtained, verifying the utility of Eq. (9.22) in regions where interactions are conservative. These results show that an enhanced signal-to-noise ratio is obtained from the stiffness data. These data clearly contain information beyond that

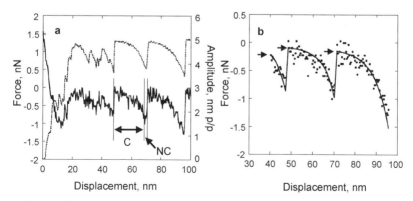

Figure 9.20. a, Force (*heavy line*) and DFM amplitude (*light dashed line*) for pulling a chromatin molecule off a glass substrate. Regions labeled *C* are conservative. **b**, Agreement between force calculated from the amplitude curve in the conservative regions (*solid lines*) and the measured force (*dots*). *Arrows*, the points at which the integrations of stiffness data were started. (Reprinted with permission from Ref. 102.)

encapsulated in the relation Eq. (9.22). In the nonconservative regions, much would be gained by recording *both* the amplitude and the phase of the DFM signal and using these to determine a complex modulus for the molecule. The dissipative part is of particular interest because it would include information about bond breaking and viscosity. I found it useful to record both ac and dc data to separate elastic from dissipative data.

9.10   CHEMICAL BONDS AND ANTIBODY–ANTIGEN INTERACTIONS

The AFM lacks an intrinsic, controlled chemical sensitivity, but great progress has been made in functionalizing the probe tip to make it selectively sticky. In one approach, friction is measured through twisting of the cantilever as a functionalized tip is dragged over a heterogeneous surface.[103] This method has even been extended to carbon nanotube tips.[104] Another way of detecting interactions is to record structure in the force-distance curve as a tip is pulled away from a surface, a technique often referred to as force spectroscopy. Hoh and co-workers[105] used this to detect hydrogen bonds between the tip and a glass surface in water.

More recently, techniques have been developed for attaching large molecules to either the tip or the substrate (or both) and detecting the bonding between them as the tip is pulled away from the substrate. These measurements are important because they hold out the prospect of comparing binding parameters for single pairs of molecules with calorimetric data obtained on large populations. They are becoming easier to carry out and interpret and may offer a better path to systematic development of drugs, in as much as total free energies and some kinetic data are available. Examples of such experiments are protein-ligand interactions[106–110] and antibody–antigen binding.[111–115]

The basic arrangement of such experiments is outline in Figure 9.21a. The receptor (shown as a circle with a wedge missing) may be bound to a tip, and the ligand (shown as a triangle) bound to a surface. Hinterdorfer et al.[112] showed

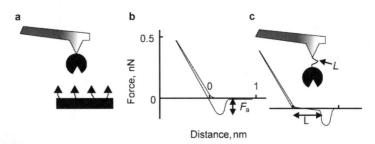

Figure 9.21. **a**, The arrangement of a bond rupture measurement. **b**, The bond rupture force is obtained from the adhesion portion of force curves (F_a). **c**, When a flexible linker (length L) retains one of the ligands, the adhesion peak is delayed, giving a characteristic recognition signal.

that measurements were greatly facilitated when a flexible linker was used to bind at least one of the receptor–ligand pair, because it relaxes constraints on their alignment as the tip is docked on the target molecule. These workers used a flexible polymer chain (polyethylene glycol) of about 6 nm in length. When the tip is pulled back from the surface (Fig. 9.21b) a characteristic recognition signal is seen in the retrace force curve (magnitude F_a). A flexible linker would result in a shift of the adhesion peak in the retraction curve to the right (Fig. 9.21c). A histogram of the values of F_a obtained from many such measurements may show a distinct peak that is taken to be the rapture force of the bond. Multiple bonds give rise to multiple peaks in the histogram, in which case the separation of the individual peaks is used as a measure of the bond force.

The rapture force is not unique. It depends on the dynamics of the process, as discussed by Evans and Ritchie[116] and more recently by Moore et al.[110] There are several contributions to the rate dependence. At high rates of loading (rate of increase of force) molecular friction generates a force proportional to the velocity of the tip.[111] At smaller rates, the time dependence of the entropy term in the free energy dominates. This is because time is required for the system to explore different configurations, some of which will facilitate unbinding more readily. Moore et al.[110] used the following semiempirical equation to fit force data taken as a function of loading rate over a wide range of scan speeds

$$F_a = \frac{1}{r}\left(\Delta H - \alpha T \Delta S - k_B T \ln\frac{r}{v\tau_0}\right) + \xi v \qquad (9.23)$$

where, ΔH and ΔS are the transition state enthalpy and entropy associated with the bond, r is the size of the binding pocket, v is the velocity of the tip, and τ_0 is an inverse attempt frequency, on the order of 10^{-8} s for biotin-streptavidin. The factor $0 \leq \alpha \leq 1$ accounts for the reduced contribution of entropy to lowering the overall barrier at higher loading rates. Moore et al.[110] obtain excellent fits to experimental data using Eq. (9.23), finding logarithmic dependence on loading rates at small rates and linear behavior at higher rates. The critical rate at which behavior turned from linear to logarithmic was on the order of 100 nN/s, related to the diffusivity of the ligand in the binding pocket. At small loading rates (0.1 nN/s) the effective barrier fell to a value close to that obtained from thermodynamic measurements.

A further development of these methods is functional imaging. Here, a functionalized tip is used to locate specific moieties in an image. The DFM intrinsically incorporates high-speed force curve measurements into imaging, because the tip is oscillated many times over a given pixel. This is useful for mapping complicated surfaces. For example, it could be used to locate specific receptor proteins on a cell surface. Raab et al.[118] used magnetic DFM with an antibody functionalized tip to image the distribution of antigen (lysozyme) on a mica substrate. The magnetic DFM permitted the tip to be engaged gently so that the antibody (and its flexible tether) were not damaged on approach to the surface. The use of magnetic DFM facilitates amplitude and frequency control, which in turn permits control of the antibody–antigen binding during imaging. Operated at a small amplitude A with a

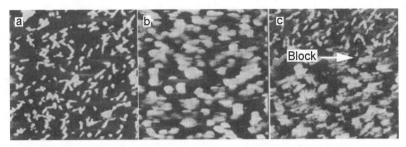

Figure 9.22. Antibody functional imaging. **a,** Taken with a bare tip over lysozyme on mica. **b,** Taken with a monoclonal antibody on the tip. **c,** The action of free antigen in blocking the sensing antibody and restoring contrast similar to the bare tip. All scans are 500 nm square. (Reprinted with permission from Ref. 118.)

tether length L, the effective molar concentration of antibody under the tip is $c = (\pi L^2 A N_0)^{-1}$. The corresponding binding time is $\tau = (ck_{on})^{-1}$, where k_{on} is the binding constant (in $M^{-1}s^{-1}$). The scan speed and oscillation amplitude must be chosen so that the tip dwells over the target for times that are long compared to τ. Using $k_{on} = 1.65 \times 10^5 M^{-1}s^{-1}$, $L = 6$ nm and $A = 2.5$ nm, Raab et al.[118] obtained τ on the order of milliseconds, permitting micron-size areas to be scanned in minutes.

Results of a functional imaging experiment are shown in Figure 9.22. Figure 9.22**a** shows a 500 nm^2 area of a mica substrate imaged under buffer and covered by a submonolayer of lysozyme. It is imaged by magnetic DFM with a bare tip. Figure 9.22**b** shows a similar area imaged with a functionalized tip. The features are characteristically higher and broader (broadened by approximately twice the tether length of 6 nm). The specific nature of this enhanced contrast is demonstrated in Figure 9.22**c**. Free antigen had been injected into the liquid cell before acquisition of this image. As the tip scanned up, it became blocked by antigen binding from solution at the point labeled by the arrow. The contrast reverts to that found with a bare tip. Recognition imaging was restored when the tip was lifted, effectively diluting the antigen concentration and permitting dissociation of the antigen on the tip. It is interesting to note that the blocked tip gives images identical to the bare tip. Evidently, the blocked antibody diffuses around freely and does not interfere with the image, much as was found for the soluble protein domains of the membrane protein discussed in Section 9.8 (Plate 9.2). The inherent variation in tip geometry and chemistry and the heterogeneity of cell surfaces may well preclude functional imaging in a single sweep. The most promising approach is to carry out imaging before and after adding free antigen, having previously established that this does not affect the target image when a bare tip is used.

9.11 STM IMAGES AND ELECTRICAL MEASUREMENTS ON SINGLE MOLECULES

In the first edition of this book, this chapter[2] was devoted almost entirely to STM, a technique that has all but vanished in biologic applications. What happened to it?

Some sensational images of DNA were published,[119–121] but these were obtained on graphite substrates. It was subsequently shown that electronic artifacts on the surface of bare graphite could mimic DNA,[122] even down to the level of atomic detail.[123] This did a great deal to discredit the technique. As an alternative to graphite and imaging in air, my group used electrochemical methods to deposit DNA onto a gold electrode where it was imaged in situ. This approach gave images that clearly showed the helical repeat of the deposited DNA,[124, 125] but it was difficult to reproduce and relied on codeposition of oxidized salts to embed the DNA. The technique was validated in a blind trial,[126] but it was clear that the DNA had to be strongly attached to an electrode surface to be imaged with an STM.[127] Image interpretation was also difficult. It appeared that the STM image corresponded to the footprint of the molecule on the substrate, so that large molecules with small contact areas on the substrate would give smaller images than small molecules with larger contact areas. Recently, Kawai's group used a pulse-injection method to deposit DNA onto a clean copper surface in ultrahigh vacuum, producing beautiful images of plasmid DNA. These show the helix repeat quite clearly.[128] No doubt this will regenerate interest in STM, if only because the electronic sensitivity of the STM opens up the possibility of chemical identification. Molecular identification by electronic measurement is the subject of this section.

Initial reports of STM imaging of large molecules like DNA met with skepticism from the physics community because the molecule is too big from a vacuum tunneling point of view.[2] However, Eigler et al.[129] showed that the contrast of an excellent atomic insulator (xenon) on a metal surface could be explained entirely by the (small) degree of hybridization between Xe atomic states and the metal surface when the xenon physisorbed. A small amount of itinerant charge from the metal is propagated out into space at the Fermi energy (which is well below the energy of the atomic state in this case). This accounts for the contrast of the Xe atom in STM images. Such calculations are difficult to do for more complex molecules. It is clear from images of molecules that have obvious feet attached to them that the STM sees the contact points with the surface, because this is where the hybridization occurs.[130] Thus, in the case of DNA, one expects that the high spots in the image are actually the low points on the molecule that lie in contact with the metal. This is something of a caricature of a complex process, but it illustrates the points that identification of molecular species is difficult and that the image need not be related to molecular geometry in any simple manner.

Other electron transfer processes are chemically selective. In the case of molecules with electron acceptor or donor states (redox active molecules) chemical identification is possible. Studies of some model compounds find a correlation between electrochemistry and STM image contrast, a connection first pointed out by Schmickler[131a] and Mazur and Hipps.[131] Using a series of modified porphyrins, Han et al.[132] demonstrated that the easily reduced molecules showed strongly enhanced contrast on a negative electrode, whereas the nonreducible molecules did not; this permitted identification of the molecular species in a heterogeneous sample. Tao[133] carried out a particularly elegant experiment using a monolayer film of protoporphyrin on graphite. This material forms a well-organized monolayer (Fig. 9.23). Furthermore, molecules are available with and without a central iron

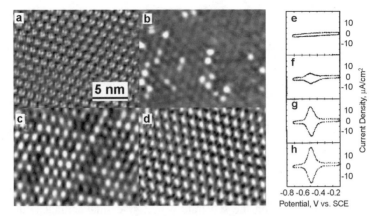

Figure 9.23. STM images taken over a monolayer of protoporphyrin on graphite. The fraction of iron-containing molecules is (**a**) 0, (**b**) 0.2, (**c**) 0.8, and (**d**) 1.0. The largest enhancement of contrast for the iron-containing molecules occurs at the redox potential for the iron oxidation–reduction process corresponding to the peak in the cyclic voltammograms on the right (**e**, 0; **f**, 0.2; **g**, 0.8; **h**, 1.0 fraction of iron containing protoporphyrins). (Reprinted with permission from Ref. 133.)

atom; and the two species are entirely miscible, forming a chemically heterogeneous but otherwise uniform monolayer. Tao[133] imaged this monolayer in an electrochemistry cell under potential control, tuning the surface potential through the redox potential for the $Fe^{++} \leftrightarrow Fe^{+++}$ process. He found that the image of the iron containing molecules was brightest at the point where the redox current was maximum (shown as the peak in the thin-film voltammograms in Fig. 9.23).

There are currently two models of this process (Fig. 9.24). Schmickler and Widrig[131a, 134] proposed that the applied electric field pulls down the LUMO energy until it coincides with the Fermi energy of the tip–substrate combination and resonant tunneling occurs. Because the state is coupled strongly to solvent and environmental fluctuations (Q in Fig. 9.24a), the electronic energy has a parabolic distribution, leading to a Gaussian distribution of thermally occupied states ($P(E)$). Schmickler and Widrig[131a, 134] proposed that STM is a way to determine this density of unoccupied states of electroactive molecules. Quite another mechanism is proposed by others[135, 136] (Fig. 9.24**b**). In this view, the electron is coupled so weakly to the tip and substrate that there is time for significant charge to build up on the LUMO and, more important, for environmental relaxation to reduce the energy of the charged state, resulting in trapping of the charge. This is nothing other than the process of electrochemical reduction, and the trapped state is lowered in energy by a factor 2λ, where λ is the relaxation parameter of the Marcus theory.[132, 136] The charge will reside in this state until a subsequent thermal fluctuation brings the relaxed state (E_L^*) up to the Fermi level of the second electrode at which point it can tunnel out. This process results in a maximum net current halfway between these states, and it defines the redox potential for the process (it is a free energy so the potential is properly defined only for a standard concentration). Because λ is on

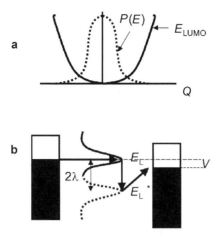

Figure 9.24. **a,** Molecular LUMOs are coupled to phonons (generalized displacement Q), giving rise to a parabolic variation (*solid line*) in electronic energy with a corresponding Gaussian probability $P(E)$ of thermal activation of a given state (*dotted line*). **b,** Transport through an electron-accepting state proceeds by resonant tunneling into a thermally accessible part of the LUMO distribution followed by relaxation after charging by an amount 2λ. The charge remains trapped in the acceptor state until another thermal fluctuation permits resonant tunneling out to the Fermi level of the second electrode.

the order of an electron volts, it would seem to be trivial to distinguish the two processes; but as of the time of writing, there is no unambiguous experimental result.

The STM suffers from the drawback that only small molecules may be used. The nature of the contact is also uncontrolled in as much as the tip will be pushed into the molecule by whatever amount is needed to achieve the set point tunnel current. Leatherman et al.[137] used an AFM with a platinum-coated probe to carry out simultaneous imaging and conductivity measurements. They studied a rather large synthetic carotene molecule embedded in a 22-carbon atom alkane-thiol film (Fig. 9.25**a**). To avoid electrochemical processes that damage the film (giving unrepeatable data), the experiment was carried out under oxygen-free and water-free toluene in an environmental chamber. Images of many molecules were analyzed to produce the current voltage data for a single molecule (Fig. 9.25**b**). The various curves are models based on the charge transfer theories described earlier, and it can be seen that the data are not yet good enough to resolve the theoretical debate. In fact, the curves are quite well described by a straight line with a resistance of $4.2\,G\Omega$. It is not clear at this stage how much of this is intrinsic and how much is related to the contact resistance between the probe and molecule. No current could be detected through the alkane film in any conditions, and its resistance was estimated to be on the order of $10^{16}\,\Omega$. However, it has recently been demonstrated that much of the early data on the resistance of metallic carbon nanotubes were dominated by contact resistance by the elegant expedient of dipping one end of a

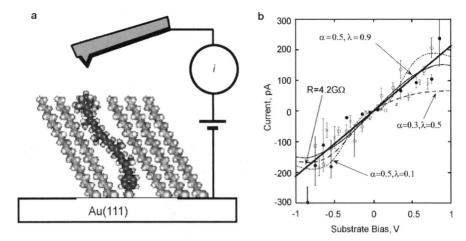

Figure 9.25. a, Arrangement of a conducting AFM measurement of conduction in a single molecule. Here a carotene molecule is shown embedded in a 22-atom alkane thiol layer. **b**, Current-voltage data for single carotene molecules. The behavior is ohmic and the corresponding resistance is 4.2 GΩ, over a million times more conductive than the surrounding alkane matrix. (Data from Ref. 137.)

nanotube into a mercury pool.[138] Contact resistance may also be important in these studies of molecules.

Certain DNA bases are, in fact electrochemically active, albeit at potentials that preclude simple experiments.[139] So perhaps the early dream of sequencing DNA by chemical imaging is not an impossible one.

9.12 FUTURE DEVELOPMENTS

The field is in a state of rapid growth, so speculations about future developments are risky. Nonetheless, it seems inevitable that the various methods for probing single molecules will be combined. The most obvious development would be to join together the optical methods for single molecule detection with SPM.[3] For example, protein-pulling experiments might be carried out on proteins labeled with fluorescent molecules chosen to signal the distance between certain parts of the protein (the FRET technique). In this way, the forced unfolding of a protein might be dissected at a submolecular level. If the background problem in live cells can be overcome, it would be of great interest to trigger signaling events mechanically, by, for example, placing a signaling protein onto its receptor on the cell surface. The subsequent chain of chemical events inside the cell could be followed with the appropriate fluorescent markers. The functional imaging methods involving antibodies would be much more powerful if carried out in conjunction with high-resolution fluorescence microscopy. Conceivably one could fish for some chemically identified (but otherwise unknown) receptor with optical probes and then, once mechanically isolated, image it with the AFM. The possibilities continue to grow and are seemingly endless.

ACKNOWLEDGMENTS

I am grateful to the following people for sharing data with me, some of it before publication elsewhere: Balagurumurthy, Xiadong Cui, XiZheng Feng, Wenhai Han, Peter Hinterdorfer, Mark Lantz, Gerry Leatherman, Sanford Leuba, Yangzhang Liu, Nongjian Tao, Analise Raab, Ron Vale, Phil Williams, and Judy Zhu. I appreciate comments on the manuscript from Nongjian Tao, Wenhai Han, Lars Chapsky, Peter Hinterdorfer, and Dmitry Cherny. The laboratory has been supported by grants from the NSF, NIH, and Molecular Imaging Corporation.

REFERENCES

1. R. P. Feynman, *J. Micromechanical Systems* **1**, 60 (1992).
2. S. M. Lindsay, in D. Bonnell, ed., *Scanning Tunneling Microscopy: Theory, Techniques and Applications*, VCH Press, New York, 1993.
3. S. Weiss, *Science* **283**, 1676 (1999).
4. F. Giessibl, *Science* **267**, 68 (1995).
5. F. J. Giessibl, *Phys. Rev.* B **56**, 16010 (1997).
6. F. Ohnesorge, *J. Surface and Interface Analysis*, in press.
7. F. Ohnesorge and G. Binnig, *Science* **260**, 1451 (1993).
8. Z. Shao and J. Yang, *Q. Rev. Biophys.* **28**, 195 (1995).
9. Z. Shao, J. Yang, and A. P. Somlyo, *Annu. Rev. Cell Dev. Biol.* **11**, 241 (1995).
10. J. Yang and Z. Shao, *Micron* **26**, 35 (1995).
11. A. Engel, C. A. Schoenberger, and D. J. Muller, *Curr. Opin. Struct. Biol.* **7**, 279 (1997).
12. C. Bustamante, C. Rivetti, and D. J. Keller, *Curr. Opin. Struct. Biol.* **7**, 709-716 (1997).
13. S. Kasas, N. H. Thomson, B. L. Smith, et al., *Biochemistry*, **36**, 461 (1997).
14. E. Stokstad, *Science* **275**, 5308 (1997).
15. W. Han, J. Mou, J. Sheng, J. Yang, and Z. Shao, *Biochemistry* **34**, 8215 (1995).
16. Z. Shao and Y. Zhang, *Ultramicroscopy* **66**, 141 (1996).
17. M. Rief, M. Gautel, F. Oesterhelt, J. M. Fernandez, and H. E. Gaub, *Science* **276**, 1109 (1997).
18. I. Rousso, E. Khachatryan, Y. Gat, I. Brodsky, M. Ottolenghi, M. Sheves, and A. Lewis, *Proc. Natl. Acad. Sci. U. S. A.* **94**, 7937 (1997).
19. C. Bustamante, D. Keller, and G. Yang, *Curr. Opin. Struct. Biol.* **3**, 363 (1993).
20. C. Bustamante, D. Erie, and D. Keller, *Curr. Opin. Struct. Biol.* **4**, 750 (1994).
21. H. G. Hansma and J. Hoh, *Ann. Rev. Biophys. Biomol. Struct.* **23**, 115 (1994).

22. C. Bustamante and D. Keller, *Physics Today* **48**, 32 (1995).

23. C. Bustamante and C. Rivetti, *Annu. Rev. Biophys. Biomol. Struct.* **25**, 395 (1996).

24. H. G. Hansma, *J. Vac. Sci. Technol.* **B14**, 1390 (1996).

25. T. Ushiki, J. Hitomi, S. Ogura, T. Umemoto, and M. Shigeno, *Arch. Histol. Cytol.* **59**, 421 (1996).

26. M. Miles, *Science* **277**, 1845 (1997).

27. D. J. Muller, C. A. Schoenberger, F. Schabert, and A. Engel, *J. Structu. Biol.* **119**, 149 (1997).

28. M. Guthold, M. Bezanilla, D. Erie, B. Jenkins, H. G. Hansma, and C. Bustamante, *Proc. Nat. Acad. Sci. U. S. A.* **91**, 12927 (1994).

29. M. Bezanilla, B. Drake, M. Nudler, M. Kashlev, P. K. Hansma, and H. G. Hansma, *Biophys. J.* **67**, 2454 (1994).

30. T. R. Albrecht, P. Grutter, D. Horne, and D. Rugar, *J. Appl. Phys.* **69**, 668 (1991).

31. J. L. Hutter and J. Bechhoefer, *Rev. Sci. Instrum.* **64**, 1868 (1993).

32. J. L. Hutter and J. Bechhoefer, *Rev. Sci. Instrum.* **64**, 3342 (1993).

33. H. J. Butt and M. Jaschke, *Nanotechnology* **6**, 1 (1995).

34. H. J. Butt, P. Siedle, K. Seifert, et al., *J. Microsc.* **169**, 75 (1993).

35. D. A. Walters, J. P. Cleveland, N. H. Thomson, and P. K. Hansma, *Rev. Sci. Instrum.* **67**, 3583 (1996).

36. F. A. Schabert and A. Engel, *Biophys. J.* **67**, 2394 (1994).

37. D. Keller, *Surf. Sci.* **253**, 353 (1991).

38. J. Vesenka, R. Miller, and E. Henderson, *Rev. Sci. Instrum.* **65**, 2249 (1994).

39. P. M. Williams, M. C. Davies, D. E. Jackson, C. J. Roberts, and S. J. B. Tendler, *J. Vac. Sci. Technol.* **B12**, 1456 (1994).

40. J. S. Villarubia, *Surf. Sci.* **321**, 287 (1994).

41. C. A. J. Putman, M. Igarashi, and R. Kaneko, *Appl. Phys. Lett.* **66**, 32221 (1995).

42. H. Dai, J. H. Hafner, A. G. Rinzler, D. T. Colbert, and R. E. Smalley, *Nature* **384**, 147 (1996).

43. J. Yang, J. Mou, J. Y. Yuan, and Z. Shao, *J. Microscopy* **182**, 106 (1996).

44. D. J. Muller and A. Engel, *Biophys. J.* **73**, 1633 (1997).

45. J. Mou, D. M. Czajkowsky, Y. Zhang, and Z. Shao, *FEBS Lett.* **371**, 279 (1995).

46. W. Han, M. Dlakic, R. E. Harrington, J. Zhu, and S. M. Lindsay, *Proc. Natl. Sci. U. S. A.* **94**, 10565 (1997).

47. W. Han, S. M. Lindsay, M. Dlakic, and R. E. Harrington, *Nature* **386**, 563 (1997).

48. J. P. Hansen and I. R. McDonald, *Theory of Simple Fluids*, Academic Press, London, 1986.

49. J. N. Israelachvilli, *Intermolecular and Surface Forces*, Academic Press, New York, 1991.

50. S. J. O'Shea and M. E. Welland, *Langmuir* **14**, 4186 (1998).

51. W. F. Heinz and J. H. Hoh, *Biophys. J.* **76**, 528 (1999).

52. D. M. Czajkowsky, M. J. Allen, V. Elings, and Z. Shao, *Ultramicroscopy* **74**, 1 (1998).

53. W. Han and S. M. Lindsay, *Appl. Phys. Lett.* **72**, 1656 (1998).

53a. D. Keller and C. C. Chou, *Surf. Sci.* **268**, 333 (1992).

54. H. F. Knapp, R. Guckenberger, and A. Stemmer, *Surf. Interface Anal.*, in press.

54a. J. B. Pethica and W. C. Oliver, *Physica Scripta* **T19**, 61 (1987).

55. R. M. Overney, E. Meyer, J. Frommer, H. J. Guntherodt, M. Fujihara, H. Takano, and Y. Gotoh, *Langmuir* **10**, 1281 (1994).

56. S. J. O'Shea, M. A. Lantz, and H. Tokomoto, *Langmuir*, **15** (4), 922–925 (1999).

57. B. Anczykowski, D. Kruger, K. L. Babcock, and H. Fuchs, *Ultramicroscopy* **66**, 251-259 (1996).

58. J. Tamayo and R. Garcia, *Langmuir* **12**, 4430 (1996).

59. J. Tamayo and R. Garcia, *Appl. Phys. Lett.* **71**, 2394 (1997).

60. S. J. O'Shea, M. E. Welland, and J. B. Pethica, *Chem. Phys. Lett.* **223**, 336 (1994).

60a. A. Roters, M. Gelbert, M. Schimmel, J. Ruhe, and D. Johannsmann, *Phy. Rev.* **E56**, 3256 (1997).

61. M. A. Lantz, S. J. O'Shea, and M. E. Welland, *Appl. Phys. Lett.* **65**, 409 (1994).

62. W. Han, S. M. Lindsay, and T. Jing, *Appl. Phys. Lett.* **69**, 4111 (1996).

63. M. Lantz, Y. Z. Liu, X. D. Cui, H. Tokumoto, and S. M. Lindsay, *Surf. Interface Anal.* **27**, 354 (1999).

64. J. E. Sader, I. Larson, P. Mulvaney, and L. White, *Rev. Sci. Instrum.* **66**, 3789 (1995).

65. P. K. Hansma, J. P. Cleveland, M. Radmacher, et al., *Appl. Phys. Lett.* **64**, 1738 (1994).

66. C. A. J. Putman, K. O. V. d. Werf, B. G. deGrooth, N. F. V. Hulst, and J. Greve, *Appl. Phys. Lett.* **64**, 2454 (1994).

67. S. M. Lindsay, Y. L. Lyubchenko, N. J. Tao, Y. Q. Li, P. I. Oden, J. A. DeRose, and J. Pan, *J. Vac. Sci. Technol.* **11**, 808 (1993).

68. T. E. Schaffer, J. P. Cleveland, F. Ohnesorge, D. A. Walters, and P. K. Hansma, *J. Appl. Phys.* **80**, 3622 (1996).

69. D. Sarid, *Atomic Force Microscopy*, Oxford University Press, New York, 1992.

70. A. Dhinojwala and S. Granick, *J. Chem. Soc. Faraday Trans.* **92**, 619 (1996).

71. T. Smith, *J. Colloid Interface Sci.* **75**, 51 (1980).

72. C. J. Chen, *Introduction to Scanning Tunneling Microscopy*, Oxford University Press, New York, 1993.

73. N. H. Thomson, S. Kasas, B. Smith, P. K. Hansma, and P. K. Hansma, *Langmuir* **12**, 5905 (1996).

74. M. Amrein and D. J. Muller, *Nanobiology* **4**, 229 (1999).

75. C. Rivetti, M. Guthold, and C. Bustamnte, *J. Mol. Biol.* **264**, 919 (1996).

76. D. Lang and P. Coates, *J. Mol. Biol.* **36**, 137 (1968).

77. M. Tanigawa and T. Okada, *Analytica Chimica Acta* **365**, 19 (1998).

78. R. M. Pashley, *J. Colloid Interface Sci.* **83**, 531 (1981).

79. D. J. Muller, M. Amrein, and A. Engel, *J. Struct. Biol.* **119**, 172 (1997).

80. H. J. Butt, *Biophys. J.* **60**, 1438 (1991).

81. H. G. Hansma and D. E. Laney, *Biophys. J.* **70**, 1933 (1996).

82. G. Zuccheri, R. T. Dame, M. Aquila, and B. Samori, *Appl. Phys.* **A66**, S585 (1998).

83. S. M. Lindsay, Y. L. Lyubchenko, A. A. Gall, L. Shlyaktenhko, and R. E. Harrington, *SPIE Proc.* **1639**, 84 (1992).

84. Y. L. Lyubchenko, A. A. Gall, L. Shlyakhtenko, P. I. Oden, S. M. Lindsay, and R. E. Harrington, *J. Biomol. Struct. Dyn.* **10**, 589 (1992).

85. Y. L. Lyubchenko, B. L. Jacobs, and S. M. Lindsay, *Nuc. Acids. Res.* **20**, 3983 (1992).

86. Y. L. Lyubchenko, L. Shlyakhtenko, R. E. Harrington, P. I. Oden, and S. M. Lindsay, *Proc. Natl. Acad. Sci. U. S. A.* **90**, 2137 (1993).

87. M. J. Allen, E. M. Bradbury, and R. Balhorn, *Scanning Microsc.* **10**, 994 (1996).

88. S. G. Roscoe and K. L. Fuller, *J. Colloid Interface Sci.* **152**, 429 (1992).

89. S. M. Lindsay and N. J. Tao, in M. Amreim and O. Marti, eds., *STM and SFM in Biology*, Academic Press, London, 1993.

90. S. M. Lindsay, N. J. Tao, J. A. DeRose, P. I. Oden, Y. L. Lyubchenko, R. E. Harrington, and L. Shlyakhtenko, *Biophys. J.* **61**, 1570 (1992).

91. P. Wagner, J. A. Spudich, W. D. Volkmuth, et al., *J. Struct. Biol.* **119**, 189 (1997).

92. C. A. McDermott, M. T. McDermott, J. B. Green, and M. D. Porter, *J. Phys. Chem.* **99**, 13257 (1995).

93. P. Wagner, P. Kernen, M. Hegner, E. Ungewickell, and G. Semenza, *FEBS Lett.* **356**, 267 (1994).

94. S. O. Kelley, J. K. Barton, N. M. Jackson, et al., *Langmuir* **14**, 6781 (1998).

95. A. A. Baker, W. Helbert, J. Sugiyama, and M. J. Miles, *J. Struct. Biol.* **119**, 121 (1997).

96. C. M. Yip, M. L. Brader, DeFelippis, and M. D. Ward, *Biophys. J.* **74**, 2199 (1998).

97. C. M. Yip and D. Ward, *Biophys. J.* **71**, 1071 (1996).

98. D. J. Muller and A. Engel, *J. Mol. Biol.*, in press.

99. N. J. Tao, S. M. Lindsay, and S. Lees, *Biophys. J.* **63**, 1165 (1992).

100. E. A-Hassan, W. F. Heinz, M. D. Antonik, N. P. D'Costa, S. Nageswaran, C. A. Schoenenberger, and J. H. Hoh, *Biophys. J.* **74**, 1564 (1998).

101. M. Rief, F. Oesterhelt, B. Heymann, and H. E. Gaub, *Science* **275**, 1295 (1997).

102. Y. Z. Liu, S. Leuba, and S. M. Lindsay, *Langmuir* **15**, 8547–8548 (1999).

103. C. D. Frisbie, F. Rozsnyai, A. Noy, M. S. Wrighton, and C. M. Lieber, *Science* **265**, 2071 (1994).

104. S. Wong, E. Joselevich, A. T. Woolley, C. L. Cheung, and C. M. Lieber, *Nature* **394**, 52 (1998).

105. J. H. Hoh, J. P. Cleveland, C. B. Prater, J.-P. Revel, and P. K. Hansma, *J. Am. Chem. Soc.* **114**, 4917 (1992).

106. E. L. Florin, V. T. Moy, and H. E. Gaub, *Science* **264**, 415 (1994).

107. G. Lee, D. A. Kidwell, and R. J. Colton, *Langmuir* **10**, 354 (1994).

108. U. Dammer, O. Popescu, P. Wagner, D. Anselmetti, H. J. Guntherodt, and G. N. Misevic, *Science* **267**, 1173 (1995).

109. H. Nakajima, Y. Kunioka, K. Nakano, K. Shimizu, M. Seto, and T. Ando, *Biochem. Biophys. Res. Comm.* **234**, 178 (1997).

110. A. Moore, M. Stevens, S. Allen, M. Davies, C. Roberts, S. Tendler, and P. M. Williams, *Biophys. J.*, in press.

111. U. Dammer, M. Hegner, D. Anselmetti, P. Wagner, M. Dreier, W. Huber, and H. J. Guntherodt, *Biophys. J.* **70**, 2580 (1996).

112. P. Hinterdorfer, W. Baumgartner, H. J. Gruber, K. Schilcher, and H. Schindler, *Proc. Natl. Acad. Sci. U. S. A.* **93**, 3477 (1996).

113. S. Allen, X. Y. Chen, J. Davies, et al., *Biochemistry* **36**, 7457 (1997).

114. P. Hinterdorfer, K. Schilcher, W. Baumgartner, H. J. Gruber, and H. Schindler, *Nanobiology* **4**, 39 (1998).

115. P. Hinterdorfer, A. Raab, D. Badt, S. J. SmithGill, and H. Schindler, *Biophys. J.* **74**, 186 (1998).

116. E. Evans and K. Ritchie, *Biophys. J.* **72**, 1541 (1997).

117. H. Grubmuller, B. Heymann, and P. Tavan, *Science* **271**, 997 (1996).

118. A. Raab, W. Han, D. Badt, S. J. Smith-Gill, S. M. Lindsay, H. Schindler, and P. Hinterdorfer, *Nature Biotechnology* **17**, 901–905 (1999).

119. P. G. Arscott, G. Lee, V. A. Bloomfield, and D. F. Evans, *Nature* **339**, 484 (1989).

120. T. P. Beebe, T. E. Wilson, D. F. Ogletree, et al., *Science* **243**, 370 (1989).

121. R. J. Driscoll, M. G. Youngquist, and J. D. Baldeschweiler, *Nature* **346**, 294 (1990).

122. C. R. Clemmer and T. P. Beebe, *Science* **251**, 640 (1991).

123. W. M. Heckl and G. Binnig, *Ultramicroscopy* **42–44**, 1073 (1992).

124. S. M. Lindsay, T. Thundat, and L. A. Nagahara, *J. Microscopy* **152**, 213 (1988).

125. S. M. Lindsay, T. Thundat, L. A. Nagahara, U. Knipping, and R. L. Rill, *Science* **244**, 1063 (1989).

126. A. M. Jeffrey, T. W. Jing, J. A. DeRose, et al., *Nucleic Acids Res.* **21**, 5896 (1993).

127. D. Rekesh, Y. Lyubchenko, L. S. Shlyakhtenko, and S. M. Lindsay, *Biophys. J.* **71**, 1079 (1996).

128. H. Tanaka, paper presented at the First International Symposium on Atomic Scale Processing and Novel Properties in Nanoscopic Materials, Osaka, Nov. 9-11, 1998.

129. D. M. Eigler, P. S. Weiss, E. K. Schweizer, and N. D. Lang, *Phys. Rev. Lett.* **66**, 1189 (1991).

130. T. A. Jung, R. R. Schlittler, J. K. Gimzewski, H. Tang, and C. Joachim, *Science* **271**, 181 (1996).

131. U. Mazur and K. W. Hipps, *J. Phys. Chem.* **99**, 6684 (1995).

131a. W. Schmickler, *J. Electroanal. Chem. Interfacial Electrochem.* **296**, 283 (1996).

132. W. Han, E. N. Durantini, T. A. Moore, A. L. Moore, D. Gust, P. Rez, G. Leatherman, G. R. Seely, N. Tao, and S. M. Lindsay, *J. Phys. Chem.* **101**, 10719 (1997).

133. N. Tao, *Phys. Rev. Lett.* **76**, 4066 (1996).

134. W. Schmickler and C. Widrig, *J. Electroanal. Chem.* **336**, 213 (1992).

135. A. Kuznetsov, P. Somner-Larsen, and J. Ulstrup, *Surf. Sci.* **275**, 52–64 (1992).

136. H. Sumi, *J. Phys. Chem.* **102**, 1833 (1998).

137. G. Leatherman, E. N. Durantini, D. Gust, et al., *J. Phys. Chem. B* **103**, 4006 (1999).

138. S. Frank, P. Poncharal, Z. L. Wang, and W. A. de Heer, *Science* **280**, 1744 (1998).

139. C. Hinnen, A. Rousseau, R. Parsons, and J. A. Reynaud, *J. Electroanal. Chem.* **125**, 193 (1981).

NANOMECHANICS

Nancy A. Burnham and Richard J. Colton

10.1 INTRODUCTION

In this chapter, we discuss the use of scanning tunneling and especially atomic force microscopy (AFM) to study the mechanical properties of materials at the nanoscale. Nanomechanics is of growing importance within the scanning probe microscopy community. The atomic force microscope shares some characteristics with other instruments that confer nanometer resolution, and we will compare them as a way of introducing the topic.

The hardness tester or microindenter characterizes mechanical properties—modulus, hardness, creep, and stress relaxation—of materials by varying the load or displacement of a tip during indentation. The displacement of the tip is usually determined capacitively, and its penetration depth is recorded as a function of load. Typical commercial instruments have force and depth resolutions of 75 nN and 0.04 nm, respectively.[1]

The surface force apparatus (SFA) measures forces—van der Waals, double-layer, viscous, hydration, and adhesion—between two crossed cylinders covered with mica or silica.[2] The forces are measured in vacuum or in the presence of vapors or liquids as a function of the separation between the smooth surfaces. The cylinders usually have a radius of curvature of 1 cm. One cylinder is mounted on a cantilever so that the force between the two cylinders can be determined by measuring the cantilever's deflection using optical interferometry. When touching, their contact area is of the order of 10 to 30 μm^2. The distance between the two surfaces can be controlled to 0.1 nm, and the force resolution is approximately 20 nN.[3]

The force microscope can be configured with the capabilities of these instruments—the surface force apparatus and the microindenter—with far greater performance. For example, the force microscope can image the surface topography, measure surface forces and adhesion, and determine mechanical properties. Our earliest instrument at the Naval Research Laboratory (NRL) had 1 nN force resolution and 0.02 nm depth resolution.[4] In addition, force microscopes can image with nanometer resolution in real time,[5] and purposefully modify structures on a nanometer scale.[6] They can be configured to study magnetic forces,[7] electrostatic forces,[8] forces at the liquid–solid interface,[9] thin liquid films,[10] and friction.[11] An unexpected bonus is that a force microscope is relatively inexpensive to buy or build.

Even though AFM was used early after its invention to investigate the nanomechanical properties of materials,[4, 12] the conventional force detection techniques (inferred from the known spring constant of the lever), small tip size, unknown tip shape, and unknown contact area lead to measurements that were qualitative at best. Recent developments in scanning probe microscopy (SPM), combining depth-sensing nanoindentation[13] with imaging capability,[14, 15] reduces these problems significantly. Accurate determination of both contact area and displacement can be achieved by coupling depth sensing nanoindentation with force modulation,[16] allowing quantitative measurement of nanoscale mechanical properties (hardness, modulus, viscoelasticity, and creep).

We proceed with a summary of nanomechanics. The field continues to develop, so that what is presented here should be considered merely an introduction to the topic. We will then discuss some interpretation problems. Finally, we will raise some important current challenges and speculate on the exciting future of this technique.

10.2 NANOMECHANICS

10.2.1 Repulsive Contact

Repulsion is a strong, short-ranged interaction between atoms owing to the electrostatic forces between like charges (i.e., electrons very close to each other). Depending on the mathematical form chosen to represent the repulsion, it can be known as hard-core or hard-sphere repulsion; power-law repulsion; or exchange, steric, or Born repulsion. If the atoms are in a lattice, the balance between interatomic repulsion and attraction determines the materials properties such as the elastic modulus. To learn about the fundamental properties of repulsive energies and forces in solids, we must turn to the classical mechanics of Hertz who, while working as a research assistant to Helmholtz, developed a theory on the deformation of elastic bodies.[17, 18] His work was motivated by his experiments on optical interference between glass lenses in contact (Newton's rings); he was concerned about the elastic deformation of the lenses.

To understand the functional dependencies of the variables involved in elastic deformation theory, consider Figure 10.1, where a sphere is pressed into a flat to depth d. The radius of the sphere is R; the radius of the contact area is a. For $d \ll R$, using the Pythagorean theorem, it is easy to show that $a^2 = Rd$. For small displacements, the relative deformation or strain in the material varies as d/a. The load required to achieve a certain strain is the stress, which varies as the effective elastic modulus K times the strain, or

$$\text{stress} \sim Kd/a$$

where

$$\frac{1}{K} = \frac{3}{4}\left(\frac{1 - \nu_1^2}{E_1} + \frac{1 - \nu_2^2}{E_2} \right)$$

for Young moduli E_1 and E_2 and Poisson ratios ν_1 and ν_2.

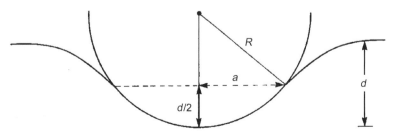

Figure 10.1. A sphere with radius R in contact with a flat surface. The radius of the contact area is a and the penetration depth is d.

The stress is expressed as a force per unit area; thus to determine the total load required to deform the system by d, we simply multiply by the contact area πa^2, or πRd. We see that the functional relationship between the load F and elastic deformation for a sphere and a flat is (for $d \ll R)F \sim Kda$. This is the same relationship as obtained by Hertz using a more precise derivation. The expression $a^2 = Rd$ also allows us to write

$$F = Kda = \frac{Ka^3}{R} = (K^2 Rd^3)^{1/2}$$

Hertzian mechanics assumes that there is no adhesion or surface forces (i.e., no attraction whatsoever) between the contacting surfaces.

In the 1960s, Sneddon[19] solved the deformation problem for systems of common geometry. The functional dependence of load on deformation (for $d \ll R$) for a flat-ended punch indenting a flat surface is d, for a spherical or parabolic indenter $d^{3/2}$, and for a conical indenter d^2. These deformation equations are used to analyze data from hardness testers or microindentation instruments. Data usually consist of a curve that plots load as a function of penetration depth of the indenter. Figure 10.2 shows a typical indentation curve. From these curves, the elastic modulus and hardness of the sample can be determined.[1, 20–23] An ideally elastic material would exhibit no hysteresis between loading and unloading. An ideally plastic material would have an unloading curve with infinite slope. Most materials are elastoplastic, as shown in Figure 10.2.

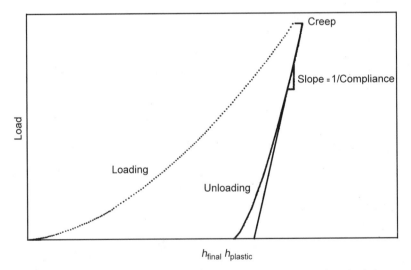

Figure 10.2. Typical indentation curve, which is a plot of the load placed on the indenter as a function of penetration depth of the indenter. From these curves, the materials properties of the sample may be determined. (Reprinted with permission from Ref. 21.)

The elastic modulus can be determined from the initial slope of the unloading curve (straight line in Fig. 10.2). The intercept of this line with the depth axis is defined as the plastic depth.[1, 20–23] Deviations from this straight line are caused by the indenter slowly losing contact with the sample during unloading. If the geometry of the indenter is known, then the projected contact area at the plastic depth can be calculated. The hardness is then the applied load divided by the projected contact area at the plastic depth. The procedure that Oliver and Pharr[22] adopted is more sophisticated, in that they fit the unloading curve to a power law relationship.

With the surge in interest in thin films, the ability to measure material properties at low loads is now important. At low loads, the end of the indenter tip must be well characterized.[21] Also, substrate effects may be involved if the indenter penetrates more than a certain critical distance. In the case of films harder than the substrate, this distance is much less than the film thickness, but for soft films it can equal the film thickness.[24] More significant is that surface forces become important at low loads. The force microscope can detect surface forces as well as exceed the force and depth resolution of a commercial indenter.

Early in the development of nanomechanics, we used AFM to investigate the mechanical properties of some elastic and plastic materials.[4] The AFM nanoindentation results are shown in Figure 10.3 for an elastomer, a graphite sample, and a gold foil. The differences in the indentation curves are striking. The elastomer and graphite behave elastically, and the gold behaves plastically. Results at lighter loads also indicated to us the significant influence of surface forces on the deformation processes.

Although Hertz and Sneddon did their work many decades ago, it was not until the 1970s that surface forces were included in theoretical models of elastic deformation. The four prominent models that include surface forces are the Johnson-Kendall-Roberts-Sperling[25, 26] (JKRS), whose work was motivated by the study of windshield wipers); the Derjaguin-Muller-Toporov[27] (DMT); the Maugis-Dugdale[28–30](MD); and the Muller-Yushchenko-Derjaguin / Hughes-White[31, 32] (MYD/HW) models. Lucid comparisons of the models are available.[33–35] Because MYD/HW is a numerical approach, we shall exclude it from our discussion, for in practice only the analytical methods are used. A semiempirical approach, Burnham-Colton-Pollock[35](BCP) mechanics, allows rapid calculation at the expense of accuracy at high repulsive loads. Table 10.1 summarizes the salient features of the five models that we discuss here.

JKRS theory includes surface forces in its derivation by considering the energy of adhesion within the contact area between solids. As shown in Figure 10.4, the solids deform locally into a connective neck. Their results give a radius of contact area at zero applied load of

$$a_o = \left(\frac{6\pi w R^2}{K} \right)^{1/3}$$

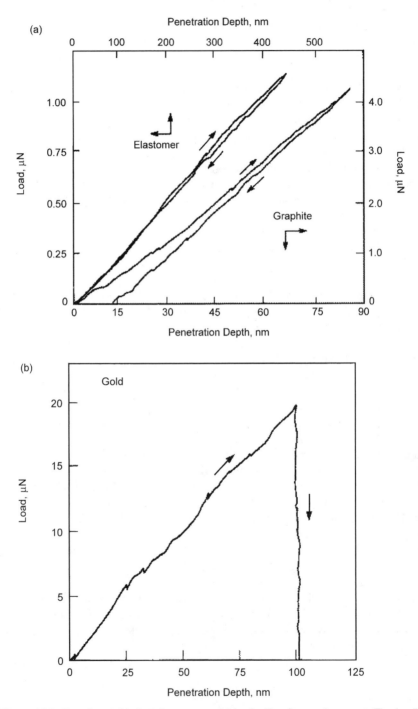

Figure 10.3. Experimental indentation curves obtained with a force microscope. The loading and unloading behavior of (**a**) an elastomer and highly oriented pyrolytic graphite and (**b**) a gold foil. (Reprinted with permission from Ref. 4.)

Table 10.1 Comparison of contact deformation theories

Theory	Assumptions	Limitations
Hertz	No surface forces	Not appropriate for small loads
DMT	Long-range forces act outside contact area; contact geometry remains Hertzian	May underestimate contact area
BCP	Long-range forces act outside contact area; surface allowed to bulge out	May overestimate contact area and underestimate adhesion
JKRS	Short-range forces act inside contact area	Many underestimate loading owing to forces
MD	Periphery of tip–sample interface modeled as a crack	Parametric equations

with the work of adhesion w, defined by the Dupré equation

$$w = \gamma_1 + \gamma_2 - \gamma_{12}$$

where γ_1 and γ_2 are the surface energies of the sphere and flat, and γ_{12} is the interfacial energy. R and K are as defined earlier.

The force required to separate the tip and sample after contact is the pull-off force or adhesion, F_{ad}. It is equal to $-1.5\pi wR$. In contrast, Hertzian theory predicts no contact area, unless an external load is applied, and no pull-off force.

The JKRS assumptions lead to unphysical values in the stress distribution at the edge of the connective neck. In comparison, DMT theory assumes that the surface forces extend over a finite range, and act in the area just outside contact. This removes the infinite stress at the contact boundary. However, DMT assumes Hertzian deformation only (i.e., no neck at the interface), which may underestimate the actual contact area. The DMT results for contact radius at zero applied load and pull-off force are

$$a_o = \left(\frac{2\pi wR^2}{K} \right)^{1/3}$$

and

$$F_{ad} = -2\pi wR$$

A more rigorous method, the MYD/HW model[31,32] uses a Lennard-Jones potential for attraction and allows the deformation geometry to be non-Hertzian. Unfortunately, its solutions can be reached only numerically.

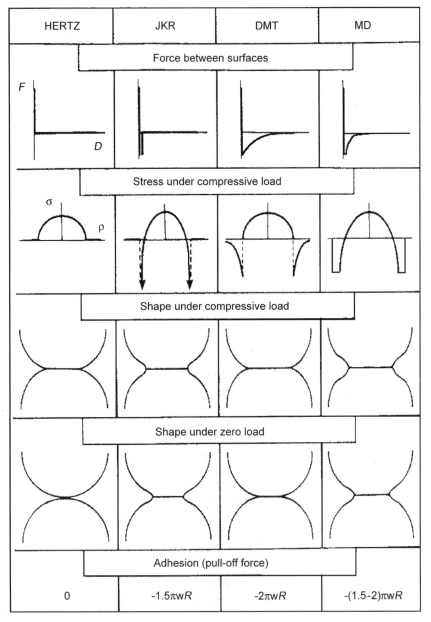

Figure 10.4. A comparison of the main features of the four theories of the deformation and adhesion of two elastic spheres. (Adapted from Ref. 33.)

DMT is more appropriate for systems with small radii of curvature, low work of adhesion, and high elastic moduli (e.g., a diamond tip on a diamond surface). JKRS applies to large radii, high work of adhesion, and low moduli (e.g., the crossed cylinders of mica in a surface force apparatus[2]). The experimental configurations typical in FM unfortunately fall into a region where either DMT or JKRS could apply. To decide which theory is more valid, the expected neck height l should be evaluated.

$$l = \left(\frac{2^{20} w^2 R}{3^5 \pi^4 K^2} \right)^{1/3} = \left(\frac{44.3 w^2 R}{K^2} \right)^{1/3}$$

where l is the extent of the connective neck shown in Figure 10.4 for the JKRS profile. If l exceeds a few angstroms, an appreciable neck is formed, and JKRS theory should be used. If l is of the order of a few picometers, DMT is a viable option. For the situations in which the neck height lies between these extremes, either BCP or MD is appropriate. BCP is an approximation[12, 35] but mathematically simple. It predicts that the contact radius at zero applied load and pull-off force are

$$a_o = \left(\frac{3.18 \pi w R^2}{K} \right)^{1/3}$$

and

$$F_{ad} = -1.35 \pi w R$$

A more appropriate solution (but more complex owing to parametric equations) is the Maugis-Dugdale theory.[28–30] In its extremes, it reduces to JKRS and DMT. For the interim cases, the mathematics and assumptions are more rigorous than for BCP. MD theory accounts for the changing geometry of the contact in that in the DMT limit the contact geometry remains Hertzian, and in the JKRS limit the contact relaxes into a neck shape. Consequently, the contact radius at zero applied load and the pull-off force span the range from the JKRS to the DMT values. MD theory also gives predictions for the conditions under which jump to contact (next section) can be expected. For a comparison of the equations governing the relationships between force, contact radius, and indentation depth, see Burnham and Kulik.[35]

To summarize this section, we emphasize the following points: (1) the theories of Hertz and Sneddon can be used to evaluate elastic and plastic deformation of materials by indentation at high loads; (2) the JKRS, DMT, BCP, and MD theories are more appropriate for analysis at low loads in which surface forces become important; and (3) although there are limitations to the Hertz, JKRS, DMT, and BCP theories, they are mathematically tractable.

10.2.2 Jump to Contact

One of the predictions of the JKRS and MD theories is that at sufficiently small separations the two surfaces jump together. This effect was also predicted by

Pethica and co-workers,[36] Smith and co-workers,[37] and Landman and co-workers[38] using a variety of interaction potentials. Experimental evidence for this phenomenon exists in papers by Pollock[39] and Gimzewski and Möller,[40] but has not yet been fully investigated. When the gradient of the tip–sample force equals the second derivative of the potential of one or both of the solids (which will be a function of the elastic moduli and the local geometry of the solids), an instability occurs, and the surfaces jump together.

Figure 10.5a is from Landman et al.[38] and shows the deformation in a gold surface as it jumps to contact with a nickel tip mounted on a rigid support. Most of the deformation occurs in the gold because its modulus is much lower than that of

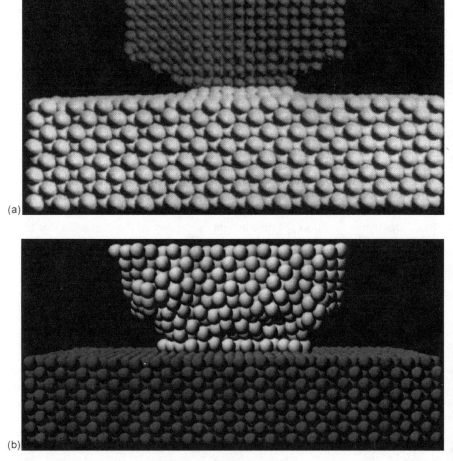

(a)

(b)

Figure 10.5. Atomic configurations generated by molecular dynamics simulations of (a) the cross section of a nickel tip in contact with a gold surface just after the jump to contact (note the bulging of the gold substrate under the nickel tip), and (b) a gold tip in contact with a nickel surface just after the jump to contact. (Reprinted with permission from Ref. 38.)

nickel. If the tip and sample materials are interchanged (i.e., a gold tip and nickel surface), then extreme deformation occurs in the tip (Fig. 10.5**b**). This example illustrates the combined effects of materials properties and the local geometry on the behavior of the system. The simulations were obtained by molecular dynamics calculations using embedded atom potentials.

There are actually two forms of jump behavior in AFM. We have just described jump to contact as it is related to the local stiffness of the tip when equaled by the sample force gradient. In addition, there can be a cantilever instability, which will occur when the sample force gradient equals the cantilever's stiffness k. (In Fig. 10.5 the effective spring constant of the rigid support is infinite.) The cantilever instability can occur at any distance from the sample. These two forms of jump behavior should be carefully distinguished, and the force microscope should be designed so as to accentuate the desired behavior. We prefer the terminology *cantilever instability* for the macroscopic instrumental artifact, and *jump to contact* for the interesting nanoscale phenomenon.

The general behavior of a spring under the influence of an arbitrary surface force is depicted in Figure 10.6. The spring (which could represent either the entire cantilever or the local stiffness of the tip) has a spring constant k and is a distance D away from the sample. The arrows are used to guide the eye throughout the full interaction cycle. The cycle starts with the sample far away and the spring in its rest position. As D decreases, the spring bends toward the sample so that at any equilibrium separation D the attractive force F balances the spring's restoring force defined by k times its deflection. The measured slope of the force vs. separation curve will be a convolution of the tip–sample force gradient $\partial^2 U/\partial D^2$ (where U is the potential) with the spring constant k[35] and is

$$\text{measured slope} = \frac{(k)\left(\frac{\partial^2 U}{\partial D^2}\right)}{\left(k + \frac{\partial^2 U}{\partial D^2}\right)}$$

When the total force gradient (the denominator) equals zero ($\partial^2 U/\partial D^2$ is negative for attractive potentials), the slope becomes infinite and there is an instability. The spring jumps from A to A'. On reversing the direction of the sample, the spring will extend from A' to B until it jumps away from the sample at B to some point B', giving rise to hysteresis and discontinuities in the measured force curve. If the instabilities are caused by the cantilever, then the resulting hysteresis should not be confused with the hysteresis of the tip–sample interaction (jump to contact).

10.2.3 Adhesion

The maximum negative force on loading the sample may be different from that on unloading the sample. The forces bringing the tip and sample together are called attractive forces, whereas the forces keeping the tip in contact with the sample are owing to adhesion. Adhesion usually exceeds attraction in magnitude. If adhesion equals attraction and there is no hysteresis in the force-distance data, then no energy is lost during the loading–unloading cycle.

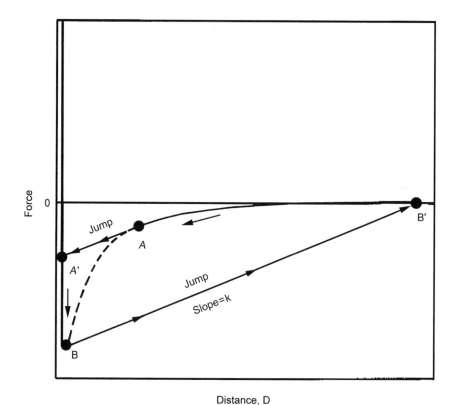

Figure 10.6. The general behavior of a spring under the influence of an arbitrary surface force. The spring has a spring constant k and is a distance D from the sample surface. (Reprinted with permission from Ref. 38.)

Adhesion is of great technological importance. Until recently, owing to the difficulty of characterizing nanometer-scale topography, the fundamental processes of adhesion were a challenging study. This problem motivated adhesion studies of single asperity contact. First experiments of a single asperity on a flat surface were done in the 1970s.[41] In these earlier studies, force in the 100 nN range was determined using capacitance techniques to measure the deflection of a cantilever, and contact area was inferred from electrical resistance measurements (from the flat through the asperity). Today, surfaces can be mapped with force resolution down to 10 pN.[42] The researcher may also choose to "invert" the force measurement by putting the sample on a cantilever and detecting the sample force gradient with STM.[43]

These new types of investigations have just begun. Here, some recent experimental conclusions pertaining to adhesion are summarized: (1) Hysteresis in the force versus displacement curves found during tip–sample separation after contact is attributed to adhesion and inelastic processes, provided there is no cantilever instability.[38] (2) The magnitude of the adhesion is sensitive to approximately a monolayer of contamination.[44] (3) Surface forces alone may initiate plastic behavior for tip–sample systems with high surface energy and low hardness.[39] (4) The hardness approaches the critical yield stress of materials at small penetration depths.[45] (5) The surface energy may be determined regardless of the deformation mechanism.[46] (6) Purely elastic and purely plastic behavior are successfully modeled using continuum theory, but elastoplastic behavior still eludes understanding.[47] (7) Clean metal-metal systems may separate by ductile extension of the softer material or by brittle separation near the interface.[48] (8) Not every clean metal-metal system in UHV exhibits jump behavior.[49]

Molecular dynamics simulations have modeled the tip–surface interaction, with the hope of both incorporating existing continuum theory and gaining new insights into atomistic behavior. The nickel tip and gold surface interaction studied by Landman et al.[38] falls just into the JKRS regime, according to the parameter l (see Section 10.2.1). The surfaces are found to jump into contact (Section 10.2.2), forming a connective neck, as is expected with JKRS theory. The JKRS predictions of contact radius at zero applied load and the adhesion compare quite well with the simulations. In addition, the simulations show gold transferred to the nickel tip, wetting of the tip by the gold atoms, and ductile extension of the gold, followed by fracture. A fascinating prediction, recently experimentally verified, of the calculation is that the ductile extension occurs in discrete steps. As the elongated neck is stretched, the strain increases, until it is energetically favorable for the neck to relax by forming an additional layer of atoms. This is the basis of the structure seen in the force versus displacement curve shown in Figure 10.7.

10.2.4 Friction

Mate et al.[11] were the first to exploit AFM's capability to study friction. They detected stick-slip motion between a tungsten tip and a graphite surface with the periodicity of the graphite lattice at loads over 10^{-5} N. (This is a high load that necessitates a large contact area; see Sections 10.2.2 and 10.3.4) Their data are displayed in Figure 10.8.

Historically, the need to control and understand friction was driven by applications.[51] By the time of the Pharoahs, fluids were being added to interfaces so that they could more easily slide. Ships that stuck to their docks during launch was the problem that Coulomb addressed when he developed the idea of interlocking asperities. Johnson[18] was motivated by the adhesion of automobile windshield wipers to glass for the JKRS contact mechanics described in Section 10.2.1.

A significant change has taken place since the mid-1980s. There is now a surge of activity in the fundamental aspects of friction owing to two technical developments. The first is the ability to establish single contacts, be they over large areas of atomically smooth surfaces, such as in the surface forces apparatus,[2,3] or at the end of a pointed tip, as in the mechanical and electrical contact experiments of Pollock et al.[41] The second is the force and distance sensitivity of an atomic force microscope (Fig. 10.8). The former simplifies the interpretation of the experimental data,

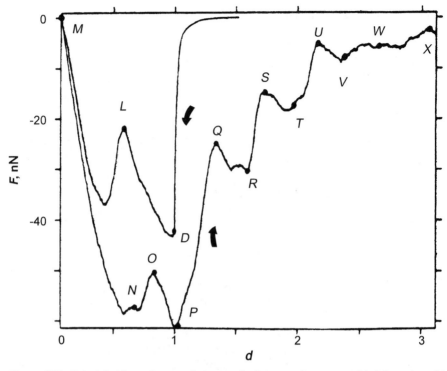

Figure 10.7. Calculated force F versus tip-to-sample distance d between a nickel tip and a gold sample for approach, jump to contact, indentation, and subsequent separation, d denotes the distance between the rigid tip-holder assembly and the static substrate of the gold surface ($d = 0$ at the jump-to-contact point, marked D). The letters on the curve signify the actual distances between the bottom of the nickel tip and the top of the gold surface: D, 3.8 Å; L, 2.4 Å; M, 0.8 Å; N, 2.6 Å; O, 3.0 Å; P, 3.8 Å; Q, 5.4 Å; R, 6.4 Å; S, 7.0 Å; T, 7.7 Å; U, 9.1 Å; V, 9.6 Å; W, 10.5 Å; and X, 12.8 Å.

the latter allows the periodicity of the atomic lattice to be resolved. A hint of the amount of intellectual activity in this decade is available.[52, 53] Here we shall briefly summarize one promising way of thinking about friction.

Amontons' law, $F_f = \mu N$, i.e., the friction force is a function of the applied normal load N times a coefficient, was formulated and tested 300 years ago.[51] And for centuries, the origin of friction was thought to be owing to the interlocking of asperities and their mutual wear.[51, 54] With the development of the surface forces apparatus,[3] wearless friction was observed.[2] An alternative approach to understanding friction between atomically smooth surfaces became necessary.[53] If one recalls that atomically "flat" surfaces are still corrugated on the atomic scale, then one can imagine how the sheets of atoms may fall into registry. Some effort is needed to slide one plane of atoms over the other as the crests of the atoms mutually pass, then fall back into registry. A descriptive name for this model is the cobblestone model, as if the atoms of one surface were the wheels of a cart, and the atoms of the other surface cobblestones in a street.

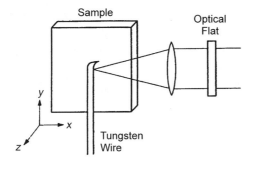

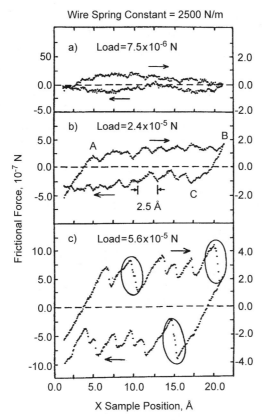

Figure 10.8. The frictional force on the tungsten tip in contact with the surface of graphite under three different loads and as a function of sample position. The frictional force is determined by measuring the deflection of the cantilever beam parallel to the sample surface as shown at the top of the figure. The observed corrugation at higher loads corresponds to slip-stick behavior with atomic periodicity. *Circled sections*, double slips. (Reprinted with permission from Ref. 11.)

There are several interesting aspects to the cobblestone model. As one might expect, the amount of effort required to lift the cart over the stones is related to the adhesion between and load applied to the surfaces. But the behavior of the system is different in response to adhesion and load.[55] For adhesion, it is not the absolute value of the forces that correlates with friction, but rather the adhesion hysteresis (the work done in a loading-unloading cycle) that is directly proportional to the friction between the surfaces.[56] The adhesion hysteresis times the distance the cart is lifted per pair of interacting atoms is summed over the atoms in contact to give an indication of the friction. Thus adhesion-controlled friction is proportional to the area of contact. This hypothesis has been experimentally verified.[57] For load-controlled friction, the applied load is distributed over n individual contacts, each contributing $1/n$ of the total friction. The sum over all of the contacts yields a friction force directly proportional to load. Ergo, Amontons' law is modified at the molecular level.[55] The friction force equals a coefficient times the contact area A plus an Amontons-like contribution, or

$$F_f = C_1 A + C_2 N$$

Area A and load N are explicit in this equation. Adhesion hysteresis is implicit in the constants C_i. What contributes to adhesion hysteresis? Both surface and bulk effects. The mechanical structure of the surfaces gives rise to friction anisotropy, depending on the direction of the scan with respect to the surface crystallographic direction.[58] The tilt of molecules at the sample surface can cause asymmetry in friction data, in other words, a left-to-right scan has a slightly different friction than a right-to-left one.[59] Surface chemistry can change the amount of adhesion,[60] implying that the load range over which the friction is adhesion controlled can also be changed, and that adhesion hysteresis may vary as well. The reduced elastic modulus K, a bulk mechanical property, has a similar influence on adhesion hysteresis.[35] Another bulk effect that may or may not fit into the adhesion hysteresis approach is the electronic contribution to friction. In this prediction and its experimental demonstration,[61] the electrons in one body dissipate energy as they are dragged along by the presence of the moving second body. The jury is still out on how well the newer data fit into the cobblestone picture.

10.2.5   Molecular Mechanics

The three-dimensional (3-D) structures of complex biomolecules are determined not only by the strong covalent bonds between atoms but also by relatively weak intramolecular forces such as electrostatic, van der Waals, hydrogen bonding, and hydrophobic forces. Similarly, when these forces act together between molecules and in a manner dependent on the specific 3-D structure of a molecule, they are collectively termed *molecular recognition forces*. Until recently, our understanding of molecular recognition was based on indirect physical and thermodynamic measurements such as X-ray crystallography, light scattering, nuclear magnetic resonance spectroscopy, and calorimetry. More recently scientists have used micropipette techniques, optical tweezers, SFA, and AFM to directly measure these molecular recognition forces.

These force measurement techniques must record both force and position with high resolution and accuracy. The AFM has both high force sensitivity (10^{-13} N or two orders of magnitude more sensitivity than the force of a single hydrogen bond) and position control (0.01 nm) and can probe areas as small as $10 \, nm^2$ under physiological conditions. To measure the intermolecular forces between specific molecules, one molecule is attached to the end (tip) of a cantilever beam and its complement to a flat substrate. Bringing the two molecules (surfaces) together and then pulling them apart allows the specific binding interactions to be measured. To date, many types of interactions have been measured including hydrogen[62] and noncovalent[63] bonds. The early studies used the biotin-streptavidin ligand-receptor pair that mimics an antibody–antigen interaction but with higher binding affinity. Molecular dynamics simulations of the bond rupture mechanism[64] help interpret the experimental results. Variations in the range of forces measured by different laboratories led to the idea that the rupture force may depend on the rate at which the bond is ruptured. Evans and co-workers[65] found that the bond strengths do indeed depend upon how fast the ligand is pulled away from the receptor.

The AFM and other techniques have also been used to measure the elastic properties of single molecules such as DNA,[66, 67] polysaccharides,[68] and titin proteins.[69] The extension of these molecules may be understood by examining the statistical distribution of the random coils. There are three models based on entropic elasticity theory that have been used:[70, 71] the Gaussian chain, the freely jointed chain (FJC), and the worm-like chain (WLC). The full force-extension curve can be calculated numerically from the WLC model, but the general behavior is more easily obtained by approximation in the Gaussian and FJC models.

Applying a weak tensile force F at the ends of the molecule causes it to extend by a fraction of its contour length:

$$\frac{h}{L} = \frac{2}{3}\left(\frac{FP}{k_B T}\right), \quad h \ll L$$

where h is the extension of the molecule along the direction of the applied force, L is the unstrained contour length, P is the persistence length, which is the length scale over which the directionality of the chain is maintained ($P = \mathbf{k}/(k_B T)$ where $\mathbf{k}$ is the elastic bending modulus), k_B is the Boltzman constant, and T is absolute temperature. The force required to stretch the molecule to a significant fraction of its full contour length would be $F_s \sim (k_B T)/P$. For DNA under physiological conditions, F_s is approximately 1 pN.

When the applied force is larger than F_s, the elasticity becomes nonlinear, i.e., it becomes harder and harder to stretch the polymer as it straightens and approaches its contour length. The FJC model is a simplified picture of molecular stretching where the polymer is represented as a chain of rigid links connected by pivots that can swivel. Each link has a length b, known as the Kuhn statistical length, where $b = 2P$. A force applied to the ends of the chain tends to align each link. In the model the molecular extension can be calculated analytically. At small extensions, the result is equivalent to the Gaussian model, but the force curve diverges as the extension approaches the contour length. The fractional elongation of the polymer can be represented by

$$\frac{h}{L} = \Gamma\left(\frac{Fb}{k_BT}\right)$$

where Γ is the Langevin function $\Gamma(x) = \coth(x) - 1/x$.

In the WLC model, the force diverges less strongly than in the FJC model because the polymer is able to bend continuously. As full extension is approached, the elongation can be described by

$$\frac{h}{L} = 1 - \left(\frac{2FP}{k_BT}\right)^{-1/2}, \qquad L - h \ll L$$

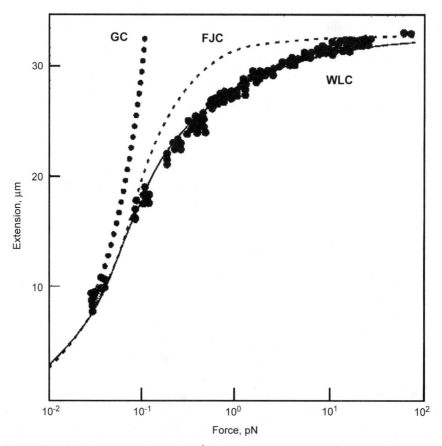

Figure 10.9. The force-extension curve of a single DNA molecule under tension applied to its ends. The data points are compared with the theoretical curves for the entropic elasticity of three polymer models: the Gaussian chain (*GC*), the freely jointed chain (*FJC*), and the worm-like Chain (*WLC*). (Reprinted with permission from Ref. 71; data from Ref. 66.)

For DNA under physiological conditions, the WLC model fits the experimental data quite well (Fig. 10.9).[71]

Stretching the polymer beyond its natural length strains the chemical bonds along its backbone. The elasticity is no longer entropic (related to its spatial conformation), but enthalpic (related to the chemical bond's heat of formation) in origin. Several groups have begun to measure the forces need to break single chemical bonds.

10.3 INTERPRETING THE EXPERIMENT

Force microscopy continues to mature. We are still learning how to interpret our data. There are many fundamental questions to answer. In this section, some important interpretation issues are addressed: interpreting quasistatic and dynamic force curves, the influence of surface forces on imaging, image resolution and imaging mechanisms, and surface forces and STM.

10.3.1 Interpreting Quasistatic and Dynamic Force Curves

In this section, we discuss the major features of quasistatic and dynamic force curves in the absence of liquid films and adsorbates. We discuss why the cantilever bends toward the sample, where contact occurs with either the sample or a thin liquid film on the sample, how the shape of the curve after contact is related to the surface forces and mechanical properties of the tip and sample, and what are the origins of hysteresis in the quasistatic force curves. The challenges associated with interpreting dynamic force curves are mentioned.

During attraction, the cantilever may bend toward the sample in a smooth and continuous motion or by small or large discontinuous jumps. The instability of the cantilever may be because the formation of a meniscus around the tip, which pulls the tip toward the sample, or because the magnitude of the sample force gradient has equaled the effective spring constant of the cantilever.

In the case of a liquid film, contact with the liquid begins at the point of instability when the meniscus draws the cantilever toward the sample. For the case in which the instability is owing to the total force gradient going to zero, contact is established after the point of instability where the curvature changes from negative to positive.

When no instabilities are evident, determining the contact point is more challenging. One method is to modulate the sample and observe the modulation amplitude of the cantilever. Contact can be assigned to the point where the detected cantilever modulation amplitude starts to increase.[12] This normally occurs at the inflection point of the approach (point A in Fig. 10.10). The curvature change corresponds to a growing repulsive force between atoms on the end of the tip and the sample. The point of maximum attraction occurs when the gradient of the repulsive interaction equals the gradient of the attractive interaction(s).

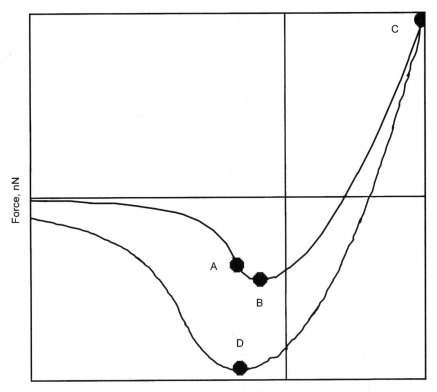

Separation or Indentation Depth, nm

Figure 10.10. Typical AFM force curve showing the force on a cantilever tip as a function of tip–sample distance (or sample position). Below the x-axis, the net force is attractive (*negative*); above the x-axis the net force is repulsive (*positive*). A, first contact; B, maximum attractive force (pull-on force); C, maximum applied load; D, maximum adhesion (pull-off force).

After measuring the force curves, we must choose the proper mechanical relationships with which to evaluate the data to determine the materials properties of the sample as well as the tip–sample contact area. In Section 10.2.1 we summarized the salient features of Hertzian mechanics, as well as the JKRS, DMT, BCP, and MD theories. All of the theories reduce to the traditional Hertzian mechanics at high loads. Choosing the appropriate theory depends on the relative magnitudes of the materials properties and surface forces. In general, MD is the most inclusive, but its parametric equations may intimidate some.

Surface forces tend to increase the tip–sample contact area via attraction in the early stages of contact. In the JKRS theory, attractive forces operative within the contact area cause the surfaces to form a connective neck between the tip and sample. In the DMT theory, the attractive interactions are assumed to occur outside the contact area, increasing the interaction volumes as the tip comes closer to the sample as indentation begins. In reality, both processes occur to some extent.

The total force interaction curve can be modeled as

$$F = \begin{cases} \sum F_{\text{attr}} & D \leq -D_{\text{o}} \\ \sum F_{\text{attr}} + \sum F_{\text{rep}} & D \geq -D_{\text{o}} \end{cases}$$

In this model, D is determined by the long-range attractive forces, such that at contact $D = D_{\text{o}}$ rather than $D = 0$. This prevents the force relations of the form $F \sim -1/D^n$ from becoming infinite.

The total force before contacting either a liquid or a solid can be attractive or repulsive and can be owing to electrostatic, capacitive, magnetic, van der Waals, or other long-range forces. The force after contact consists of both attractive and repulsive components.

The general mathematical form of the force equation up to contact (assuming no meniscus or cantilever instability effects) is the sum of all the long-range forces. Always present are van der Waals forces, but they may be small compared to the other forces discussed in Chapters 2 and 8. The distance dependence of the force may give clues to its origin, because the functional dependence on distance varies, depending on the nature of the force.[12]

In the simplest sense, the elastic modulus of the tip–sample system can be related to how quickly the force curve "turns around" after making contact. For example, in Figure 10.11a, the curvature in the region where the tungsten tip makes contact with alumina is fairly high. In parts **b** and **c** alumina has been covered with monolayers of stearic acid, a compliant material relative to the alumina. Compared to Figure 10.11a, the curvature at contact is low.[73]

Assuming a Hertzian contact radius between the tip and sample, $a^2 = Rd$, the contact radius can easily be determined at any point after contact. We stress that the Hertzian radius is a minimum estimate for the contact radius, as it ignores any tip–sample deformation that occurs as a result of surface forces.

Another major feature is the hysteresis in the curves. There are several origins of hysteresis in AFM, two of which—hysteresis in the piezoelectric ceramics and cantilever instability—are considered artifacts (Section 10.2.2). However, hysteresis in the force curve may also be owing to a strong tip–sample interaction associated with adhesion and/or plastic deformation.[38]

So far, we have discussed only quasistatic force curves, where all imposed forces or movements of the tip and sample are performed slowly in comparison to the response time of the system. There are many alternative methods to generate nanomechanical data, some of which are mentioned in Figure 10.12. They are explained in detail in Chapter 2. A popular point measurement that complements quasistatic force curve acquisition is dynamic force curve acquisition,[75] in which the cantilever is driven near its resonant frequency with some tens or even hundreds of nanometers of vibration amplitude. The sample and tip approach each other, then retract, during which the cantilever amplitude and phase are recorded.

In contrast to quasistatic force curves, where the cantilever bends forward in response to attractive forces and backward in response to repulsive ones, dynamic

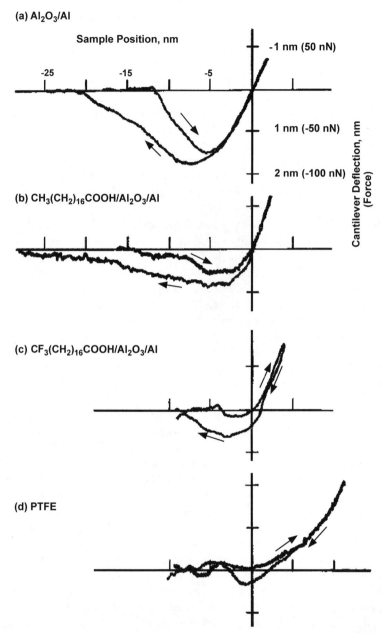

Figure 10.11. Representative data obtained with the force microscope showing surface–force interaction between a tungsten tip and (**a**) the native oxide of aluminum (Al_2O_3/Al), (**b**) stearic acid [$CH_3(CH_2)_{16}COOH$], (**c**) ω,ω,ω-trifluorostearic acid [$CF_3(CH_2)_{16}COOH$], and (**d**) polytetrafluoroethylene (PTFE). The x-axis is the sample position in nanometers; the y-axis is the cantilever deflection in nanometers relative to its rest position, or the force in nanonewtons between the cantilever and sample. The scales for all the curves are the same. (Reprinted with permission from Ref. 73.)

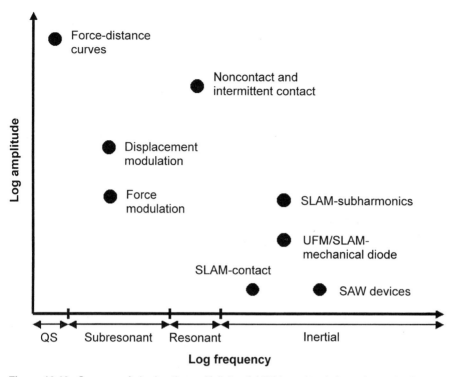

Figure 10.12. One way of viewing the multiplicity of AFM-based techniques is to plot them in frequency-amplitude space, depending on the frequency and amplitude of the excitation. The frequency axis is divided into four regimes: *quasistatic*, where the damping and acceleration of the cantilever and sample are negligible; *subresonant*, where only the acceleration of the cantilever is ignored; *resonant*, where damping, acceleration, and acceleration must be accounted for; and *inertial*, where static loads are in general small compared to the acceleration. The amplitude axis covers the range from a few angstroms for inertial modulation amplitudes, to microns, if quasistatic force curves are generated. The techniques themselves are described in Chapter 2 and the literature.[74]

force curves are far less intuitive to understand.[76] Depending on if the excitation frequency is slightly above or below the resonant frequency of the cantilever, the amplitude of the cantilever can either increase or decrease in response to attractive forces. If the tip–sample contact stiffness approaches that of the cantilever stiffness, the system becomes nonlinear, and it may no longer be treated with the harmonic approximation.[77]

In summary, force curves contain the following information: the magnitude and distance dependence of the long-range forces, the point of contact, the modulus, adhesion, and plastic deformation of the tip–sample system, and an estimate of the tip–sample contact area. We stress that the details of interpretation, particularly for dynamic force curves, are still open to discussion.

10.3.2 Influence of Surface Forces on Imaging

Surface forces are important in image interpretation. Surface forces alone create large contact areas (Sections 10.2.3 and 10.3.2), even before the cantilever applies any external load. The size of the contact area also depends on the mechanical properties of the tip and sample. Assuming that the contact area between tip and sample determines the image resolution, an awareness of surface forces and mechanical properties is important to interpret contact mode images. In noncontact and intermittent contact imaging, if the surface forces vary from spot to spot on a sample through changes in the chemistry of the surface, the detected image will be a convolution of the topography and the surface forces.

Contact mode imaging, although offering higher resolution than noncontact imaging, may deform or damage the tip or sample. The loads generated by the surface forces alone may cause very high pressures underneath the tip. An estimate of the average pressure under the tip at zero applied load may be obtained by dividing the maximum attractive force or adhesion by the circular contact area at zero applied load πa_o^2. In our early experiments, which assumed the macroscopic tip radius determined by SEM,[73] pressures varied from 300 MPa for Al_2O_3 to 4 MPa for teflon. Small asperities on the tip or sample would increase the estimated mean pressure for a given load. These pressures should be considered when imaging delicate samples. In addition, high adhesion may transfer material between the sample and tip;[38] a cantilever instability can accelerate the tip into the sample, increasing the likelihood of damage.

The pressure and deformation beneath the tip and the tip–sample adhesion may be reduced by conducting the imaging experiment in an appropriate environment. The sample may be exposed to air or a gas that lowers the surface energy and adhesion, at the expense of the sample cleanliness. The sample and cantilever may be placed in a liquid that lowers the surface force interaction.[78, 79] (Here, the entire cantilever must be immersed in the fluid, or meniscus effects will increase the interaction strength; see Chapter 2). Liquids with high dielectric constants will reduce van der Waals forces and increase capacitive interactions. In some cases, the van der Waals interaction can be made repulsive. This was nicely illustrated by the work of Hartmann[79] who calculated the van der Waals forces between a sharp SiO_2 tip and a metal surface in polar media. Not only does the polar liquid reduce the force but it may also make the interaction repulsive. An estimate of the effect can be made using the principle of combining relations[2] used to obtain approximate values for unknown Hamaker constants from known ones, i.e.,

$$A_{132} \approx \left(\sqrt{A_{11}} - \sqrt{A_{33}} \right) \left(\sqrt{A_{22}} - \sqrt{A_{33}} \right)$$

where A_{132} is defined as the nonretarded Hamaker constant for media 1 (tip) and 2 (sample) interacting across medium 3 (vacuum, air, liquid). Charged and hydrated surfaces can also render the interaction repulsive. All these interactions can be used to decrease the tip–sample interaction to allow nondestructive contact with minimal deformation.

Using the same principle, Hutter and Bechhoefer[80] manipulated van der Waals forces to improve image resolution. The relationship derived from JKRS theory is given as

$$\Re = \left(\frac{AR^2}{8KD_o^2} \right)^{1/3}$$

where $\Re$ is the resolution, A is the Hamaker constant, R is the radius of the tip, K is the reduced modulus of the tip and sample, and D_o is the tip–sample separation at contact.

10.3.3 Image Resolution and Imaging Mechanisms in Contact Mode

Force microscopy is capable of atomic resolution images in the contact mode of imaging. Atomic resolution has been demonstrated on Au,[81] graphite, MoS_2, and BN,[82] LiF,[83] and NaCl.[84] The high resolution of these images suggests that the imaging mechanism consists of a lone atom at the end of the tip that gently traces out the contours of the sample. However, as we have discussed in this chapter, the contact mechanism and surface forces associated with the tip–sample interaction make atomic contact difficult to achieve. In this section, we summarize the arguments against single atom contact in contact mode imaging, and mention the possible alternative imaging mechanisms.

Both the tip–sample geometry and the materials properties govern the tip–sample contact area. The tip–sample geometry as described by Hertzian mechanics determines the radius of the contact area as $a^2 = Rd$, where R is the tip radius and d is the deformation after contact (Section 10.2.1). For a minimum detectable deformation of 0.1 nm (typical AFM noise level), and a small tip of radius 10 nm, the contact radius a equals 1 nm. The Hertzian contact radius is actually a lower limit for the radius as it ignores both short- and long-range surface forces.

The actual contact area becomes even larger if one remembers that Hertzian mechanics neglects the effect of surface forces. Furthermore, it has been predicted by a number of authors that at some point during the tip–sample approach,[36–38] the tip–sample force gradient equals the second derivative of the potential of one or both of the solids, and an instability occurs causing the surfaces to jump together (see Section 10.2.2). For high-modulus, low-surface energy materials this effect should be minimized.

Even high-moduli materials such as diamond will deform under the high pressures beneath force microscope tips (Section 10.3.3). For example, a load of 1 nN on an area 10 pm in radius (atomic resolution) gives a mean pressure of 3000 GPa. This is approximately 30 times the yield stress of diamond.[85] So even in the case of the lightest measurable loads on the highest modulus material, it is not reasonable to assume single atom contact. Nevertheless, we do expect that two diamond surfaces in contact give a minimum obtainable contact area for a given load and tip radius owing to diamond's extremely high modulus.

If single atom contact is not feasible, then what mechanism is responsible for atomic resolution images? A few papers have been published addressing AFM imaging mechanisms, mostly focusing on lamellar materials such as graphite.

Pethica's[86] model consists of sliding planes of graphite. A piece of graphite sticks to the tip and slides over the layer below it. The high and low spots in the resulting image correspond to when the atoms slide in and out of registry. Rotation of one plane with respect to another can account for images without the expected hexagonal or trigonal symmetry.[87–89] This model is consistent with Mate's[11] friction experiments on graphite in that atomic scale periodicity in the data was observed despite the high loads and large contact areas (Section 10.2.4).

Landman et al.[90] used molecular dynamics simulations to predict the scan contours of ordered and disordered tips on $Si\langle 111 \rangle$ surfaces. The ends of the tips were composed of 16 atoms that were in contact with the sample. Both the ordered and disordered tips exhibited periodic variations in force as the tip was scanned over the surface.

The interpretation of atomic resolution images in the context of sliding contact involving large areas has, so far, been applied to graphite. Similar analysis could apply to the other lamellar materials (such as BN or MoS_2) for which atomic resolution has been achieved. Although no theory has yet been rigorously applied to $Au\langle 111 \rangle$, $LiF\langle 001 \rangle$, or $NaCl\langle 001 \rangle$, we would like to point out that these surfaces are the easy shear planes on these materials.

So far we have been concerned with mechanisms and resolution associated with contact mode imaging. In noncontact mode AFM, resolution is fortunately a little easier to understand. The resolution is determined by the size of the tip and its distance from the sample.[91] The effective size of the tip is determined by how quickly the forces drop off. For example, magnetic forces have been detected more than 200 nm away from the sample,[92] whereas van der Waals forces are rarely detectable more than 10 nm away. The resolution of the magnetic tip will then be more dependent on the macroscopic shape of the tip. The best FM resolution reported to date for noncontact mode imaging of van der Waals forces is 5 nm.[93] For magnetic force microscopy a resolution of 10 nm has been reported.[94, 95]

Work from Ohnesorge and Binnig[42] explicitly revealed the experimental conditions under which true atomic resolution can be achieved. The sample was the $\langle 10\underline{1}4 \rangle$ surface of calcite, which has a large unit cell of 5.0×8.1 Å. Imaging under water to lower the attractive forces and using a weak cantilever in the attractive part of the force curve before contact, they were able to detect single atomic defects. Recent noncontact studies pioneered by Giessibl[96] employ an AFM in ultra-high vacuum (UHV) where the cantilever is driven at high amplitudes (10 to 150 nm). At the end of its oscillation, the cantilever tip is close to the surface, yet the cantilever is not subject to an instability because of its stored elastic energy. The Si $\langle 111 \rangle$-(7×7) surface was successfully imaged using this technique, and subsequently individual point defects on insulating surfaces have been detected.[97]

These last two examples show that the words *atomic resolution* have been erroneously applied to many contact mode experimental results, for which the terminology *lattice resolution* or *atomic periodicity* would be much more appropriate.

10.3.4 Surface Forces and STM

In STM measurements involving clean surfaces prepared, for example, in UHV, the tunneling tip is held approximately 1 nm away from the surface. The proximity of the tip and surface causes a significant surface force interaction. Operating the STM in a dirtier environment often requires the tip to penetrate contamination layers to establish tunneling. In this realm, a repulsive interaction may exist between the tip and the sample. As a result in STM imaging, it is important to consider both surface forces and materials properties to interpret STM images. In this section, we give some examples of the overlap between AFM and STM.

Eigler and Schweizer[98] used attractive interactions between an STM tip and xenon atoms on nickel and platinum surfaces at 4 K to move the xenon atoms into patterns. Their procedure is to scan the surface, locate a xenon atom and lower the STM tip over the atom such that the attractive interaction is high enough to drag the atom along the surface underneath the tip. When the atom is in the desired location, the tip is raised so as to lower the magnitude of the interaction. In this way, atomic structures of xenon atoms can be fabricated using surface forces.

Ideally in STM, the measured current will change an order of magnitude for each angstrom change in separation between the tip and sample.[99] However, contamination layers can interfere with this relationship.[100] If a nonconducting layer is between the tip and the sample, then the tip will have to move a large distance to cause a significant change in current. This effect gives rise to anomalously low barrier heights. As the tip pushes through the contamination layer, the repulsive force between sample and tip rises.[100] In graphite, the repulsive interaction is the source of the enhanced corrugation measured for the graphite surface.[101] Even on clean surfaces, the barrier is expected to collapse at distances of <0.2 nm.[102]

STM has been used to study adhesion. Gimzewski and Möller[40] demonstrated the effect of a thin layer of contamination on current-voltage (*I-V*) curves and the mutual adherence of the tip–sample materials. The *I-V* curves for an iridium tip on a polycrystalline surface of Ag in UHV differ markedly between the freshly cleaned surfaces and surfaces that were cleaned 10 h before the experiment. Images of the Ag surface before and after contact were also substantially different depending on cleanliness. For a clean tip and sample, a 3 nm excursion of the tip toward the sample resulted in a new hillock on the sample of approximately 10 nm in diameter. For a tip exposed to the UHV environment for a few hours, the same tip excursion indented the surface. In the case of the clean system, the adhesion was high enough to produce a neck of material that broke upon tip retraction to form the hillock. In the contaminated case, the adhesion was lowered to the point where no neck was formed. Similar hillock and hole forming techniques have been used recently in STM lithography.

Dürig et al.[43, 44] used STM to measure the sample force gradient during tunneling and electrical contact. They mounted the sample on a cantilevered beam and brought an STM tip to within tunneling distance. By adding a small modulation voltage to the tunnel piezoceramic and monitoring the observed frequency dependence of the tunnel current, it is possible to determine the force gradient between the tip and sample . The observed behavior depends on the materials used. For an Ir tip on an Ir surface,[104] the force gradient becomes increasingly negative (attractive) until the point at which electrical contact is established. For Ir tips on Au and Al,[105] a repulsive component to the force gradient was observed several angstroms before the point of electrical contact.

From these examples, one can conclude that both materials properties and surface forces play a significant role in STM as well as AFM.

10.4 SUMMARY

AFM offers the ability to measure a whole spectrum of forces, allowing mechanical properties to be deduced from the force–distance relationships. We hope that we have given the reader an adequate introduction to this field.

Despite the remarkable progress in such a short time, work has just begun. Here we list some challenges and questions for force microscopists as of 1999. It represents only our perspective; no doubt other researchers will be able to augment this list.

- What are the effects of water vapor and surface adsorbates in imaging and force measurements?
- What are the imaging mechanism(s) for all of the different forces in both contact and noncontact modes of operation?
- Can we develop a theory that accounts quantitatively for all of the features in the force curves?
- Can traditional continuum mechanics theories be used to explain nanometer-scale contact and indentation processes, and how well do they agree with atomistic simulations?
- Can we develop fundamental understanding of such complex phenomena as friction and adhesion?

Nanomechanics had its beginnings in the late 1980s. Today, conferences often feature one or more sessions on nanomechanics and nanotribology. As research in nanomechanics progresses, we have no doubt that there will be greater insight into biological processes, surface structure, surface forces, and mechanical properties of materials on the nanometer scale. We believe that AFM will play an ever larger role in research, development, and production owing to its ability to study both insulators and conductors with nanometer lateral and nanonewton force resolution.

ACKNOWLEDGMENTS

RJC acknowledges the support of the Office of Naval Research and the many friends and colleagues—S. Asif, N. Burnham, S. Corcoran, C. Draper, J. Harrison, S. Hues, K. Lee, D. Schaefer, S. Sinnott, and K. Wahl—who have contributed to the nanomechanics program at NRL. Similarly, NAB emphasizes that the cumulated knowledge presented here is a result of interacting with an international collection of talented researchers at the EPF Lausanne; Forschungszentrum Jülich; Naval Research Laboratory; Georgia Institute of Technology; Ecole Centrale de Lyon; Paul Drude Institute; and Lancaster, Loughborough, Oxford, and Saarbrücken Universities. NAB also appreciates the use of the facilities at the EPF Lausanne for some of the preparation of this manuscript.

REFERENCES

1. N. R. Moody, W. W. Gerberich, N. Burnham, and S. P. Baker, eds., *Mat. Res. Soc. Symp. Proc.* **522** (1998).

2. J. N. Israelachvili, *Intermolecular and Surface Forces*, Academic Press, New York, 1992.

3. J. N. Israelachvili and D. Tabor, *Proc. R. Soc. Lond. A* **331**, 19 (1972).

4. N. A. Burnham and R. J. Colton, *J. Vac. Sci. Technol.* **A7**, 2906 (1989).

5. B. Drake, C. B. Prater, A. L. Weisenhorn, et al., *Science* **243**, 1586 (1989).

6. A. L. Weisenhorn, J. E. MacDougall, S. A. C. Gould, et al., *Science* **247**, 1330 (1990).

7. D. Rugar, H. J. Mamin, P. Guethner, S. E. Lambert, J. E. Stern, I. McFadyen, and T. Yogi, *J. Appl. Phys.* **68**, 1169 (1990).

8. H. K. Wickramasinghe, *Sci. Am.* 98 (1989).

9. W. A. Ducker and R. F. Cook, *Appl. Phys. Lett.* **56**, 2408 (1990).

10. C. M. Mate, M. R. Lorenz, and V. J. Novotny, *J. Chem. Phys.* **90**, 7550 (1989).

11. C. M. Mate, G. M. McClelland, R. Erlandsson, and S. Chiang, *Phys. Rev. Lett.* **59**, 1942 (1987).

12. N. A. Burnham, R. J. Colton, and H. M. Pollock, *Nanotechnol.* **4**, 64 (1993).

13. J. B. Pethica, R. Hutchings, and W. C. Oliver, *Phil. Mag. A* **48**, 593 (1983).

14. G. L. Miller, J. E. Griffith, E. R. Wagner, and D. A. Grigg, *Rev. Sci. Instrum.* **62**, 705 (1991); A. J. Stephen and J. E. Houston, *Rev. Sci. Instrum.* **62**, 710 (1991).

15. B. Bhushan, A. V. Kulkarni, W. Bonin, and J. T. Wyrobek, *Phil. Mag. A* **74**, 1117 (1996).

16. S. A. Syed Asif, K. J. Wahl, and R. J. Colton, *Rev. Sci. Instrum.* **70**, 2408 (1999).

17. H. Hertz and J. Reine, *Angew. Math.* **92**, 156 (1882); K. L. Johnson, *Proc. Instn. Mech. Engrs.* **196**, 363 (1982).

18. K. L. Johnson, *Contact Mechanics*, Cambridge University Press, Cambridge, UK, 1985.

19. I. N. Sneddon, *Int. J. Engng. Sci.* **3**, 47 (1965).

20. J. L. Loubet, J. M. Georges, O. Marchesni, and G. Meille, *Trans. ASME J. Tribol.* **106**, 43 (1984).

21. M. F. Doerner and W. D. Nix, *J. Mater. Res.* **1**, 601 (1986).

22. W. C. Oliver and G. M. Pharr, *J. Mater. Res.* **7**, 1564 (1992).

23. J. Meneve, J. F. Smith, N. M. Jennett, and S. R. J. Saunders, *Appl. Surf. Sci.* **100–101**, 64 (1996).

24. J. C. Pivin, D. Lebouvier, H. M. Pollock, and E. Felder, *J. Phys. D* **22**, 1443 (1989).

25. G. Sperling, dissertation, Karlsruhe Technical High School, 1964.

26. K. L. Johnson, K. Kendall, and A. D. Roberts, *Proc. R. Soc. Lond. A* **324**, 301 (1971).

27. B. V. Derjaguin, V. M. Muller, and Y. P. Toporov, *J. Coll. Interface Sci.* **53**, 314 (1975).

28. D. Maugis, *J. Coll. Interface Sci.* **150**, 243 (1992).

29. D. Maugis and B. Gauthier-Manuel, *J. Adhesion Sci. Technol.* **8**, 1311 (1994).

30. D. S. Dugdale, *J. Mech. Phys. Solids* **8**, 100 (1960).

31. V. M. Muller, V. S. Yushchenko, and B. V. Derjaguin, *J. Coll. Interface Sci.* **77**, 91 (1980).

32. B. D. Hughes and L. R. White, *J. Mech. Appl. Math.* **32**, 445 (1979).

33. R. G. Horn, J. N. Israelachvili, and F. Pribac, *J. Coll. Interface Sci.* **115**, 480 (1987).

34. M. D. Pashley, *Colloids Surf.* **12**, 69 (1984).

35. N. A. Burnham and A. J. Kulik, in B. Bhushan, ed., *Handbook of Micro/Nanotribology*, 2nd ed., CRC Press, Boca Raton, Fla., 1999.

36. J. B. Pethica and W. C. Oliver, *Physica Scripta* **T19**, 61 (1987); J. B. Pethica and A. P. Sutton, *J. Vac. Sci. Technol.* **A6**, 2494 (1988).

37. J. R. Smith, G. Bozzolo, A. Banerjea, and J. Ferrante, *Phys. Rev. Lett.* **63**, 1269 (1989).

38. U. Landman, W. D. Luedtke, N. A. Burnham, and R. J. Colton, *Science* **248**, 454 (1990).

39. H. M. Pollock, *Vacuum* **31**, 609 (1981).

40. J. K. Gimzewski and R. Möller, *Phys. Rev. B* **36**, 1284 (1987).

41. H. M. Pollock, P. Shufflebottom, and J. Skinner, *J. Phys. D: Appl. Phys.* **10**, 127 (1977); H. M. Pollock, *J. Phys. D* **11**, 39 (1978).

42. F. Ohnesorge and G. Binnig, *Science* **260**, 1451 (1993).

43. U. Dürig, J. K. Gimzewski, and D. W. Pohl, *Phys. Rev. Lett.* **57**, 2403 (1986).

44. U. Dürig, O. Züger, and D. W. Pohl, *J. Microsc.* **152**, 259 (1988).

45. M. D. Pashley, J. B. Pethica, and D. Tabor, *Wear* **100**, 7 (1984).

46. M. D. Pashley and J. B. Pethica, *J. Vac. Sci. Technol. A* **3**, 757 (1985).

47. D. Maugis and H. M. Pollock, *Acta Metallurgia* **32**, 1323 (1984); H. M. Pollock, D. Maugis and M. Barquins, in P. J. Blau and B. R. Lawn, eds., *Microindentation Techniques in Materials Science and Engineering*, American Society for Testing and Materials, Philadelphia, 1986.

48. S. K. R. Chowdhury, N. E. W. Hartley, H. M. Pollock, and M. A. Wilkens, *J. Phys. D* **13**, 1761 (1980); S. K. R. Chowdhury and H. M. Pollock, *Wear* **66**, 307 (1981).

49. G. Cross, A. Schirmeisen, A. Stalder, P. Grütter, M. Tschudy, and U. Dürig, *Phys. Rev. Lett.* **80**, 4685 (1998).

50. N. Agraït, G. Rubio, and S. Vieira, *Phys. Rev. Lett.* **74**, 3995 (1995); G. Rubio, N. Agraït, S. Vieira, *Phys. Rev. Lett.* **76**, 2302 (1996); J. I. Pascual, J. Méndez, J. Gómez-Herrero, A. M. Baró, N. Garcia, U. Landman, W. D. Luedtke, E. N. Bogachek, and H.-P. Chen, *Science* **267**, 1793 (1995).

51. D. Dowson, *History of Tribology*, Longman Press, London, 1979.

52. I. L. Singer and H. M. Pollock, eds., *Fundamentals of Friction: Macroscopic and Microscopic Processes*, NATO ASI Series, Vol. **E220**, Kluwer Academic Publishers, Dordecht, 1992; I. L. Singer, *J. Vac. Sci. Technol.* **A12**, 2605 (1994); B. Bhushan,

J. N. Israelachvili, and U. Landman, *Nature* **374**, 607 (1995); R. J. Colton, Langmuir **12**, 4574 (1996); J. Krim, *Sci. Am.* **275**, 74 (1996); B. Bhushan ed., *Micro/Nanotribology and Its Applications*, NATO ASI Series, Vol. **E330**, Kluwer Academic Publishers, Dordecht, 1997; B. N. J. Persson and E. Tosatti, eds., *Physics of Sliding Friction*, Kluwer Academic, Dordrecht, 1996; R. W. Carpick, and M. Salmeron, *Chem. Rev.* **97**, 1163 (1997), B. N. J. Persson, *Sliding Friction: Physical Principles and Applications*, Springer-Verlag, Heidelberg, 1998.

53. B. Bhushan, ed., *Handbook of Micro/Nanotribology*, 2nd ed., CRC Press, Boca Raton, Fla., 1999.

54. F. P. Bowden and D. Tabor, *Friction and Lubrication*, 2nd ed., Methuen, London, 1967.

55. A. Berman, C. Drummond, and J. Israelachvili, *Tribol. Lett.* **4**, 95 (1998).

56. H. Yoshizawa, Y.-L. Chen, and J. Israelachvili, *J. Phys. Chem.* **97**, 4128 (1993); **97**, 11300 (1993).

57. R. W. Carpick, N. Agraït, D. F. Ogletree, and M. Salmeron, *J. Vac. Sci. Technol.* **B14**, 1289 (1996); M. Enachescu, R. J. A. van den Oetelaar, R. W. Carpick, D. F. Ogletree, C. F. J. Flipse, and M. Salmeron, *Phys. Rev. Lett.* **81**, 1877 (1998).

58. M. Hirano, K. Shinjo, R. Kaneko, and Y. Murata, *Phys. Rev. Lett.* **67**, 2642 (1991); J. A. Harrison, C. T. White, R. J. Colton, and D. W. Brenner, *Phys. Rev. B* **46**, 9700 (1992); R. Overney, M. Fujihira, H. Takano, W. Paulus, and H. Ringsdorf, *Phys. Rev. Lett.* **72**, 3546 (1994); R. Nisman, P. Smith, and G. J. Vancso, *Langmuir* **10**, 1667 (1994); H. Bluhm, U. D. Schwarz, K. P. Meyer, and R. Wiesendanger, *Appl. Phys.* **A61**, 525 (1995); K. Meine, D. Vollhardt, and G. Weidemann, *Langmuir* **14**, 1815 (1998); U. Gehlert, J. Fang, and C. M. Knobler, *J. Phys. Chem.* **B102**, 2614 (1998); L. F. Chi, M. Gleiche, and H. Fuchs, *Langmuir* **14**, 875 (1998).

59. M. Liley, D. Gourdon, D. Stamou, U. Meseth, T. M. Fischer, C. Lautz, H. Stahlberg, H. Vogel, N. A. Burnham, and C. Duschl, *Science* **280**, 273 (1998).

60. G. Hähner, A. Marti, and N. D. Spencer, *Tribol. Lett.* **3**, 359 (1997).

61. A. Dayo, W. Alnasrallah, and J. Krim, *Phys. Rev. Lett.* **80**, 1690 (1998).

62. J. H. Hoh, J. P. Cleveland, C. B. Prater, J. P. Revel, and P. K. Hansma, *J. Am. Chem. Soc.* **114**, 4917 (1992).

63. G. U. Lee, D. A. Kidwell, and R. J. Colton, *Langmuir* **10**, 354 (1994); E. L. Florin, V. T. Moy, and H. E. Gaub, *Science* **264**, 415 (1994); V. T. Moy, E. L. Florin, and H. E. Gaub, *Science* **266**, 257 (1994); A. Chilkoti, T. Boland, B. D. Ratner, and P. S. Stayton, *Biophys. J.* **69**, 2125 (1995); U. Dammer, M. Hegner, D. Anselmetti, P. Wagner, M. Dreier, W. Huber, and H.-J. Güntherodt, *Biophys. J.* **70**, 2437 (1996); P. Hinterdorfer, W. Baumgartner, H. J. Gruber, K. Schilcher, and H. Schindler, *Proc. Natl. Acad. Sci. U. S. A.* **93**, 3477 (1996).

64. H. Grubmuller, B. Heymann, and P. Tavan, *Science* **271**, 997 (1996); S. Izrailev, S. Stepaniants, M. Balsera, Y. Oono, and K. Schulten, *Biophys. J.* **72**, 1568 (1997).

65. E. Evans and K. Ritchie, *Biophys. J.* **72**, 1541 (1997); R. Merkel, P. Nassoy, A. Leung, K. Ritchie, and E. Evans, *Nature* **397**, 50 (1999).

66. S. B. Smith, L. Finzi, and C. Bustamante, *Science* **258**, 1122 (1992).

67. A. Bensimon, A. Simon, A. Chiffaudel, V. Croquette, F. Heslot, and D. Bensimon, *Science* **265**, 2096 (1994); G. U. Lee, L. A. Chrisey, and R. J. Colton, *Science* **266**, 771 (1994); T. T. Perkins, D. E. Smith, R. G. Larson, and S. Chu, *Science* **268**, 83 (1995); P. Cluzel, A. Lebrun, C. Heller, R. Lavery, J.-L. Viovy, D. Chatenay, and F. Caron,

Science **271**, 792 (1996); S. B. Smith, Y. Cui, and C. Bustamante, *Science* **271**, 795 (1996).

68. M. Reif, F. Oesterhelt, B. Heymann, and H. E. Gaub, *Science* **275**, 1295 (1997).

69. M. Reif, M. Gautel, F. Oesterhelt, J. M. Fernandez, and H. E. Gaub, *Science* **276**, 1109 (1997); M. S. Z. Kellermayer, S. B. Smith, H. L. Granzier, and C. Bustamante, *Science* **276**, 1112 (1997); L. Tskhovrebova, J. Trinick, J. A. Sleep, and R. M. Simmons, *Nature* **387**, 308 (1997).

70. M. Doi and S. F. Edwards, *The Theory of Polymer Dynamics*, Oxford University Press, Oxford, 1986; P. J. Flory, *Statistical Mechanics of Chain Molecules*, Hanser Publishers, Munich, 1989.

71. R. H. Austin, J. P. Brody, E. C. Cox, T. Duke, and W. Volkmuth, *Phys. Today,* **Feb**, 32 (1997).

72. M. Grandbois, M. Beyer, M. Rief, H. Clausen-Schaumann, and H. E. Gaub, *Science* **283**, 1727 (1999).

73. N. A. Burnham, D. D. Dominguez, R. L. Mowery, and R. J. Colton, *Phys. Rev. Lett.* **64**, 1931 (1990).

74. N. A. Burnham, A. J. Kulik, F. Oulevey, C. Mayencourt, D. Gourdon, E. Dupas, and G. Gremaud, in B. Bhushan, ed., *Micro/Nanotribology and Its Applications* NATO ASI Series **E330**, Kluwer Academic, 1997.

75. Q. Zhong, D. Inniss, K. Kjoller, and V. B. Elings, *Surf. Sci. Lett.* **290**, L688 (1993).

76. N. A. Burnham, O. P. Behrend, F. Oulevey, et al., *Nanotechnol*, **8**, 67 (1997); www. nanomechanics.com.

77. www.nanomechanics.com/.

78. A. L. Weisenhorn, P. Maivald, H.-J. Butt, and P. K. Hansma, *Phys. Rev. B* **45**, 11226 (1992).

79. U. Hartmann, *Phys. Rev. B* **43**, 2404 (1991); U. Hartmann, *Ultramicroscopy* **42–44**, 59 (1992); U. Hartmann in R. Wisendanger and H.-J. Güntherodt, eds., *Scanning Tunneling Microscopy III–Theory of STM and Related Probe Methods*, Springer-Verlag, Berlin, 1993.

80. J. L. Hutter and J. Bechhoefer, *J. Appl. Phys.* **73**, 4123 (1993).

81. A. L. Weisenhorn, P. K. Hansma, T. R. Albrecht, and C. F. Quate, *Appl. Phys. Lett.* **54**, 2651 (1989).

82. T. R. Albrecht and C. F. Quate, *J. Appl. Phys.* **62**, 2599 (1987).

83. E. Meyer, H. Heinzelmann, H. Rudin, and H.-J. Güntherodt, *Z. Phys.* **B79**, 3 (1991).

84. G. Meyer and N. M. Amer, *Appl. Phys. Lett.* **56**, 2100 (1990).

85. C. A. Brookes, *Properties of Diamond*, Academic Press, London, 1979; I. P. Hayward, Ph.D. thesis, Cambridge University, 1987.

86. J. B. Pethica, *Phys. Rev. Lett.* **57**, 3235 (1986).

87. R. J. Colton, S. M. Baker, R. J. Driscoll, M. G. Youngquist, J. D. Baldeschwieler, and W. J. Kaiser, *J. Vac. Sci. Technol.* **A6**, 349 (1988).

88. H. A. Mizes, S. Park, and W. A. Harrison, *Phys. Rev. B* **36**, 4491 (1987).

89. F. F. Abraham and I. P. Batra, *Surf. Sci.* **209**, L125 (1989).

90. U. Landman, W. D. Luedtke, and M. W. Ribarsky, *J. Vac. Sci. Technol.* **A7**, 2829 (1989).

91. Yu. N. Moiseev, V. M. Mostepanenko, V. I. Panov, and I. Yu. Sokolov, *Phys. Lett. A* **132**, 354 (1988).

92. H. J. Mamin, D. Rugar, J. E. Stern, R. E. Fontana Jr., and P. Kasiraj, *Appl. Phys. Lett.* **55**, 318 (1989).

93. Y. Martin, C. C. Williams, and H. K. Wickramasinghe, *J. Appl. Phys.* **61**, 4723 (1987).

94. Y. Martin, D. Rugar, and H. K. Wickramasinghe, *Appl. Phys. Lett.* **52**, 244 (1988).

95. P. Grütter, A. Wadas, E. Meyer, H. Heinzelmann, H.-R. Hidber, and H.-J. Güntherodt, *J. Vac. Sci. Technol.* **A8**, 406 (1990); P. Grütter, T. Jung, H. Heinzelmann, A. Wadas, E. Meyer, H.-R. Hidber, and H.-J. Güntherodt, *J. Appl. Phys.* **67**, 1437 (1990).

96. F. J. Giessibl, *Science* **267**, 69 (1995).

97. *Proceedings of the 1st NC-AFM Workshop, Appl. Surf. Sci.* **137** (1999).

98. D. M. Eigler and E. K. Schweizer, *Nature* **344**, 524 (1990).

99. G. Binnig, H. Rohrer, Ch. Gerber, and E. Weibel, *Appl. Phys. Lett.* **40**, 178 (1982).

100. J. H. Coombs and J. B. Pethica, *IBM J. Res. Develop.* **30**, 455 (1986).

101. J. M. Soler, A. M. Baro, N. García, and H. Rohrer, *Phys. Rev. Lett.* **57**, 444 (1986).

102. S. Ciraci and I. P. Batra, *Phys. Rev. B* **36**, 6194 (1987); N. D. Lang, *Phys. Rev. B* **37**, 10395 (1988).

103. C. R. K. Marrian, ed., *Technology of Proximal Probe Lithography*, SPIE Press, Bellingham, 1993; P. Avouris, ed., *Atomic and Nanometer-Scale Modification of Materials: Fundamentals and Applications*, Kluwer Academic, Dordrecht, 1993; R. Wiesendanger, *Scanning Probe Microscopy and Spectroscopy: Methods and Applications*, Cambridge University Press, 1994.

104. U. Dürig, O. Zuger, and D. W. Pohl, *Phys. Rev. Lett.* **56**, 2045 (1990).

105. U. Dürig and O. Züger, paper presented at the Eleventh International Vacuum Congress, Cologne, Germany, 1989.

11

NEAR-FIELD SCANNING
OPTICAL MICROSCOPY

Daniel A. Higgins and Erwen Mei

11.1 INTRODUCTION

Conventional or "far-field" light microscopy is by far the oldest form of microscopy. The standard methods used to produce magnified optical images of microscopic objects are familiar to all and use lenses, mirrors, and the like.[1,2] Very early in the development of these methods, it was realized that the ultimate spatial resolution that could be achieved was limited by diffraction phenomena. As determined by Abbe,[3] the theoretical resolution limit for an objective lens of numerical aperture, (NA) is approximately half the wavelength λ of light employed:

$$R = \frac{0.61\lambda}{NA} \tag{11.1}$$

Resolution in far-field optical microscopes is, therefore, intimately tied to the wavelength of light used.

To improve optical resolution, Eq. (11.1) suggests that one only need use light of shorter wavelength. However, significant difficulties arise when UV (and higher energy) photons are employed to obtain nanometer-scale resolution. These wavelengths are often strongly absorbed by atmospheric gases; by liquids; and of particular concern, by many of the materials from which conventional optical elements are constructed. As a result, exotic optical materials must be used in some cases. Often, such imaging experiments must also be performed in high vacuum environments. Today, these experiments are indeed routinely performed,[4,5] although samples are restricted to those that are vacuum compatible. Unfortunately, the wavelength of light employed in such experiments is often dictated by the desired spatial resolution rather than by properties of the sample. The desire to achieve high resolution may, therefore, result in a reduction in the amount of spectroscopic information that can be obtained. Hence, the experimentalist must often choose between experiments that provide high resolution and those that yield valuable spectral data.

The desire to overcome the limitations of far-field optics, and hence greatly improve optical resolution, has led to the development of near-field scanning optical microscopy (NSOM). The development of NSOM as a scanned probe microscopy began in the early part of the twentieth century, when it was first proposed by Synge.[6,7] Although Synge did not have the means to actually construct such a microscope, his proposals demonstrated remarkable foresight into the instrumental methods and applications of modern near-field experiments. Synge described how optical imaging could be performed in a manner such that the spatial resolution would no longer be limited by diffraction phenomena. He proposed the use of a tiny light source formed from a subwavelength-size aperture in a conducting metal screen. His proposed microscope is shown in Figure 11.1. With light incident on the backside of the aperture, fields that are laterally confined to a cross-sectional area determined by the aperture dimensions are produced in and near the front side of the hole. Field confinement occurs only within a distance of about one aperture diameter from the hole and is restricted to the region now known as the near field.

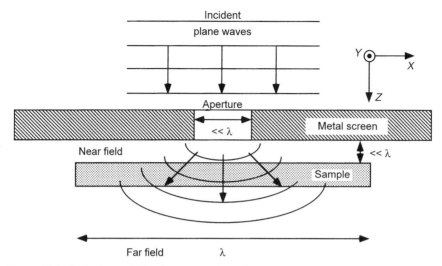

Figure 11.1. A simple model for the aperture-based near-field optical microscope. A subwavelength-size aperture is produced in a perfectly conducting screen of infinite extent and infinitesimal thickness. The screen and aperture are shown in cross section. Plane waves of light are incident on the back side of the screen. The sample is placed close to the aperture at a distance much less than the wavelength of light employed.

By holding the sample in this near-field regime, only the sample region directly beneath the aperture is illuminated. Hence, the optical properties of subwavelength-size sample regions can be directly probed. As proposed by Synge, images may be recorded by simply moving the aperture and sample relative to one another, using a single-element detector to observe the optical signal. Images are recorded one element at a time. The spatial resolution of such a near-field imaging system is limited by the size of the aperture and its distance from the sample, rather than by the wavelength of light employed.

The concept proposed by Synge was independently rediscovered by other scientists numerous times in subsequent years.[8-11] Again, because of experimental difficulties, little progress was made toward the experimental realization of subwavelength-resolved imaging. In 1972, Ash and Nichols[11] conclusively demonstrated subwavelength imaging by near-field methods for the first time. In their studies, an apparatus similar to that originally described by Synge[6] was employed. This microscope operated in the microwave spectral region ($\lambda = 3$ cm), allowing for the use of macroscopic apertures (0.5 mm diameter) and relatively large aperture–sample separations. Resolution of 0.5-mm-wide lines on a metallic grating was achieved using this system.

Because of their much more rigorous dimensional requirements for both aperture size and probe–sample separation, visible light NSOM experiments did not appear in the literature until the early 1980s, paralleling the development of other scanned-probed microscopies.[12] Although its development was distinct from these other methods, NSOM uses similar instrumentation for controlling tip and sample motions. Therefore, the scanned-probe technology developed in this era was readily incorporated into NSOM methods, leading to the explosive growth of near-field microscopy, beginning in the mid-1980s.

Research groups from Cornell University and IBM Zurich independently initiated the modern era of NSOM,[13-18] while others made significant early contributions.[19-21] These groups were the first to construct and demonstrate the operation of near-field microscopes using visible light. A key aspect of this work was the demonstration that subwavelength resolution could be obtained in the visible. Early, both the Cornell and IBM groups realized the benefits of using sharp probes with apertures fabricated at their ends. The Cornell group produced probes from hollow aluminum-coated glass pipettes[17] and the IBM group fabricated probes by chemically etching (in HF) rod-shaped quartz crystals.[15] Such probes allow for samples of moderate topography to be imaged with high spatial resolution.

A particularly important body of literature, which built on the early work, was contributed by the group at AT&T Bell Labs.[22,23] The work from AT&T included the development and demonstration of numerous new methods for improving NSOM imaging and spectroscopy, as well as demonstrating their applications. Examples of their work are referenced throughout this chapter. Of general importance to all NSOM applications were the development of tapered, fiber-optic probes (Fig. 11.2),[22] new feedback methods,[24] and the demonstration that NSOM was sensitive enough to allow for single molecule detection.[25] During the early 1990s a large number of research groups around the world initiated projects devoted to developing new NSOM methods and to implementing these methods in experimental studies.[26-52]

Based on the work of a large number of researchers, modern NSOM instruments now routinely provide ≈ 50 nm spatial resolution using light in the midvisible range. The ultimate practical resolution of contemporary aperture-based NSOM probes has been shown to be about 12 nm,[22] and apertureless NSOM methods provide resolution in the 1 nm range.[47] A particularly important attribute of most modern NSOM instruments is their ability to record high-resolution (<1 nm in the vertical direction) topographic images simultaneously with near-field optical images.

Modern state-of-the-art NSOM instruments are now applied in a broad range of fields, including chemistry, physics, biology, materials science, and engineering. A diverse range of NSOM methods exists as well. Perhaps the most common microscopes are those employing aperture-based probes. A wide variety of apertureless NSOM methods find broad application as well.[46,47,53,54] Myriad methods and optical contrast mechanisms have been employed to obtain optical images and spectroscopic information by NSOM. It is difficult to even mention all of them and impossible to describe them in any detail. The scope of this chapter is, therefore, limited to an overview of the theory, instrumentation, and methods of aperture-based NSOM. In addition, no attempt is made to present a comprehensive review of the methods and applications of aperture-based NSOM. Although the methods and applications not covered are no less important, their detailed description is not possible in a single chapter. The interested reader can find such information in numerous reviews,[16,44,55-64] books,[65,66] and the primary literature.

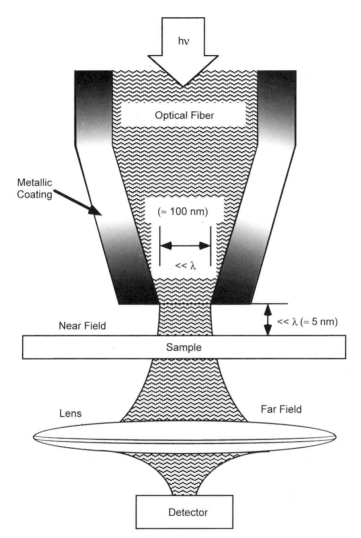

Figure 11.2. The basic elements of a fiber probe based transmission NSOM, not drawn to scale. A tapered, metallized optical fiber is used. The near-field light exits the optical aperture at the end of the probe and is coupled through the sample to an objective lens placed in the far field. The far-field lens sends the light on to an optical detector.

Table 11.1 gives a listing of numerous near-field imaging mechanisms, types of samples studied, and corresponding references.

11.2 THEORY

The theory behind high-resolution NSOM imaging and spectroscopy is evolving rapidly. Valuable theoretical descriptions of model systems have been published

Table 11.1. References for near-field imaging mechanisms

Samples	Transmission	Absorption	Reflection	Luminescence	Polarization	Photovoltage	Infrared/Raman	Apertureless	Plasmon/Tunneling
Molecular films		67	68, 69	17, 25, 34, 35, 42, 52, 70–74	25, 34, 50, 67, 69		75		76, 77
Polymers	24, 27, 50, 78		45	23, 31, 37, 45, 50, 79, 80	50, 81			47	21, 54, 82, 83
Organic composites				25, 28, 30, 42, 72, 84–88	89–91		92–95		
Metallic films	11, 15–19, 22, 23, 48, 50, 79, 96–103		23, 26, 45, 68, 100, 102, 104		22, 23, 26, 81, 105		106	53	107–109
Inorganic semiconductors			40, 68, 110	41, 45, 111–114	112, 113	39, 43, 115–118	110		
Inorganic dielectrics	19, 50, 96, 119, 120		45, 121		50, 122		92, 110, 123	46, 47, 53	20, 21, 54, 83, 92, 124–126
Biologic materials	79, 127, 128	23, 27		36, 51, 80, 128–131					126, 132
Magnetic materials	133			112, 113	23, 81, 112, 113, 134				

and provide an excellent picture of the various important attributes of NSOM experiments.[33, 38, 62, 82, 135–151] Several of these are critically important to understanding how high-resolution information is obtained. Perhaps most obvious is a concrete understanding of the optical field distribution in the near field of the probe. Such theories become quite complex when it is realized that the sample itself alters the fields expected from an isolated aperture. In addition, an understanding of how high-resolution near-field information is transmitted to a far-field detector is also important. These issues are discussed here in detail.

11.2.1   Electric Fields in the Near Field

A complete understanding of the electric fields in the near field of the probe aperture requires that a solution to Maxwell's equations be found, taking into account the detailed geometry of each individual probe and sample. Such exact solutions are difficult to obtain and, in fact, may be useless in the development of a general understanding of NSOM. A much more theoretically tractable model was presented in 1944 by Bethe.[135] Bethe found the fields for the case of a circular hole in a perfectly conducting screen of infinite lateral extent and infinitesimal thickness, as in Synge's[6] model. The coordinate system and model used are presented in Figure 11.1. The screen and aperture are shown in cross-section. In this model, plane waves are assumed to be normally incident on the back side of the screen, polarized along the x-direction. The electric and magnetic fields are required to satisfy certain boundary conditions in and near the aperture. First, the tangential electric field (E_X and E_Y) must vanish on the screen, as must the normal component of the magnetic field (H_Z). In addition, the field must be continuous within the hole. To satisfy these conditions, fictitious magnetic currents and charges resulting from the incident field may be defined in the hole. The fields on the opposite side of the aperture are then determined from the following expressions:[136]

$$\mathbf{E} = \nabla \times \mathbf{F}, \quad \mathbf{H} = -ik\mathbf{F} - \nabla\Psi \tag{11.2}$$

where $\mathbf{F}$ and Ψ are the magnetic vector and scalar potentials, and k is the magnitude of the familiar wave vector $\mathbf{k}$, where $k = 2\pi/\lambda$. From his model, Bethe derived expressions for the fields in and near the hole, as well as at large distances. He also gave the first prediction of the dependence of aperture transmission on aperture size: Throughput should scale as the sixth power of the aperture diameter.[135]

However, in subsequent work by Bouwkamp,[136, 137] it was demonstrated that an error had been made in the original derivation. Although Bethe's solution is correct for fields at a large distance from the hole (in the far field), it gives inaccurate predictions of the fields within the near-field regime. Using the same model, Bouwkamp rederived the expressions for the fields by finding a power series solution in ka (a is the radius of the aperture) to the above equations and boundary conditions. The expressions published provide the correct first-order approximation to the fields. The following expressions give the electric fields in the plane of the aperture

$(z = 0)$:[136]

$$E_X = -\frac{4ik}{3\pi}\left(\frac{2a^2 - x^2 - 2y^2}{\sqrt{a^2 - x^2 - y^2}}\right), \quad E_Y = -\frac{4ik}{3\pi}\left(\frac{xy}{\sqrt{a^2 - x^2 - y^2}}\right), \quad E_Z = 0$$

$$(11.3)$$

Significantly more complicated expressions were also given for E_X, E_Y, and E_Z as a function of distance from the aperture. These expressions were used to calculate the fields at a distance of 5 nm from a 100 nm aperture for x-polarized incident fields; the results are presented in Figure 11.3 as $|E_X|^2$, $|E_Y|^2$, $|E_Z|^2$. As can be seen, the fields remain laterally confined to the region directly beneath the aperture. That is, at this distance, they have not yet spread significantly through diffraction. At greater distances, the fields spread rapidly, demonstrating that field confinement is limited to a distance corresponding approximately to one aperture diameter from the probe (along z).

The extent to which the near-field probe modifies the original polarization of the light is of importance for polarization-dependent work. As expected, for a subwavelength-size metallic aperture, the polarization is significantly altered. The data presented in Figure 11.3 show that both y- and z-polarized components

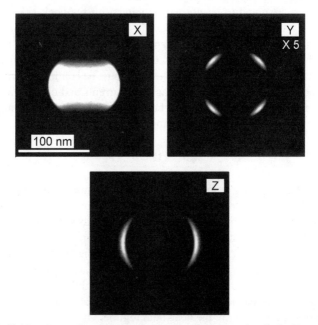

Figure 11.3. Field patterns from an ideal aperture, based on the Bethe-Bouwkamp model shown in Figure 11.1.[135, 136] The brightness scale depicts the squared magnitude of the electric field along x, y, and z directions, with brighter regions corresponding to larger fields. The fields were obtained numerically from equations for the field components given by Bouwkamp. The calculation was performed for 488 nm light, at 5 nm from a 100-nm-diameter aperture in a perfectly conducting metal screen.

are produced from light of initially pure x-polarization. From the data presented in Figure 11.3, the integrated power in the z component is approximately one-third the x component. The y component is much smaller. The magnitudes of these other components vary with the aperture size.

Although the Bethe-Bouwkamp model represents a gross oversimplification of real NSOM experiments, more recent experimental[25] and theoretical[38, 144-146] work has shown that it provides a remarkably accurate view of the near field. However, in extensive theoretical work, Novotny and co-workers[38, 144-146] sought to improve the current understanding. They presented field calculations for model systems that much more closely approximate modern, metallized, fiber-based probes. In this work, the probe was assumed to have a geometry similar to that of a tapered fiber probe. A wedge-shaped probe was employed in two-dimensional (2-D) simulations[38] and a conical probe geometry was used in later three-dimensional (3-D) work.[138, 145, 146] Realistic empirical values for the dielectric properties of the glass and metallic portions of the model probes were employed. The properties of the metal were taken to be those of aluminum. The multiple multipole method was then used to calculate the fields within the probe, at the aperture, and in the near-field regime. The results of one such 2-D simulation are shown in Figure 11.4.[38] The color scale in this figure represents the strength of the electric field. The arrows depict the Poynting vector (defined as the direction and magnitude of electromagnetic energy flow per unit area) at each location. Data for light polarized normal to

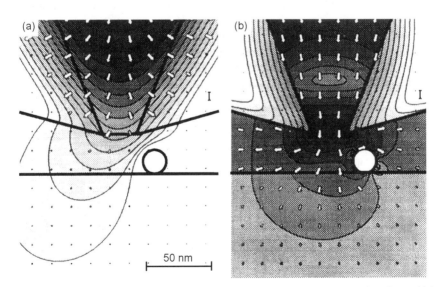

Figure 11.4. Field patterns in a 2-D model of an NSOM probe calculated using the multiple multipole method. The probe is modeled as a glass wedge coated with aluminum. The dielectric properties were selected to be those of glass and aluminum in the visible spectral region. A metallic particle is positioned above the substrate (glass) surface. **a**, The squared magnitude of the field polarized normal to the plane of the page. **b**, The squared magnitude of the field polarized parallel to the page. *Arrows*, the Poynting vector at each location. (Reprinted with permission from Ref. 38.)

the plane of the page and parallel to the page are presented. Most strikingly different from early calculations is the observation that the fields are not strictly confined to the probe aperture dimensions, as predicted by the Bethe-Bouwkamp theory. Real metals have a finite skin depth over which the electric fields decay. Therefore, nonzero field strength is found within the aluminum coating, effectively broadening the size of the aperture by a few nanometers. The skin depth for aluminum at 500 nm is ≈ 13 nm. Subsequent 3-D calculations show similar results.[138, 145, 146] Although all of these results yield fields qualitatively similar to those predicted by the Bethe-Bouwkamp theory, the more realistic nature of the model indicates that the practical limit to NSOM resolution is determined by the optical properties of the metal coating. In the visible spectrum, resolution is, therefore, limited to ≈ 13 nm for an aluminum-coated probe.

11.2.2 Field Perturbation by the Sample

The discussion presented above is not meant to imply that the near fields are determined independently of the sample. In fact, the sample may significantly perturb the fields in and near the probe, leading to possibly complex contrast mechanisms (see below). Again, Novotny and co-workers[38, 138, 145, 146] addressed this issue in detail, using the same 2-D and 3-D models presented above. The effects of small metallic and dielectric particles positioned near the probe and those of smooth dielectric surfaces were studied.

The results of their 2-D calculations are shown in Figure 11.4 for a metallic object of a few nanometers size. In the absence of such a particle, the fields from the aperture are symmetric about the probe axis. As is readily apparent, the particle produces a distinctly nonsymmetrical, polarization-dependent field distribution. The fields polarized parallel to the surface of the particle are driven toward zero, whereas those polarized perpendicular to it are enhanced. Dielectric particles also affect the fields, although in a somewhat different manner.[38] As a result of these effects, the amount of light detected in an imaging experiment varies as the probe and object are moved relative to each other. Metallic objects may appear darker or brighter than the surrounding material, depending on the polarization and materials properties.[38] Dielectric objects show different intensity variations that are also quite complicated. These results are consistent with prior near-field imaging experiments on metallic and dielectric objects.[81] Similar phenomena are observed when a planar dielectric sample is brought into the near-field of the probe.[38, 145] The strength of the fields in the near-field regime are found to increase, as is the extent to which light is coupled to the far field. Again, these effects have been observed experimentally as well.[82] All such phenomena must be taken into account when interpreting contrast in NSOM images.

11.2.3 Resolution and Near-Field to Far-Field Coupling

From the theory described above, it is clear that a subwavelength-scale region of the sample may be illuminated exclusively of all surrounding regions. Hence, if the near-field light can be collected from this region, high-resolution optical information may be obtained. However, the vast majority of near-field light does not propagate through space. This is especially true for the components carrying the highest resolution information. Because it is experimentally impractical to place a detector in the near field, these fields must be converted into propagating waves. The high-resolution information may then be carried to the far-field detector.

The theory of Fourier optics is often employed to understand these concepts. Following the model of Novotny et al.,[143] one can describe the spatial field pattern at some distance z from the probe by $E(x, y, z)$. The spatial field pattern may then be written as the inverse Fourier transform of the field pattern at the aperture, in the frequency domain. The frequency-domain field at the aperture is represented by $E_0(k_x, k_y)$. The spatial field is then given by Eq. (11.4).

$$E(x, y, z) = \frac{1}{2\pi} \int\int_{-\infty}^{+\infty} E_0(k_x, k_y) e^{i(k_x x + k_y y + k_z z)} dk_x dk_y \qquad (11.4)$$

The variables k_x, k_y, and k_z are the components of the wave vector $\mathbf{k}$ and represent the spatial frequencies carried by the light. Again, the wave number k is simply the magnitude of $\mathbf{k}$:

$$k = \frac{2\pi}{\lambda} = \sqrt{k_x^2 + k_y^2 + k_z^2} = \sqrt{k_t^2 + k_z^2} \qquad (11.5)$$

where k_t and k_z are the transverse and longitudinal wave numbers. When $k_t^2 > (2\pi/\lambda)^2$, k_z must be imaginary. In this case, $E(x, y, z)$ decays exponentially in z (i.e., the fields are evanescent).

The values for k_t are determined by the field distribution at the NSOM probe aperture. Using a simplistic model in which the fields are exactly confined to the dimensions of the aperture,

$$k_t = j\left(\frac{\eta\pi}{d}\right) \qquad (11.6)$$

where $j = 1, 2, \ldots, 8$; d is the aperture diameter; and η is the refractive index of the optical material in the aperture. It is now readily apparent that a particular Fourier component of the field will be evanescent when the following inequality holds:

$$j > \frac{2d}{\eta\lambda} \qquad (11.7)$$

From this simplistic model, even the lowest frequency mode ($j = 1$) is predicted to be evanescent for a typical fused silica probe with a 100-nm aperture, using 500 nm light. However, in a real system, the fields are not as rigidly confined. Therefore, some propagating waves are launched from real 100-nm apertures, although a large fraction of the fields are evanescent.

Some of the evanescent fields are converted to propagating modes when a sample is brought near the probe. Conversion occurs primarily via refraction and / or light scattering and is caused by all materials.[55, 66, 139] In aperture-based NSOM methods, it is possible that the fiber probe itself behaves as an important scattering source, converting a significant fraction of the evanescent waves. Hence, the highest resolution information is not lost and may be recorded by simply detecting propagating modes emanating from the probe–sample junction. The conversion of evanescent modes to propagating modes is also the basis for photon tunneling microscopy,[20, 58, 76, 83, 124, 141, 152–155] as well as the many varieties of apertureless NSOM.[46, 47, 53]

As has been shown in theoretical models,[38] the converted, originally evanescent modes propagate away from the probe–sample junction at rather large angles from the optical axis of the microscope (z-direction). These oblique rays carry the highest frequency information. To enhance sensitivity to near-field phenomena and this high-frequency information, these rays must be efficiently collected. Methods for doing so are described in the next section. These effects may place some constraints on the selection of the collection optics.

11.3   INSTRUMENTATION

The primary components of a typical aperture-based, transmitted-light near-field optical microscope are shown in Figure 11.5. The near field probe is the most important component in all near-field microscopes. Therefore, the production, characterization, and other important aspects of near-field probes are presented here in great detail. Because probe–sample distance control is also of great importance to achieving high optical resolution, the methods used for sensing and maintaining probe–sample separation are also discussed at length. The conventional optical elements used to deliver light to the microscope and to collect the light from the probe are described briefly, as are typical optical detectors employed in NSOM experiments. Along with these components, a piezo electric scanning stage is employed to position the sample within the near-field of the probe. This stage is also used to raster scan the sample (in the x and y directions) during image acquisition. The technology employed is identical to that used in other scanned-probe microscopes and is described in detail in earlier chapters. However, it should be realized that the sample, rather than the probe, is scanned in NSOM to avoid the obvious optical alignment problems that arise when the probe is scanned.

11.3.1   Probes for Near-Field Microscopy

The detailed characteristics of the near-field probe play a dominant role in determining spatial resolution and signal levels in all NSOM experiments. In his initial articles, Synge[6] described many of the attributes of a good probe. Since the first NSOM publications, intensive effort has been directed toward understanding probe properties and improving their optical and mechanical characteristics.

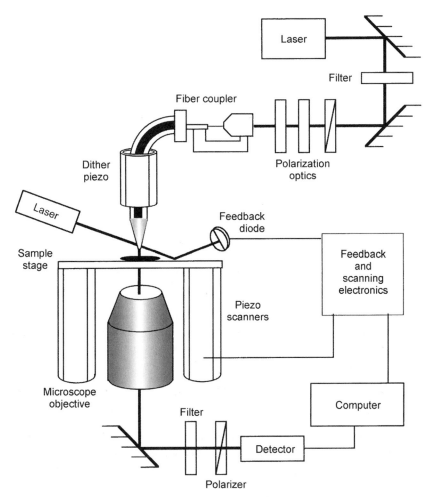

Figure 11.5. Typical instrument for transmitted light NSOM experiments. Two lasers are employed: one for near-field imaging, the other for shear-force feedback (tip–sample distance regulation). Optical components include polarizers and half- and quarter-wave plates, through which the light is sent before coupling into the probe fiber. The sample is placed on a piezo electric stage, which provides x, y, and z motion of the sample. A conventional microscope objective is used to collect near field light coupled to the far field. An analyzer and filters are used to select the polarization and wavelength of light to be detected.

The probes employed in early visible-light NSOM experiments, again, were constructed from glass pipettes and etched quartz crystals.[17, 29, 55] Variations of these probes continue to be employed in certain applications today.[96, 97] However, in the majority of systems, these probes have since been replaced by tapered, aluminum-coated, single-mode optical-fiber probes, originally developed by Betzig et al.[22] Their popularity stems from their simplicity, ease of fabrication (relatively speaking), and the high efficiency with which they deliver light to the aperture. As described below, these probes have been extensively studied and modified over recent years. Because of their ubiquitous use, primary emphasis will be given here to the most common forms of aperture-based fiber-optic probes; however, numerous alternative classes of probes exist. For example, a variety of different probes have been implemented in apertureless near-field microscopy.[46, 47, 53, 54] These probes are similar to conventional AFM and / or STM probes and are used as subwavelength-size light scattering devices.

Fiber Optic Probes   For visible-light NSOM, fiber-optic probes are most often produced from fused-silica-based single mode optical fiber.[32, 49, 66, 125, 156, 157] These probes may be used from the blue to near infrared (IR). In the recent literature, NSOM methods have also been extended into the UV[158] and near-mid IR.[92, 106, 110] These spectral regions are somewhat more difficult to work in and require the use of fiber fabricated from different materials.[92] Alternatively, apertureless methods have been employed in these spectral regions.[110]

The optical fibers used in visible-light imaging are usually composed of a 125-μm-diameter graded-index fiber with a nominal core diameter of 5 μm. The total length of fiber employed is usually around 1 m / tip. The fiber is coated with polymer during manufacturing to improve mechanical strength. This polymer coating must be removed from one end of the fiber before probe fabrication and may be accomplished with a conventional fiber stripper or by simply swelling the polymer with a solvent (i.e., methylene chloride) followed by mechanical stripping by hand.

Probe Tapering   A number of properties determine the utility of a particular probe for high-resolution NSOM imaging.[32, 49, 66, 157] Of most obvious importance is the diameter of the optical aperture. Also of importance is the optical throughput of the probe. As noted earlier, probe throughput is predicted to scale as the sixth power of the aperture radius.[135] However, the probe taper also greatly affects throughput, because the light must first be efficiently delivered to the aperture.[49, 142] For high throughput, the probe diameter must remain greater than the waveguide mode cutoff (i.e., ≈0.4λ) up to a point near to the final aperture.[49, 66, 125, 142] Light no longer propagates through the fiber once it has tapered to such dimensions. Probes with fairly steep final tapers (i.e., between 30° and 70°) provide both the highest resolution and greatest throughput.[138, 142] However, the overall taper shape has also been shown to play a significant role in determining throughput.[49] An optical micrograph of a typical NSOM probe is shown in Figure 11.6. A SEM micrograph of a similar tip is shown in Figure 11.7. Probes with similar shapes generally have good transmission characteristics. In all cases, throughput efficiency is at best 1×10^{-4}.

The overall probe taper also effects the mechanical properties of the probe. These are of particular importance for tip–sample distance regulation (see below). A probe with a lengthy narrow portion well above the final aperture will be mechanically weak and susceptible to damage. As a general rule, the overall length of the tapered region is usually maintained at about 1 mm. Probes with shapes similar

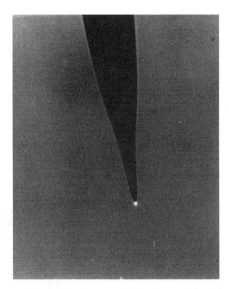

Figure 11.6. Optical micrograph of a tapered, aluminum-coated single-mode optical fiber NSOM probe photographed using a 50X objective with numerical aperture 0.55. Light was coupled into the fiber for this photograph. A single diffraction-limited spot of light is observed at the end of the probe. The probe shape shown here provides good optical throughput and good mechanical properties. The aperture on this probe is nominally 125 nm in diameter.

to that shown in Figure 11.6 have sufficient mechanical strength for use with conventional feedback methods.

The most common method for producing the taper is to use a commercially available laser-based pipette puller. A CO_2 laser is used to rapidly and reproducibly heat the fiber above its melting point.[22] The fiber is then pulled apart by both gravitational and instrument-controlled mechanical pulling cycles. Modern versions of this instrument are microprocessor controlled, allowing for control of a number of heating, timing, and pulling force parameters. Using the numerous parameters available, probes with a wide range of end diameters may be produced.[49] Again, to achieve greater spatial resolution than conventional optical microscopy, the end diameter must be much smaller than half the wavelength of light to be employed. Probe diameters of 50–100 nm are routinely achieved but can be made much smaller.[22] The uncoated probe shown in Figure 11.7 has an end diameter of 125 nm. The ends of probes produced by these methods are usually quite flat as well,[49] as shown in the figure. Probe flatness is also of importance in achieving the highest possible resolution. A highly curved probe face may prevent close approach of the entire aperture to the sample surface. More recently, probe tapers have been produced by other means, such as chemical etching[78, 98, 99, 125, 132, 158–160] and nanophotolithography.[161] Probes produced by these methods have the advantage that the taper angle and end diameter may be more readily controlled than in conventional thermal methods.

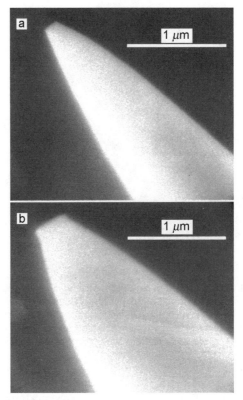

Figure 11.7. a, Scanning electron micrograph of an uncoated tapered fiber-optic probe. The end of the probe is observed to be flat and has a diameter of 125 nm. **b,** Scanning electron micrograph of a probe after coating with ≈100 nm of aluminum.

Metallic Coating To produce a well-defined aperture and to achieve high optical resolution, the tapered fiber probes produced above are coated with a thin metallic film. The coatings also play an important role in confining light within the tapered region of the probe. When the probe diameter decreases to subwavelength dimensions, the light is no longer confined to the fiber. If no coating is employed, a majority of the light simply exits the fiber at this point. Some light still reaches the end, but it is difficult to separate the near-field component of this light from the extremely large far-field background. It has, however, been proposed that uncoated probes may work for near-field imaging,[68, 151] although great care must be exercised in doing so.[100]

The vast majority of probes employed are, therefore, coated with ≈100 nm of metal. These coatings are most often produced by thermal evaporation performed under vacuum.[162, 163] Recent publications have described probe coating-methods and materials and characterization of the resulting films.[49, 70] The procedures used for coating NSOM probes are somewhat involved. The probes must be coated uniformly on all sides, but the end of the probe is left uncoated, forming the aperture. The mechanical apparatus required to hold the probes and rotate them about their axial dimension above the metal source is complex. The probes are often tilted off horizontal by about $30°$ so that a shadow is produced on the end of the probe, preventing metal deposition in this area. Either vacuum-compatible motors, or a drive mechanism coupled to a motor outside the vacuum is used to rotate the probes. The stringent requirements for film quality also make the coating procedure somewhat difficult. Once probe tapering has been mastered, it is often the metallic coating that limits the final optical quality of the probe.

Metal selection is based on a number of parameters, including reflectivity, optical skin depth, and quality of the final film. From an optical perspective, the metals that work best include silver, gold, and aluminum, each with its own advantages. High-quality silver and aluminum films have high reflectivities in the visible spectrum, $\approx95\%$ and $\approx90\%$, respectively, for light near 500 nm.[162] However, aluminum has the smallest skin depth (which limits practical resolution) in the visible range (≈13 nm at 500 nm). The reflectivity of gold films is much lower and is $<70\%$ throughout much of the visible. The large losses in gold occur by absorption in the metallic film; as a result, the probe is heated to a greater extent. As discussed below, probe heating can cause changes in the sample and, in severe circumstances, can lead to destruction of the metal coating on the probe.[111, 164] As a result of the above considerations, aluminum is almost always employed as the coating material in visible-light NSOM. From a chemical and film quality perspective, aluminum may seem like a poor selection. It readily oxidizes, often forming a rough, nonreflective surface. Oxidation that occurs after coating is complete is confined to the outer, less critical surface[162] and does not pose a significant problem. In contrast, silver films continue to oxidize over a much longer time period. In all cases, the coating parameters must be rigorously controlled to prevent the formation of a rough film with large aggregates.

Hollars and Dunn[70] described the important aspects of metal deposition for the production of high-quality coatings on NSOM fiber probes. Figure 11.8 presents aluminum film topographic data, recorded by AFM, showing the quality of films that may be achieved under different conditions. To minimize roughness caused by oxidation and aggregate formation, the rate of aluminum deposition should be as large as possible. Coatings are often produced at a rate of 100 to 300 Å/s. Base pressure in the vacuum system should also be kept below 10^{-5} torr. Coating rates and pressures in this range yield an aluminum deposition rate that is greater than the rate of collisions between metal atoms and other trace gases. Besides oxygen, the presence of trace water was also found to be problematic.

Once the probes have been coated, they must be inspected before use. This is usually performed by observation under optical and electron microscopes (Figs. 11.6 and 11.7). For optical characterization, light is coupled into the probe. A single diffraction-limited spot of light is observed at the very end of a good probe (Fig. 11.6). Rough metallic coatings yield probes with numerous pinholes, which can lead to a significant increase in background signal. Rough coatings, which incorporate protrusions near the end of the probe, may also yield poor optical resolution.

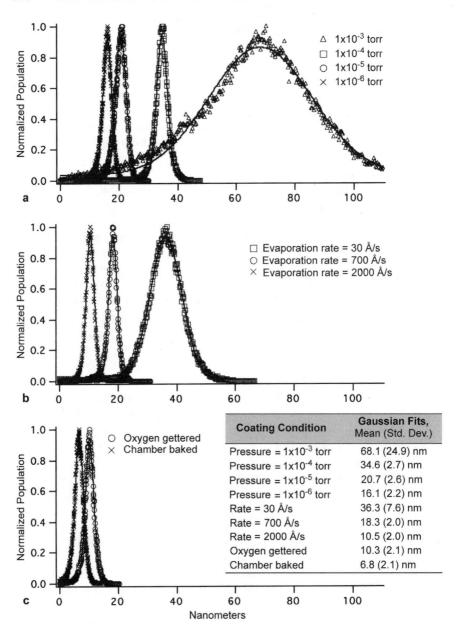

Figure 11.8. Histograms of aluminum coating roughness on NSOM probes measured by AFM as a function of (**a**) base pressure, (**b**) evaporation rate, and (**c**) oxygen gettering and chamber baking. (Reprinted with permission from Ref. 70.)

In this case, the aperture–sample distance may be too large for near-field imaging. Metallic protrusions may also severely distort the optical fields, complicating optical contrast.

Cantilevered or Bent Probes While the majority of NSOM tips currently employed are straight probes, many applications now use bent fiber probes. An optical micrograph of a bent probe is shown in Figure 11.9. These probes have been shown to provide resolution and sensitivity identical to that of straight probes. Indeed, bent probes have been demonstrated to provide sufficient sensitivity for single molecule detection.[165] These probes are used in a manner similar to the cantilevered probes employed in AFM. They have the distinct advantage that tapping mode methodology can be used for sensing the sample surface.[79, 80, 101, 112, 119, 127, 129, 165] Such methods are readily implemented in alternative environments, as has been shown in a number of tapping-mode NSOM studies performed in liquids.[79, 101] These methods also allow for the acquisition of important data on surfaces forces, along the direction normal to the surface.[71]

Although the probes are in intermittent contact with the sample surface in tapping mode NSOM, only weak forces are imparted on the sample. Cantilevered probes have a small force constant in the dimension normal to the surface. Hence, in most cases, both probe and sample survive such contact without modification. In contrast, straight probes must be prevented from contacting the surface because they are extremely rigid along this direction; intermittent contact in this case usually leads to sample and / or probe modification.

Production of cantilevered probes follows the same procedure described above for straight probes. However, a bend must be formed in the fiber after initial tapering. This may be accomplished by using the same laser-based pipette puller to remelt the fiber a short distance above the end of the probe so that it droops. Rotation of these probes during coating requires a slightly more complicated mechanical apparatus.

Figure 11.9. Optical micrograph of a bent, aluminum-coated fiber-optic probe used for tapping mode NSOM. (Reprinted with permission from Ref. 129.)

One possible disadvantage of cantilevered probes is that a large fraction of the light coupled into the fiber is either reflected or transmitted into the surrounding environment from the bend region. Hence, much less light is delivered to the probe aperture. To achieve light levels similar to those obtained with straight probes, the amount of light in the fiber must be increased proportionately. Although this does not necessarily lead to significant probe heating (see below), it may result in significantly more stray light. Care must be taken to prevent this light from reaching the detector. Conventional confocal detection schemes work well in this regard.

Alternative NSOM Probes A wide variety of other probes have also been developed. Some may be classified as aperture-based probes and others are useful in apertureless NSOM methods.[44, 57, 63, 65, 66] Some particularly interesting probes use nanoscale emitters attached to, or in place of, the aperture.[29, 166] A wide array of both optically opaque Si and optically transparent Si_3N_4 probes similar to those used in AFM have also been employed.[46, 76] A particularly interesting Si-based apertureless NSOM probe was fabricated with a photodiode integrated into the cantilever.[54, 126] In other studies, apertureless metallic probes were employed.[53] Again, uncoated fiber-optic probes have been used in a number of NSOM experiments.[68, 167] Although it is difficult to separate the near-field and far-field components when such probes are used in transmitted-light experiments,[100, 151] they are useful in photon tunneling and surface plasmon microscopes, which largely avoid such problems.[168]

11.3.2 Feedback Methods

For high-resolution near-field imaging the probe–sample separation must be maintained at a distance small relative to the aperture diameter throughout image collection. In a typical NSOM experiment, the probe–sample separation is held constant. The sample or tip must move up and down in response to sample topography. However, in experiments in which the sample possesses little topography, constant probe height experiments have been performed as well.[26, 82] Because the electronic circuitry and piezoelectric positioners employed in this process are identical to those used in other scanned probe microscopies, these subjects are not addressed here. However, the methods for sensing tip–sample separation are somewhat specific to NSOM. Recent developments in this area are described in detail.

Early Methods A number of methods for sensing probe–sample proximity have been employed. Initially, it was proposed that optical effects observed as the probe is brought near the sample could be employed.[26] As the tip and sample approach one another, the intensity of light transmitted through (or reflected from) the sample rises and falls owing to far-field interference phenomena.[26, 102] However, this method has significant limitations. Perhaps most important, the optical properties of the sample must be fairly constant, a requirement that severely restricts the types of samples that can be studied.

Electronic tunneling between the tip and sample was also used.[16, 17, 48, 97] Because of the rapid decay of tunneling processes with probe–sample separation, such methods are, of course, highly sensitive. However, they are restricted to use in situations in which the sample is somewhat conductive and cannot be employed on the insulating glass surfaces that find ubiquitous use as substrates in NSOM experiments.

Shear-Force Feedback   Perhaps the most common method of feedback uses so-called shear-forces. The methods involved were developed simultaneously by Betzig and co-workers[24] and Vaez-Iravani's group.[27] As shown in Figure 11.10, a straight NSOM probe is vibrated (dithered) laterally, parallel to the sample surface, at one of the probe's mechanical resonances. A piezoelectric device provides the driving force.[24, 27] Dither amplitude is maintained at or below 10 nm peak to peak. A small amplitude is required, in part to prevent degradation of the optical resolution.

The probe resonance is sensitive to minute changes in any forces acting on the probe. As a result, weak lateral interactions of the probe with the sample cause

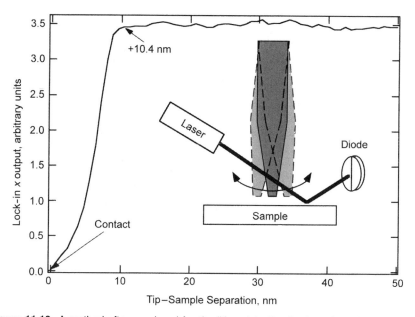

Figure 11.10. A method often employed for tip dither detection in shear force feedback. The probe is dithered laterally 10 nm peak to peak. A laser is focused near the end of the probe. Residual light after probe incidence is reflected from the sample surface and on to a position-sensitive split photodiode. A lock-in amplifier is used to detect the amplitude and phase characteristics of the probe dither as the probe approaches the surface. Plotted is the x output of the lock-in amplifier as a probe is brought from infinite distance to contact with the sample surface. Interaction between the probe and sample extends over a distance of about 10 nm before contact.

changes in the amplitude and phase characteristics of probe motion. These changes are monitored and employed as a measure of probe–surface proximity. Experimental studies performed in a number of environments have demonstrated the general applicability of shear-force methods.[37, 169] However, the actual mechanism by which the tip and sample interact is somewhat ill-defined at present. Gregor et al.[170] suggested that it actually involves probe–sample contact. It is, however, generally accepted that in air, it is the drag induced on the dithering probe by a "viscous" contamination layer that provides surface sensitivity.[171] Figure 11.10 also shows a plot demonstrating that, in air, the interaction distance between probe and sample extends over a distance of about 10 nm.

The properties of the mechanical resonance employed depend on how the probe is mounted. The bare, metal-coated fiber probe is usually cemented into a glass capillary during fabrication and is subsequently mounted to a piezoelectric transducer. Between 1 and 3 mm of the fiber protrudes from the end of the capillary. The mechanical resonances of the probe are primarily determined by this length.[49] A simple physical model of a bar driven at one end and left free to vibrate at the other may be used to predict the resonance frequencies,[172] which are typically in the 10 to 100 KHz range.[49] The quality factor of the probe resonance is particularly important and is given by $Q = f/\Delta f$, where f is the resonance frequency and Δf its width. The quality factor determines how well the entire feedback system responds to changes in sample topography.[173] A large Q is desirable for the purposes of achieving high sensitivity. When Q is >500, weak interaction forces between the probe and sample cause a relatively large change in the probe resonance. However, large Q values also imply a long response time. The amplitude response time τ of the dithering probe is given by:[173, 174]

$$\tau = \frac{Q\sqrt{3}}{\pi f} \tag{11.8}$$

The response time for a probe with $Q = 1000$ and $f = 30$ KHz is 18 ms, too long to provide a reasonable feedback response. Although small values of Q (i.e., $Q \ll 100$) yield rapid responses, they are much less sensitive to minute changes in the tip–sample interaction. The optimum Q is in the intermediate range, typically a little greater than 100 for $f = 30$ KHz.

As noted above, dither amplitude is typically maintained at or below 10 nm peak to peak. This small amplitude minimizes probe–sample interaction forces and hence the potential for sample damage. With larger amplitudes, sample and/or probe damage may occur by tip–sample contact as the probe passes through regions of rapidly changing topography. However, for samples with significant topography, careful optimization of dither amplitude, tip–sample separation, and raster-scanning rate can prevent such problems.[175] For samples with little topography, shear force feedback has been demonstrated to work well, and extremely fragile samples, such as Langmuir-Blodgett monolayers, have been imaged.[72]

Sensitive dither detection is accomplished most often by optical means. A diagram of a typical experimental configuration is shown in Figure 11.10. A laser is focused on the external surface of the aluminum-coated NSOM probe, near its end. For visible-light NSOM experiments, lasers operating to the red of 860 nm are typically employed for feedback purposes. The dithering probe produces a time-dependent shadow in the optical path of the feedback laser light. This shadow is directed onto a photodiode (usually multielement). A lock-in amplifier is used to sense the amplitude and phase characteristics of the probe dither and to produce an error signal for use in the feedback electronics. An example of the signal obtained is shown in Figure 11.10.

In the most common optical dither detection scheme, the feedback laser light must reflect from (or transmit through) the sample surface after striking the NSOM probe.[24] Such methods sometimes suffer from problems caused by changes in sample optical and topographic properties. Alternative methods circumventing this problem have been developed. In one such method, light reflected from two distinct portions of the probe is collected, and the resulting interference effects are observed upon recombination of the two reflections. This method was originally developed by Vaez-Iravani and Toledo-Crow.[103] Recently, similar interferometric detection systems were developed.[176] Although optically complicated, these methods are highly sensitive and avoid interactions of the feedback light with the sample surface.

All of the optical shear-force detection methods described require careful alignment of external optics and are, therefore, somewhat difficult. Further alignment difficulties may arise when imaging under liquids. Researchers have thus developed a number of nonoptical methods for detecting probe motion. In one system, a Wheatstone-Bridge circuit is used to detect changes in the power dissipated (via impedance changes) by the dither piezo as the resonance properties of the probe change owing to probe–sample interaction.[177–179]

One method, described by Karrai and Grober[173] and Muramatsu et al.,[180] uses a piezo-electric quartz tuning fork for shear-force detection. Here, the fiber probe is mounted directly to the tuning fork (Fig. 11.11). A conventional piezo tube is employed to drive probe dither; however, the resonance employed is that of the tuning fork. Therefore, probe dithering is a result of tuning fork motion, rather than a probe resonance. Interactions between the probe and sample surface alter the tuning fork resonance. The feedback signal is acquired electronically by detecting the time-dependent voltage developed across the tuning fork. A similar shear-force detection system was demonstrated that used two piezoplates: one for excitation and one for detection of probe motion.[169]

Tapping Mode   Perhaps newest to NSOM applications is the tapping mode method of feedback. For tapping mode NSOM, the bent, tapered, metallized fiber probes described above are employed.[79, 129] The fiber is mounted to a piezoelectric bimorph for probe modulation normal to the sample surface. Only small amplitude motion is employed so that the sample is within the near field of the probe at all times and the probe makes weak, intermittent contact with the sample surface. Probe or cantilever motion is detected by optical means. A laser beam is reflected from the (approximately) horizontal portion of the fiber (before the bend) to a position-sensitive photodiode. The probe is driven at a mechanical

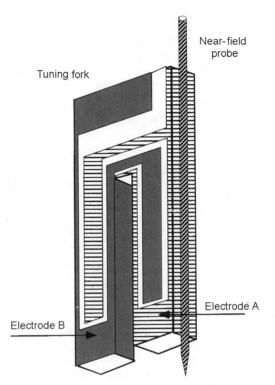

Figure 11.11. NSOM probe mount for nonoptical shear-force feedback, based on the work of Grober and co-workers.[173] The probe is cemented to one leg of a quartz crystal tuning fork. Probe motion is driven at a tuning fork resonance, which is used to sense proximity of the probe to the sample surface.

resonance, and its motion is observed via lock-in detection of the signal from the diode. One advantage of this method is that tip–sample interaction forces may be kept to a minimum, perhaps in the piconewton to nanonewton range,[129] and even controlled, so that they may be used for physical measurements of sample properties.[71]

This method of feedback was demonstrated in a number of experiments performed in different environments.[80, 127, 130, 160] Hollars and Dunn[71] at the University of Kansas used tapping mode NSOM to simultaneously record topographic, optical, and compliance images of Langmuir-Blodgett monolayers prepared from fluorescent phospholipid dyes and dipalmitoylphosphatidylcholine (DPPC). Representative images from this study are shown in Figure 11.12. At the surface pressure under which these monolayers were transferred to the substrate, both liquid-condensed- and solid-condensed-phase regions were formed. The fluorescent dye dopants were found to preferentially position themselves in the liquid-phase regions.[72] This conclusion was proven through the acquisition of the compliance images that showed the regions containing the dye to be much more fluid than the surrounding environment, as expected for liquid regions.

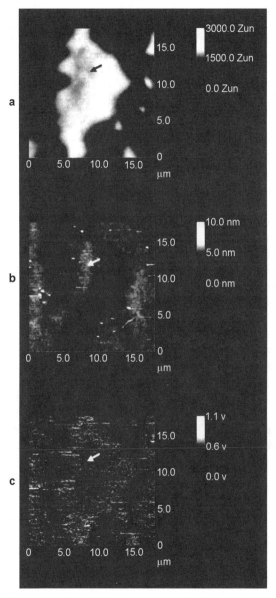

Figure 11.12. (a) Tapping mode NSOM fluorescence, (b) topography, and (c) compliance images of a monolayer prepared from a mixture of phospholipid and a surface-active fluorescent dye. The monolayer was prepared by the Langmuir-Blodgett method and was transferred to the substrate at a surface pressure of 15 mN/m. The large fluorescent region in the center of the images (*arrows*) is a liquid-phase region surrounded by a solid phase region. The dye is soluble in the liquid phase. The topographic image shows the height difference between the two. The compliance image shows that the fluorescent region is less rigid, confirming its assignment as a liquid phase region. (Reprinted with permission from Ref. 71.)

11.3.3 Imaging Modes and Instrument Design

The details of NSOM instrument design depend on the imaging mode employed. A number of such modes exist and are outlined below. For organizational purposes, the optical elements incorporated in modern NSOM microscopes are presented here primarily for transmitted-light, illumination-mode NSOM systems. These same elements are used in the other modes as well, but in different configurations. The changes required for implementation in the other imaging modes are also outlined.

Illumilnation Mode

Transmitted-Light Configuration   NSOM experiments in which the probe is used as the illumination source and light is collected in transmission through the sample are perhaps simplest instrumentally (Fig. 11.5). Light from a laser source is first passed through a number of conventional optical filters and polarization optics and is then coupled into the near-field probe fiber. The probe is mounted so that its axial dimension is normal to the sample surface. A conventional microscope objective, placed on the opposite side of the sample, is used to collect either light coupled from the probe through the sample or light generated by the sample in response to the near-field light (e.g., fluorescence[181] or Raman scattering[75]). This light is then sent through a second set of filters and polarization optics and on to the detector.

Lasers are used almost exclusively as light sources in near-field applications because of their low noise, well-defined optical modes, and ease of manipulation (e.g., focusing and polarization control). Laser sources producing light from the blue to the infrared regions have been used. For near-field imaging, continuous-wave lasers are most often used, whereas pulsed lasers (e.g., Nd:YAG-pumped dye lasers or Ti:sapphire lasers) are employed in most time-resolved experiments (see below). For imaging applications, the sample usually dictates the wavelength of light needed, and hence, laser selection. The pulsed lasers are usually tunable to some extent and may be used with various nonlinear optical methods for generating light of the required wavelength.

Before coupling into the near-field probe, conventional optical filters are employed to remove potential interferences in the source light, especially when gas-phase ion lasers are used. The light is then usually passed through a polarizer, followed by quarter-wave and half-wave retardation plates. Near-field probes often impart ellipticity to light of originally linear polarization. The polarization control elements allow for compensation of such effects. These optics also allow for an arbitrary polarization state to be selected at the NSOM probe aperture by the microscope operator.

Once the light has passed from the probe aperture and has interacted with the sample, the light is collected in the far field. The specific properties of the objective employed for collection purposes are not always of critical importance. Again, resolution is determined by the probe aperture size and its distance from the sample. In simple visible transmitted-light experiments, relatively inexpensive objectives may be used. However, in all cases, the working distance of the objective must obviously be greater than the combined thickness of the sample and substrate. For low light level experiments, high numerical aperture, oil-immersion objectives are best, although the total sample–substrate thickness must then be $<\approx 200\,\mu m$. As noted, the numerical aperture of the objective does play an important role in determining the collection efficiency of the light carrying the highest resolution information. Again, the rays carrying this information propagate at an oblique angle from the optical axis.[38] Efficient collection of these rays is best achieved by using a high numerical aperture, oil-immersion objective in all cases.

After the light is collected, optical filters are used to isolate the desired spectral region for detection. In the case of fluorescence or Raman scattering experiments, notch filters are usually employed to block residual laser light. Finally, a polarization analyzer is used to select the polarization of light to be detected.

Conventional optical detectors are used to record the optical signal. The output of the detector is fed into a computer for image and/or spectroscopic data collection. Conventional uncooled photomultiplier tubes (PMTs) work well for high-light-level imaging applications. Maximum optical powers reaching the detector are usually $\approx 100\,pW$. At this level, the signal-to-noise ratio is typically limited by signal-shot noise. Because PMTs incorporate large-area photocathodes, optical alignment is relatively simple. Reduced sensitivity to optical alignment also minimizes artifacts caused by deviations in the optical path during imaging (i.e., from topographic and/or optical effects). The origins of such artifacts are discussed below. Such large-area detectors, however, provide little background rejection. As noted, NSOM probes sometimes incorporate pinholes well above the near-field aperture. When a large-area detector is employed, a significant fraction of the stray light originating from these pinholes will be detected. A small aperture is often placed in front of the photocathode to minimize sensitivity to stray light. The wide analog signal bandwidths that PMTs provide also make them useful in certain NSOM experiments. Such signals are required in millisecond to microsecond time-resolved methods and in various optical modulation experiments.[67, 89, 105, 122, 164]

For samples producing weak signals (i.e., fluorescence), photon counting methods must often be employed. Although cooled PMTs may be used, silicon avalanche photodiode (SiAPD) detectors are best suited to these applications. The quantum efficiency in the red visible region of the spectrum with SiAPD detectors is far superior to PMTs. These detectors also have extremely small dark count rates (nominally < 100 dark counts/s) and a large dynamic range, allowing for megahertz photon counting rates. One particular advantage is their small photosensitive area (typically $150\,\mu m$ in diameter). By producing a magnified image of the probe aperture on such a small detector, the advantages of confocal detection are achieved, minimizing sensitivity to stray light. However, this system is sensitive to optical alignment, increasing the potential for incorporation of artifacts in the optical images (see below).

Finally, charge coupled device (CCD) detectors, coupled to imaging spectrographs (typically one-eighth to one-third meter focal length), have been used extensively for the recording of fluorescence and Raman spectra by NSOM methods.[28, 75, 181]

Reflected Light Configuration   NSOM experiments on optically opaque samples must be performed in some form of reflected-light geometry. In one particular configuration, the probe aperture is used only as the illumination source and the reflected light is collected by external far field optics. Figure 11.13 shows examples of such an instrument.[104, 182–184] Again, a transmission objective may be employed for light collection; however, because of space limitations, the objective cannot be positioned close to the sample. A long working distance, low numerical aperture objective must be employed, resulting in a greatly reduced collection efficiency. In alternative designs, a reflection objective may be used.[41]

Collection Mode: Reflection and Transmission Configurations   In an alternative optical arrangement for NSOM imaging, similar to the systems shown in Figures 11.5 and 11.13, the position of the light source and the collection optics are swapped.[18] That is, the sample is illuminated by light focused through the objective, and the NSOM probe is used to collect the near-field light. This method is similar to photon scanning tunneling optical microscopy (PSTM),[76, 107] although light collection is not necessarily restricted to evanescent waves. Collection-mode NSOM has also been performed on self-luminous samples (e.g., electroluminescent samples).[185] In such experiments, the external light source and illumination optics may be eliminated from the microscope design.

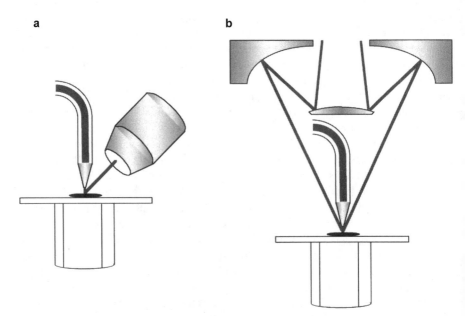

a   **b**

Figure 11.13. Reflection mode NSOM instruments. **a**, A conventional long working distance objective is employed. **b**, A reflective objective is employed, which may greatly increase optical collection efficiency.

Combined Mode Imaging A number of NSOM microscopes have been designed to combine various attributes of some of the above-mentioned imaging modes. These systems provide the advantage that multiple imaging mechanisms may be employed, accessing different contrast mechanisms, and therefore, providing additional optical information. In one particular dual-mode NSOM, the probe is used simultaneously as the illumination source and the collection optic.[39, 45, 84] A bidirectional fiber coupler is used in this microscope to separate the light traveling in opposite directions through the fiber.

Novel Imaging Modes Along with the more well known methods of near-field imaging, several novel methods also exist. For example, a similar optical arrangement to that used for reflected-light NSOM imaging is used in numerous apertureless NSOM systems.[53, 121] In these microscopes, the probe is used solely as a light scattering source; an objective collects the scattered component of the near-field light. Similarly, methods such as photon tunneling microscopy,[20, 58, 76, 83, 124, 152–155] and surface plasmon NSOM[77, 107–109, 150, 154, 186] use various combinations of the configurations described above and provide enhanced contrast at sample surfaces via the scattering and subsequent collection of evanescent waves and plasmons.

11.3.4   Imaging Artifacts and Probe–Sample Interactions

One advantage of NSOM over the other scanned probe microscopies is that it allows for direct optical characterization of materials. The optical contrast mechanisms that are available are all well known from their extensive implementation in conventional far-field optical microscopy and spectroscopy. These contrast mechanisms use transmission, reflection, absorption, emission (fluorescence, phosphorescence, electroluminescence, etc.), light scattering, refractive index, polarization (birefringence, dichroism) and interference phenomena to produce optical images. Both static and dynamic NSOM imaging methods have been developed based on several of these. Most forms of optical spectroscopy may also be coupled with NSOM to obtain highly detailed information on the chemical and physical nature of a sample. Images and spectra recorded by NSOM are, for the most part, readily interpretable based on the well-developed prior knowledge of similar far-field techniques; however, the information obtained may, at times, be influenced by effects not found in far-field methods. Effects specific to near-field optics may lead to erroneous claims of high optical resolution and novel optical contrast mechanisms. Care must be taken to ensure that contrast observed in NSOM images and spectroscopic data is indeed real.

Optical–Topographic Coupling The primary mechanism for artifact generation in near-field optical images is via coupling of optical and topographic information. Pohl and co-workers[82] identified a number of important factors in this regard. For conventional, aperture-based transmitted-light or luminescence NSOM, the most important mechanisms for optical–topographic coupling are those owing to changes in the optical alignment and those owing to optical property variations in the near-field of the probe as topographic features are encountered.

The first effect may arise from slight changes in the path the light takes to the detector as a topographic feature is scanned. Such effects may be dramatic when a confocal detector is employed. Because the sample is scanned rather than the probe, such effects are usually small. However, changes in the focal properties of the collected light may still occur. When a high numerical aperture oil-immersion objective is employed, changes in the sample position (height) lead to slight changes in the optical path taken by the light (specifically, the ratio of oil to sample and substrate in the optical path changes). As has been shown, variations in the optical properties of the sample may also result in variations of the focal properties of the collected light, even in the absence of any topography.[120] A slight change in focus owing to such effects may lead to a noticeable change in the signal from a small area, confocal detector.

The second effect is caused by the dependence of the near-field to far-field coupling efficiency on the optical properties of the material directly beneath the probe. On a uniform sample, such optical variations provide desired optical contrast. However, topographic features may produce artificial contrast by a similar mechanism. Figure 11.14 depicts this type of artifact generation. As the probe scans across the edge of a topographic feature, the properties of the region nearest the tip

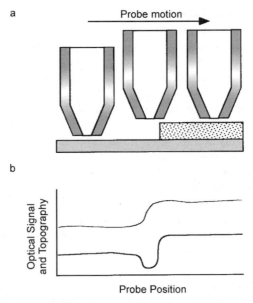

Figure 11.14. Coupling of optical and topographic information in transmitted-light NSOM. **a**, A near-field probe passing over a topographic feature of similar optical properties to that of the surrounding sample. **b**, Sketch of the topographic (*gray*) and optical (*black*) data obtained as the edge of the feature is traversed.

change from those of the sample to those of the ambient environment and back to those of the sample, resulting in variations in the detected optical signal. For other methods, such as photon tunneling microscopy and other forms of apertureless NSOM, similar effects also contribute.[82] In all such cases, the coupling of optical and topographic information leads to strong correlations between the optical and the topographic images.

In all but the most obvious cases, the presence of such artifacts may be difficult to prove. However, because they are extremely difficult to prevent, images of samples that are prone to such effects (i.e., samples with large topographic features) must be carefully analyzed. Pohl and co-workers[82] suggested a number of methods for elucidating the presence of artifacts. Comparison of optical and topographic images recorded simultaneously provides the best check. If the two images appear to be totally uncorrelated, it is likely that purely optical contrast mechanisms dominate. Strong optical–topographic coupling is suggested if strong correlations exist between the two; however, stronger evidence is required to prove causality. The authors[82] proposed that such evidence may be obtained by recording optical images in constant height mode and comparing them to those acquired in constant tip–sample separation mode. They suggested that if strong correlations are not observed between the optical images, optical–topographic coupling is likely a problem in the images recorded with constant tip–sample separation.

Probe–sample Interactions   Aside from mechanical perturbations of the sample, the potential for electronic perturbation of the sample by the metallized NSOM probe is also a concern in certain situations.[31] Such effects were first observed in time-resolved NSOM measurements of single-molecule excited-state lifetimes.[73] Although NSOM represents a highly sensitive method for studying single molecules,[25, 34, 35, 52, 73, 84–87, 181, 187–189] care must be taken in interpreting time-resolved emission data recorded by NSOM.[88] Representative single molecule data are shown in Figure 11.15.[52] From these and other studies,[190, 191] it is now clear that the excited state lifetime measured for a single molecule depends on the exact position of the molecule relative to the metal and glass portions of the probe. The measured lifetime also depends on the orientation of the molecule (dipole) beneath the probe.[52, 190, 191] Although important for samples with nanosecond (or longer) excited-state lifetimes, such effects have been shown to be negligible in other circumstances.[88] The extent to which such effects contribute is easily determined by recording data with the probe alternately positioned in the near and far fields of the sample.[88]

Probe Heating   Thermally induced changes in the probe and sample may occur during NSOM imaging.[192] As addressed in numerous publications,[111, 164, 192, 193] the NSOM probe may be heated to relatively high temperatures via absorption of light by the aluminum coating. Aluminum typically absorbs $\approx 10\%$ of incident light. This becomes a critical issue only near the end of the probe, where the intensity striking the aluminum is greatest and the thermal mass smallest. When high powers are coupled into straight fiber-optic probes (i.e., more than several

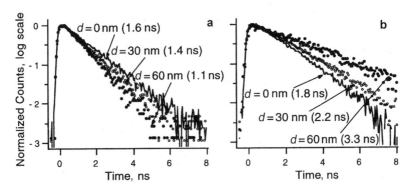

Figure 11.15. Single molecule fluorescence decays recorded as a function of position with respect to the aluminum-coated NSOM probe center. **a,** With a 5-nm tip–sample gap, the lifetime is longest when the molecule is beneath the glass portion of the probe and shortest when it is beneath the metallized portion. **b,** The trend reverses when the tip–sample separation is increased to 20 nm. (Reprinted with permission from Ref. 52.)

milliwatts), the probe itself may be damaged by thermally induced destruction of the aluminum coating.[164] Under more mild conditions, the probe may reach temperatures great enough to alter the sample.

Hallen and co-workers[164] studied the effects of probe heating. The authors point out that probe heating may alter the intensity of light coupled through the probe aperture. Three dominant mechanisms that may lead to an increase or decrease in probe throughput were identified. First, thermal expansion of the probe aperture leads to a highly nonlinear temperature dependence in the optical throughput. Again, Bethe's[135] theory suggests that throughput scales as the sixth power of the aperture diameter. Second, thermal elongation of the end of the probe leads to the attenuation of wave guide modes, thus reducing throughput as the probe temperature increases. Third, the reflectivity of aluminum decreases with increasing temperature as well. The later effect is most problematic because it provides a mechanism for positive temperature feedback, possibly causing a rapid rise in probe temperature.

In the experiments by Hallen et al.,[164] the optical throughput of the probe in the near IR region (1.15 μm) was measured as the power of visible light simultaneously coupled into the probe was modulated. Increases in the IR throughput with increasing visible light intensity were attributed to temperature-dependent aperture size effects. The temperature of the probe was estimated from the results. For one particular probe, it was determined that 0.5 mW of 514-nm light resulted in a temperature rise of 50°C near the aperture. A much higher power of 4.5 mW coupled into the probe led to a temperature rise of 390°C and permanent probe damage. Of course such effects vary significantly with the properties of a given probe.

11.4 EMERGING NSOM METHODS AND APPLICATIONS

Aside from the powerful near-field optical imaging and spectroscopic methods presented above, a number of new methods have been described, including methods for performing NSOM experiments in alternative environments and at low temperatures. Others allow for time-resolved information to be obtained on time scales ranging from minutes to femtoseconds. Still others allow for acquisition of detailed chemical and physical information using diverse spectroscopic techniques. These relatively new methods are described here; descriptions of routine applications of NSOM are available elsewhere.[16, 44, 55–64, 66]

11.4.1 Single Molecule Studies

Static molecular organization and the translational and orientational motions of molecular species play critical roles in governing the properties of thin-film materials. Because these properties likely vary on short distance scales (i.e., in the submicron range), NSOM studies of such phenomena are of particular importance from both fundamental and technological perspectives. Much of the work in this area has used samples composed of well-dispersed single molecules as probes of the local environment. The orientations of single molecules have been determined,[25] and their orientational and translational motions observed.[74, 86] In addition, their fluorescence spectra have been recorded,[84, 181] and their excited state lifetimes measured.[34, 52, 84] Although most single molecule experiments are performed at room temperature, low-temperature NSOM methods have also been developed.[42, 194]

NSOM was not the first method to detect single molecules, but it was the first microscopic method used to spatially locate single molecules by fluorescence imaging. Betzig and Chichester[25] at AT&T Bell Laboratories first demonstrated the utility of NSOM in such studies. The authors observed fluorescent spots in near-field images of polymer films on which dilute (i.e., nanomolar) solutions of dye had been coated. Using several experimental observables, the authors were able to deduce that the fluorescent spots resulted from the excitation and detection of single molecule fluorescence. About 10 nW of light from the near-field probe was used to excite the molecules, resulting in thousands of photons per second of fluorescence from each, against a background typically 10 times smaller.

These initial single molecule experiments unequivocally demonstrated the sensitivity of NSOM methods for characterization of weakly fluorescent materials. They also point to the fact that NSOM sensitivity does not result from a greater signal level than is expected in far-field measurements. Rather, it is the reduction in background that provides this sensitivity. The extremely small volumes in which excitation occurs and from which the signal is collected effectively prevents the detection of fluorescence and scattered light from surrounding regions. As a result, background levels are typically <1000 counts/s.

Intriguing aspects of this initial work include the observation of distinct polarization-dependent "shapes" for the individual fluorescent spots. This effect arises from the fact that the absorption and emission processes involve a single transition dipole. As shown in Figure 11.3, the electric fields at the end of the NSOM probe are distributed in distinctive patterns for the various polarization components. Each molecule absorbs light polarized parallel to its transition dipole, effectively sampling a combination of the various components. The shapes depend on the molecular orientation and reflect these field distributions. Betzig and Chichester[25] showed that these shapes, taken with their polarization dependence, could be used to determine the orientation of individual molecules.

Since this work, numerous single molecule studies have been published. Single molecule fluorescence spectra have yielded information on molecular-scale inhomogeneities in thin film materials.[181] As noted, NSOM fluorescence lifetime measurements are somewhat complicated.[52, 73, 195] More recently, it was discovered that far-field confocal microscopic methods may provide advantages in this regard and may be superior for single molecule studies.[84] The advantages of greater spatial resolution and the simultaneous acquisition of film morphologic data often outweigh the difficulties encountered with NSOM methods, making such methods valuable for dynamics studies in a variety of mesostructured materials.[88]

In two single molecule studies, Bopp and co-workers[85] and Ruiter and co-workers[74] have used NSOM to record diffusional trajectories (i.e., Brownian motion and/or probe-induced motion) for individual molecules (Fig. 11.16). The same area of a glass sample onto which rhodamine 6G had been coated was imaged repeatedly by Ruiter et al.[74] Differences in the position of each fluorescent spot from one image to the next indicates that the molecules were moving. The map shown in

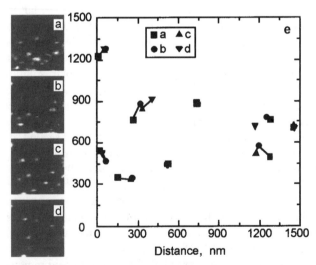

Figure 11.16. a–d, Consecutive fluorescence images of a glass substrate surface coated with well-dispersed rhodamine 6G dye molecules. **e**, Single molecule motion (translational trajectories) determined from these images. The images were corrected for drift. Each image required 10 min to record. (Reprinted with permission from Ref. 74.)

Figure 11.16 depicts the motions of each molecule. By calculating the mean square displacement for each molecule in time, the authors were able to determine single molecule diffusion coefficients. Although powerful, such methods allow for only relatively slow molecular motions to be observed, owing to the long time required for image collection ($\approx$10 min).

In the same work, Ruiter et al.[74] used polarization-dependent NSOM methods to observe the orientational motions of single molecules. Two detectors were employed in these studies, each detecting fluorescence polarized along orthogonal directions. In a series of images, certain molecules appeared to slowly rotated (on a minute time scale), as evidenced by slow variations in the polarization of emission. Others were found to rotate on a much faster time scale, and still others were found to remain fixed on the sample surface.

11.4.2 Polarization-Modulation Methods

Polarization-dependent methods are now routinely employed in NSOM experiments and new methods are constantly being developed. A subset of these new methods uses rapid modulation of the polarization of light from the probe to provide for the detailed characterization of a variety of sample optical properties. The first microscopic polarization-modulation methods were incorporated in far-field microscopes.[196–198] Similar NSOM methods have since been described by several different research groups.[67, 105, 122, 134]

In the work by Higgins et al.,[67] polarization modulation was performed by passing linearly polarized light through an electro-optic (EO) modulator and quarter wave plate combination before coupling the light into the near-field probe fiber. A time-varying potential was applied across the EO modulator, producing a time-varying ellipticity in the polarization state of the light. The quarter wave plate was used to convert the polarization to a time-varying linear polarization (i.e., the polarization vector rotates) from the NSOM probe. A sawtooth waveform was used to drive the EO modulator so that the polarization angle varied linearly through 180° in time. No output polarizer was used. The near-field light coupled through the sample was then monitored using a PMT. The output of the PMT was fed into the input of a lock-in amplifier for detection of the modulated optical signal produced by dichroism and/or birefringence effects in the sample.

This method was used to study micron-size single crystals of a rhodamine 110[67] (Fig. 11.17). With the NSOM probe positioned over an optically isotropic region of the sample, no signal modulation was observed, after removal of background caused by the polarization-dependent transmission of the aperture and microscope optics.[122] With an optically anisotropic sample region beneath the probe, a strong polarization-dependent modulation in the optical signal was observed. Signal modulation occurred in both resonant and nonresonant experiments. For the resonant case, the experiments are similar to linear dichroism experiments performed using bulk spectroscopic methods. The modulated signal results from polarization-dependent absorption in the sample. The modulation amplitude provides information on the absorption strength. The phase angle (with respect to the field applied to

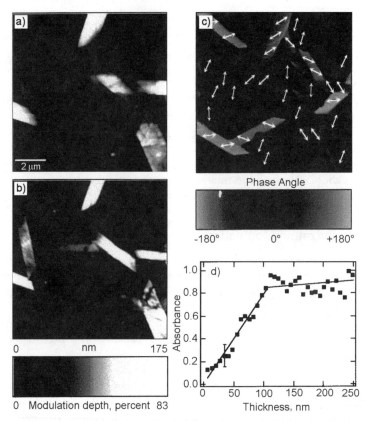

Figure 11.17. Images of rhodamine 110 dye crystals recorded by polarization-modulation NSOM. The three images shown correspond to (**a**) topography, (**b**) amplitude of polarization-dependent intensity modulation, and (**c**) phase of the intensity modulation. Contrast in the optical images is owing to dichroism; 514-nm light was employed in these experiments and is strongly absorbed by the crystals. The intensity scale in part **b** provides the magnitude of absorption for each crystal. (**d**) A plot of absorbance as a function of crystal thickness showing Beer's law behavior for thin crystals. The phase data provide the direction of polarization most strongly absorbed. *Arrows*, appended to depict this polarization. (Reprinted with permission from Ref. 67.)

the EO modulator) provides the direction of polarization most strongly absorbed and hence the orientation of the chromophores. In the off-resonance case, the birefringence of the sample leads to a modulation in the transmitted light intensity as well. This effect arises from the dependence of the near-field to far-field coupling efficiency on the polarization-dependent dielectric constant (optical) of the material beneath the probe. Here, the phase angle provides the direction of the optically fast and slow axes.

Variations of the above method have been described in both early work by Vaez-Iravani and Toledo-Crow[105] and later studies by McDonald et al.[122] and Lacoste et al.[134] Vaez-Iravani's group used the EO modulator to produce light of sinusoidally varying (in time) ellipticity. An output polarizer was employed so that a single polarization state of the light coupled to the far field could be monitored. Detailed equations describing the signal from the detector at both the first and the second harmonics were used to distinguish between birefringence and transmitivity effects. The method was demonstrated for chrome patterns produced on quartz substrates.

The method developed by Hsu's group[122] used a photoelastic modulator and quarter wave plate to produce linearly polarized light of time-dependent orientation. The polarization angle was modulated sinusoidally in time in these experiments. In studies of nonabsorbing birefringent materials, optical anisotropy in the sample plane created a time-varying ellipticity in the polarization state of transmitted light. A circular analyzer placed before the detector converted this to a time-varying intensity. The resulting polarization-dependent optical signal was made up of a series of harmonics and a dc component. By measuring the fundamental and second harmonic components using two lock-in amplifiers, images of the local retardance were obtained. By mathematical combination of the signals, the magnitude of the birefringence (i.e., the difference between the ordinary and extraordinary refractive indices) and the orientation of the optical axis were determined. This method was used to observe strain-induced birefringence in otherwise isotropic $SrTiO_3$ crystals.

11.4.3 Time-Resolved NSOM

Ultrafast Methods The possibility for studying dynamics on nanometer-length scales with femtosecond to picosecond time resolution will lead to a much better understanding of thin-film materials. Much of the time-resolved NSOM work published to date has used time-correlated, single-photon counting (TCSPC) methods to measure excited-state lifetimes in local sample regions.[31, 34, 73, 84, 88] These NSOM experiments simultaneously provide picosecond time resolution and better than 100-nm spatial resolution. For such studies, either femtosecond or picosecond pulses from a cavity-dumped or pulse-picked mode-locked Ti:sapphire or Nd:YAG-pumped dye laser are coupled directly into the probe fiber. The length of the fiber employed is important only for the shortest femtosecond pulses, when dispersion becomes an issue. Tapered, metallized fiber probes and optics identical to those described above are used. However, the optical detector and signal processing electronics are much different from those used for conventional NSOM. The detector employed is usually a microchannel-plate PMT,[31, 88] although short-pulse SiAPDs have been used as well.[34] The most important parameters are the rise time of the detector signal and its timing reproducibility (timing jitter). The output signal from the detector is then amplified and passed through a discriminator, the output of which is used as the start pulse in a time-to-amplitude converter (TAC). A fast photodiode (nanosec rise time) is used to directly monitor the output pulses from the laser. The signal from the photodiode is passed through a second discriminator. The output of this discriminator is used as the stop pulse for the TAC. Data from the TAC is then used to construct a histogram of the time delays between excitation and single-photon emission events, thus yielding the excited-state lifetime. Such methods have been used to determine excited-state lifetimes for aggregates,[88] polymers,[31] molecular semiconductors,[199] and single molecules.[34, 73, 84]

Ultrafast pump-probe studies have also been performed in near-field microscopes. The basic experimental principle is similar to that employed in far-field pump-probe spectroscopy. The output of a mode-locked Ti:sapphire laser is usually used and is split into two separate beams (e.g., 50:50), one of which may be variably delayed. The two beams are then recombined and coupled into the fiber, yielding variably delayed pump and probe pulses at the near-field aperture. External (to the NSOM fiber) pump and probe geometries may also be employed.[104] A femtosecond-resolved signal is obtained from the sample, characteristic of the desired time-dependent phenomena, via any of a number of nonlinear processes. Methods used include transient absorption and luminescence autocorrelation.[112, 113] Because the pump and probe are often at the same wavelength and because the signal from each follows an identical path to the detector, a large background, independent of the pump–probe delay, is often present. As a result, the two beams are usually modulated in the kilohertz range with optical choppers. The signal is then measured by lock-in detection at the sum or difference frequency of the two. For the shortest pulses (i.e., <100 fsec), dispersion in the fiber may cause significant temporal pulse broadening. Dispersion precompensation must therefore be employed to maintain a short pulse width at the probe aperture.[40]

Ultrafast NSOM was used to study magnetic heterostructures,[112, 113] and semiconductors.[200] In addition, strain-induced changes in sample reflectivity were studied in patterned gold nanostructures.[104]

Other Time-Resolved Methods The influence of an externally applied perturbation (i.e., an electric field) on the local properties of thin film materials may also be studied by newly developed NSOM methods. For example, molecules within thin films may change their orientation under the influence of an electric field. Such effects are indeed the basis for liquid crystal flat-panel display technology. In these systems, the dynamics of molecular reorientation are important for their technological applications, yet may vary dramatically on nanometer-length scales within individual samples owing to the presence of film defects and other features.

A new NSOM method for the study of such phenomena was developed by Higgins's group.[89, 90] In this method, an electric field (either dc or ac) is applied across the sample by applying a small voltage between the metallized NSOM fiber probe and a conductive, optically transparent substrate, on which the sample is supported. As has been shown,[89] with application of only a few volts between tip and sample, a highly concentrated field is produced directly beneath the probe, limiting the sample region that is perturbed by the field. The molecular response is then detected by near-field optical methods. The dynamic signal may be recorded in either the frequency or the time domain. For frequency-domain studies, a lock-in amplifier is used for phase-sensitive detection of the optical signal resulting from application of a sinusoidally modulated electric field. In time-domain studies, the signal is simply recorded after a rapid change in the applied field strength.

The samples studied to date are polymer-dispersed liquid crystal films. These films are made up of micron-size droplets of birefringent nematic liquid crystal dispersed in a polymer matrix. To understand the dynamic response in these materials, a detailed picture of the static (i.e., relaxed) molecular orientation must first be obtained. Static liquid crystal organization is determined by imaging the samples under crossed-polarization, transmitted-light conditions, in the absence of an applied field[91] (Fig. 11.18).

Upon application of an electric field, the liquid crystal beneath the probe reorients and the optical signal level changes. The optical response observed is due solely to a change in the local birefringence properties in the sample plane owing to molecular reorientation. Dynamics images are recorded by modulating the applied electric field with a sinusoidal waveform. A lock-in amplifier is used to measure both the amplitude and the phase of the modulated signal. These signals are used in the recording of images, as shown in Figure 11.18.[89] The amplitude images provide information of the trajectories through which the molecules reorient, and the phase images provide information on how the rate of reorientation varies between individual sample regions. By slowly switching the field on and off using a square waveform, the rate of molecular reorientation may be directly determined.[90] The signal rise and decay are fit (approximately) with an exponential function to measure both the rise time (after application of an electric field) and the relaxation time (after field removal).

The reorientation of liquid crystal molecules by mechanical and/or optical perturbation of the sample was reported by Moyer et al.[69] The authors observed changes in the optical properties of a thin film of smectic liquid crystal resulting from prior NSOM imaging. The changes in optical properties were attributed to reorientation of the liquid crystal during imaging, although no time-resolved information was obtained.

11.4.4 Low-Temperature NSOM

The ability to perform NSOM experiments at cryogenic temperatures has many benefits. Luminescence intensities may be enhanced by the elimination of nonradiative decay mechanisms. Spectral lines may be significantly narrowed as well. However, such experiments are quite difficult to perform because of the need to operate the microscope inside a cryostat. Grober et al.[41] designed such an instrument. The microscope itself is quite similar to ambient-temperature near-field microscopes. The same optical and mechanical elements are employed. Shear force feedback is used to maintain tip–sample separation. However, all of this is performed with the microscope submersed in either cryogenic gas or liquid. This type of microscope may be operated in either illumination or collection modes. A Schwartzchild (reflection) objective is used to collect light reflected from the sample surface in the illumination mode (Fig. 11.13**b**).

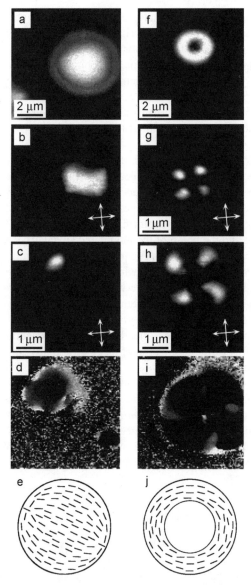

Figure 11.18. Topographic, static, and dynamic optical images of nematic liquid crystal droplets embedded in a polymer film. A mixture of cyanobiphenyls was employed, the polymer is poly(-vinyl alcohol). **a and f**, Topographic images. **b and g**, Transmitted-light optical images recorded under crossed polarization conditions. **c and h**, Dynamic optical images depicting liquid crystal reorientation under the influence of an electric field. The sinusoidally modulated field was applied between the aluminum-coated near-field probe and the substrate. The optical data were recorded by measuring the resulting amplitude of the transmitted-light signal modulation with a lock-in amplifier. Signal modulation is the result of liquid crystal reorientation. **d and i**, Dynamic optical images obtained by recording the phase angle of the modulated transmitted-light signal. **e and j**, Models for the liquid crystal organization of the droplets shown in each column of figures.

Operation of this microscope was demonstrated in studies of GaAs and AlGaAs quantum wire structures.[114] In more recent work on transparent samples, a conventional microscope objective was employed in a cryogenic NSOM of similar design.[194] This microscope was used to study CdS nanocrystals. Cryogenic near-field microscopes similar to those described here have also been used in single molecule studies.[42, 194]

11.4.5   Near-Field Raman and Infrared Microscopy

New NSOM methods based on Raman and IR spectroscopies were recently developed. Vibrational spectroscopic methods provide much greater chemical selectivity than room-temperature fluorescence methods. The fact that only a few research groups have developed such systems indicates the experimental difficulties involved. However, the advantages gained by the ability to characterize molecular order, organization, interactions, and a variety of other properties with nanometer-scale resolution and vibrational-mode specificity is driving significant effort in this area.

Although near-field Raman instrumentation is similar to that used in fluorescence experiments (see above), the signal levels for unenhanced Raman scattering are several orders of magnitude smaller than those obtained in fluorescence experiments. Near-field Raman imaging is thus made difficult by the need to integrate the signal for long periods of time for each image pixel. This process is made difficult by the inherent drift of piezo-electric devices. As a result, near-field Raman spectra are frequently reported without Raman images. In most near-field Raman experiments, some form of enhancement of the Raman signal must be used. Near-field Raman spectra reported previously include those of organic molecules deposited on silver substrates, in which both resonance- and surface-enhancement effects aided in improving signal levels.[75, 93, 95]

Despite the associated difficulties, Raman imaging by NSOM methods was described Jahncke and co-workers[123] at North Carolina State University. In these experiments, scattered light from the near-field probe (514 nm) was collected in both transmitted-light and backscattered geometries. Raman scattering at a specific Stokes shift was detected by passing the light coupled from the near field through a double monochromator and onto a PMT. The samples studied were Rb-doped KTiOPO$_4$. Rb-doped and undoped regions of the material were readily distinguished.

Partly because the small signal levels obtained in near-field Raman experiments and partly because of the desire to obtain complimentary spectroscopic information, IR NSOM is now being developed. In addition, IR methods provide a significant advantage over conventional far-field methods in terms of spatial resolution. Whereas visible-light NSOM provides resolution on the order of $\lambda/10$, IR NSOM potentially yields resolution closer to $\lambda/100$ and smaller.

IR near-field microscopes incorporate available infrared light sources, optics, and detectors. Instrument designs are similar to those of visible NSOM systems. However, difficulties do exist. First, broad-band, tunable IR lasers are not common.

However, free electron lasers are available in certain locations and have been used as light sources in IR NSOM experiments.[92] In alternative systems, conventional light sources were employed.[92] Single-line lasers, such as the CO_2 laser, have also been used.[110] Furthermore, conventional fused silica fibers do not work well in much of the IR region, because of strong Si-O and Si-OH absorption. Researchers are working to overcome the problems of fiber transparency by constructing fibers from other materials. New fibers based on chalcogenide glasses may be excellent alternatives.[201, 202] Fluoride glass fibers may be employed as well.[92] As a result of the present material limitations, apertureless IR NSOM has proven powerful for avoiding such difficulties altogether.[110] Finally, the selection of IR detectors is somewhat limited to those that provide sufficient sensitivity. Typically HgCdTe detectors are used for near-mid IR applications.[110]

11.4.6 Near-Field Photocurrent and Photovoltage Measurements

Of particular importance in many semiconductor-based EO devices is the generation of charges via optical absorption. Such processes also provide a unique view of the local physical properties of a particular semiconducting material. With the development of nanostructured and mesostructured devices, measurements of such phenomena must be performed with nanometer-scale resolution and have been performed by STM. Recently, a number of new NSOM methods were developed to obtain complementary information in similar experiments.

Photovoltage measurements may be made using the near-field aperture as the optical source. The signal of interest in such experiments arises from the photo-induced separation of charges in the sample, rather than from detection of the reflected or transmitted light. Hsu's group[43, 115] described a method for measuring such signals in detail. They used this method to study charge-carrier effects near threading dislocations in compositionally graded Ge_xSi_{1-x}. In these studies, light from an argon ion laser was coupled into the fiber-optic probe. The illumination intensity was modulated using an optical chopper. During raster scanning of the sample, the photovoltage produced across the sample was recorded via lock-in detection at the optical modulation frequency. These same methods have since been employed in studies of $Cu(In,Ga)Se_2$ solar cells.[116] Figure 11.19 shows that small particles tend to cluster together to form larger grains in such a sample. The boundaries between grains show a reduced photovoltage, possibly owing to reduced charge generation efficiency and/or migration rates, or enhanced charge recombination rates.

Of course, photocurrent measurements may be made as well. Buratto et al.[39] demonstrated this method in detailed studies of carrier-transport phenomena in InGaAsP quantum well lasers. In these studies, NSOM photocurrent maps were used to characterize the p-n junctions in these devices. The results allowed for variations in the quality of the p-n junctions to be observed. The variations were attributed to the presence of defects and growth inhomogeneities in the various

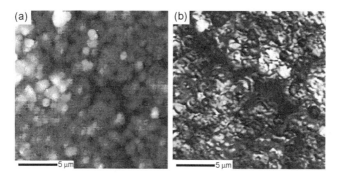

Figure 11.19. Simultaneously recorded (**a**) topographic and (**b**) near-field photovoltage images. The sample is a Cu(In,Ga)Se$_2$ thin film. Topography in part **a** spans 1.8 μm and the photovoltage in part **b** varies by 113 nV. Contrast in part **b** may be owing to variations in the rate of charge-carrier recombination. (Reprinted with permission from Ref. 116.)

regions. Since this initial work, NSOM photocurrent imaging has found widespread application in the characterization of a number of semiconductor systems.[117, 118]

11.5 CONCLUSIONS

The driving force behind the rapid advance of NSOM as a scanned probe microscopy is the need to better understand the optical properties of mesostructured materials on a length scale not attainable by conventional optical methods. As a result, the wide array of near-field methods presented here have been developed. Both static and dynamic optical spectroscopic characteristics of materials may now be directly studied on the sub-100-nm length scale. All such optical and spectroscopic information is obtained simultaneously with film topographic data. The coupling of this information provides a powerful means to address a number of materials issues. As demonstrated in the array of emerging methods, new types of information are now being obtained by coupling other methods into NSOM experiments. In the future, applications of NSOM in other spectral regions will further enhance the amount of information obtained.

The already broad array of NSOM imaging methods and applications will continue to grow. A large number of research groups throughout the world are currently working on many new NSOM methods and applications, spanning a number of scientific and engineering disciplines. This chapter has covered only a few of the current and emerging methods of near-field microscopy. The primary focus was on thin-film materials characterization by a limited number of NSOM techniques (e.g., aperture-based NSOM). Extensive literature exists in subject areas that did not fit into the scope of this chapter. For example, the numerous and rapidly expanding biologic applications[128, 131] of NSOM were not covered. In addition, there are many applications of NSOM in optical data storage.[133] The wide array of NSOM applications found in the literature indicates its general utility. This field will continue to expand, yielding a better understanding of a variety of materials and aiding in the development of new technologies.

ACKNOWLEDGMENTS

The authors thank the National Science Foundation for support of this work through the CAREER Award Program. The members of the Higgins group are acknowledged for their helpful comments and aid in preparing this manuscript.

REFERENCES

1. L. C. Martin, *The Theory of the Microscope*, Elsevier, New York, 1966.
2. T. Wilson and C. Sheppard, *Theory and Practice of Scanning Optical Microscopy*, Academic Press, Orlando, Fla., 1984.
3. E. Abbe, *Archiv. Mikros. Anat.* **9**, 413 (1873).
4. D. F. Parsons, ed., *Short Wavelength Microscopy*, New York Academy of Sciences, New York, 1978.
5. D. Sayre, ed., *.X-Ray Microscopy II: Proceedings of the Second International Symposium*, Springer-Verlag, New York, 1988.
6. E. H. Synge, *Phil. Mag.* **6**, 356 (1928).
7. E. H. Synge, *Phil. Mag.* **13**, 297 (1932).
8. H. H. Pattee, *J. Opt. Soc. Am.* **43**, 61 (1953).
9. J. A. O'Keefe, *J. Opt. Soc. Am.* **46**, 359 (1956).
10. A. V. Baez, *J. Opt. Soc. Am.* **46**, 901 (1956).
11. E. A. Ash and G. Nicholls, *Nature* **237**, 510 (1972).
12. G. Binnig, H. Rohrer, C. Gerber, and E. Weibel, *Phys. Rev. Lett.* **49**, 57 (1982).
13. A. Lewis, M. Isaacson, A. Muray, and A. Harootunian, *Biophys. J.* **41**, 405a (1983).
14. A. Lewis, M. Isaacson, A. Harootunian, and A. Muray, *Ultramicroscopy* **13**, 227 (1984).
15. D. W. Pohl, W. Denk, and M. Lanz, *Appl. Phys. Lett.* **44**, 651 (1984).
16. U. Durig, D. W. Pohl, and F. Rohner, *J. Appl. Phys.* **59**, 3318 (1986).
17. A. Harootunian, E. Betzig, M. Isaacson, and A. Lewis, *Appl. Phys. Lett.* **49**, 674 (1986).
18. E. Betzig, M. Isaacson, and A. Lewis, *Appl. Phys. Lett.* **51**, 2088 (1987).
19. U. C. Fischer, *J. Vac. Sci. Technol. B* **3**, 386 (1985).
20. D. Courjon, K. Sarayeddine, and M. Spajer, *Opt. Commun.* **71**, 23 (1989).
21. D. Courjon, J.-M. Vigoureux, M. Spajer, K. Sarayeddine, and S. Leblanc, *Appl. Opt.* **29**, 3734 (1990).
22. E. Betzig, J. K. Trautman, T. D. Harris, J. S. Weiner, and R. L. Kostelak, *Science* **251**, 1468 (1991).
23. E. Betzig and J. K. Trautman, *Science* **257**, 189 (1992).
24. E. Betzig, P. L. Finn, and J. S. Weiner, *Appl. Phys. Lett.* **60**, 2484 (1992).
25. E. Betzig and R. J. Chichester, *Science* **262**, 1422 (1993).

26. J. A. Cline, H. Barshatzky, and M. Isaacson, *Ultramicroscopy* **38**, 299 (1991).

27. R. Toledo-Crow, P. C. Yang, Y. Chen, and M. Vaez-Iravani, *Appl. Phys. Lett.* **60**, 2957 (1992).

28. D. A. Higgins and P. F. Barbara, *J. Phys. Chem.* **99**, 3 (1995).

29. K. Lieberman, S. Harush, A. Lewis, and R. Kopelman, *Science* **247**, 59 (1990).

30. D. Birnbaum, S.-K. Kook, and R. Kopelman, *J. Phys. Chem.* **97**, 3091 (1993).

31. D. A. M. Smith, S. A. Williams, R. D. Miller, and R. M. Hochstrasser, *J. Fluor.* **4**, 137 (1994).

32. B. I. Yakobson, P. J. Moyer, and M. A. Paesler, *J. Appl. Phys.* **73**, 7984 (1993).

33. O. Keller, M. Xiao, and S. Bozhevolnyi, *Surf. Sci.* **280**, 217 (1993).

34. W. P. Ambrose, P. M. Goodwin, J. C. Martin, and R. A. Keller, *Science* **265**, 364 (1994).

35. W. P. Ambrose, P. M. Goodwin, J. C. Martin, and R. A. Keller, *Phys. Rev. Lett.* **72**, 160 (1994).

36. R. C. Dunn, G. R. Holtom, L. Mets, and X. S. Xie, *J. Phys. Chem.* **98**, 3094 (1994).

37. M. H. P. Moers, H. E. Gaub, and N. F. van Hulst, *Langmuir* **10**, 2774 (1994).

38. L. Novotny, D. W. Pohl, and P. Regli, *J. Opt. Soc. Am. A* **11**, 1768 (1994).

39. S. K. Buratto, J. W. P. Hsu, E. Betzig, et al., *Appl. Phys. Lett.* **65**, 2654 (1994).

40. J. B. Stark, U. Mohideen, E. Betzig, and R. E. Slusher, in P. F. Barbara, W. H. Know, G. A. Mourou, and A. H. Zewail, eds., *Ultrafast Phenomena IX*, Springer New York, 1994.

41. R. D. Grober, T. D. Harris, J. K. Trautman, and E. Betzig, *Rev. Sci. Instrum.* **65**, 626 (1994).

42. W. E. Moerner, T. Plakhotnik, T. Irngartinger, U. P. Wild, D. W. Pohl, and B. Hecht, *Phys. Rev. Lett.* **73**, 2764 (1994).

43. J. W. P. Hsu, E. A. Fitzgerald, Y. H. Xie, and P. J. Silverman, *Appl. Phys. Lett.* **65**, 344 (1994).

44. H. Heinzelmann and D. W. Pohl, *Appl. Phys. A* **59**, 89 (1994).

45. H. Bielefeldt, I. Hörsch, G. Krausch, M. Lux-Steiner, J. Mlynek, and O. Marti, *Appl. Phys. A* **59**, 103 (1994).

46. F. Zenhausern, M. P. O'Boyle, and H. K. Wickramasinghe, *Appl. Phys. Lett.* **65**, 1623 (1994).

47. F. Zenhausern, Y. Martin, and H. K. Wickramasinghe, *Science* **269**, 1083 (1995).

48. M. Garcia-Parajo, E. Cambril, and Y. Chen, *Appl. Phys. Lett.* **65**, 1498 (1994).

49. G. A. Valaskovic, M. Holton, and G. H. Morrison, *Appl. Opt.* **34**, 1215 (1995).

50. G. A. Valaskovic, M. Holton, and G. H. Morrison, *J. Microscopy* **179**, 29 (1995).

51. R. C. Dunn, E. V. Allen, S. A. Joyce, G. A. Anderson, and X. S. Xie, *Ultramicroscopy* **57**, 113 (1995).

52. R. X. Bian, R. C. Dunn, X. S. Xie, and P. T. Leung, *Phys. Rev. Lett.* **75**, 4772 (1995).

53. R. Bachelot, P. Gleyzes, and A. C. Boccara, *Appl. Opt.* **36**, 2160 (1997).

54. S. Akamine, H. Kuwano, and H. Yamada, *Appl. Phys. Lett.* **68**, 579 (1996).

55. D. W. Pohl, in C. J. R. Sheppard and T. Mulvey, ed., *Advances in Optical and Electron Microscopy*, Academic Press, London, 1990.

56. H. Heinzelmann, T. Huser, T. Lacoste, et al., *Opt. Eng.* **34**, 2441 (1995).

57. R. Kopelman and W. Tan, *Appl. Spectrosc. Rev.* **29**, 39 (1994).

58. H. Heinzelmann, T. Lacost, T. Huser, H. J. Güntherodt, B. Hecht, and D. W. Pohl, *Thin Solid Films* **273**, 149 (1996).

59. R. J. Hamers, *J. Phys. Chem.* **100**, 13103 (1996).

60. L. A. Bottomley, J. E. Coury, and P. N. First, *Anal. Chem.* **68**, 185R (1996).

61. D. A. Vanden Bout, J. Kerimo, D. A. Higgins, and P. F. Barbara, *Acc. Chem. Res.* **30**, 204 (1997).

62. J.-J. Greffet and R. Carminati, *Prog. in Surf. Sci.* **56**, 133 (1997).

63. D. Courjon, M. Spajer, F. Baida, C. Bainier, and S. Davy, *Condensed Matter News* **6**, 14 (1998).

64. E. Oesterschulze, *Surf. Coat. Technol.* **97**, 694 (1997).

65. D. W. Pohl and D. Courjon, eds., *Near Field Optics*, Vol. 242, Kluwer Academic, Dordrecht, 1993.

66. M. A. Paesler and P. J. Moyer, *Near-Field Optics: Theory, Instrumentation, and Applications*; Wiley-Interscience, New York, 1996.

67. D. A. Higgins, D. A. Vanden Bout, J. Kerimo, and P. F. Barbara, *J. Phys. Chem.* **100**, 13794 (1996).

68. G. Kaupp, A. Herrmann, and M. Haak, *J. Vac. Sci. Technol. B* **15**, 1521 (1997).

69. P. J. Moyer, K. Walzer, and M. Hietschold, *Appl. Phys. Lett.* **67**, 2129 (1995).

70. C. W. Hollars and R. C. Dunn, *Rev. Sci. Instrum.* **69**, 1747 (1998).

71. C. W. Hollars and R. C. Dunn, *J. Phys. Chem. B* **101**, 6313 (1997).

72. J. Hwang, L. K. Tamm, C. Böhm, T. S. Ramalingam, E. Betzig, and M. Edidin, *Science* **270**, 610 (1995).

73. X. S. Xie and R. C. Dunn, *Science* **265**, 361 (1994).

74. A. G. T. Ruiter, J. A. Veerman, M. F. Garcia-Parajo, and N. F. van Hulst, *J. Phys. Chem. A* **101**, 7318 (1997).

75. S. R. Emory and S. Nie, *Anal. Chem.* **69**, 2631 (1997).

76. M. H. P. Moers, R. G. Tack, N. F. van Hulst, and B. Bölger, *J. Appl. Phys.* **75**, 1254 (1994).

77. J. D. Pedarnig, M. Specht, W. M. Heckl, and T. W. Hänsch, *Appl. Phys. A* **55**, 476 (1992).

78. Y. Chuang, K. Sun, C. Wang, J. Y. Huang, and C. Pan, *Rev. Sci. Instrum.* **69**, 437 (1998).

79. T. Ataka, H. Muramatsu, K. Nakajima, N. Chiba, K. Homma, and M. Fujihara, *Thin Solid Films* **273**, 154 (1996).

80. E. Tamiya, S. Iwabuchi, N. Nagatani, et al., *Anal. Chem.* **69**, 3697 (1997).

81. E. Betzig, J. K. Trautman, J. S. Weiner, T. D. Harris, and R. Wolfe, *Appl. Opt.* **31**, 4563 (1992).

82. B. Hecht, H. Bielefeldt, Y. Inouye, D. W. Pohl, and L. Novotny, *J. Appl. Phys.* **81**, 2492 (1997).

83. N. F. van Hulst, F. B. Segerink, F. Achten, and B. Bölger, *Ultramicroscopy* **42–44**, 416 (1992).

84. J. K. Trautman and J. J. Macklin, *Chem. Phys.* **205**, 221 (1996).

85. M. A. Bopp, A. J. Meixner, G. Tarrach, I. Zschokke-Gränacher, and L. Novotny, *Chem. Phys. Lett.* **263**, 721 (1996).

86. M. A. Bopp, G. Tarrach, M. A. Lieb, and A. J. Meixner, *J. Vac. Sci. Technol. A* **15**, 1423 (1997).

87. T. Ha, T. Enderle, D. S. Chemla, P. R. Selvin, and S. Weiss, *Phys. Rev. Lett.* **77**, 3979 (1996).

88. P. J. Reid, D. A. Higgins, and P. F. Barbara, *J. Phys. Chem.* **100**, 3892 (1996).

89. E. Mei and D. A. Higgins, *J. Phys. Chem. A* **102**, 7558 (1998).

90. E. Mei and D. A. Higgins, *Appl. Phys. Lett.* **73**, 3515 (1998).

91. E. Mei and D. A. Higgins, *Langmuir* **14**, 1945 (1998).

92. A. Piednoir, C. Licoppe, and F. Creuzet, *Opt. Commun.* **129**, 414 (1996).

93. D. Zeisel, B. Dutoit, V. Deckert, T. Roth, and R. Zenobi, *Anal. Chem.* **69**, 749 (1997).

94. D. Zeisel, V. Deckert, R. Zenobi, and T. Vo-Dinh, *Chem. Phys. Lett.* **283**, 381 (1998).

95. D. A. Smith, S. Webster, M. Zyad, S. D. Evans, D. Fogherty, and D. Batchelder, *Ultramicroscopy* **61**, 247 (1995).

96. S. Münster, S. Werner, C. Mihalcea, W. Scholz, and E. Oesterschulze, *J. Microsc.* **186**, 17 (1997).

97. K. Lieberman and A. Lewis, *Appl. Phys. Lett.* **62**, 1335 (1993).

98. H. Muramatsu, N. Chiba, and M. Fujihira, *Appl. Phys. Lett.* **71**, 2061 (1997).

99. M. Muranishi, K. Sato, S. Hosaka, A. Kikukawa, T. Shintani, and K. Ito, *Jpn. J. Appl. Phys.* **36**, L942 (1997).

100. V. Sandoghdar, S. Wegscheider, G. Krausch, and J. Mlynek, *J. Appl. Phys.* **81**, 2499 (1997).

101. N. Chiba, H. Muramatsu, K. Nakajima, K. Homma, T. Ataka, and M. Fujihira, *Thin Solid Films* **273**, 331 (1996).

102. D. W. Pohl, U. C. Fischer, and U. T. Dürig, *J. Microsc.* **152**, 853 (1988).

103. M. Vaez-Iravani and R. Toledo-Crow, *Appl. Phys. Lett.* **62**, 1044 (1993).

104. A. Vertikov, M. Kuball, A. V. Nurmikko, and H. J. Maris, *Appl. Phys. Lett.* **69**, 2465 (1996).

105. M. Vaez-Iravani and R. Toledo-Crow, *Appl. Phys. Lett.* **63**, 138 (1993).

106. B. D. Boudreau, J. Raja, R. J. Hocken, S. R. Patterson, and J. Patten, *Rev. Sci. Instrum.* **68**, 3096 (1997).

107. B. Hecht, H. Bielefeldt, L. Novotny, Y. Inouye, and D. W. Pohl, *Phys. Rev. Lett.* **77**, 1889 (1996).

108. M. Specht, J. D. Pedarnig, W. M. Heckl, and T. W. Hänsch, *Phys. Rev. Lett.* **68**, 476 (1992).

109. Y.-K. Kim, P. M. Lundquist, J. A. Helfrich, et al., *Appl. Phys. Lett.* **66**, 3407 (1995).

110. A. Lahrech, R. Bachelot, P. Gleyzes, and A. C. Boccara, *Appl. Phys. Lett.* **71**, 575 (1997).

111. C. Lienau, A. Richter, and T. Elsaesser, *Appl. Phys. Lett.* **69**, 325 (1996).

112. J. Levy, V. Nikitin, J. M. Kikkawa, D. D. Awschalom, and N. Samarth, *J. Appl. Phys.* **79**, 6095 (1996).

113. J. Levy, V. Nikitin, J. M. Kikkawa, et al., *Phys. Rev. Lett.* **76**, 1948 (1996).

114. R. D. Grober, T. D. Harris, J. K. Trautman, et al., *Appl. Phys. Lett.* **64**, 1421 (1994).

115. J. W. P. Hsu, E. A. Fitzgerald, Y. H. Xie, and P. J. Silverman, *J. Appl. Phys.* **79**, 7743 (1996).

116. A. A. McDaniel, J. W. P. Hsu, and A. M. Gabor, *Appl. Phys. Lett.* **70**, 3555 (1997).

117. M. S. Unlu, B. B. Goldberg, and W. D. Herzog, *Appl. Phys. Lett.* **67**, 1862 (1995).

118. Q. Xu, M. H. Gray, and J. W. P. Hsu, *J. Appl. Phys.* **82**, 748 (1997).

119. D. P. Tsai and W. K. Li, *J. Vac. Sci. Technol. A* **15**, 1427 (1997).

120. E. B. McDaniel and J. W. P. Hsu, *J. Appl. Phys.* **81**, 2488 (1997).

121. Y. Inouye and S. Kawata, *Opt. Commun.* **134**, 31 (1997).

122. E. B. McDaniel, S. C. McClain, and J. W. P. Hsu, *Appl. Opt.* **37**, 84 (1998).

123. C. L. Jahncke, M. A. Paesler, and H. D. Hallen, *Appl. Phys. Lett.* **67**, 2483 (1995).

124. C. Reddick, R. J. Warmack, and T. L. Ferrel, *Phys. Rev. B* **39**, 767 (1989).

125. N. Essaidi, Y. Chen, V. Kottler, E. Cambril, C. Mayeux, N. Ronarch, and C. Vieu, *Appl. Opt.* **37**, 609 (1998).

126. K. Fukuzawa, Y. Tanaka, S. Akamine, H. Kuwano, and H. Yamada, *J. Appl. Phys.* **78**, 7376 (1995).

127. H. Muramatsu, N. Chiba, K. Homma, et al., *Thin Solid Films* **273**, 335 (1996).

128. W. Wiegrabe, S. Monajembashi, H. Dittmar, et al., *Surf. Interface Anal.* **25**, 510 (1997).

129. C. E. Talley, G. A. Cooksey, and R. C. Dunn, *Appl. Phys. Lett.* **69**, 3809 (1996).

130. H. Muramatsu, N. Chiba, T. Ataka, S. Iwabuchi, N. Nagatani, E. Tamiya, and M. Fujihira, *Opt. Rev.* **3**, 470 (1996).

131. H. Shiku, C. W. Hollars, M. A. Lee, C. E. Talley, G. Cooksey, and R. C. Dunn, *Proc. SPIE* **3273**, 156 (1998).

132. S. Jiang, H. Ohsawa, K. Yamada, T. Pangaribuan, M. Ohtsu, K. Imai, and A. Ikai, *Jpn. J. Appl. Phys.* **31**, 2282 (1992).

133. E. Betzig, J. K. Trautman, R. Wolfe, E. M. Gyorgy, P. L. Finn, M. H. Kryder, and C. Chang, *Appl. Phys. Lett.* **61**, 142 (1992).

134. T. Lacoste, T. Huser, and H. Heinzelmann, *Z. Phys. B* **104**, 183 (1997).

135. H. A. Bethe, *Phys. Rev.* **66**, 163 (1944).

136. C. J. Bouwkamp, *Philips Res. Rep.* **5**, 321 (1950).

137. C. J. Bouwkamp, *Philips Res. Rep.* **5**, 401 (1950).

138. D. W. Pohl, L. Novotny, B. Hecht, and H. Heinzelmann, *Thin Solid Films* **273**, 161 (1996).

139. J. M. Vigoureux, F. Depasse, and C. Girard, *Appl. Opt.* **31**, 3036 (1992).

140. J. M. Vigoureux and D. Courjon, *Appl. Opt.* **31**, 3170 (1992).

141. C. Girard and D. Courjon, *Phys. Rev. B* **42**, 9340 (1990).

142. L. Novotny, D. W. Pohl, and B. Hecht, *Opt. Lett.* **20**, 970 (1995).

143. L. Novotny, B. Hecht, and D. W. Pohl, *Ultramicroscopy* **71**, 341 (1998).

144. L. Novotny and C. Hafner, *Phys. Rev. E* **50**, 4094 (1994).

145. L. Novotny and D. W. Pohl, in O. Marti and R. Möller, eds., *Photons and Local Probes*, Kluwer Academic Publishers, 1995.

146. L. Novotny, D. W. Pohl, and B. Hecht, *Ultramicroscopy* **61**, 1 (1995).

147. L. Novotny, *J. Opt. Soc. Am. A* **14**, 91 (1997).

148. L. Novotny, *J. Opt. Soc. Am. A* **14**, 105 (1997).

149. H. Furukawa and S. Kawata, *Opt. Commun.* **132**, 170 (1996).

150. M. Xiao, A. Zayats, and J. Siqueiro, *Phys. Rev. B* **55**, 1824 (1997).

151. S. I. Bozhevolnyi and B. Vohnsen, *J. Opt. Soc. Am. B* **14**, 1656 (1997).

152. J. M. Vigoureux, C. Girard, and D. Courjon, *Opt. Lett.* **14**, 1039 (1989).

153. N. F. van Hulst, N. P. De Boer, and B. Bölger, *J. Microsc.* **163**, 117 (1991).

154. S. I. Bozhevolnyi, A. V. Zayats, and B. Vohnsen, in M. Nieto-Vesperinas, ed., *Optics at the Nanometer Scale*, Vol. 319, Kluwer Academic Publishers, Dordrecht, 1996.

155. T. Okamoto and I. Yamaguchi, *Jpn. J. Appl. Phys. 2* **36**, L166 (1997).

156. R. Chang, P. Wei, W. S. Fann, M. Hayashi, and S. H. Lin, *J. Appl. Phys.* **81**, 3369 (1997).

157. R. L. Williamson and M. J. Miles, *J. Appl. Phys.* **80**, 4804 (1996).

158. T. Pagnot and C. Pierali, *Opt. Commun.* **132**, 161 (1996).

159. D. Mulin, D. Courjon, J. Malugani, and B. Gauthier-Manuel, *Appl. Phys. Lett.* **71**, 437 (1997).

160. M. Xiao, J. Nieto, R. Machorro, J. Siqueiros, and H. Escamilla, *J. Vac. Sci. Technol. B* **15**, 1516 (1997).

161. T. Matsumoto and M. Ohtsu, *J. Lightwave Technol.* **14**, 2224 (1996).

162. L. Holland, *Vacuum Deposition of Thin Films*, John Wiley & Sons, New York, 1956.

163. C. F. Powell, J. H. Oxley, and J. M. Blocher, eds., *Vapor Deposition*, John Wiley & Sons: New York, 1966.

164. A. H. La Rosa, B. I. Yakobson, and H. D. Hallen, *Appl. Phys. Lett.* **67**, 2597 (1995).

165. C. E. Talley, M. A. Lee, and R. C. Dunn, *Appl. Phys. Lett.* **72**, 2954 (1998).

166. A. Lewis and K. Lieberman, *Nature* **354**, 214 (1991).

167. S. Madsen, S. I. Bozhevolnyi, K. Birkelund, M. Müllenborn, J. M. Hvam, and F. Grey, *J. Appl. Phys.* **82**, 49 (1997).

168. J. P. Fillard, M. Castagne, M. Benfedda, S. Lahimer, and H. U. Danzebrink, *Appl. Phys. A* **63**, 421 (1996).

169. R. Brunner, A. Bietsch, O. Hollricher, and O. Marti, *Rev. Sci. Instrum.* **68**, 1769 (1997).

170. M. J. Gregor, P. G. Blome, J. Schöfer, and R. G. Ulbrich, *Appl. Phys. Lett.* **68**, 307 (1996).

171. F. F. Froehlich and T. D. Milster, *Appl. Phys. Lett.* **70**, 1500 (1997).

172. D. Sarid, *Scanning Force Microscopy*, Oxford University Press, New York, 1994.

173. K. Karrai and R. D. Grober, *Appl. Phys. Lett.* **66**, 1842 (1995).

174. W. A. Atia and C. C. Davis, *Appl. Phys. Lett.* **70**, 405 (1997).

175. C. Durkan and I. V. Shvets, *J. Appl. Phys.* **79**, 1219 (1996).

176. M. Pfeffer, P. Lambelet, and F. Marquis-Weible, *Rev. Sci. Instrum.* **68**, 4478 (1997).

177. J. W. P. Hsu, M. Lee, and B. S. Deaver, *Rev. Sci. Instrum.* **66**, 3177 (1995).

178. M. Lee, E. B. McDaniel, and J. W. P. Hsu, *Rev. Sci. Instrum.* **67**, 1468 (1996).

179. J. W. P. Hsu, A. A. McDaniel, and H. D. Hallen, *Rev. Sci. Instrum.* **68**, 3093 (1997).

180. H. Muramatsu, N. Yamamoto, T. Umemoto, K. Homma, N. Chiba, and M. Fujihira, *Jpn. J. Appl. Phys.* **36**, 5753 (1997).

181. J. K. Trautman, J. J. Macklin, L. E. Brus, and E. Betzig, *Nature* **369**, 40 (1994).

182. T. D. Harris, R. D. Grober, J. K. Trautman, and E. Betzig, *Appl. Spectrosc.* **48**, 14A (1994).

183. K. D. Weston, J. A. DeAro, and S. K. Buratto, *Rev. Sci. Instrum.* **67**, 2924 (1996).

184. K. D. Weston and S. K. Buratto, *J. Phys. Chem. B* **101**, 5684 (1997).

185. M. Isaacson, J. A. Cline, and H. Barshatzky, *J. Vac. Sci. Tech. B* **9**, 3103 (1991).

186. I. I. Smolyaninov, *Int. J. Modern Phys. B* **11**, 2465 (1997).

187. H. P. Lu and X. S. Xie, *Nature* **385**, 143 (1997).

188. X. S. Xie, *Acc. Chem. Res.* **29**, 598 (1996).

189. J. J. Macklin, J. K. Trautman, T. D. Harris, and L. E. Brus, *Science* **272**, 255 (1996).

190. R. Chang, W. Fann, and S. H. Lin, *Appl. Phys. Lett.* **69**, 2338 (1996).

191. C. Girard, O. J. F. Martin, and A. Dereux, *Phys. Rev. Lett.* **75**, 3098 (1995).

192. J. L. Kann, T. D. Milster, F. F. Froehlich, R. W. Ziolkowski, and J. B. Judkins, *Appl. Opt.* **36**, 5951 (1997).

193. D. I. Kavaldjiev, R. Toledo-Crow, and M. Vaez-Iravani, *Appl. Phys. Lett.* **67**, 2771 (1995).

194. W. Göhde, J. Tittel, T. Basche, C. Bräuchle, U. C. Fischer, and H. Fuchs, *Rev. Sci. Instrum.* **68**, 2466 (1997).

195. A. Rahmani, P. C. Chaumet, F. de Fornel, and C. Girard, *Phys. Rev. A* **56**, 3245 (1997).

196. C. W. See and M. Vaez-Iravani, *Electron. Lett.* **22**, 1079 (1986).

197. V. K. Gupta and J. A. Kornfield, *Rev. Sci. Instrum.* **65**, 2823 (1994).

198. C. Juang, L. Finzi, and C. J. Bustamante, *Rev. Sci. Instrum.* **59**, 2399 (1988).

199. D. M. Adams, J. Kerimo, E. J. C. Olson, A. Zaban, B. A. Gregg, and P. F. Barbara, *J. Am. Chem. Soc.* **119**, 10608 (1997).

200. S. Smith, N. C. R. Holme, M. Kwok, B. G. Orr, R. Kopelman, and T. B. Norris, eds., *Ultrafast Equal Pulse Correlation Measurements in GaAs Structures with a Near-Field Microscope*, Springer, New York, 1996.

201. J. Heo, J. S. Sanghera, and J. D. Mackenzie, *Opt. Eng.* **30**, 470 (1991).

202. M. F. Churbanov, *J. Non-Cryst. Solids* **184**, 25 (1995).

12

APPLICATIONS IN ELECTROCHEMISTRY

Allen J. Bard and Fu-Ren F. Fan

Electrochemistry is concerned with the behavior of electronically conductive electrodes (metals, carbon, semiconductors) immersed in ionically conductive liquids. Processes of interest include electrodeposition of metals, corrosion, adsorption of species from solution, and changes in the structure of the electrode surface caused by the passage of current. Information about such processes is usually obtained from the electrochemical response of the system.[1] The electrode is some-times observed spectroscopically while it is immersed in the solution (in situ), e.g., by Raman or infrared (IR) spectroscopic methods, but more frequently it is re-moved from solution and characterized by an ex situ technique (e.g., scanning electron microscopy, X-ray photo-emission spectroscopy).[2] Although ex situ tech-niques have been valuable in providing information about changes in surface structure caused by electrode processes, it is necessary to remove the electrode from the solution environment, often to a high vacuum, which changes the surface species, e.g., by desolvation. Moreover, removal of the electrode prevents one from observing progressive changes in the electrode surface during an electrochemical process as a function of time or potential. The scanning tunneling microscope (STM) has provided an important new tool for both in situ and ex situ character-ization of electrodes, i.e., the imaging of surface structures at a scale ranging from submicron to atomic resolution and the acquisition of information about the elec-tronic properties of the surface.

Particular problems arise when the STM is used in an electrochemical environ-ment. For example, the STM tip must be insulated to minimize the faradaic current and the instrumentation must provide for independent control of the tip and sub-strate potential. Tip insulation is not a problem when AFM is applied in electro-chemical cells. Because the tips used in AFM are usually electrical insulators, there are fewer problems with independent control of substrate potential in AFM studies of electrochemical systems. In this chapter the special considerations that are in-volved with the use of STM in in situ electrochemical studies are discussed. We then consider particular applications of in situ and ex situ STM to several different types of electrochemical systems.

12.1   APPARATUS FOR ELECTROCHEMICAL STM

12.1.1   Electrochemical Cells

To appreciate the nature of the special considerations for performing STM in an operating electrochemical cell, it is useful to understand the basics of electrochemi-cal cell design and operation.[1] An electrochemical cell usually contains three elec-trodes (Fig. 12.1). The working electrode is the electrode whose properties are being studied. Current is passed between the working electrode and another elec-trode immersed in the solution, called the auxiliary or counterelectrode, by connec-tion of an external power supply. The auxiliary electrode is sometimes separated from the working electrode chamber by a porous separator (e.g., a glass frit). Current flows along the working electrode–solution interface by the passage of a charge across the interface, which causes chemical reactions (oxidation or reduc-tion), for example, the reduction of water to hydrogen or the oxidation of a solution component. A chemical reaction also occurs at the auxiliary electrode surface, but as long as this takes place well away from the working electrode and does not cause an appreciable chemical change in the solution composition, it is usually ignored.

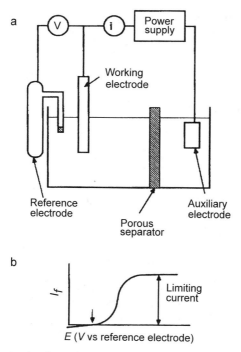

Figure 12.1. **a**, An electrochemical cell. **b**, Typical current-potential curve (voltammogram) at a microelectrode.

This current flow at an electrode is called a faradaic current I_f, because the amount of chemical products generated by the passage of current can be calculated by Faraday's law. The reaction that occurs at an electrode is governed by the potential of that electrode with respect to a given fixed reference electrode (i.e., by the energy of electrons or the Fermi level of the electrode). The reference electrode in an electrochemical cell is usually one whose potential remains constant with time and can be prepared reproducibly. Common reference electrodes are the saturated calomel electrode (SCE) prepared from mercury in contact with mercurous chloride and saturated KCl solution, or a silver/silver chloride electrode in contact with saturated KCl. Sometimes, as a matter of convenience, a quasi-reference electrode is used. Such an electrode is usually simply a piece of platinum or silver wire. Such electrodes are less stable and reproducible and must be calibrated against a true reference electrode if meaningful potentials are to be obtained.

The electrochemical characteristics of a working electrode are frequently obtained by observing how the faradaic current varies with the potential of the electrode vs. the reference electrode (E_w), i.e., by recording an I_f vs. E_w curve, or voltammogram. A typical voltammogram is shown in Figure 12.1. At a given electrode for a given solution composition there is a potential, or sometimes a range of potentials, where no faradaic reactions will occur and thus I_f is at or near zero. Electrochemical STM cells are actually four-electrode cells, where both the tip and the substrate are working electrodes. Thus it is convenient to consider and control the potentials of both the tip E_t and the substrate E_s with respect to the reference electrode.

12.1.2 Tip Considerations

When the STM tip is immersed in the electrochemical cell and is far (i.e., a few micrometers) from the substrate, the only current that will flow through the tip is the faradaic current I_f. The magnitude of I_f depends on E_t, the solution composition, and the exposed area of the tip. For example, if the solution contains a species O at a concentration, C_O, which can be reduced in an n-electron reaction at E_t, at a rate that depends only on the rate of diffusion of O to the tip, the steady-state limiting current is given with good approximation by Eq. (12.1):[3]

$$I_f = 4nFD_OC_Or \tag{12.1}$$

where D_O is the diffusion coefficient of O, F is Faraday's constant, and r is the radius of a disk of area equal to the effective exposed area of the tip. Note that when potential is first applied to the tip, the current will be larger than this value because of a transient faradaic current and a nonfaradaic component for charging of the double layer.[1] This transient response is usually not important in STM studies. When the tip approaches the substrate, the faradaic current will be affected by the presence of the substrate (the feedback effect, discussed later) and at very close separations the tunneling current I_t will begin to flow. The total tip current I is given by $I = I_f + I_t$. Because I_f can be large compared to I_t, it is necessary in electrochemical STM studies to minimize I_f. This is usually accomplished by decreasing the exposed area of the tip by coating it with an insulator. The ideal STM tip for these studies would be a single exposed atom of the tip metal with the remainder of the tip insulated. In practice, of course, a considerably larger exposed tip area exists; but even so, with the proper choice of operating conditions, I_f can be reduced to negligible levels with respect to I_t.

A number of different approaches have been used in the preparation of insulated STM tips. The simplest approach involves coating or painting a Pt-Ir, Au, or W tip that has been etched to a point electrochemically with epoxy varnish,[4,5] silicone polymer,[6] Apiezon wax,[7] or nailpolish.[8] The coating method must not insulate the whole tip or leave a residue that extends below the exposed area that will contact the substrate before the exposed area attains tunneling distance. Controlled coating is attained by careful choice of the viscosity of the coating material or by control of the coating method[4] (Fig. 12.2). Such tips have been used in a number of studies with aqueous solutions but are probably less useful when nonaqueous solvents, such as acetonitrile, are used, because they tend to swell and dissolve in organic solvents. Glass-coated tips are probably the most stable, but are more difficult to fabricate. These can be constructed by sealing of etched wires, usually Pt-Ir, in small glass capillaries[3,9] or by pushing an etched wire through a drop of molten

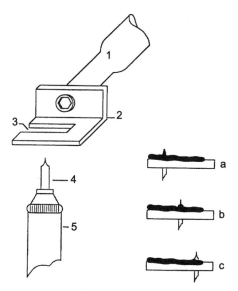

Figure 12.2. The apparatus and technique for tip insulation. Where the tip penetrates the Apiezon wax along the slit determines how much of the tip gets insulated. **a**, At too cold a region, the tip is completely covered with Apiezon wax. **b**, At the optimum point, only the extreme end of the tip is exposed. **c**, At too hot a region, the tip receives little insulation and is thus mostly bare. *1*, soldering iron; *2*, copper plate; *3*, 1-mm-wide slit; *4*, STM tip; *5*, holder for manipulator. (Reprinted with permission from Ref. 7.)

glass at a controlled rate at a controlled temperature.[10] Glass-coated Pt-Ir STM tips are also commercially available from several sources. Improved insulation of glass-coated tips by vacuum deposition of SiO_2 on the uncovered end of the tip has been reported.[11] This treatment, as well as some of the other coating methods, leaves the tip completely covered. In these cases, the end of the tip can be exposed by placing it in the STM with a high bias voltage (e.g., 10 V) applied between tip and substrate. The onset of tunneling or field emission blows a hole in the insulation at the point of closest approach of tip to substrate, while leaving most of the tip insulated. Tungsten and tantalum tips have also been used for in situ STM studies.[12, 13] With these the thick oxide coating that forms on the W and Ta surfaces provides the insulation. Note, however, that for W the oxide (WO_3) can be reduced at negative potentials, and an appreciable faradaic current can flow under these conditions. It is not clear how tunneling is established at these oxide-covered tips in solution, although atomic resolution has been attained with W tips under these conditions.[8]

The total exposed area of the tip can be estimated via Eq. (12.1) by performing an electrochemical experiment in an external electrochemical cell with a solution containing a known concentration of an electroactive species (e.g., 10 mM $Fe(CN)_6^{3-}$ in 0.2 M KCl for Pt-Ir). Alternatively, the area can be estimated by measuring the charge required for adsorption of hydrogen (on Pt-Ir) or oxygen (on Pt-Ir or Au) or for the underpotential deposition of a metal.[1, 14] These estimates of tip area are rarely done routinely, however; and the faradaic contribution to the total current is usually found with the tip mounted in the STM when it is far from the substrate. I_f is then minimized by adjusting E_t. When operating the STM in the constant current mode, the current set point must be larger than I_f to obtain adequate control.

12.1.3 Electrochemical STM Cells and Instrumentation

A number of papers have reported electrochemical cells designed for use with the STM and commercial electrochemical STMs are now available. The substrate is usually placed horizontally at the bottom of the cell with the tip approaching from above. Because a long tip can lead to vibration problems, the amount of solution above the sample is minimized. Thus most cells have quite small volumes and only limited space for the counter and reference electrodes. Provision must also be made for firmly clamping the substrate and making electrical contact to it. Ideally, the cell should be closed to prevent evaporation of solution, which causes changes in the solution composition and promotes thermal drifts. Moreover, it is undesirable to have solution vapors contacting the piezoelectric elements of the STM. However, a tight solution cover is incompatible with the use of an STM scanning head; and in most cases, the cell cover has a rather large hole for the tip. The solution should not contact the tip holder, which is sometimes coated with an insulating material (epoxy cement, silicone cement, Teflon spray). Most electrochemical experiments are carried out under an inert atmosphere by bubbling nitrogen or argon through the solution before an experiment and flowing these inert gases over the solution during the studies. This prevents oxygen, which can be reduced at electrodes, and impurities in the atmosphere, which can be adsorbed on electrode surfaces, from entering the cell solution. However, deaeration is difficult in electrochemical STM cells because of the limited space, as well as the enhanced evaporation and possible vibrations caused by the flowing gas streams. Most electrochemical STM experiments reported to date have used solutions exposed to the atmosphere, although in some experiments the cover over the STM head (the Faraday shield) has been filled with inert gas or the whole STM head has been placed in a glove bag or glove box.

Different cell designs for STM are shown in Figures 12.3 and 12.4. Different means are used to fix the substrate working electrode and make contact to it. Frequently, the electrode material is mounted by cementing it with a conductive cement (e.g., Ag epoxy) to a metal contact in an assembly that screws into the bottom of the cell (Fig. 12.4). The edges of the substrate must then be sealed with epoxy or silicone cement to prevent leakage of solution to the electrical contact.[5, 15, 16] The substrate can also be cast into a polymeric support[17] or in shrinkable Teflon tubing.[18, 19] This method provides good support and contact to the substrate but does not allow quick mounting and replacement of new substrate electrodes. The alternative approach is to use the substrate electrode as the base of the electrochemical cell (Fig. 12.3).[9, 20] A plastic (Kel-F or Teflon) piece forms the walls of the cell, which is pressed against the substrate surface via an O-ring. This approach allows rapid replacement and contacting of the substrate but requires a larger

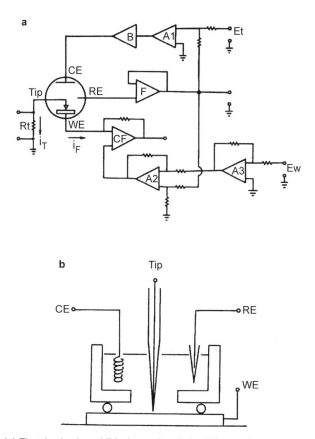

Figure 12.3. (a) Electric circuit and (b) electrochemical cell for the four-electrode configuration. *WE*, working electrode; *CE*, counter electrode; *RE*, reference electrode; *Tip*, tunneling tip; i_F, electrochemical current; i_T, tunneling current. (Reprinted with permission from Ref. 9.)

substrate. Moreover, tight clamping of the substrate may lead to drifts as the substrate mechanically relaxes and flexes under the applied stress.

The counterelectrode and reference electrode are often simply wires (i.e., a quasi-reference electrode) brought in through the cell cover or the top of the cell[9, 20] (Fig. 12.3). A better, but more elaborate arrangement involves bringing these in through the walls of the cell (Fig. 12.4). Ideally, to provide a uniform current density and thus a uniform potential drop between substrate electrode and solution, the counterelectrode should be arranged symmetrically with respect to the substrate.[17] However, the small size of most cells, the shape of the substrate electrode, and the effect of the scanning tip prevent true uniformity of current density. Except for resistive solutions (e.g., those with low dielectric constant solvents or low electrolyte concentrations), nonuniform current density effects are probably minor. When possible, the use of true (rather than quasi) reference electrodes is preferable.

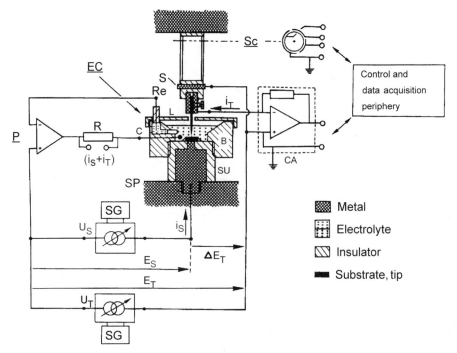

Figure 12.4. The main parts of the experimental STM assembly and the potentiostatic circuitry for individual potential control of substrate and tunneling tip. The piezoelectric single-tube scanner (Sc) has a titanium spacer plate (S); the electrochemical cell (EC) consists of a Plexiglas beaker (B) with Pt counterelectrode (C) and Ag/AgCl reference electrode (RE) in 0.1 M NaCl, a Plexiglas lid (L), and a PETF support unit (SU) with an epoxy-sealed substrate mounted on support plate (SP). The low-noise potentiostat (P) has low-impedance voltage units (U_s, U_T) equipped with a low-pass filter and signal generator (SG), a precision resistor (R) for measuring $i_S + i_T$, and a low-noise current amplifier (CA) for measuring i_T. (Reprinted with permission from Ref. 5.)

Small reference electrodes can contact the cell through small salt bridges, e.g., of Teflon tubing or a porous Vycor glass tip.[6, 19]

The earliest electrochemical STM studies involved only the tip and substrate as electrodes with no independent control of the tip and substrate potentials.[21, 22] Although this configuration is simple and uses only the usual STM biasing arrangement, it is not the best arrangement for electrochemical studies. The potential of the substrate with respect to a reference electrode is not controlled and the tip may be in a potential region where undesirable reactions (e.g., hydrogen evolution or metal plating) occur. Most electrochemical studies are now performed in a four-electrode cell (tip, substrate, auxiliary, reference) with potential control with a bipotentiostat.[1] Bipotentiostats allow independent control of two working electrodes in the same electrochemical cell and were developed mainly for use with rotating ring-disk electrodes. They are usually based on operational amplifier circuits and allow independent adjustment of E_t and E_s and measurement of both the tip current (I_t)

and the substrate current (I_s). Note that the magnitudes of I_t and I_s are usually not equal in this arrangement and that the current flow through the auxiliary electrode is $I_t + I_s$. Typical bipotentiostat arrangements used in STM are shown in Figures 12.3 and 12.4. Commercial STMs, such as the electrochemical version of the Nanoscope II, also incorporate bipotentiostats that can be controlled by the computer software. Reports on modifying earlier versions of commercial STMs to allow electrochemical studies have also appeared.[23] An alternative, but less useful, approach is to use an external battery system to control the substrate potential with respect to a reference electrode.[24, 25]

12.2 APPLICATIONS OF ELECTROCHEMICAL STM

12.2.1 Ex Situ Studies

Arvia's group[26–30] first studied the surface morphology changes of Pt electrodes after electrochemical treatment consisting of repetitive square wave potential signals (RSWPS) and of Au electrodes after electrochemical reduction of oxide covered surfaces. The treated Pt specimens showed clear reoriented patches on the surface in the form of steps with well-defined orientations separated by terraces of different sizes. The high-resolution STM images showed corrugated regions made of small spheres of ca. 10 nm average diameter. The small spheres were separated by channels of ca. 10 nm width. The increase in real surface area produced by electrochemical activation was explained through a model consisting of a metal overlayer made of small adhering spheres of ca. 10 nm diameter leaving interconnected channels of nearly the same average diameter. Ex situ STM has also been used to study electroreduction of [Re(CO)$_3$(vbpy)Cl] on HOPG,[31] chloride ion–induced reconstruction of an Ag electrode,[32] and the surface microstructure of electrodeposited α- and β-PbO$_2$.[33, 34] Schardt's group[8, 35] studied iodine adsorption on Pt(111) with atomic resolution both in air and in 0.1 M aqueous HClO$_4$ solution. Uosaki et al.[36] examined the surface morphology of Pt chemically deposited on an ion exchange membrane, Nafion 120, and found that there is a relation between the surface roughness determined by electrochemical methods and that observed by STM. Yoshihara et al.[37] correlated the surface structure of a nickel film plated on Cu substrates with that obtained by STM. Szklarczyk et al.[38] studied Al(111) and polycrystalline Al samples directly after polishing or further annealing under vacuum at 10 to 5 bar at 750 K for 15 h.

12.2.2 In Situ Studies

A difficulty with studying surfaces by STM is that only a tiny area of the electrode surface is examined in any one image, and it is difficult to establish whether that area is a representative portion of the surface. Large displacements of the tip between treatments or removal of the sample from the environment (as in ex situ studies) prevents one from noting progressive changes of exactly the same area of the electrode by STM. An important aspect of in situ monitoring of a substrate is that it allows one to examine the same area of the electrode surface as the electrode is subjected to different treatment. Another important aspect of in situ electrochemical STM is that both substrate and tip can be potentiostatically or galvanostatically controlled and thus maintain a well-defined electrochemical environment for STM measurement.

Considerable progress has been made over the past few years toward the development of STM as an in situ technique for the study of the (electrode/electrolyte) interface.[11] Early studies used a conventional STM instrument without additional potential or current control of the substrate and tip during imaging. This method can provide sufficient control of the current to yield structural images of the substrate as long as the superimposed electrochemical currents remain steady during imaging. It has been successfully applied to image a Pt-overlaid integrated circuit in H_2O or a 1 mM $K_4Fe(CN)_6$/1 mM $K_3Fe(CN)_6$ solution;[22] to image HOPG in H_2O, aqueous electrolyte, or nonpolar liquids;[10, 15, 21, 39, 40, 41] to study Ag,[42] Au,[43] and Pt[39,40] depositions on HOPG; to study Au[44] or Cu[45] deposition on Au; and to study the morphology change of Ag[18] or Pt[46, 47] electrodes after electrochemical activation. It has also been applied to image electrophoretically deposited nucleic acids in buffered aqueous solution.[48–50]

More recently, the interfacing of STM to electrochemical techniques has become feasible; the substrate and the tip can be potentiostatically or galvanostaticaly controlled during imaging. These techniques have since then been widely used for real time in situ studies of the electrodeposition of metals and conductive polymers; morphology change of the substrate surface caused by electrochemical oxidation–reduction treatment; and corrosion of metals, alloys, and semiconductors. We will briefly describe the more recent studies. Readers who are interested in the earlier developments can find detailed descriptions elsewhere.[11]

Changes in Electrode Surfaces by Electrochemical Treatment The potentiostatic EC STM technique was applied by Wiechers et al.[12, 51] to a flame-annealed Au(111) substrate in 50 mM H_2SO_4/5 mM NaCl, by Nichols et al.[34] to Au(100) electrodes in 50 mM H_2SO_4/5 mM HCl, by Lustenberger et al.[13] to HOPG in 0.1 M aqueous $NaClO_4$ solution, by Otsuka and Iwasaki[19] to study polycrystalline Ag and Au substrates after oxidation–reduction cycling in 0.1 M KCl solution, by Gewirth and Bard[16] to the anodic oxidation of HOPG in 0.1 M H_2SO_4. A similar technique was applied to monitor surface modification of Pt electrodes in H_2SO_4,[52, 53] Pt and Pd electrodes in 2 mM $HClO_4$,[54] Ag electrodes in 0.1 M KCl,[55, 56] and Au(111) on mica in 0.1 M $HClO_4$ with and without 5×10^{-5} M HCl.[6] The galvanostatic EC STM technique was also used to study the electrochemical dissolution of a Ag film that was electrodeposited on HOPG from a Ag plating solution.[57]

Although STM resolution at or above the nanometer scale is routinely obtained, atomic height steps and substrate changes in the subnanometer range can also be resolved.[6, 34] Figure 12.5 illustrates the topographic changes of Au(111) films on mica in 0.1 M $HClO_4$/5×10^{-5} M HCl accompanying electrochemical oxidation–

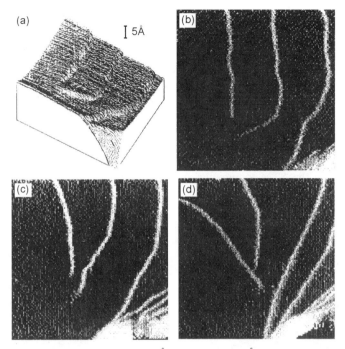

Figure 12.5. STM Images of a 2000 × 2000 Å region of a 1000-Å-thick Au(111) film on mica at + 0.7 V vs. NHE in 5 × 10^{-5} M HCl, 0.1 M $HClO_4$, showing the emergent points of two dislocations: (**a**) after cycling to + 1.6 V and back to + 0.7 V at 20 mV/s, (**b**) same image presented as a gray scale map of dz/dx, (**c**) after cycling to + 1.9 V, (**d**) after three cycles to + 2.0 V. (Reprinted with permission from Ref. 6.)

re-reduction cycles. Figure 12.5c shows the result of a potential cycle to 1.9 V. The three steps in Figure 12.5b have each moved a bit to the left. Figure 12.5d shows the result of three cycles to 2.0 V; note that the higher step now originates from the dislocation lower in the picture as a consequence of the dissolution of the middle terrace. Within the region shown in the figure, about half a monolayer of Au appears to have been lost by dissolution. In contrast, similar losses of Au have not been noted in the absence of Cl^-. A further difference that distinguishes the STM images in Cl^--containing electrolytes from those in Cl^--free electrolytes is the lack of evidence of roughening of the terrace after electrochemical oxidation–re-reduction cycles. The apparent lack of surface roughening has been attributed to the enhanced mobility of surface Au atoms by Cl^- adsorption. Specific effect of chloride ion on the topographic modification of Au electrodes under potential control was also demonstrated by Wiechers et al.[13] and Otsuka and Iwasaki.[19] Small ripple structures were observed when the electrode was biased at potentials at which Cl^- adsorption took place, and the topograph was irreversibly changed when the electrode was scanned back to potentials at which Cl^- desorption occurred. Such a profile change was been attributed to enhanced mobility of surface Au atoms by Cl^- adsorption. Similar mobility of Au atoms is found for Au(111) on mica during oxygen etching in a CN^- medium, where etch pits are found to heal with time during scanning.[58]

Gewirth and Bard[16] applied a constant-current imaging mode coupled with barrier height (i.e., $d(\log I)/ds$) imaging to study the anodic oxidation of HOPG in 0.1 M H_2SO_4. As shown in Figure 12.6, before oxidation, both topograph and barrier-height images are featureless at HOPG potential of 0.05 V vs. a Ag quasi-reference electrode (AgQRE). This implies the HOPG surface is quite inert and does not absorb or intercalate material from the solution at this potential. The barrier height to tunneling was 0.9 eV. After 20 potential cycles between 0.0 and 1.8 V vs. AgQRE at 200 mV/s, the island of surface corrugation grew together anisotropically giving the STM topographs shown in Figure 12.6c. The flat area of the HOPG surface still had a barrier height of ca. 0.9 eV, whereas the protruded regions showed reduced barrier heights, suggesting the formation of a more insulating layer that might correspond to oxidized areas. Finally another 10 potential sweeps gave

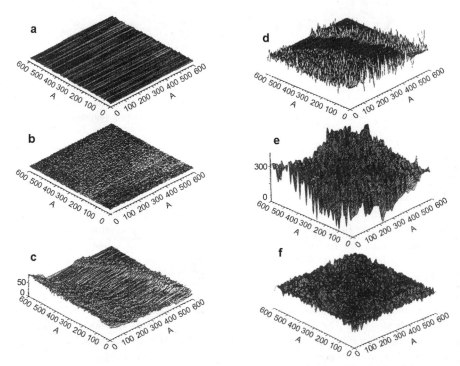

Figure 12.6. (**a, c, e**) Constant-current STM images and (**b, d, f**) $d(\log i)/ds$ scans of the same areas of the anodic oxidation of highly oriented pyrolytic graphite in 0.1 M H_2SO_4 at 0.05 V vs. AgQRE. V_{tip}, −50 mV; i_{tunnel}, 7 nA. Scan speed was 260 Å/s. (**a**) HOPG topograph before oxidation and (**b**) corresponding barrier height measurement (barrier height, 0.9 eV). (**c**) HOPG topograph after 20 potential cycles and (**d**) corresponding barrier height. (Barrier height decreases in the upward direction.) (**e**) HOPG topograph after further oxidation of the surface and (**f**) corresponding barrier height. (Reprinted with permission from Ref. 16.)

rise to rough topographs (Fig. 12.6e) that showed relatively uniform, but low, barrier heights (Fig. 12.6f). The surface was now apparently completely oxidized, and it was covered with a grayish film after removal from the solution.

The potentiostatic EC STM technique was also applied to flame-annealed Pt[53] and Pt(111)[52] surfaces. Both before and after electrochemical potential cycling in aqueous H_2SO_4 solution, significant roughening of the electrode surface was induced by extended cycling of the electrode in H_2SO_4 solutions between the regions where an adsorbed oxide film forms and is removed. It was suggested that the Pt atoms underwent substantial transport and rearrangement during the anodic–cathodic potentiostatic control cycle. For a Pt(111) surface, randomly oriented islands a few atoms in height were observed on the terraces after five potential cycles, suggesting that Pt adatoms produced by the reduction of the oxide aggregated on the terraces. More recently, surface reconstruction of metals in solution was revealed by in situ STM studies. For example, Gao et al.[59, 60] reported that the Au(110)−(1×1) structure transformed to a (1×2) structure in the double-layer region, when the electrode potential was lowered to -0.3 V vs. SCE in $HClO_4$ solution. In situ STM combined with IR spectroscopy was applied to characterize the CO adlayer on Rh(111) and Pt(100) electrodes in aqueous solutions.[61, 62]

Corrosion and Etching Drake et al.[43] first performed real-time STM imaging of the corrosion of an iron film (400 Å) on 300 Å of Al in 0.2 mM NaCl solution. They saw substantial roughening on the nanometer scale within minutes of immersion. Zhang and Stimming[63] studied copper dissolution in aqueous $HClO_4$ solutions of different concentrations. The STM images showed that the surface changes became more pronounced with increasing acid concentration. The local corrosion rates estimated from the STM images were comparable to those obtained from SEM micrographs.

An early report from this laboratory demonstrated the possibility of employing STM for in situ investigation of potential-driven phase transition of a nickel electrode in 0.5 M sulfuric acid.[24] When applying STM to a nickel electrode at its rest potential or in the active region (e.g., -0.1 V vs. SCE), the images were clean and reproducible. The current–tip displacement (*I-s*) curves were symmetric and had an exponential relationship as expected for electron tunneling. When the nickel surface was scanned with the electrode polarized in the passivating region (e.g., 0.7 V vs. SCE) under the same tunneling conditions as those in the active region, erratic motion of the tip was observed and the *I-s* response was asymmetric, spread over a large distance, and did not show a form typical of tunneling.

Following the study on Ni electrode in 0.5 M H_2SO_4, we applied STM for real-time study of the corrosion of type 304L stainless steel in aqueous chloride media.[64] Chloride ion accelerated the corrosion rate, presumably by removing the surface insulating oxide layer through complexation. The corrosion rate of 304L stainless steel increased with decreasing pH, which was consistent with the pH dependence of their breakdown potential. The introduction of an organic corrosion inhibitor, *N*-lauroylsarcosine (NLS), into buffered Cl^- solutions produced a highly disordered surface structure that exhibited nonideal tunneling behavior. The tunneling current resembled that seen with the formation of an insulating interfacial layer, which was probably caused by strong adsorption of NLS.

Moffat et al.[65] carried out a real-time STM examination of dealloying of a Cu-Au alloy in 0.01 M H_2SO_4, 0.99 M Na_2SO_4. The voltammetric curves of ordered and disordered Cu_3Au, Cu, and Au in 0.01 M H_2SO_4 + 0.99 M Na_2SO_4 at a scan rate of 0.1 mV/s are shown in Figure 12.7. The curves for Cu_3Au can be divided into three regimes. From the open circuit potential to −0.4 V vs. a Hg/Hg_2SO_4 reference electrode (SSE), the rate of dissolution increases with potential (regime I). This is attributed to oxidation of Cu from sites of low coordination. The current reaches a local maximum and then decreases reaching a plateau up to a potential of ca. 0.2 V (regime II). Upon further polarization the current increases with a significant rise apparent as the potential exceeds a critical potential, E_c, of ca. 0.3 V (regime III). In Figure 12.8 a series of real-time STM images are recorded in regime I at different substrate potentials for given time periods. The tip potential was always kept at −0.2 V vs. SSE. The development of localized rough structures is apparent (notice the ripple structure shown on the right-hand side domain). These ripples are thought to represent the clustering of Au atoms at sites of extensive Cu dissolution. An independent real-time STM experiment following a potential step from −0.5 to −0.35 V also shows localized rough structures. The rough region grows laterally as a function of time. After a few minutes, further growth is insignificant. These dissolution nuclei appear to be tied to solid-state defects. Pickering and Byrne[66] showed that the charge associated with transient dealloying in this

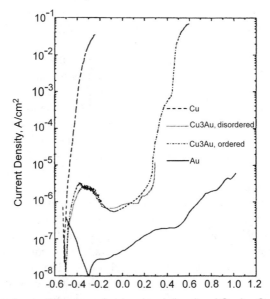

Figure 12.7. Current-potential curves of ordered and disordered Cu_3Au, Cu, and Au in 0.01 M H_2SO_4 + 0.99 M Na_2SO_4. (pH = 2.8, T = 30°C, deaerated with argon, scan rate = 0.1 mV/s). (Reprinted with permission from Ref. 65.)

potential region is a function of the defect density in the solid. The ripples on the surface appear to be mobile (notice the change of the step edge, which runs from the top to the bottom near the central portion of the images). Ripples developed on the step edge at an earlier time (Fig. 12.8c) are annealed in minutes at the same substrate potential (Fig. 12.8d). As the potential was increased above the critical potential, new regions of high activity appeared. At 0.4 V, the surface roughened over a significant fraction of the area. Clearly, the critical potential is associated with a roughening transition occurring on a global scale. Stepping the potential back to 0 V led to significant annealing of the roughening caused by the dealloying at 0.4 V. These observations are consistent with those reported on Au(111)[6, 13, 19, 34] and illustrate the smoothing action of the surface diffusion of gold atoms. Effects of surface atom mobility have also been observed in the de-alloying of Ag-Au alloys[67] and the surface reactions of Au(111) with oxygen in an aqueous cyanide solution.[58]

Electrodeposition of Metals and Conductive Polymers The potentiostatic EC STM technique has also been applied for in situ studies of electrodeposition of various metals, e.g., Ag on HOPG,[9] Pb on Au(111),[68] Ag on

Figure 12.8. A sequence of topographic, constant-current STM images of disordered Cu₃Au progressively held in the low overpotential regime: **a**, −0.480 V for 2 min; **b**, for 4 min; **c**, at −0.450 V for 2 min; **d**, for 4 min; **e**, at −0.400 for 2 min. (Reprinted with permission from Ref. 65.)

Ag(100),[17,69,70] Ag on Au(111),[71] Cu on Pt,[23,53] Pb on HOPG,[25,72] and Cu on Ag(100),[34] and Au(111).[34,73]

Green et al.[68] studied underpotential deposition (UPD) of Pb on Au(111). The cyclic voltammogram for Pb UPD on Au(111) is given in Figure 12.9. Typical features attributed to the deposition and stripping of the Pb overlayer (0.15 V vs. Pb/Pb^{2+}) are superimposed on a negative current background, which has been attributed to the reduction of O_2 from the atmosphere.

Figure 12.10 shows a series of images taken with a Au electrode about 2 h after removal from the evaporator and immediately after immersion in the electrolyte, during voltage sweeps between 600 and 50 mV. A 250×250 nm portion of the bare surface located on top of a grain is shown in Figure 12.10a. A pair of spiral dislocations can be seen; each atomic step is 0.25 ± 0.03 nm high. Figure 12.10b shows the same region covered with a monolayer of Pb after a potential sweep to 50 mV. The overlayer is an almost perfect replica of the Au topograph under these conditions, except that the step edges of the Pb covered surface are slightly more ragged. A reverse potential sweep removes the adlayer from the electrode surface, giving the surface in Figure 12.10c. Note that the deposition and removal of a single Pb layer causes substantial roughening of the original terraced surface. This type of roughening was observed for all samples subjected to a fill plate strip cycle, for sweep rates from 0.5 to 1000 mV/s. Figure 12.10d shows a second deposition onto the roughened substrate; the topographic structure is almost identical to that in part **a** instead of a reproduction of the roughened surface.

Real-time STM images on bulk deposition of Cu on Pt and Pd in 2 mM $HClO_4/5$ mM $CuSO_4$[23,54] have shown that Cu deposition started referentially at the step edges and extended to form large plateaus composed of grains of various sizes. Cu deposition on a flat Pt surface was also shown to take place first at specific site rather than as uniform layers.[53] The Pt surface was roughened appreciably by a Cu deposition–stripping cycle. A control experiment with a Pt substrate in the same supporting electrolyte in the absence of Cu^{2+} showed no appreciable surface change after the same electrochemical treatment. This irreversible change in the morphology of the substrate by the deposition–stripping cycle of a second metal was also observed for Pb deposition–stripping on Ag(100) and has been interpreted as recrystallization of selected parts of the substrate caused by the formation of metastable alloy phases or by exchange processes within the range of several substrate or deposit layers.[5]

The morphology of electrodeposited Ag on HOPG, investigated with STM under potentiostatic conditions by Itaya and Tomita,[9] indicated that Ag deposition on HOPG occurs by an island growth mechanism, as in Pt deposition. Ag deposition on HOPG was studied under galvanostatic conditions.[74] Nucleation of Ag was initiated at a current density of 60 μA cm^{-2} for 20 s, then real-time STM was used to monitor the growth of Ag nuclei at a lower current density (ca. 20 μA cm^{-2}), which permitted STM observation over a relatively wide scan area. Szklarczyk et al.[25,72] deposited a bulk layer of Pb on HOPG. Part of the Pb deposit was then removed by electrochemical dissolution to electropolish the surface of the deposit and to thin the Pb layer at the same time. The thin electropolished Pb layer was then

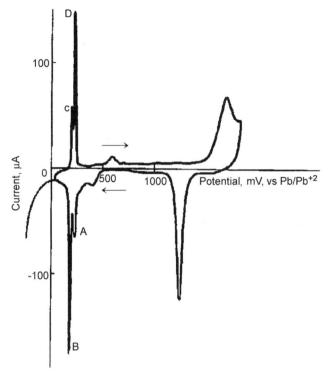

Figure 12.9. Cyclic voltammogram taken on Au(111) in the STM electrochemical cell. The Pb underpotential peaks for deposition (*A and B*) and stripping (*C and D*) are indicated. If the sweep were continued below 0 mV, bulk deposition of Pb would occur (*dotted line*). Solution, 5 mM PbO, 0.05 M HClO$_4$; sweep rate, 5 mV/s. (Adapted from Ref. 68.)

investigated by STM both in air and in solution. Most recently, Magnussen et al.[73] first reported the atomic structure of the copper UPD on Au(111) and Au(100) in 0.05 M H$_2$SO$_4$ and 5 mM CuSO$_4$. The atomic structure of copper UPD on Pt(111) was also reported by Sashikata et al.[75] Atomic force microscopy (AFM) was also applied by Gewirth's group to study the UPD of Cu[76] and Hg[77] on Au(111). As shown in Figure 12.11a, AFM images taken at $+0.7$ V vs. a Cu wire in either 0.1 M HClO$_4$ or 0.1 M H$_2$SO$_4$ containing 1 mM Cu^{2+} before Cu deposition showed large areas exhibiting Au(111) atomic resolution. When the potential was swept to -0.100 V at 10 mV/s to initiate bulk Cu deposition, the surface structure quickly (ca. 10 s) settled into images of a close-packed surface with interatomic distance of 0.26 ± 0.02 nm (Fig. 12.11b). Sweeping the potential to 0.110 V removed the bulk-deposited Cu but left the UPD monolayer in place. Here, different images were found in different electrolytes. In perchloric acid, images appeared with the same spacing (0.29 ± 0.02 nm) found for the Au(111) surface at $+0.7$ V (Fig. 12.11c). However, the lattice direction was rotated by $30° \pm 10°$ relative to the underlying Au lattice. The Cu adatoms are arranged in an incommensurate fashion with respect

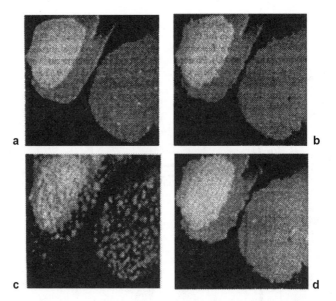

Figure 12.10. In situ images of the deposition and stripping of Pb overlayer on a 250×250 nm^2 region of pristine Au. Shading is keyed to height and five atomic levels are shown. **a**, Bare Au. **b**, After deposition of one monolayer of Pb. **c**, After the Pb monolayer is stripped. **d**, Second deposition of Pb layer. (Adapted from Ref. 68.)

to the Au(111) surface (Fig. 12.11). In sulfuric acid, a $(\sqrt{3} \times \sqrt{3})$R30° overlayer structure was observed (Figs. 12.11e and 12.11f). The distance of the nearest neighbors of Cu adatoms is 0.49 ± 0.02 nm and the atomic rows of Cu are rotated $30° \pm 10°$ relative to the underlying Au lattice. This $(\sqrt{3} \times \sqrt{3})$R30° structure was also found by STM.[73,78]

We have also studied metal deposition or dissolution on a HOPG substrate that was initially treated to produce monolayer deep pits by heating in air at 650°C.[79, 80] These pits are useful as markers and nucleation sites for in situ study of the deposition of metals on HOPG. Metal deposition or dissolution takes place preferentially at the step edge.[81] STM imaging after Pb deposition, when the substrate potential is stepped to −600 mV vs. SCE, and its subsequent stripping, when the HOPG potential is returned to the initial potential, −79 mV vs. SCE, shows that Pb is preferentially grown at the step edges of HOPG when the substrate potential is stepped into the Pb deposition region. Tip–substrate interactions also occur. A possible cause is the movement of Pb on the HOPG surface as the tip pushes the metal particles. One should be aware of other possible tip effects in electrochemical studies. For example, the tip can shield the area directly below it, as is well known from analogous effects found with Luggin capillaries. An even more significant effect arises when the potential applied to the tip affects the potential distribution

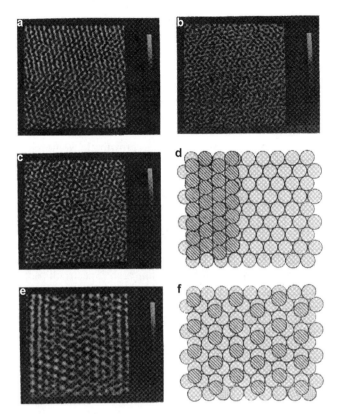

Figure 12.11. AFM images (6 × 6 nm) of a Au(111) electrode surface in electrolyte $+10^{-3}$ M Cu^{2+} under potential control at different voltages. **a**, V $= +0.7$ V showing Au(111) corrugation. **b**, Bulk-deposited Cu at -0.1 V. The atom–atom distance is 0.26 nm. **c**, Close-packed overlayer of Cu observed at $+0.114$ V in 0.1 M perchloric acid. The atom–atom distance is 0.29 nm. **d**, The incommensurate close-packed overlayer of Cu on Au. *Open circles,* Au atoms; *striped circles,* Cu atoms. Only part of the monolayer is exhibited to demonstrate the overlayer–underlayer orientation. **e**, ($\sqrt{3} \times \sqrt{3}$)R30° overlayer of Cu on Au observed at $+0.144$ V in 0.1 M sulfate. Atom–atom distance is 0.49 nm. **f**, The ($\sqrt{3} \times \sqrt{3}$)R30° overlayer of Cu (*striped circles*) on Au (*open circles*). (Reprinted with permission from Ref. 76.)

on the substrate being examined. For example, in the Pb deposition on HOPG study, if the tip potential was too positive, Pb would not deposit below it on the HOPG, although deposition on the remainder of the HOPG surface was clearly seen. We have also investigated the applicability of STM for in situ study of the growth of a conductive polymer, polypyrrole (PP) in nonaqueous solvent. Pyrrole was electro-polymerized in acetonitrile containing 0.2 M tetrabutylammonium fluoroborate (TBABF$_4$) and 20 to 50 mM pyrrole on a fairly flat Pt film on mica. During electropolymerization, the tip was pulled a few thousand angstroms away from the substrate to minimize possible localized shielding effects of the substrate in the immediate surroundings of the tip. The STM images clearly show that PP is not uniformly deposited on Pt but involves formations of clusters on the bare substrate and follows a nucleation and growth mechanism. This is consistent with the ex situ STM observation by Yang et al.[82, 83] Examination of PP on HOPG by STM in air showed different structures, depending on whether the sample was imaged on the thick continuous polymer or in the boundary region between the continuous polymer film and the graphite substrate. The continuous region showed a nodular structure, and the discontinuous–interfacial region showed a micro island structure, reflecting the nucleation and initial polymer growth. In some cases, the micro islands were interconnected by polymer strands with average heights and widths of ca. 1.8 nm.

Semiconductor Electrodes, STM, and Tunneling Spectroscopy Studies
Sonnenfeld et al.[84] investigated chemomechanically polished, heavily doped (1 to $4 \times 10^{19}\,cm^{-3}$) (001) GaAs samples immersed in 0.01 M aqueous KOH solution. The STM images revealed 100-nm features co-existing with large regions of ripples approximately 4 nm wide and 1 nm high. Spectroellipsometric data of similarly prepared surfaces on the same samples indicated a 0.5-nm-thick layer with a packing fraction of 50% relative to a lightly doped (001) GaAs standard, suggesting that the ellipsometric measurement is primarily sensitive to the ripples, not to the larger surface irregularities. Real-time STM images could also be observed in a chemically active solution, such as 0.01 M KOH solution.

Thundat et al.[85] studied photoelectrodeposition of Au on n- and p-GaAs(100) surfaces in a buffered acidic (pH = 3.5) plating solution and photocorrosion of p-GaAs(100) surface in NaOH solutions. Localized gold dots (ca. 200 nm in diameter) can be formed underneath the tip by photoelectrodeposition on p-GaAs samples; whereas for n-GaAs samples, dots were not formed directly underneath the tip. Instead, small Au islands (ca. 15 nm in diameter) were found in an annular region away from the tip. The dimensions of the Au dots depended on the length and height of the voltage pulse used for deposition. The extent of photocorrosion of p-GaAs depended on the tip bias, intensity of light and the pH of the solution.

Gilbert and Kennedy[86] studied the n-TiO$_2$(001) surface in air and found that the photoelectrochemical (PEC) characteristics of the n-TiO$_2$(001) electrode before and after PEC etching in 0.9 M NaF correlated well with the surface structure determined by STM. Tunneling spectroscopy (TS) was also performed on n-TiO$_2$ (001)[87, 88] and α-Fe$_2$O$_3$(0001) exposed to air at room temperature. Significant hysteresis in the tunneling spectra (Fig. 12.12) was observed on reversal of bias scan direction. The observed hysteresis is reproducible over many scan cycles, and its magnitude depends on scan amplitude, scan frequency, and bias polarity. The origin of this phenomenon has been attributed to field-induced gap states.

Itaya and Tomita[89, 90] studied the n-TiO$_2$ and n-ZnO/aqueous solution interface and the effect of substrate potential on the STM behavior. Electron transfer from the conduction band of the semiconductor to the tip was found to be possible, if the electrode potential was negative of the flat-band potential V_{fb} for n-type semiconductors.

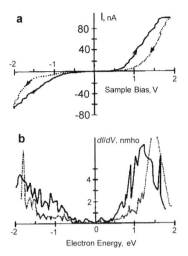

Figure 12.12. Example of hysteresis in STM tunneling spectra of TiO_2(00l) in air. **a**, Cyclic I/V characteristic. **b**, Differential conductivity spectra taken from above data. *Solid lines*, forward positive-direction scan; *dotted lines*, reverse negative-direction scan. (Reprinted with permission from Ref. 87.)

The topography and the surface electronic structure of p- and n-Si(100) surfaces were examined in 0.05 M H_2SO_4 aqueous solution by Tomita and co-workers.[91] Tunneling current-electrode potential (I/V) characteristics were measured with constant tunneling distance. The electrode potential was scanned from -0.5 to 0.4 V vs. SCE with a scan rate of 1 V/s while the tip potential was held at 0.45 V vs. SCE. The current first decreased to a plateau and then monotonically decreased when the electrode potential approached the tip potential. This I/V behavior was different from that observed for n-TiO_2 and ZnO, in which the current sharply decreased at V_{fb}, and has been attributed to the existence of surface states. Stable and reproducible STM images could also be obtained in 0.05 M H_2SO_4 solution before and after photo-oxidation of Si surfaces.

Sakamaki et al.[92,93] investigated the surface density of states (SDOS) of a hydroxylated and reduced n-TiO_2(110) surface and showed that the SDOS were different from the states in the bulk material. A surface band gap of 1.6 eV, which is different from the bulk value (3.05 eV), was reported. They also observed a gap state with an energy located at 1.35 eV below the Fermi level of the n-TiO_2 sample and attributed this to the OH-Ti^{3+} O vacancy state.

We[94] studied a strongly reduced n-type TiO_2(001) surface and examined its surface electronic states and topograph images in air. Good images of the surface could be obtained with the sample held at a fairly large negative voltage (e.g., -1.0 V). Two types of tunneling spectroscopic measurements were made with the tip held fixed at a given position above the TiO_2 surface. In I/V measurements, the current was recorded as a function of bias voltage between sample and tip. In conductance-voltage $(dI/dV/V)$ measurements, the magnitude of the 10 kHz modulation in-phase current was determined as a function of voltage. The conductance measurement reveals electron tunneling into the conduction band and the valence band, as well as a prominent peak identified as a surface state ca. 0.3 eV below the conduction band edge.

Similar techniques were applied to study heavily doped n-type FeS_2 (001) surfaces.[95] Figure 12.13 shows a typical conductance (dI/dV) spectrum for a heavily doped (ca. 2.7×10^{18} cm^{-3} doping density) n-FeS_2 single crystal sample over the bias voltage range of -1.8 to 1.8 V. As shown, there is a wide region of low conductance, referred to as the conductance well, terminating at both ends in regions of sharply rising conductance. The conductance well has a width close to the indirect band gap of FeS_2 (ca. 0.95 eV). The normalized conductance spectrum in the positive bias region is a broad peak of width ca. 1.5 eV that has several fine features superimposed, tentatively attributed to electron transfer between the tip and the broad, delocalized conduction band of FeS_2. In the negative bias region the conductance spectrum shows a narrow peak (0.5 eV wide) centered around ca. -1.6 V, which is superimposed on a high increasing background. This narrow peak has been attributed to electron transfer between the tip and the fairly localized uppermost valence level mainly consisting of Fe 3 d_{t2g} states. Also shown in Figure 12.13 is a small, but distinct, peak located at ca. -0.25 V, which might be associated with an impurity state.

More recently, atomic images of a Si(111) surface have been obtained in solution.[96, 97] Long-range ordered structure can be imaged in 1% HF after the Si(111) surface is pretreated with a mixed solution of H_2O_2 and HF. Several atomically flat terraces appear that are separated by mono atomic steps intersecting in a sawtooth pattern. This surface feature was also observed in UHV by Hessel et al.[98] and was attributed to the surface misorientation of the Si sample to the structures of the steps. An ordered structure, similar to that observed by STM in UHV[99] and by

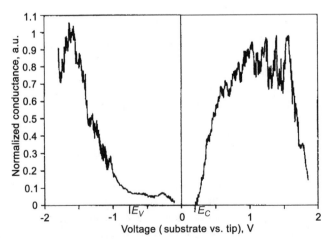

Figure 12.13. Normalized conductance $(dI/dV)/(I/V)$ as a function of bias voltage under the experimental conditions noted for Figure 12.15. (Reprinted with permission from Ref. 95.)

AFM in air,[100] is attributed to the $Si(111)-(1 \times 1){:}H$ phase, with a nearest atomic spacing of 0.38 ± 0.02 nm. STM has also been used to photoetch a CdSe thin film[101] and to study CdTe UPD on Au(100) by using the electrochemical atomic layer epitaxy technique. Parkinson[102] used STM to etch patterns on two-dimensional (Z-D) layer-type compounds (e.g., $SnSe_2$, $TiSe_2$), and Zhao and Fendler[103] characterized Zn and CdS particles grown from Langmuir-Blodgett monolayers.

12.3 SCANNING ELECTROCHEMICAL MICROSCOPY

12.3.1 Principles

Scanning electrochemical microscopy (SECM) involves measurements of the faradaic current that flows at the tip electrode to deduce information about substrate materials. Although instrumentation and electrochemical cells for SECM are similar to those used in STM, the principles of operation and the scope of applications are different. As discussed, the steady-state current that flows at a microdisk electrode when it is far from a substrate because of a faradaic reaction $(O + ne \rightarrow R)$ is given by Eq. (12.1). This faradaic current $(i_{T,\infty})$ represents the flux of O to the electrode surface by hemispherical diffusion (Fig. 12.14a). This current is perturbed when the

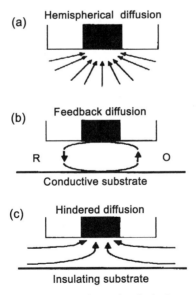

Figure 12.14. Basic principles of scanning electrochemical microscopy. **a,** The microelectrode tip is far from the substrate, where diffusion leads to a steady-state current $i_{T,\infty}$ (eq. (12.1)). **b,** The tip is near the conductive substrate, where feedback diffusion leads to $i_T > i_{T,\infty}$. **c,** The tip is near an insulating substrate, where hindered diffusion leads to $i_T < i_{T,\infty}$. (Reprinted with permission from Ref. 104.)

tip is brought close (i.e., on the order of the tip diameter) to a substrate. If the substrate is an insulator (Fig. 2.14**b**), diffusion of O to the tip is blocked, and the tip current i_T will be smaller than $i_{T,\infty}$. The closer the tip is to the substrate the smaller i_T is. On the other hand, if the tip is brought close to a conductive substrate that is at a potential at which the product of the tip reaction R can be oxidized to O, i_T will be larger than $i_{T,\infty}$, because of feedback of O from substrate to tip (Fig. 12.14**c**). For a conductive substrate, the closer the tip is to the substrate, the larger i_T is. Thus, as in STM, the tip current can be used to image substrates (e.g., by measuring i_T as a function of the xy position). The magnitude of i_T compared to $i_{T,\infty}$ indicates the nature of the surface (i.e., electrically conducting or insulating) and the distance of the tip from the surface. As described in review articles[104, 105] SECM has been used to image metals, polymers, biologic materials, and semiconductor materials.

12.3.2 SECM Theory

The magnitude of i_T as a function of tip–substrate spacing d can be calculated for both insulating and conductive substrates by using well-established models of electrochemical systems.[106] When i_T is normalized with respect to $i_{T,\infty}$, and d is normalized with respect to the tip disk radius a, general dimensionless plots result that can be used to find d from a measured value of i_T. Such a plot for the steady-state current above a conductor and insulator is given in Figure 12.15. Note that

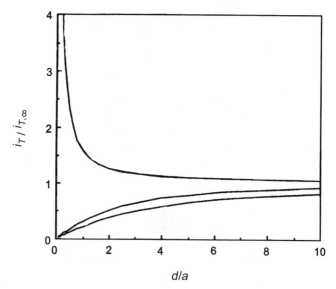

Figure 12.15. Plot of i_T normalized with respect to $i_{T,\infty}$ as a function of the tip–substrate distance normalized with respect to the radius a of the tip. Above a conducting substrate (*top curves*) different sheath sizes exhibit no difference in tip current. Above an insulating substrate (*bottom curves*) different sheath sizes lead to different tip currents (the larger the sheath, the smaller the tip current).

these curves are not based on any adjustable or unknown parameters and are independent of such factors in Eq. (12.1) as C and D. The i_T value above an insulator does depend on the radius of the insulating sheath around the conductive disk portion of the tip, however; and the model does assume an SECM tip geometry of a planar disk surrounded by a planar insulating sheath, as shown in Figure 12.14. Indeed the tip geometry can be an important factor in determining the SECM response.[107] The model also assumes that the rate of the regeneration reaction ($R \rightarrow O + ne$) is either zero (insulator) or infinite (conductor). Results are also available for intermediate rates of this reaction at a substrate,[108, 109] and for the transient current that flows in SECM (which governs, for example, the xy scan rate that can be employed in imaging).[110] A typical example of imaging an interdigitated array is shown in Figures 12.16 and 12.17.[111] Note that this array contains both insulating and conductive zones. The SECM is also useful in extracting thermodynamic and kinetic parameters for reactions at insulator surfaces, e.g., those of minerals[112, 113] or membranes containing enzymes.[114] Additional applications involve the study of polymer films[115, 116, 117, 118] and diffusion through membranes.[119, 120]

12.3.3 Fabrication with the SECM

Electrochemical methods are widely used for fabrication and surface modification; typical examples include metal electrodeposition and etching, electromachining, and formation of electronically conductive polymers. Such methods can be readily transported into SECM, and several examples of surface modification at high resolution by controlled reactions at tip or substrate have been reported. An early example was the photoetching of single crystal GaAs.[121] In this case, the GaAs was immersed in an aqueous solution containing 5 mM NaOH and 1 mM EDTA. When the sample was irradiated as the tip was scanned across the surface, a line of submicrometer width was etched in the GaAs surface. The direct deposition of submicrometer silver lines within a polymer (Nafion) matrix with the SECM was also reported.[122] Details concerning the direct deposition and etching of metals have been discussed.[123, 124]

The feedback approach can be used for these purposes. Here the substrate can be a polymer film containing the desired metal; a typical example would be a poly (vinylpyridine) film in the protonated form containing gold as the $AuCl_4^-$ complex.[125] A reductant, such as $Ru(NH_3)_6^{2+}$, is generated at the tip and diffuses to the substrate where the complex is reduced to the metal. A similar feedback approach can be used to etch a semiconductor, such as GaAs.[126] In this case a strong oxidant is generated at the tip, which is held in close proximity to the semiconductor surface. The feedback approach to fabrication is especially useful because the feedback current can be monitored and controlled to maintain tip–substrate distance. Moreover, in the feedback mode the substrate need not be a conductive material. In an extension of the etching of semiconductors, the SECM was used as an analytical tool to study the mechanism of the etching processes.[127]

(a)

(b)

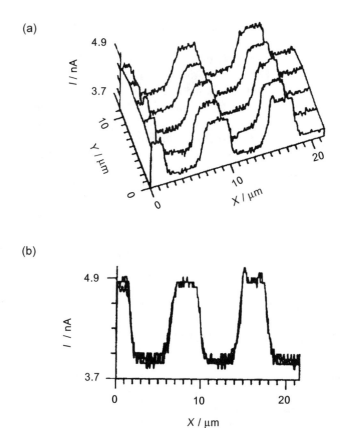

Figure 12.16. Scan of an interdigitated array consisting of 3-μm-wide Pt bands alternating with 5-μm wide SiO_2 spaces: **a**, 3-D view; **b**, side view. Tip, ca. 0.2 μm diameter Pt microdisk; solution, 40 mM methyl viologen dichloride in 2 M KCl; $i_{T,\infty} = 3.7$ nA; $E_T = -0.76$ V vs. SCE. (Reprinted with permission from Ref. 110.)

12.3.4 SECM in Thin Water Films in Air

SECM imaging in thin films of water that form on the surface of a hydrophilic substrate like mica is also possible. In the first reports of this technique,[128] an STM with pA current capability was used to image species on a mica substrate with a metal contact and the imaging was explained in terms of tunneling. However, electron tunneling through the liquid film between the tip and the distant contact is unlikely and an electrochemical mode involving redox processes in the water film at the tip has been proposed.[129, 130] Imaging in air allows much higher resolution than liquid phase SECM, since a sharp tip (W or Pt), whose area is defined by the small contact with the water film, can be employed. Imaging of DNA and protein molecules by this approach has been described.[128, 131]

(a)

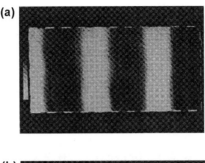

(b)

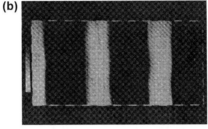

Figure 12.17. Gray-scaled image of Figure 12.16 (a) from minimum to maximum current and (b) from average to maximum current. The *white lines* around the edge denote a 1-μm distance, and the scale at the left shows gray scale variation: *light color*, large cathodic current; *dark color*, small cathodic current. (Reprinted with permission from Ref. 110.)

12.4 SUMMARY

As with all chapters in this book, this one emphasizes fundamental aspects of scanning probes. The results chosen to illustrate concepts in electrochemistry are not always the most recent. STM has continuously been used as a very effective and efficient tool for real-time, in-situ characterization of both single-crystalline and mesoscopic structures at the solid/electrolyte interface. Defects such as vacancies, dislocations, surface steps, kink sites, adatoms and their dynamics have been studied on an atomic scale basis. Nucleation and growth processes associated with surface reconstruction, adsorption/desorption, deposition/dissolution, passivation and other phenomena have also been examined. Quantitative evaluation of the rate parameters associated with various surface processes in combination with simulation and theory has started to emerge and may allow electrochemists to examine many long-standing questions concerning the interplay between electrode structures and electrochemical kinetics. Readers interested in more examples or in the most recent research results may consult conference proceedings[132–135] on STM in electrochemistry as well as some excellent recent reviews of STM studies on semiconductor[136] and metal[137] electrodes. Detailed reviews on SECM have also appeared.[138–143]

REFERENCES

1. A. J. Bard and L. R. Faulkner, *Electrochemical Methods*, John Wiley & Sons, Inc., New York, 2nd ed. 2000.

2. D. M. Kolb, *Ber. Bunsenges. Phys. Chem.* **92**, 1175 (1988) and references therein.

3. R. M. Wightmin and D. O. Wipf, in A. J. Bard, ed., *Electroanalytical Chemistry*, Vol. 15, Marcel Dekker, New York, 1988.

4. A. A. Gewirth, D. H. Craston, and A. J. Bard, *J. Electroanal. Chem. Interfacial Electrochem.* **261**, 477 (1989).

5. R. Christoph, H. Siegenthaler, H. Rohrer, and H. Wiese, *Electrochim. Acta* **34**, 1011 (1989).

6. D. J. Trevor, C. E. D. Chidsey, and D. N. Loiacono, *Phys. Rev. Lett.* **62**, 929 (1989).

7. L. A. Nagahara, T. Thundat, and S. M. Lindsay, *Rev. Sci. Instrum.* **60**, 3128 (1989).

8. S.-L. Yau, C. M. Vitus, and B. C. Schardt, *J. Am. Chem. Soc.* **112**, 3577 (1990).

9. K. Itaya and E. Tomita, *Surf. Sci.* **201**, L507 (1988).

10. M. J. Heben, M. M. Dovek, N. S. Lewis, R. M. Penner, and C. F. Quate, *J. Microsc.* **152**, 651 (1988).

11. R. Sonnenfeld, J. Schneir, and P. K. Hansma, in R. E. White, J. O'M. Bockris, and B. E. Conway, eds., *Modern Aspect of Electrochemistry*, no. 21, Plenum, New York, 1990.

12. J. Wiechers, T. Twomey, D. N. Kolb, and R. J. Behm, *J. Electroanal. Chem. Interfacial Electrochem.* **248**, 451 (1988).

13. P. Lustenberger, H. Rohrer, R. Christoph, and H. Siegenthaler, *J. Electroanal. Chem. Interfacial Electrochem.* **243**, 225 (1988).

14. R. Woods in A. J. Bard, ed., *Electroanalytical Chemistry*, Vol. 9, Marcel Dekker, New York, 1976.

15. M. M. Dovek, M. J. Heben, C. A. Lang, N. S. Lewis, and C. F. Quate, *Rev. Sci. Instrum.* **59**, 2333 (1988).

16. A. A. Gewirth and A. J. Bard, *J. Phys. Chem.* **92**, 5363 (1988).

17. M. H. J. Hottenhuis, M. A. H. Mickers, J. W. Gerritsen, and J. P. van der Eerden, *Surf. Sci.* **206**, 259 (1988).

18. S. Morita, I. Otsuka, T. Okada, H. Yokoyama, T. Iwasaki, and N. Mikoshiba, *Jpn. J. Appl. Phys.* **26**, L1853 (1987).

19. I. Otsuka and T. Iwasaki, *J. Microsc.* **152**, 289 (1988).

20. Digital Instruments, *Operating Manual. Nanoscope II Electrochemical STM.* Digital Instruments, Santa Barbara, Calif., 1990.

21. R. Sonnenfeld and P. K. Hansma, *Science* **232**, 211 (1986).

22. H.-Y. Liu, F.-R. F. Fan, C. W. Lin, and A. J. Bard, *J. Am. Chem. Soc.* **108**, 3838 (1986).

23. K. Uosaki and H. Kita, *J. Electroanal. Chem. Interfacial Electrochem.* **239**, 301 (1989).

24. O. Lev, F.-R. Fan, and A. J. Bard, *J. Electrochem. Soc.* **135**, 783 (1988).

25. M. Szklarczyk and J. O'M. Bockris, *J. Electrochem. Soc.* **137**, 452 (1990).

26. J. Gómez, L. Vázuez, A. M. Baró, N. García, C. L. Perdriel, W. E. Triaca, and A. J. Arvía, *Nature* **323**, 612 (1986).

27. L. Vázquez, J. Gómez, A. M. Baró, et al., *J. Am. Chem. Soc.* **109**, 1730 (1987).

28. L. Vázquez, J. M. Gómez Rodriguez, J. Gómez Herrero, A. M. Baró, N. García, J. C. Canullo, and A. J. Arvía, *Surf. Sci.* **181**, 98 (1987).

29. A. J. Arvía, R. C. Salvarezza, and W. E. Triaca, *Electrochim. Acta* **34**, 1057 (1989).

30. J. Gómez, L. Vázquez, A. M. Baró, et al., *J. Electroanal. Chem. Interfacial Electrochem.* **240**, 77 (1988).

31. S. R. Snyder, H. S. White, S. López, and Héctor D. Abruña, *J. Am. Chem. Soc.* **112**, 1333 (1990).

32. I. Otsuka and T. Iwasaki, *J. Vac. Sci. Technol. A* **8**, 530 (1990).

33. B. A. Sexton, G. F. Cotterill, S. Fletcher, and M. D. Horne, *J. Vac. Sci. Technol.* 544.

34. R. J. Nichols, O. M. Magnussen, J. Hotlos, T. Twomey, R. J. Behm, and D. M. Kolb, *J. Electroanal. Chem. Interfacial Electrochem.* **290**, 21 (1990).

35. B. C. Schardt, S.-L. Yau, and F. Rinaldi, *Science* **243**, 1050 (1989).

36. K. Uosaki, J. Wang, and H. Kita, *J. Electroanal. Chem. Interfacial Electrochem.* **273**, 275 (1989).

37. S. Yoshihara, E.-I. Sato, and A. Fujishima, *Ber. Bunsenges. Phys. Chem.* **94**, 603 (1990).

38. M. Szklarczyk, Lj. Minevski, and J. O'M Bockris, *J. Electroanal. Chem. Interfacial Electrochem.* **289**, 279 (1990).

39. K. Itaya and S. Sugawara, *Chem. Lett.* 1927 (1987).

40. K. Itaya, S. Sugawara, and K. Higaki, *J. Phys. Chem.* **92**, 6714 (1988).

41. J. Schneir and P. K. Hansma, *Langmuir* **3**, 1025 (1987).

42. R. Sonnenfeld and B. C. Schardt, *Appl. Phys. Lett.* **49**, 1172 (1986).

43. B. Drake, R. Sonnenfeld, J. Schneir, and P. K. Hansma, *Surf. Sci.* **181**, 92 (1987).

44. J. Schneir, V. Elings, and P. K. Hansma, *J. Electrochem. Soc.* **135**, 2774 (1988).

45. L. A. Nagahara, S. M. Lindsay, T. Thundat, and U. J. Knipping, *J. Microsc.* **152**, 145 (1988).

46. F.-R. Fan and A. J. Bard, *Anal. Chem.* **60**, 751 (1988).

47. K. Itaya, K. Higaki, and S. Sugawara, *J. Chem. Soc. Faraday Trans. I* **85**, 1351 (1989).

48. S. M. Lindsay, L. A. Nagahara, T. Thundat, and P. Oden, *J. Biomol. Struct. Dynam.* **7**, 289 (1989).

49. S. M. Lindsay, T. Thundat, and L. Nagahara, *J. Microsc.* **152**, Pt. 1, 213 (1988).

50. T. Thundat, L. A. Nagahara, P. Oden, and S. M. Lindsay, *J. Vac. Sci. Technol. A* **8**, 645 (1990).

51. T. Twomey, J. Wiechers, D. M. Kolb, and R. J. Behm, *J. Microsc.* **152**, 537 (1988).

52. K. Itaya, S. Sugawara, K. Sashikata, and N. Furuya, *J. Vac. Sci. Technol. A* **8**, 515 (1990).

53. F.-R. Fan and A. J. Bard, *J. Electrochem. Soc.* **136**, 3216 (1989).

54. K. Uosaki and H. Kita, *J. Electroanal. Chem. Interfacial Electrochem.* **259**, 301 (1989).

55. K. Sakamaki, K. Itoh, A. Fujishima, and Y. Gohshi, *J. Vac. Sci. Technol. A* **8**, 525 (1990).

56. K. Sakamaki, K. Itoh, A. Fujishima, Y. Gohshi, and H. Nakagawa, *Bull. Chem. Soc. Jpn.* **62**, 2890 (1989).

57. R. S. Robinson, *J. Electrochem. Soc.* **136**, 3145 (1989).

58. R. L. McCarley and A. J. Bard, *J. Phys. Chem.* **95**, 9618 (1991).

59. X. Gao, A. Hamelin, and M. J. Weaver, *J. Chem. Phys.* **95**, 6993 (1991).

60. X. Gao, A. Hamelin, and M. J. Weaver, *Phys. Rev.* **B44**, 10983 (1991).

61. S.-L. Yau, X. Gao, S.-C. Chang, B. C. Schardt, and M. J. Weaver, *J. Am. Chem. Soc.* **113**, 6049 (1991).

62. C. M. Vitus, S.-C. Chang, B. C. Schardt, and M. J. Weaver, *J. Phys. Chem.* **95**, 7559 (1991).

63. X. G. Zhang and U. Stimming, *Corrosion Sci.* **30**, 951 (1990).

64. F. Fan and A. Bard, *J. Electrochem. Soc.* **136**, 166 (1989).

65. T. P. Moffat, F.-R. Fan, and A. J. Bard, *J. Electrochem. Soc.* **138**, 3224 (1991).

66. H. W. Pickering and P. J. Byrne, *J. Electrochem. Soc.* **118**, 209 (1971).

67. I. C. Oppenheim, D. J. Trevor, C. E. D. Chidsey, P. L. Trevor, and K. Sieradzki, *Science* **254**, 687 (1991).

68. M. P. Green, K. J. Hanson, D. A. Scherson, X. Xing, M. Richter, P. N. Ross, R. Carr, and I. Lindau, *J. Phys. Chem.* **93**, 2181 (1989).

69. J. P. van der Eerden, M. A. Mickers, J. W. Gerritsen, and M. H. J. Hottenhuis, *Electrochim. Acta* **34**, 1141 (1989).

70. B. Wichman, J. P. van der Eerden, and J. W. Gerritsen, *J. Cryst. Growth* **99**, 1333 (1990).

71. R. S. Robinson, *J. Microsc.* **152**, 541 (1988).

72. M. Szklarczyk, O. Velev, and J. O'M. Bockris, *J. Electrochem Soc.* **136**, 2433 (1989).

73. O. M. Magnussen, J. Hotlos, R. J. Nichols, D. M. Kolb, and R. J. Behm, *Phys. Rev. Lett.* **64**, 2929 (1990).

74. R. S. Robinson, *J. Vac. Sci. Technol. A* **8**, 511 (1990).

75. K. Sashikata, N. Furuya, and K. Itaya, *J. Electroanal. Chem. Interfacial Electrochem.* **316**, 361 (1991).

76. S. Manne, P. K. Hansma, J. Massie, V. B. Elings, and A. A. Gewirth, *Science* **251**, 183 (1991).

77. C. H. Chen and A. A. Gewirth, *Phys. Rev. Lett.* **68**, 1571 (1992).

78. T. Hachiya, H. Honbo, and K. Ithaya, *J. Electroana. Chem.* **315**, 275 (1991).

79. H. Chang and A. J. Bard, *J. Am. Chem. Soc.* **112**, 4598 (1990).

80. H. Chang and A. J. Bard, *J. Am. Chem. Soc.* **113**, 5598 (1991).

81. S. A. Hendricks, Y.-T. Kim, and A. J. Bard, *J. Electrochem. Soc.* **139**, 2818 (1992).

82. R. Yang, K. M. Dalsin, D. F. Evans, L. Christensen, and W. A. Hendrickson, *J. Phys. Chem.* **93**, 511 (1989).

83. L. Christensen, W. A. Hendrickson, R. Yang, and D. F. Evans, *Polymer Prepr.* **30**, 100 (1989).

84. R. Sonnenfeld, J. Schneir, B. Drake, P. K. Hansma, and D. E. Aspnes, *Appl. Phys. Lett.* **50**, 1742 (1987).

85. T. Thundat, L. A. Nagahara, and S. M. Lindsay, *J. Vac. Sci. Technol. A* **8**, 539 (1990).

86. S. Gilbert and J. H. Kennedy, *J. Electrochem. Soc.* **135**, 2385 (1990).

87. S. Gilbert and J. H. Kennedy, *Surf. Sci.* **225**, L1 (1990).

88. S. Gilbert and J. H. Kennedy, *Langmuir* **5**, 1412 (1989).

89. K. Itaya and E. Tomita, *Chem. Lett.* 1989, 285.

90. K. Itaya and E. Tomita, *Surf. Sci.* **219**, L515 (1989).

91. E. Tomita, N. Matsuda, and K. Itaya, *J. Vac. Sci. Technol. A* **8**, 534 (1990).

92. K. Sakamaki, S. Matsunaga, K. Itoh, A. Fujishima, and Y. Gohshi, *Surf. Sci.* **219**, L531 (1989).

93. K. Sakamaki, K. Itoh, A. Fujishima, and Y. Gohshi, *J. Vac. Sci. Technol. A* **8**, 614 (1990).

94. F.-R. Fan and A. J. Bard, *J. Phys. Chem.* **94**, 3761 (1990).

95. F.-R. Fan and A. J. Bard, *J. Phys. Chem.* **95**, 1969 (1991).

96. S.-L. Yau, F.-R. F. Fan, and A. J. Bard, *J. Electrochem. Soc.* **139**, 2825 (1992).

97. K. Itaya, S. Sugawara, Y. Morita, and H. Tolumoto, *Appl. Phys. Lett.* **60**, 2534 (1992).

98. H. Hessel, A. Felts, M. Reiter, U. Memment, and R. J. Behm, *Chem. Phys. Lett.* **186**, 275 (1991).

99. R. S. Becker, G. S. Higashi, Y. J. Chabal, and A. J. Becker, *Phys. Rev. Lett.* **65**, 1917 (1990).

100. Y. Kim and C. M. Lieber, *J. Am. Chem. Soc.* **113**, 2333 (1991).

101. C.-Y. Liu and A. J. Bard, *Chem. Phys. Lett.* **174**, 162 (1990).

102. B. Parkinson, *J. Am. Chem. Soc.* **112**, 7498 (1990).

103. X. K. Zhao and J. H. Fendler, *J. Phys. Chem.* **95**, 3716 (1991).

104. A. J. Bard, G. Denuault, C. Lee, D. Mandler, and D. O. Wipf, *Acc. Chem. Res.* **23**, 357 (1990).

105. A. J. Bard, F.-R. F. Fan, D. T. Pierce, P. R. Unwin, D. O. Wipf, and F. Zhou, *Science* **254**, 68 (1991).

106. J. Kwak and A. J. Bard, *Anal. Chem.* **61**, 1221 (1989).

107. J. M. Davis, F.-R. F. Fan, and A. J. Bard, *J. Electroanal. Chem.* **238**, 9 (1987).

108. M. V. Mirkin and A. J. Bard, *J. Electroanal. Chem.* **323**, 1 (1992).

109. D. O. Wipf and A. J. Bard, *J. Electrochem. Soc.* **138**, 469 (1991).

110. A. J. Bard, G. Dennuault, B. C. Dormblaser, R. A. Friesner, and L. S. Tuckerman, *Anal. Chem.* **63**, 1282 (1991).

111. V. Lee, C. J. Miller, and A. J. Bard, *Anal. Chem.* **63**, 78 (1991).

112. P. R. Unwin and A. J. Bard, *Anal. Chem.* **64**, 113 (1992).

113. P. R. Unwin and A. J. Bard, *J. Phys. Chem.* **96**, 5035 (1992).

114. D. T. Pierce, P. R. Unwin, and A. J. Bard, *Anal. Chem.* **64**, 1795 (1992).

115. C. Lee and A. J. Bard, *Anal. Chem.* **62**, 1906 (1990).

116. M. V. Mirkin, F.-R. F. Fan, and A. J. Bard, *Science* **257**, 364 (1992).

117. C. Lee and F. C. Anson, *Anal. Chem.* **64**, 528 (1992).

118. J. Kwak and F. C. Anson, *Anal. Chem.* **64** 250 (1992).

119. E. R. Scott, H. S. White, and J. B. Phipps, *J. Membrane Sci.* **58**, 71 (1991).

120. E. R. Scott, H. S. White, and J. B. Phipps, *Solid State Ionics* **53–56**, 176 (1992).

121. C. W. Lin, F.-R. F. Fan, and A. J. Bard, *J. Electrochem. Soc.* **134**, 1038 (1987).

122. D. H. Craston, C. W. Lin, and A. J. Bard, *J. Electrochem. Soc.* **135**, 785 (1988).

123. O. E. Hüsser, D. H. Craston, and A. J. Bard, *J. Electrochem. Soc.* **136**, 3222 (1989).

124. O. E. Hüsser, D. H. Craston, and A. J. Bard, *J. Vac. Sci. Technol.* **B6**, 1873 (1988).

125. D. Mandler and A. J. Bard, *J. Electrochem. Soc.* **137**, 1079 (1990).

126. D. Mandler and A. J. Bard, *J. Electrochem. Soc.* **137**, 2468 (1990).

127. D. Mandler and A. J. Bard, *Langmuir* **6**, 1489 (1990).

128. R. Guckenberger, M. Heim, G. Cevec, H. F. Knapp, W. Wiegrabe, and A. Hillebrand, *Science* **266**, 1538 (1994).

129. F.-R. F. Fan and A. J. Bard, *Science* **270**, 1849 (1995).

130. F. Forouzan and A. J. Bard, *J. Phys. Chem. B.* **101**, 10876 (1997).

131. F.-R. F. Fan and A. J. Bard, *Proc. Natl. Acad. Sci, USA* **96**, 14622 (1999).

132. "Nanoscale Probes of the Solid/Liquid Interfaces," A. A. Gewirth and H. Siegenthaler, Eds., NATO ASI Series E, Vol. 288, Kluwer Academic Publishers, Boston, 1995.

133. Proceedings of the IUVSTA Workshop on Surface Science and Electrochemistry, *Surf. Sci.* **335** (1995).

134. Proceedings of the 6th International Fischer Symposium on Nanotechniques in Electrochemistry, *Electrochim. Acta* **40**, 10 (1995).

135. "In-Situ Local Probe Techniques for Studies of Electrochemical Interface," W. J. Lorenz and W. Plieth, Eds., VCH-Wiley, New York, 1998.

136. P. Allongue in "Advances in Electrochemical Science and Engineering," H. Gerischer and C. W. Tobias, Eds., VCH Publishers, New York, 1995.

137. T. P. Moffat in "Electroanalytical Chemistry," A. J. Bard and I. Rubinstein, Eds., Vol. 21, Marcel Dekker, New York, 1999.

138. A. J. Bard, F.-R. F. Fan, and M. V. Mirkin, in "Electroanalytical Chemistry," A. J. Bard and I. Rubinstein, Eds., Marcel Dekker, New York, 1993, Vol. 18, pp. 243–373.

139. M. Arca, A. J. Bard, B. R. Horrocks, T. C. Richards, and D. A. Treichel, *Analyst* **119**, 719–26 (1994).

140. M. V. Mirkin, *Mikrochim. Acta* **130**, 127–153 (1999).

141. A. J. Bard, F.-R. F. Fan, and M. V. Mirkin, in "The Handbook of Surface Imaging and Visualization," A. T. Hubbard, Ed., CRC, Boca Raton, Fl, 1995, pp. 667–679.

142. A. J. Bard, F.-R. F. Fan, and M. V. Mirkin, in "Physical Electrochemistry: Principles, Methods and Applications." I. Rubinstein, Ed., Marcel Dekker, New York, 1995, pp. 209–242.

143. A. J. Bard and M. V. Mirkin, Eds., "Scanning Electrochemical Microscopy," Marcel Dekker, New York, 2001.

APPENDICES

APPENDIX I

DEFINITION OF VARIABLES

A	area, amplitude, Hamaker constant
A^*	Richardson constant
A_0, A_1, A_2	magnitudes of static (zero frequency), first- and second-harmonic responses
a	lattice parameter, interatomic spacing, contact area
$\mathbf{a_1, a_2, b_1, b_2}$	unit mesh vectors
α	static polarizability, critical exponent
$\mathbf{B}$	magnetic flux density
β	critical exponent
C	capacitance, London dispersion force coefficient, concentration
c	number of components, speed of light
$\chi(\mathbf{s})$	susceptibility
χ_{ij}	components of strain tensor
χ	magnetic susceptibility
D	separation distance, diffusion coefficient
$D(E_x)$	transmission probability, energy distribution
D_j	instability distance
D_o	interfacial separation at contact
d	sample–tip separation, thickness, interatomic spacing, depth, tunnel gap
d_{ijk}	components of the tensor of piezocoefficients
$\Delta(z)$	corrugation in the z-direction
E	energy, elastic modulus
$E(\mathbf{k})$	dispersion relation
E_C	critical potential
E_D	activation energy of desorption
E_F	Fermi energy level
E_g	band gap energy
$\mathbf{E}$	electric field
E_k	kth component of $\mathbf{E}$

ε	dielectric constant, interaction strength
ε_{o}	permittivity in free space, on-site energy
F	force
F_{o}	long-range force
$F_{d}(D)$	attractive force during deformation
$F_{h}(D)$	Hertzian repulsion
$F_{dc}, F_{1\omega}, F_{2\omega},$	static (zero frequency), first- and second-harmonic components of the force
f	degrees of freedom
Φ	barrier height, work function
ϕ	work function, order parameter
ϕ_{s}	work function of the sample
$\phi_{s}(\mathbf{r})$	scalar magnetostatic potential
ϕ_{t}	work function of the tip
G	Gibb's free energy
$G(D)$	sample–tip interaction force
γ	surface energy, critical exponent
H	Hamiltonian operator, magnetic field
h	Miller index along x-direction, Planck constant, height, neck height
I_{t}, i, I	tunneling current
I_{C}	collector current
I_{F}	Faradaic current
I-V	current–voltage relationship
I/V	integral conductance
dI/dV	differential conductance
φ	time-independent wave function, phase shift
φ_{1}, φ_{2}	phase shifts for first- and second-harmonic responses
$\psi_{s}(\mathbf{r})$	magnetostatic free energy
K	effective elastic modulus
$\mathbf{k}_{x}, \mathbf{k}_{y}, \mathbf{k}_{z}$	wave vectors along x, y, and z directions
k	Miller index along y-direction, Boltzmann constant, spring constant
k_{B}	Boltzmann constant
κ	decay length, spring constant
κ^{-1}	Debye length
L	length
L_{s}	latent heat of sublimation
l	length, Miller index along z-direction
λ	wavelength, attenuation length
M	magnetic permittivity, molar mass
$\mathbf{M}(\mathbf{r})$	magnetization vector
M_{ijkl}	components of the tensor of electrostriction coefficients
m	mass
m_{i}	valency of species i

m	magnetic moment
μ	dipole moment
μ_0	magnetic permittivity
μ_i	chemical potential of species i
N_i	moles of species i
n	electron density, index of refraction
n	normal unit vector
$\langle n_i \rangle$	average site occupation
$\langle n_i n_j \rangle$	pair correlation function
ν	frequency, Poisson's ratio, critical exponent
P	pressure
P	polarization vector
p	number of phases, momentum, vapor pressure
p_o	nominal vapor pressure of liquid
q	Fourier wave vector
Q_t	effective tip charge
Q_t	effective surface charge
Q_{dc}	charge induced by dc component of tip bias
Q_{ac}	charge induced ac component of tip bias
q	magnetic charge
R	resistance, radius of curvature
$R(\theta)$	angle between unit mesh vectors
R	displacement vector
r	radius
r	displacement vector
ρ	density, resistivity
ρ_s	sample electron density
ρ_t	tip electron density
$\rho(\mathbf{r}, E_F)$	local density of states at E_F
S	surface area, entropy
s	sample–tip separation
s	scattering vector
σ	surface charge density, conductivity
σ_o	hopping conductivity
Θ	coverage
θ_C	critical angle
T	translation vector
T	temperature, hopping matrix element
T_C	critical temperature
T_m	melting temperature
T_p	precipitation temperature
$T(E, V)$	tunneling probability
t	time, reduced temperature
τ	hopping matrix element
U	potential energy, internal energy

$\langle u^2 \rangle^{1/2}$	root mean square thermal vibrational amplitude
V	voltage, volume
V_{ac}	ac bias voltage
V_B	barrier height
V_{dc}	dc bias voltage
V_{stab}	stabilization voltage
W	charge transfer rate
W_{ab}	transition probability from a to b
w	work of adhesion
ω	angular frequency, resonance frequency, critical exponent
x	displacement along x-direction
ξ	correlation length
y	displacement along y-direction
Z	atomic number
z	sample–tip separation, displacement along z-direction

APPENDIX II

DEFINITION OF ACRONYMS

A–D	analog–digital
AES	Auger electron spectroscopy
AFM	atomic force microscopy(e)
AgQRE	silver quasi-reference electrode
ARUPS	angle-resolved ultraviolet photoelectron spectroscopy
BEEM	ballistic electron emission microscopy(e)
CCT	constant current topograph
CE	counterelectrode
CITS	current imaging tunneling spectroscopy
DBQW	double-barrier quantum well
DLVO	Derjaguin-Landau-Verwey-Overbeek
DMT	Derjaguin-Muller-Toporov
DOS	density of states
ECSTM	electrochemical scanning tunneling microscopy
EELS	electron energy loss spectroscopy
EFM	electrostatic force microscopy(e)
EM	electron microscopy(e)
FIM	field ion microscopy(e)
FM	force microscopy(e)
HOMO	highest occupied molecular orbital
HOPG	highly oriented pyrolytic graphite
HREELS	high-resolution electron energy loss spectroscopy
HRTEM	high-resolution transmission electron microscopy(e)
IPS	inverse photoemission spectroscopy
IR	infrared
IRRAS	infrared reflection-absorption spectroscopy
JKRS	Johnson-Kendall-Roberts-Sperling
KP(F)M	Kelvin probe (force) microscopy(e)
LB	Langmuir-Blodgett
LCM	leakage current microscopy(e)
LDOS	local density of states

LEED	low-energy electron diffraction
LEEM	low energy electron microscopy(e)
LRO	long-range order
LUMO	lowest unoccupied molecular orbital
MBE	molecular beam epitaxy
MFM	magnetic force microscopy(e)
MYD/BHW	Muller-Yushchenko-Derjaguin / Burgers-Hughes-White
NDR	negative differential resistance
PE	photoemission
PEC	photoelectrochemical
PP	polypyrrole
PR	photoresponse
PRI	piezoresponse imaging
PZT	$PbTi_xZr_{1-x}O_3$
QMR	quantum-mechanical reflection
RBS	Rutherford backscattering
RE	reference electrode
REM	reflection electron microscopy
RSWPS	repetitive square wave potential signal
SCE	standard calomel electrode
SCM	scanning capacitance microscopy(e)
SDOS	surface density of states
SECM	scanning electrochemical microscopy(e)
SEM	scanning electron microscopy(e)
SEMPA	scanning electron microscopy(e) with polarization analysis
SFA	surface force apparatus
SFM	scanning force microscopy(e)
SPM	scanning probe microscopy(e)
SRO	short-range order
SSM	scanning SQUID (superconductive quantum interference device) microscopy(e)
SSRM	scanning spreading resistance microscopy(e)
STM	scanning tunneling microscopy(e)
STS	scanning tunneling spectroscopy(e)
TEM	transmission electron microscopy(e)
TS	tunneling spectroscopy
UHV	ultra-high vacuum
UPD	underpotential deposition
UPS	ultraviolet photoemission spectroscopy
WE	working electrode
WKB	Wentzel-Kramers-Brillouin

APPENDIX III

EMPIRICAL SIMULATION OF STM IMAGES
APPLICATION TO OXIDES

J. R. Smith and Dawn A. Bonnell

A *C*-language based computer program for simulating STM images of binary compounds is presented here. The program works as follows. A surface template is input, along with the effective radii of each ion in that template. Also input are the charge density ($\sim$to STM current) at which the simulated image is to be calculated and the area of the simulated image. Then, the tip–sample distance corresponding to the charge density is calculated. Only 256×256 pixels are calculated in the image, so the scale obviously depends on the image area. The calculated image is a simulation of a constant current image. Once a data set has been obtained (an array of 256×256 pixels), the data are rescaled and converted to a standard image format (tif, etc.). The following is a brief synopsis of each of the major routines in the program. As presented, the code is configured to calculate the $TiO_2(110)$ surface.

1. Coordinate transform: a routine that changes the coordinates of any orbital in the template. This is used to change the orientation of any of the orbitals in the calculation.
2. The normalization constants for the wave functions must be calculated independently and used in the routine CalcChDen, described below.
3. RADO, RADTI, ANGO, ANGTI: routines for the radial and angular portions of orbitals in the calculation. The radii for the radial functions are adjustable.
4. CalcChDen: a routine that calculates the empty state density (ESD) owing to the template surface for any point in space.
5. CalcImage: a routine that uses CalcChDen to calculate the tip–sample distance in the direction normal to the surface (z) that gives a constant current. This calculation is performed for each pixel.

```
#include<stdio.h>
#include <math.h>
#define NoPtsGrid 65536        /* number of pixels on the output
grid */
#define XLimit 256
#define YLimit 256             /* should be sqrt of
NoPtsGrid */
#define GridLengthHeight 30.0        /* height/width of grid
 (Angstroms) */
#define Ion2Rad 1.4            /* 1 denotes oxygen, 2 denotes
Ti */
#define Ion1Rad 0.9
#define ScalePts 3000          /* # of pts on scale of output
data */
#define ZLimit 10                 /* Z limit
(Angstroms) */
#define ZIncrement 1e-2
#define aConst 5
#define bConst 0.001
#define cConst 5
#define dConst 1.0            /* angular and radial
portions */
#define eConst 1
#define fConst 1
#define gConst 1
#define hConst 1
#define dConst2 1.0
#define eConst2 1
#define fConst2 1
#define gConst2 1
#define hConst2 1
#define MaxNoPtsTmplt 1000              /* number atoms in
tmplt */
#define Pi 3.1459              /* Mathematical
Constant */
#define Ti2Alpha 3.9411
#define Ti3Alpha 3.45
#define Ti4Alpha 3.45


/*-----------------------------------------------------------------------------

* Note – DEFINITION OF CONSTANTS...
*

*

* The first in the input file is "Aalpha" which is the coefficient
```

```
* that multiplies the site (ionic) orbital contribution to the
* sites electronic wave function. It is a measure of the degree of
* mixing in the system. For simplicity:
*
* Aalpha + Bbeta = 1
*
* The second in the input file is "DOSTi". DOSTi multiplies Ti electron
* orbitals. It is meant to allow the orbital to represent the
* situation that occurs when taking an STM image in a particular
* energy window. An energy window has a particular density of Ti like
* states associated with it. What we'd like to do is to project this
* fraction of the total states on to the a spatial distribution that
* can be thought to approximate the spatial extent of these orbitals.
* DOSTi changes the Ti orbital to one that represents the Ti electrons
* on only a fraction of the energy spectrum.
*
*
*
*-------------------------------------------------------------------------------*/
double
NTi1, NTi2, NTi3, NTi4, NTi5, NO1, NO2, NO3, NO4, NO5;

float T1O_1, T1O_2, T1O_3, T2O_1, T2O_2, T2O_3, T3O_1, T3O_2, T3O_3,
T4O_1, T4O_2, T4O_3, T5O_1, T5O_2, T5O_3, O1O_1, O1O_2, O1O_3, O2O_1,
O2O_2, O2O_3, O3O_1, O3O_2, O3O_3, O4O_1, O4O_2, O4O_3, O5O_1,
O5O_2, O5O_3, RadFitConstTi, RadFitConstO,
          xSpacing, Aalpha, Bbeta, alphaTi, alphaO, betaTi, betaO,
          ySpacing, DOSTi; /* spacing between pixels in x / y direction (pts / Angstrom)
*/
int NoPtsTmplt;

/* performs double coordinate rotation */
void CoordinateTrans (float *x, float *y, float *z, float theta1,
          float theta2, float theta3) {

     float xold, yold, zold, theta_rad1, theta_rad2, theta_rad3,

     PiConst = 3.14159265358979323846;

     theta_rad1 = theta1 * (2*PiConst) / (360);
     theta_rad2 = theta2 * (2*PiConst) / (360);
     theta_rad3 = theta3 * (2*PiConst) / (360);

     xold = *x;
        yold = *y;
```

```
    zold = *z;

    *x = xold*cos (theta_rad1) + yold*sin (theta_rad1);
    *y = -xold*sin (theta_rad1) + yold*cos (theta_rad1);

  xold = *x;
    yold = *y;
    zold = *z;

    *x = xold*cos (theta_rad2) + zold*sin (theta_rad2);
    *z = -xold*sin (theta_rad2) + zold*cos (theta_rad2);

  xold = *x;
  yold = *y;
  zold = *z;

    *y = yold*cos (theta_rad3) + zold*sin (theta_rad3);
    *z = -yold*sin (theta_rad3) + zold*cos (theta_rad3);
}

/* Evaluates Normalization Constant given alphas and betas */
/* A simple (and stupid) function used to copy a file onto another. */
void copyFile (FILE *f1, FILE *f2) {

    int c;

    c = fgetc (f1);
    while (c ! = EOF) {
        fputc (c, f2);
            c = fgetc (f1);
        }
}

/* Function that calculates the Charge Density at one pixel on the
 * output grid. The output of the function is the array called
"ChargeDensity"
 * and the template data is the input. The function returns the value of
 * the charge density at the pixel whose coordinates (x, y, z) are given
 * as input. NOTE: This is NOT the same function as the one in
surface. c
 * with the same name.
 */

/*
```

Note: This program was used to test the
charge density calculation in order to come up with a new
CDCalc routine.

2/9/98

```
*/

double RADO (float r) {
    extern float RadFitConstO;
    double a, b, c, d, e, f;

    a = (1 / (2*sqrt (6)))*r*sqrt (pow (RadFitConstO, 5))*exp
        (−RadFitConstO*r/2);
    return (a);

}

double RADTI (float r) {
    extern float RadFitConstTi;
    double a;
    a = (1 / (sqrt (720)))*sqrt (pow (RadFitConstTi, 7))
        *r*r*exp (−RadFitConstTi*r/2);

    return (a);

}

double ANGTI (float x, float y, float z, float r) {

    double a;

    a =
    pow ((1 / (r*r)), 2) * (dConst*pow (sqrt (15) * (x*x
        −y*y)/(4*sqrt (Pi)), 2)
        +eConst*pow (sqrt (30)* (x*z)/ (2*sqrt (2*Pi)), 2)
        +fConst*pow (sqrt (5)* (3*z*z −
            r*r)/(4*sqrt (Pi)), 2)
        +gConst*pow (sqrt (30)* (y*z)/ (2*sqrt (2*Pi)), 2)
        +hConst*pow (sqrt (15)* (x*y)/ (2*sqrt (Pi)), 2));
    return (a);
}
double ANGO (float x, float y, float z, float r) {

    double a;
```

```
a =
    (sqrt (3) / (2*sqrt (Pi)*r))* (aConst*pow (x, 2)
    + bConst*pow (z, 2) + cConst*pow (y, 2));

return (a);
}

double CalcChDen
(double z, float template [] [4], float x, float y,
    float IonRad []) {

    extern float Aalpha, Bbeta, T1O_1, T1O_2, T1O_3, T2O_1, T2O_2, T2O_3,
        T3O_1, T3O_2, T3O_3, T4O_1, T4O_2, T4O_3,
        O1O_1, O1O_2, O1O_3, O2O_1, O2O_2, O2O_3,
        O3O_1, O3O_2, O3O_3, O4O_1, O4O_2, O4O_3,
        O5O_1, O5O_2, O5O_3;
    extern int NoPtsTmplt;
    int i, j;

    float dr, dx, dy, dz, Inx, Iny, Inz, Inr, X [2] [4], XNN [4] [2] [4],
        Theta_Oxyl = 0, Theta_Oxy2 = 0, Theta_Ti1 = 0, Theta_Ti2 =
0;
    double CD, calc, angTi, angO, psiO, psiTi, RadO, RadTi;

    CD = 0;

    /* Step through all ions in the template and make
     * sure to add the contribution of each to the value
     * charge density at (x, y, z).
     */
    for (i = 0; i < NoPtsTmplt; i++) {
        dx = (x-template [i] [1])/ 0.521977;
        dy = (y-template [i] [2])/ 0.521977;
        dz = (z-template [i] [3])/ 0.521977;
        dr = sqrt (pow (dx, 2) +
            + pow (dy, 2) + pow (dz, 2));

        /* note that while x, y, z are all in units of
         * Angstroms, r is in units of Bohr radii
         */
        if (sqrt (pow (1.1-template [i] [0], 2)) < 1e-7) {

            calc = 0;
            CoordinateTrans (& (dx), & (dy), & (dz), T1O_1,
                T1O_2, T1O_3);
```

```
dr  =  sqrt (pow (dx, 2) +
        + pow (dy, 2) + pow (dz, 2));
calc +  =  Aalpha*pow (RADTI (dr), 2)*ANGTI (dx, dy, dz, dr);

/* 2.1, row 0 */
Inx  =  dx;
Iny  =  dy−2.7716;
Inz  =  dz −1.9665;
CoordinateTrans (& (Inx), & (Iny), & (Inz), O1O_1,
                 O1O_2, O1O_3);
Inr  =  sqrt (Inx*Inx + Iny*Iny + Inz*Inz);
calc +  = Bbeta*pow (RADO (Inr), 2)
*ANGO (Inx, Iny, Inz, Inr);

/* 2.1, row−1 */
Inx  =  dx;
Iny  =  dy + 2.7716;
Inz  =  dz −1.9665;
CoordinateTrans (& (Inx), & (Iny), & (Inz), O1O_1,
                 O1O_2, O1O_3);
Inr  =  sqrt (Inx*Inx + Iny*Iny + Inz*Inz);
calc +  =  Bbeta*pow (RADO (Inr), 2)
*ANGO (Inx, Iny, Inz, Inr);

CD +  =  DOSTi*calc / sqrt (NTi1);
}

else if (sqrt (pow (1.2−template [i] [0], 2)) <1e−7) {

calc  =  0;
CoordinateTrans (& (dx), & (dy), & (dz), T2O_1,
                 T2O_2, T3O_3);
    dr  =  sqrt (pow (dx, 2) +
    + pow (dy, 2) + pow (dz, 2));

calc +  =  Aalpha*pow ((1 / (sqrt (720))), 2)

*pow (sqrt (pow (Ti2Alpha, 7)), 2)
*pow (dr*dr*exp (−Ti2Alpha*dr / 2), 2)*

pow ((1 / (dr*dr)), 2) * (dConst*pow (sqrt (15) * (dx*dx
−dy*dy) / (4*sqrt (Pi)), 2)
+ eConst*pow (sqrt (30) * (dx*dz) / (2*sqrt (2*Pi)), 2)
+ fConst*pow (sqrt (5) * (3*dz*dz −
dr*dr) / (4*sqrt (Pi)), 2)
```

```
  + gConst*pow (sqrt (30) * (dy*dz) / (2*sqrt (2*Pi)), 2)
  + hConst*pow (sqrt (15) * (dx*dy) / (2*sqrt (Pi)), 2));

  /* 2.5, row 0 */
  Inx  =  dx;
  Iny  =  dy;
  Inz  =  dz + 3.3984;
  CoordinateTrans (& (dx), & (dy), & (dz), O5O_1, O5O_2,
                    O5O_3);
  Inr  =  sqrt (Inx*Inx + Iny*Iny + Inz*Inz);
  calc +  = Bbeta*pow (RADO (Inr), 2)
  *ANGO (Inx, Iny, Inz, Inr);

  CD +  = DOSTi*calc/sqrt (NTi2);
}

else if (sqrt (pow (1.3−template [i] [0], 2)) <1e−7) {

  calc  =  0;

                    calc = 0;
  CoordinateTrans (& (dx), & (dy), & (dz), T3O_1,
                    T3O_2, T3O_3);
     dr  =  sqrt (pow (dx, 2) +
   + pow (dy, 2) + pow (dz, 2));
     calc  +  = Aalpha
  *pow ((1 / (sqrt (720))), 2)
  *pow (sqrt (pow (Ti4Alpha, 7)), 2)
  *pow (dr*dr*exp (−Ti4Alpha*dr / 2), 2)
  *ANGTI (dx, dy, dz, dr);

     /* 2.1, row −1 */
     Inx  =  dx;
     Iny  =  dy + 2.7716;
     Inz  =  dz −1.9665;
  CoordinateTrans (& (dx), & (dy), & (dz), O1O_1,
                    O1O_2, O1O_3);
     Inr  =  sqrt (Inx*Inx + Iny*Iny + Inz*Inz);
     calc +  = Bbeta*pow (RADO (Inr), 2)
     *ANGO (Inx, Iny, Inz, Inr);

  CD +  =  DOSTi*calc / sqrt (NTi3);

}
```

```
else if (sqrt (pow (1.4−template [i] [0], 2)) <1e−7) {

        calc  =  0;
    CoordinateTrans (& (dx), & (dy), & (dz), T4O_1, T4O_2,
                T4O_3);
        dr  =  sqrt (pow (dx, 2)  +
                +  pow (dy, 2) + pow (dz, 2));
        calc  + =
    Aalpha
    *pow ((1 / (sqrt (720))), 2)
    *pow (sqrt (pow (Ti4Alpha, 7)), 2)
    *pow (dr*dr*exp (−Ti4Alpha*dr / 2), 2)
    *ANGTI (dx, dy, dz, dr);
    /* 2.1, row −1 */
        Inx  =  dx;
        Iny  =  dy + 2.7716;
        Inz  =  dz −1.9665;
    CoordinateTrans (& (dx), & (dy), & (dz), O1O_1,
                O1O_2, O1O_3);
        Inr  =  sqrt (Inx*Inx + Iny*Iny + Inz*Inz);
        calc +  =  Bbeta*pow (RADO (Inr), 2)
    *ANGO (Inx, Iny, Inz, Inr);

        CD  + =  DOSTi*calc / sqrt (NTi4);

}

else if (sqrt (pow (1.5−template [i] [0], 2)) <1e−7) {

        calc  =  0;
        CoordinateTrans (& (dx), & (dy), & (dz), T5O_1, T5O_2,
                T5O_3);
        dr  =  sqrt (pow (dx, 2)  +
                +  pow (dy, 2) + pow (dz, 2));
        calc +  =  Aalpha
    *pow ((1 / (sqrt (720))), 2)
    *pow (sqrt (pow (Ti3Alpha, 7)), 2)
    *pow (dr*dr*exp (−Ti3Alpha*dr / 2), 2)
    *ANGTI (dx, dy, dz, dr);

        CD  + =  DOSTi*calc / sqrt (NTi5);

}
else if (sqrt (pow (2.1−template [i] [0], 2)) <1e−7) {

    calc  =  0;
```

```
           CoordinateTrans (& (dx), & (dy), & (dz), O1O_1,
                        O1O_2, O1O_3);
       dr = sqrt (pow (dx, 2) +
                    + pow (dy, 2) + pow (dz, 2));
           calc + = Aalpha*pow (RADO (dr), 2) *ANGO (dx, dy, dz, dr);


       /* 1.1, row 0 */
           Inx = dx;
           Iny = dy + 2.7716;
           Inz = dz + 1.9665;
       CoordinateTrans (& (dx), & (dy), & (dz), T1O_1,
                        T1O_2, T1O_3);
           Inr = sqrt (Inx*Inx + Iny*Iny + Inz*Inz);
       calc + =
Bbeta*pow (RADTI (Inr), 2) *ANGTI (Inx, Iny, Inz, Inr);


       /* 1.1, row 1 */
           Inx = dx;
           Iny = dy − 2.7716;
           Inz = dz + 1.9665;
       CoordinateTrans (& (dx), & (dy), & (dz), T1O_1,
                        T1O_2, T1O_3);

           Inr = sqrt (Inx*Inx + Iny*Iny + Inz*Inz);
           calc + =

Bbeta*pow (RADTI (Inr), 2) *ANGTI (Inx, Iny, Inz, Inr);


           CD + = (1 − DOSTi) *calc / sqrt (NO1);


       }

   else if (sqrt (pow (2.2 − template [i] [0], 2)) < 1e−7) {

       calc = 0;
       CoordinateTrans (& (dx), & (dy), & (dz), O2O_1,
                        O2O_2, O2O_3);
           dr = sqrt (pow (dx, 2) +
                    + pow (dy, 2) + pow (dz, 2));
           calc + = Aalpha*pow (RADO (dr), 2) *ANGO (dx, dy, dz, dr);


       /* 1.2, row 0 */
           Inx = dx − 2.2865;
           Iny = dy − 2.7716;
           Inz = dz + 0.56;
```

```
CoordinateTrans (& (dx), & (dy), & (dz), T2O_1,
                T2O_2, T2O_3);
    Inr = sqrt (Inx*Inx + Iny*Iny + Inz*Inz);
    calc + =
Bbeta*pow (RADTI (Inr), 2)
*ANGTI (Inx, Iny, Inz, Inr);

CD + = (1-DOSTi) *calc/sqrt (NO2);

}

else if (sqrt (pow (2.3-template [i] [0], 2)) <1e-7) {

calc = 0;

CoordinateTrans (& (dx), & (dy), & (dz), O3O_1,
                O3O_2, O3O_3);
    dr = sqrt (pow (dx, 2) +
            + pow (dy, 2) + pow (dz, 2));
calc + = Aalpha*pow (RADO (dr), 2) *ANGO (dx, dy, dz, dr);

/* 1.2, row 0 */
    Inx = dx + 2.2865;
    Iny = dy - 2.7716;
    Inz = dz + 0.56;
CoordinateTrans (& (dx), & (dy), & (dz), T2O_1,
                T2O_2, T2O_3);
    Inr = sqrt (Inx*Inx + Iny*Iny + Inz*Inz);

calc + = Bbeta*pow (RADTI (Inr), 2) *ANGTI (Inx, Iny, Inz, Inr);
CD + = (1-DOSTi) *calc / sqrt (NO3);

}

else if (sqrt (pow (2.4-template [i] [0], 2)) <1e-7) {

calc = 0;

CoordinateTrans (& (dx), & (dy), & (dz), O4O_1,
                O4O_2, O4O_3);
    dr = sqrt (pow (dx, 2) +
            + pow (dy, 2) + pow (dz, 2));
calc + = Aalpha*RADO (dr) *ANGO (dx, dy, dz, dr);

/* 1.1, row 0 */
```

```
        Inx = dx;
        Iny = dy + 2.7716;
        Inz = dz−2.7265;
    CoordinateTrans (& (dx), & (dy), & (dz), T1O_1,
                T1O_2, T1O_3);
        Inr = sqrt (Inx*Inx + Iny*Iny + Inz*Inz);

        calc + = Bbeta*RADTI (Inr) *ANGTI (Inx, Iny, Inz, Inr);

    /* 1.1, row 1 */
        Inx = dx;
        Iny = dy−2.7716;
        Inz = dz−2.7265;
    CoordinateTrans (& (dx), & (dy), & (dz), T1O_1,
                T1O_2, T1O_3);
        Inr = sqrt (Inx*Inx + Iny*Iny + Inz*Inz);

        calc + = Bbeta*RADTI (Inr) *ANGTI (Inx, Iny, Inz, Inr);

        CD + = 4* (1−DOSTi) *calc*calc / NO4;

    }

    else if (sqrt (pow (2.5−template [i] [0], 2)) <1e−7) {

        calc = 0;

    CoordinateTrans (& (dx), & (dy), & (dz), O5O_1,
                O5O_2, O5O_3);
        dr = sqrt (pow (dx, 2) +
                + pow (dy, 2) + pow (dz, 2));
        calc + = Aalpha*RADO (dr) *ANGO (dx, dy, dz, dr);

    /* 1.2, row 0 */
        Inx = dx;
        Iny = dy;
        Inz = dz−3.3984;
    CoordinateTrans (& (dx), & (dy), & (dz), T2O_1,
                T2O_2, T2O_3);
        Inr = sqrt (Inx*Inx + Iny*Iny + Inz*Inz);

        calc + = Bbeta*RADTI (Inr) *ANGTI (Inx, Iny, Inz, Inr);

        CD + = 4* (1−DOSTi) *calc*calc / NO5;
```

```
      }

      else {
         printf ("\nError: ion = %f\nCheck template! \n\n",
      template [i] [0]);
         for (i = 0; i < NoPtsTmplt; i + +) {

            printf ("\nt [%d] = %6.4e %6.4e %6.4e %6.4e",

i, template [i] [0], template [i] [1],
               template [i] [2], template [i] [3]);

         }
      exit (0);
      }

   }
   return (CD);
}

/* This functions simply writes the output data to the output file */

void WriteOutput (double Output [ ] [YLimit + 1]) {

   int xi, yi;
   float x, y;
   FILE *fp, *header, *xcel;
   extern float xSpacing, ySpacing;

   printf ("\nWriting to file 'outa'. . . ");
   /*fp = fopen ("outa", "w"); */
   xcel = fopen ("outa.xcel", "w");

   /* copy standard di input file header to output data file */
   /* NOTE: A file named "header" is assumed to exist in the
    * same directory in which this program is being run. This file
    * is expected to contain a copy of a standard header for use by
    * the di software.
    */
   /*header = fopen ("header", "r");
      copyFile (header, fp);
   fclose (header); */

   /* Write data to output file */
   for (yi = YLimit;yi> =0;yi--) {
```

```
      fprintf (xcel, "\n");
        for (xi = 0; xi< = XLimit−7; xi = xi + 7) {

          fprintf (xcel, "%f %f %f %f %f %f %f %f",
              Output [xi] [yi]
                  , Output [xi + 1] [yi]
                  , Output [xi + 2] [yi]
                  , Output [xi + 3] [yi]
                  , Output [xi + 4] [yi]
                  , Output [xi + 5] [yi]
                  , Output [xi + 6] [yi]
                  , Output [xi + 6] [yi]);
          y = yi*ySpacing;
          x = xi*xSpacing;
          /*fprintf (fp, "\n%8d%8d%8d%8d%8d%8d%8d%8d"
                  , Output [xi] [yi]
                  , Output [xi + 1] [yi]
                  , Output [xi + 2] [yi]
                  , Output [xi + 3] [yi]
                  , Output [xi + 4] [yi]
                  , Output [xi + 5] [yi]
                  , Output [xi + 6] [yi]
                  , Output [xi + 7] [yi]); */


        }


      }
      /*fprintf (fp, "\n\n");
      fclose (fp); */


}


/* This function simply rescales the Data so that it
 * can be easily read into the di software. The constant (defined at
 * the beginning of this program) called "ScalePts" sets the new scale.
 * The output data will span −ScalePts...0....ScalePts.
 */
void ReScale (double Data [ ] [XLimit + 1],
        int Output [ ] [XLimit + 1]) {

        int xi, yi, /* These variables allow iteration through the pixel
                * grid. At any time the x coordinate of the current
                * pixel is x = xi*xSpacing (similar def of y coord.)
                */
          i, j, counter, lineavcount = 0; /* iteration variable */
```

```
float a, b, c, d, sum, lineav;
   double HighValue = 0, LowValue = 0, / * highest and lowest values
                                       * of Data on
                                       * entire pixel grid
                                       */
      ScaleIncrement;                  / * "bin size" of new scale
                                       * for output data
                                       */
   printf ("\nRescaling data...");
sum = 0;
counter = 0;
   / * search for higest and lowest value */
   for (xi = 0; xi < = XLimit; xi + +) {

      for (yi = 0; yi < = YLimit; yi + +) {

         if (xi*0.1171875 > = 9.559850613 &&
            xi*0.1171875 < = 12.74646) {

            if (yi*0.1171875 > = 12.7
               && yi*0.1171875 > = 12.8) {

               lineav = lineav + Data [xi] [yi];
               lineavcount + + ;

            }
         }
         if (Data [xi] [yi] ! = 0) {

            counter + + ;
            sum + = Data [xi] [yi];

            if (Data [xi] [yi] > = HighValue)
               HighValue = Data [xi] [yi];
            if (Data [xi] [yi] < = LowValue)
               LowValue = Data [xi] [yi];

         }
      }
      printf("\n\nHV = %6.4e, LV = %6.4e, AVG = %6.4e, lineave = %6.4e",
HighValue, LowValue, sum / counter, lineav / lineavcount);

   / * Set the "bin size" in the new distribution so that all data
    * points are included.
    */
   ScaleIncrement = (HighValue−LowValue) / (2*ScalePts);
```

```
if (ScaleIncrement = = 0) {

        printf ("\nError: ScalePts too large.ScaleIncrement = 0\n");
        return;
}

/* Re-map the Data to new scale */
for (xi = 0; xi <= XLimit; xi + +) {

        for (yi = 0; yi <= YLimit; yi + +) {

        j = 0;
        for (i = −ScalePts; i < ScalePts + 1; i + +) {

                        if (j*ScaleIncrement< = Data [xi] [yi]
                        && (j + 1) * (ScaleIncrement)
                        > Data [xi] [yi])
                                Output [xi] [yi] = i;
                        j + +;
                }
        }
    }
}

/* This is the main function that calls the routines which calculate
 * the charge densities and rescale the data.
 */

void DoesALot (void) {

int     i,
        xi, /* i, xi, yi = variables used in indexing */
        yi,
        CDflag, /* indicates error in charge density calc. (1 = error) */
        Output [XLimit + 1] [YLimit + 1]; /* output data array */

extern float xSpacing, ySpacing, Aalpha, Bbeta, RadFitConstTi, RadFitConstO;
float template [MaxNoPtsTmplt] [4], CDInc, Zlimit, Zend,
        x, y, ChargeDensity,      /* coordinates of pixel in
Angstroms */
        ionRad [MaxNoPtsTmplt], /* empirical radii of ions in template (Bohr
radii) */
        r;      /* distance between ion in template and pixel
(Angstroms) */
```

```
double z [XLimit + 1] [YLimit + 1], CD, a, b, c, d, e, f, g, h, j; /* holds value of cd
for each pixel */
FILE *error;


    /* calculate spacing for both x and y coordinates */
    xSpacing = GridLengthHeight / XLimit;
    ySpacing = xSpacing; /* square grid is assumed */
    ionRad [0] = 1.1;
    ionRad [1] = 1.2;
    ionRad [2] = 2.1;
    ionRad [3] = 2.2;
    ionRad [4] = 2.3;


    error = fopen ("error", "w");
    fprintf (error, "\n\n");
    /* read in template data */

    scanf ("\n%f %f %f", &T1O_1, &T1O_2, &T1O_3);
    scanf ("\n%f %f %f", &T2O_1, &T2O_2, &T2O_3);
    scanf ("\n%f %f %f", &T3Q_1, &T3O_2, &T3O_3);
    scanf ("\n%f %f %f", &T4O_1, &T4O_2, &T4O_3);
    scanf ("\n%f %f %f", &O1O_1, &T1O_2, &O1O_3);
    scanf ("\n%f %f %f", &O2O_1, &O2O_2, &O2O_3);
    scanf ("\n%f %f %f", &O3O_1, &O3O_2, &O3O_3);
    scanf ("\n%f %f %f", &O4O_1, &O4O_2, &O4O_3);
    scanf ("\n%f %f %f", &O5O_1, &O5O_2, &O5O_3);
      scanf ("\n%f", &ChargeDensity);
      CDInc = ChargeDensity*1e-2;
      scanf ("\n%f", &Aalpha);
      scanf ("\n%f", &DOSTi);
      scanf ("\n%f", &RadFitConstTi);
    scanf ("\n%f", &RadFitConstO);
    Bbeta = 1 - Aalpha;
      printf ("\nPROGRAM TO SIMULATE STM IMAGE OF SURFACES
          (a.c)...");
      printf ("\n\nThis version does not contain the time saving routine.");

      printf ("\n----------------------------------------");
      printf ("\nRun parameters:");
          printf ("\nnNTi4 = %6.4e", NormConst (1.4));
      /* Evaluate Normalization Constants */
      NTi1 = NormConst (1.1);
      NTi2 = NormConst (1.2);
```

```
NTi3 = NormConst (1.3);
NTi4 = NormConst (1.4);
NTi5 = NormConst (1.5);
NO1 = NormConst (2.1);
NO2 = NormConst (2.2);
NO3 = NormConst (2.3);
NO4 = NormConst (2.4);
NO5 = NormConst (2.5);

/* Print the normalization constants */
printf ("\nNTi1 = %6.4e", NormConst (1.1));
printf ("\nNTi2 = %6.4e", NormConst (1.2));
printf ("\nNTi3 = %6.4e", NormConst (1.3));
printf ("\nNTi4 = %6.4e", NormConst (1.4));
printf ("\nNO1 = %6.4e", NormConst (2.1));
printf ("\nNO2 = %6.4e", NormConst (2.2));
printf ("\nNO3 = %6.4e", NormConst (2.3));
printf ("\nNO4 = %6.4e", NormConst (2.4));
printf ("\nNO5 = %6.4e", NormConst (2.5));

printf ("\n\nTemplate has %d ions.", NoPtsTmplt);
printf ("\nRadFitConstO = %6.4e", RadFitConstO);
   printf ("\nRadFitConstTi = %6.4e", RadFitConstTi);
printf ("\n\nCharge Density = (%6.4e) (electronic charge)",
   ChargeDensity);
   printf ("\n\nSharing: Alpha = %f, Beta = %f\n",
      Aalpha, Bbeta);

   printf ("\nTiDOS = %6.4e, ODOS = %6.4e", DOSTi, (1−DOSTi));
   printf ("\n----------------------------------------");

for (i = 0; i < NoPtsTmplt; i++) {
      scanf ("\n%f %f %f %f", &template [i] [0],
         &template [i] [1], &template [i] [2], &template [i] [3]);
}
/* calculate the charge density at all points outside of the ions */
for (xi = 0; xi <= XLimit; xi++) {

   if (xi == 50)
      printf ("\nxi = %d", xi);
   if (xi == 100)
      printf ("\nxi = %d", xi);
   if (xi == 150)
      printf ("\nxi = %d", xi);
   if (xi == 200)
```

```c
        printf ("\nxi = %d", xi);

    for (yi = 0; yi < = YLimit; yi+ +) {
        y = yi*ySpacing;
        x = xi *xSpacing;
        z [xi] [yi] = 0;

        Zlimit = ZLimit;

        /* redefine Z limit based on need */
        a = Zlimit;
        b = 0;
        while (a−b> = ZIncrement) {

            c = CalcChDen ((a+b)/2, template, x, y, ionRad);
            if (ChargeDensity > c) {

                a = a/2 + b/2;

            }
            else {

                b = a/2+b/2;
            }
            if (b > = a)
                    printf ("\n\nError in z det.");
        }
        z [xi] [yi] = a;
        }
    }

        /* create and write to output file */
    /* ReScale (z, Output); */
    WriteOutput (z);

    printf ("\nDone! \n");
    printf ("\n---------------------------------------");
    printf ("\n--------------------------------------- \n");
}
void main (void) {

    extern int NoPtsTmplt;
    scanf ("%d", &NoPtsTmplt);
    DoesALot ( );
}
```

CALCULATIONS OF TUNNELING CURRENT

A. Frye and Dawn A. Bonnell

The following program calculates tunneling spectra of semiconductors. It is configured to calculate the spectrum for a n-type compound with a 3.2 eV bandgap. The program is quite general and can include midgap defect states and a semiconductors with a variety of properties. It is a Mathematica program that runs on version 3.0 and executes within 10–60 minutes on a 400 MHz processor, the differences depending on the magnitude of current. The output of the code is a) the total dynamic band bending function; b) the potential distribution functions; c) the predicted experimental band bending function; d) a linear current-voltage plot; e) a semi log current-voltage plot.

```
(* Output control *)
Off[General::spell1];                 (* turn off similar spelling warning *)
Off[NIntegrate::precw];               (* turn off less than precision warning *)
Off[NIntegrate::slwcon];              (* turn off slow convergence warning *)
Off[NIntegrate::ncvb];                (* turn off non-convergence warning *)
Off[FindRoot::cvnwt];                 (* turn off non-convergence warning *)
Off[FindRoot::precw];                 (* turn off less than precision warning *)
Off[InterpolatingFunction::dmwarn];   (* turn off domain size warning *)
Off[Plot::plnr];
SetDirectory["user files:Asa:mathematica:c results"];
(* define physical constants *)
jeV = 1.60217733*10^-19;              (* joules to electron volts conversion factor *)
k = (1.380658*10^-23/jeV);            (* Boltzmann's constant; eV/K *)
hbar = (1.05457266 10^-34/jeV);       (* reduced Planck's constant; eV s *)
h = 2 Pi hbar;                        (* Planck's constant; eV s *)
m0 = 9.1093897*10^-31/jeV;            (* free electron mass; eV s^2/m^2 *)
```

e = 1.60217733*10^-19; (* fundamental unit of charge; C *)
ep0 = 8.854187817*10^-12; (* permittivity of free space; F/m = C/(V m) *)

(* define variables *)
temp = 300; (* temperature; Kelvins *)
s = 9; (* tunneling gap width; angstroms *)
r = 5; (* radius of curvature of tip; angstroms *)
mceff = 12; (* conduction band relative effective mass *)
mveff = 0.99; (* valence band relative effective mass *)
mmeff = 0.99; (* metal electrode relative effective mass *)
kappasc = 300; (* semiconductor static dielectric constant *)
kappaschf = 5; (* semiconductor high frequency dielectric constant *)
kappain = 1; (* insulator static dielectric constant *)
kappainhf = 1; (* insulator high frequency dielectric constant *)
nd = 1.0*10^19; (* free carrier density; cm^{-3} *)
initBB = 0.30; (* initial surface potential; eV *)
beta = 0.002; (* adjustable parameter for potential distribution *)
eGap = 3.2; (* semiconductor band gap; eV *)
scfermi = 0; (* semiconductor Fermi level; eV *)
wfM = 4.55; (* tungsten metal work function; eV *)
eAff = 3.0; (* semiconductor electron affinity; eV *)
emobil = 30; (* semiconductor electron mobility; cm^2/V s *)
numPoints = 100; (* number of points in IV curve *)
vaMin = -4; (* starting applied tip bias; V *)
vaMax = 4; (* ending applied tip bias; V *)
eDC1 = 0.0012; (* ionization energy of donor defect; eV *)

(* derive constants *)
ceffm := If[mceff > mmeff, mmeff, mceff]; (* eff. masses for integral limits *)
veffm := If[mveff > mmeff, mmeff, mveff];
barNc = 2* (* conduction band effective dos; m^{-3} *)
((mceff m0 k temp)/(2 Pi (hbar^2)))^1.5;
barNv = 2*
((mveff m0 k temp)/(2 Pi (hbar^2)))^1.5; (* valence band effective dos; m^{-3} *)
eConB := N[scfermi − ((3 (nd*10^6))/Pi)^(2/3)* (* conduction band edge; eV *)
(h^2/(8 mceff m0))] /; nd > = (barNc*10^-6);
eConB := N[scfermi − (k temp Log[(nd*10^6)/barNc])] /; nd < (barNc*10^ − 6);
wfSC = eAff + eConB; (* semiconductor work function; eV *)
eValB = eConB − eGap; (* valence band edge; eV *)
eIntB := N[0.5(eConB + eValB) +
  (k temp Log[(mveff/mceff)^(3/4)])]; (* intrinsic Fermi level; eV *)
eD1 = eConB − eDC1; (* energy of donor state; eV *)
debeye = Sqrt[(kappasc ep0 jeV k temp)/(e^2 (nd*10^6))]; (* Debeye length; m *)
ub = (scfermi − eIntB)/(k temp);
hevi[t_] := If[t > 0,1,0]; (* Heaviside function *)

(* derive variables *)
wdep[vs_] := Sqrt[(2 kappasc ep0 jeV Abs[bbFunc[vs]])/(nd*10^6 e^2)];
(* depletion layer width *)

phicSC[z_,vs_,longE_] :=
    N[((((wdep[vs] − z)^2/wdep[vs]^2)*bbFunc[vs]) + eConB − longE];
phivSC[z_,vs_,longE_] :=
    N[−1*((((wdep[vs] − z)^2/wdep[vs]^2)*bbFunc[vs]) + eValB − longE)];
phiVac[z_,vi_,longE_] := N[((((wfM − vi − longE)*(z/(s*10^−10)))
    + ((wfSC − longE)*(1−(z/(s 10^10))))
    − ((((0.4 e^2)/(8 Pi kappainhf jeV ep0))*((s 10^ − 10)/(z ((s 10^ − 10) − z)))))];

phiczSC[vs_,longE_] :=
    N[(((((wdep[vs] − z)^2/wdep[vs]^2)*bbFunc[vs]) + eConB − longE)];
phivzSC[vs_,longE_] :=
    N[−1*((((wdep[vs] − z)^2/wdep[vs]^2)*bbFunc[vs]) + eValB − longE)];
phizVac[vi_,longE_] := N[((((wfM − vi − longE)*(z/(s*10^−10)))
    + ((wfSC − longE)*(1−(z/(s 10^−10))))
    − ((((0.4 e^2)/(8 Pi kappainhf jeV ep0))*((s 10^−10)/(z ((s 10^−10) − z)))))];

(* generate equilibrium band bending function -> bbFunc[] *)
w[p_,q_] := (p − q)/(k temp);
fD1 = (1 + 2 Exp[x−w[eD1,eIntB]])^−1;
donorint := NIntegrate[fD1, {x,ub,us}];
xval = (eValB−energy)/(k temp);
fdival[j_,eta_] := Re[(j!^−1)*NIntegrate[(xval^j)/(Exp[xval−eta] + 1),
    {energy,−30,0}, WorkingPrecision -> 15, AccuracyGoal -> 10]];
xcond = (energy−eConB)/(k temp);
fdicond[j_,eta_] := Re[(j!^−1)*NIntegrate[(xcond^j)/(Exp[xcond−eta] + 1),
    {energy,0,30}, WorkingPrecision -> 15, AccuracyGoal -> 10]];
elecintdeg := (((2 barNc)/(3 nd*10^6))*
    ((fdicond[1.5,us−w[eConB,eIntB]])−(fdicond[1.5,ub−w[eConB,eIntB]])));
holeintdeg := (((2 barNv)/(3 nd*10^6))*
    ((fdival[1.5,w[eValB,eIntB]−us])−(fdival[1.5,w[eValB,eIntB]−ub])));
elecintnondeg =
    N[(barNc/(nd*10^6))*(Exp[us−w[eConB,eIntB]]−Exp[ub−w[eConB,eIntB]])];
holeintnondeg =
    N[(barNv/(nd*10^6))*(Exp[w[eValB,eIntB]−us]−Exp[w[eValB,eIntB]−ub])];
elecint := elecintdeg /; nd > = (barNc*10^−6);
elecint := elecintnondeg /; nd < (barNc*10^−6);
holeint := holeintdeg /; nd > = (barNc*10^−6);
holeint := holeintnondeg /; nd < (barNc*10^−6);
tvs := N[initBB − (k temp) − (((e debeye^2 nd 10^6)/(kappasc ep0))*
    (−1 donorint − elecint + holeint))];
phibbs := N[(ub − us)*(k temp)];

```
bbCurve = Table[{ −tvs,phibbs}, {us,−73,73,1.0}];
bbFunc = Interpolation[bbCurve,InterpolationOrder -> 1];
bbPlot = ListPlot[bbCurve, PlotJoined -> True, Frame -> True,
FrameLabel -> {"Sample voltage,Vs (V )",None}, AspectRatio -> 1,
PlotRange -> {{−vaMax,−vaMin}, {(−eGap + initBB),(eGap + 0.5)}},
AxesLabel -> {None, "Equilibrium Surface Potential (eV )"}];
Clear[bbCurve,us,tvs];
" = extrinsic Debeye length (meters)" debeye; " = reduced bulk potential" ub
" = equilibrium depletion width (meters)" wdep[0]
" = depletion width to Debeye length" wdep[0]/debeye


(* define limits for barrier integrals *)
zcbtemp[vs_,longE_] := NSolve[phiczSC[vs,longE] = = 0,z];
deltazcb[vs_,longE_] := z /. zcbtemp[vs,longE][[1]];
zccb[vs_,longE_] := If[Re[deltazcb[vs,longE]] > 0,Re[deltazcb[vs,longE]],0];


zvbtemp[vs_,longE_] := NSolve[phivzSC[vs,longE] = = 0,z];
deltazvb[vs_,longE_] := z /. zvbtemp[vs,longE][[1]];
zcvb[vs_,longE_] := If[Re[deltazvb[vs,longE]] > 0,Re[deltazvb[vs,longE]],0];


zvactemp[vi_,longE_] := NSolve[phizVac[vi,longE] = = 0,z];
zsoltemp[vi_,longE_] := z /. zvactemp[vi,longE][[1]];
vsol[vi_,longE_] := If[Re[zsoltemp[vi,longE]] < 0,2,1];
za[vi_,longE_] := z /. zvactemp[vi,longE][[vsol[vi,longE]]];
zb[vi_,longE_] := z /. zvactemp[vi,longE][[(vsol[vi,longE] + 1)]];


(* define transmission factor integrals *)
phibarVac[z_,vi_,longE_] := If[0 < phiVac[z,vi,longE], phiVac[z,vi,longE], 0];
phibarcSC[z_,vs_,longE_] := If[0 < phicSC[z,vs,longE], phicSC[z,vs,longE], 0];
phibarvSC[z_,vs_,longE_] := If[0 < phivSC[z,vs,longE], phivSC[z,vs,longE], 0];
etaVac[vi_,longE_] := (2*Sqrt[(2 m0)/(hbar^2)])*
NIntegrate[Sqrt[phibarVac[z,vi,longE]],{z,Re[za[vi,longE]],Re[zb[vi,longE]]},
WorkingPrecision -> 10, AccuracyGoal -> 8];


etacSC[vs_,longE_] := (2*Sqrt[(2 m0)/(hbar^2)])*
NIntegrate[Sqrt[phibarcSC[z,vs,longE]],{z,0,zccb[vs,longE]},
WorkingPrecision -> 10, AccuracyGoal -> 8];


etavSC[vs_,longE_] := (2*Sqrt[(2 m0)/(hbar^2)])*
NIntegrate[Sqrt[phibarvSC[z,vs,longE]],{z,0,zcvb[vs,longE]},
WorkingPrecision -> 10, AccuracyGoal -> 8];


(* define energy band integrals *)
dVac[vi_,longE_] := Exp[−1 etaVac[vi,longE]];
dcSC[vs_,longE_] := Exp[−1 etacSC[vs,longE]];
```

```
dvSC[vs_,longE_] := Exp[-1 etavSC[vs,longE]];

eValBs[va_] := If[bbFunc[(-1*vsFunc[va])] > 0,
bbFunc[(-1*vsFunc[va])] + eValB, eValB];

jCB[va_] := N[(4 Pi e m0 ceffm)/h^3]*
NIntegrate[hevi[ +1 (totE - eConB)]*dVac[viFunc[va],w]
*dcSC[(-1*vsFunc[va]),w], {totE,0,-va},
{w,eConB,totE}, WorkingPrecision -> 10, AccuracyGoal -> 8];

jVB[va_] := N[-(4 Pi e m0 veffm)/h^3]*
NIntegrate[hevi[-1 (totE - eValBs[va])]*dVac[viFunc[va],w]
*dvSC[(-1*vsFunc[va]),w], {totE,eValBs[va],-va},
{w,eValBs[va],totE}, WorkingPrecision -> 10, AccuracyGoal -> 8];

(* define defect induced current *)
alpha = 3*10^-4;
surfE[va_] := N[(-1*vsFunc[va])/wdep[0]];

sbl[va_] := N[((((e^3 nd 10^6)/(8 Pi^2 ep0^3 kappasc kappaschf^2))*
    (bbFunc[(-1*vsFunc[va])] + (k temp)))^0.25]; (* barrier lowering *)

dI[va_] := N[1*(alpha*temp^1.5*surfE[va]*emobil*mceff^1.5*
    (etaVac[viFunc[va],bbFunc[(-1*vsFunc[va])])]) *
Exp[(-1*(bbFunc[(-1*vsFunc[va])] + Abs[eConB] - sbl[va]))/(k temp)]*
Exp[(e/(k temp))*(e/(4 Pi ep0 kappaschf))^0.5*((surfE[va]^0.5))])];

jDI[va_] := If[va > = 0,dI[va],0];

(* generate potential distribution functions -> viFunc[] and vsFunc[] *)
vaFunc := N[((1 - (s 10^-10 kappain^-1 ep0^-1beta))^-1)*
    (vs - ((s 10^-10 kappain^-1 ep0^-1)*
    (Sqrt[(2 e kappasc ep0 nd 10^6)*(initBB - (k temp) - vs)] -
    Sqrt[(2 e kappasc ep0 nd 10^6)*(initBB - (k temp))])))];
vss[va_] := FindRoot[vaFunc = = va, {vs,10^1}];
vs[va_] := Re[vs /. vss[va]];
vi[va_] := va - vs[va];
viCurve = Table[{va,vi[va]}, {va,vaMin,vaMax,0.5}];
viFunc = Interpolation[viCurve,InterpolationOrder -> 3];
viPlot = Plot[viFunc[x], {x,vaMin,vaMax}, Frame -> True,
PlotRange -> {{vaMin,vaMax},{vaMin,vaMax}},
FrameLabel -> {"Tip Bias, Va (V )",None}, AspectRatio ->1,
AxesLabel -> {None, "Insulator voltage, Vi (V )"}];
vsCurve = Table[{va,vs[va]}, {va,vaMin,vaMax,0.5}];
vsFunc = Interpolation[vsCurve,InterpolationOrder -> 3];
```

```
vsPlot = Plot[vsFunc[x], {x,vaMin,vaMax}, Frame -> True,
PlotRange -> {{vaMin,vaMax},{vaMin,vaMax}},
   FrameLabel -> {"Tip Bias, Va (V)",None}, AspectRatio ->1,
   AxesLabel -> {None, "Sample voltage, Vs (V)"}];
Clear[vsCurve,viCurve];
Show[vsPlot,viPlot, PlotRange -> {{vaMin,vaMax},{vaMin,vaMax}},
AxesLabel -> {None,None},AspectRatio -> 1];
surfPot = Table[{ -va,bbFunc[(-1*vsFunc [va])]}, {va,vaMin,vaMax,0.05}];
surfPotPlot = ListPlot[surfPot, PlotJoined -> True, Frame->True,
FrameLabel -> {"Sample Bias, Va (V)", None}
PlotRange -> {{-vaMax,-vaMin),{(-eGap + initBB),(eGap + 0.3)}},
AxesLabel -> {None, "Surface Potential (eV)"},AspectRatio-> 1];

(* define total current density *)
current[va_] := (N[jCB[va]] + N[jVB[va]] + N[jDI[va]]) N[Pi] r^2 10^-20;
(* generate iv curve *)
i[v_] := Re[current[v]];
ivCurve = Table[{(-1*v),i[v]}, {v,vaMin,vaMax,((vaMax - vaMin)/numPoints)}];
ivCurve≫sro2_22;                    (* name of file to save results *)
Clear[ivCurve];
rawdata = ≪sro2_22;
linearspectra = Interpolation[rawdata];
vmin = (-vaMax); vmax = (-vaMin);
Plot[(linearspectra[x]*10^9),{x,vmin,vmax},PlotRange->{{vmin,vmax},
   {-10,+10}}];
Plot[Log[10,Abs[(linearspectra[x]*10^9)]],{x,vmin,vmax},
   AxesOrigin->{vmin,-1.6},
PlotRange-> {{vmin,vmax},{-1.6,1}},Frame->True,AspectRatio -> 1];
```

INDEX